THE RISE OF THE STANDARD MODEL

Editors Lillian Hoddeson, Laurie Brown, Michael Riordan, and Max Dresden have brought together a distinguished group of elementary particle physicists and historians of science to explore the recent history of particle physics. Based on a conference held at Stanford University, this is the third volume of a series recounting the history of particle physics and offers the most up-to-date account of the rise of the Standard Model, which explains the microstructure of the world in terms of quarks and leptons and their interactions

Major contributors include Murray Gell-Mann, John Heilbron, Leon Lederman, Michael Redhead, Silvan Schweber, and Steven Weinberg. The wide-ranging articles explore the detailed scientific experiments, the institutional settings in which they took place, and the ways in which the many details of the puzzle fit together to account for the Standard Model.

THE RISE OF
THE STANDARD MODEL

Particle Physics in the 1960s and 1970s

Edited by

LILLIAN HODDESON
University of Illinois at Urbana-Champaign

LAURIE BROWN
Northwestern University

MICHAEL RIORDAN
Stanford Linear Accelerator Center

MAX DRESDEN
Stanford Linear Accelerator Center

CAMBRIDGE
UNIVERSITY PRESS

CAMBRIDGE UNIVERSITY PRESS
Cambridge, New York, Melbourne, Madrid, Cape Town, Singapore, São Paulo

Cambridge University Press
The Edinburgh Building, Cambridge CB2 2RU, UK

Published in the United States of America by Cambridge University Press, New York

www.cambridge.org
Information on this title: www.cambridge.org/9780521570824

First published 1997

A catalogue record for this publication is available from the British Library

Library of Congress Cataloguing in Publication data

The rise of the standard model : particle physics in the 1960s and 1970s
/ edited by Lillian Hoddeson . . . [et al.].
p. cm.
ISBN 0-521-57082-4 (hardcover : alk. paper). – ISBN 0-521-57816-7
(pbk. : alk paper)
1. Standard model (Nuclear physics) – History.   2. Particles
(Nuclear physics) – History.   I. Hoddeson, Lillian
QC794.6.S75R57   1997
539.7´4 – dc20   96-5397
                                                CIP

ISBN-13 978-0-521-57082-4 hardback
ISBN-10 0-521-57082-4 hardback

ISBN-13 978-0-521-57816-5 paperback
ISBN-10 0-521-57816-7 paperback

Transferred to digital printing 2006

Contents

Contents

Contributors

James Bjorken
Mail Stop 81, Stanford Linear Accelerator Center, Stanford, CA 94309

Sidney Bludman
Department of Physics, University of Pennsylvania, Philadelphia, PA 19104

Mark Bodnarczuk
Breckenridge Consultants Group, P.O. Box 7399-329, Breckenridge, CO 80424-7399

Robert Brout
Universite Libre de Bruxelles, Service de Physique Theorique, Campus Plaine CP 225, Blvd. du Triomphe, 1050 Bruxelles, Belgium

Laurie M. Brown
Department of Physics and Astronomy, Northwestern University, Evanston, Illinois 60201

Tian Yu Cao
Department of Philosophy, Boston University, 745 Commonwealth Avenue, Boston, MA 02215

James Cronin
Enrico Fermi Institute, University of Chicago, 5640 South Ellis Avenue, Chicago, IL 60637

Max Dresden
Mail Stop 61, Stanford Linear Accelerator Center, Stanford, CA 94309

Jerome Friedman
Department of Physics, Massachusetts Institute of Technology, Cambridge, MA 02139

Peter Galison
Department of History of Science, Harvard University, Cambridge, MA 02138

Murray Gell–Mann
Santa Fe Institute, 1399 Hyde Park Road, Santa Fe, NM 87505

Gerson Goldhaber
Physics Department, 50A-2160, Lawrence Berkeley Laboratory, Berkeley, CA 94720

David Gross
Department of Physics, P.O. Box 708, Jadwin Hall, Princeton, NJ 08544

J. L. Heilbron
Office of the Chancellor, University of California, Berkeley, CA 94720

Peter Higgs
Department of Physics, James Clerk Maxwell Building, The University of Edinburgh, Edinburgh EH9 3JZ, UK

Lillian Hoddeson
Department of History, Gregory Hall on Wright St., University of Illinois, Urbana, Illinois 61801

Gerard 't Hooft
Institute for Theoretical Physics, Princetonplein 5, P.O. Box 80.006, 3508 TA Utrecht, The Netherlands

John Iliopoulos
Laboratoire de Physique Théorique, Ecole Normale Supérieure, 24 rue Lhomond, 75231 Paris, France

Kjell Johnsen
CERN, Building 584, CH-1211 Geneva 23, Switzerland

Makoto Kobayashi
Theory Group, National Laboratory for High Energy Physics, 1-1 Oho, Tsukuba-shi, Ibaraki-ken 305, Japan

Adrienne Kolb
Fermilab, Post Office Box 500, Batavia, IL 60510

John Krige
Centre de Recherch en Histoire des Sciences, Parc de La Villette, 211, Avenue Jean Jaures, 75019 Paris, France

Leon Lederman
Directors Office, Fermilab, Post Office Box 500, Batavia, IL 60510

Harry Lipkin
Nuclear Physics Department, Weizmann Institute of Science, Post Office Box 26, Rehovot 76100, Israel

Giacomo Morpurgo
Dipt. di Fisica, Univ. di Genova, via Dodecaneso 33, I-16146 Genova, Italy

Yoichiro Nambu
Enrico Fermi Institute, University of Chicago, 5640 South Ellis Avenue, Chicago, IL 60637

Donald Perkins
Nuclear Physics Lab, University of Oxford, Keble Road, Oxford, OX1 3RH, UK

Martin Perl
Mail Stop 61, Stanford Linear Accelerator Center, Stanford, CA 94309

Alexander Polyakov
Department of Physics, Post Office Box 708, Jadwin Hall, Princeton, NJ 08544

Charles Prescott
Mail Stop 78, Stanford Linear Accelerator Center, Stanford, CA 94309

Michael Redhead
Department of History and Philosopy of Science, University of Cambridge, Free School Lane, Cambridge CB2 3RH, UK

Burton Richter
Directors Office, Stanford Linear Accelerator Center, Stanford, CA 94309

Michael Riordan
Mail Stop 80, Stanford Linear Accelerator Center, Stanford, CA 94309

Nicholas Samios
Directors Office, Building 460, Brookhaven National Laboratory, Upton, NY 11973

Melvin Schwartz
Department of Physics, Columbia University, New York, NY 10027

Silvan Schweber
Department of Physics, Brandeis University, Post Office Box 9110, Waltham, MA 02254

Contributors

Roy Schwitters
Center for Particle Physics, University of Texas, Austin, TX 78712

Dmitrij V. Shirkov
Laboratory for Theoretical Physics, Joint Institute for Nuclear Research, Dubna, Russia

Leonard Susskind
Department of Physics, Stanford University, Stanford, CA 94305

Paul Teller
Department of Philosophy, University of California, Davis, CA 95616

Martinus Veltman
Department of Physics, University of Michigan, Ann Arbor, MI 48109

Steven Weinberg
Department of Physics, University of Texas, Austin, TX 78712

Catherine Westfall
Lyman Briggs School, Michigan State University, East Lansing, MI 48824

Robert R. Wilson
916 Stewart Avenue, Ithaca, NY 14850

Sau Lan Wu
EP Division, CERN, Building 32-RA05, CH-1211 Geneva 23, Switzerland

Editors' Acknowledgments

We are deeply grateful to all the people who contributed to the success of the Third International Symposium on the History of Particle Physics. Without their efforts it would have been a much smaller and far less enjoyable gathering. Limits of space prevent us from thanking each and every one of them by name, but certain individuals deserve special recognition.

We thank laboratory directors Burton Richter and John Peoples for their strong support of this Symposium. In assembling the program, we benefited extensively from the sage advice of Barry Barish, James Bjorken, Peter Galison, Gerson Goldhaber, Sam Schweber, and Lenny Susskind, who served with us as members of the Program Committee.

In hosting the Symposium and attending to the myriad details that contributed to making it an enjoyable and successful gathering, the untiring efforts of Nina Adelman Stolar stand out from all the rest. She arranged lodging and transportation for speakers and participants, distributed the invitations and registration forms, and supervised a small army of SLAC and Fermilab staff members who supported us for the four-day event; the Symposium would not have been such a success without her contributions. Nina was ably assisted in these tasks by Juanita O'Malley, who handled communications with speakers and participants. Herbert McIntyre coordinated all the audio-visual services, while Bernie Lighthouse took care of the specific needs of individual speakers. Rene Donaldson designed and produced the Symposium program, based on period photographs of leading physicists located by Robin Chandler;

blow-ups of these photos also graced the walls of the auditorium lobby. In addition, we wish to thank the following people for their help:

Joe Faust, for photographing Symposium speakers and participants;

Angela Gonzalez, for designing the Symposium poster and the dust jacket of this book;

Susan Grommes, Adrienne Kolb, and May West, for their gracious support and assistance during the Symposium sessions;

Jeff Machado, for catering the Symposium banquet and other social events;

Michael Peskin, for organizing and supervising scientific secretaries;

Helen Quinn, for coordinating activities for science teachers;

Pauline Wethington, for her extensive help with registration and in many other Symposium activities.

The scientific secretaries, who helped us in recording the question-and-answer sessions after the speakers' presentations, were: Jin Dai, Rob Elia, Adam Falk, Ovid Jacob, Ross King, Amit Lath, Hung Jung Lu, Carl Schmidt, Yael Shadmi, Matthew Strassler.

In addition, the following people gave generously of their time to help us before, during, and after the Symposium: Louise Addis, Gregory Bologoff, Andrea Chan, Maura Chatwell, Doug Dupen, Bette-Jane Ferandin, Diana Gregory, Bruce Hemingway, Karen Hernandez, Sharon Ivanhoe, Donà Jones, Jeff Leiter, Cortney Lighthouse, Gina Mastrantonio, Sally McFadden, Judy Meo, Brad Moore, Neal Morrison, Rocky Nilan, Robbin Nixon, Luana Plunkett, David Price, Nader Saghafi, Allison Sato, Jennifer Simmons, Ida Stelling, Neil Strand, Noreen Sugrue, and Steve Tieger.

In preparing the manuscript for this book, the efforts of Shirley Boozer, Tonya Lillie, and Dan Lewart were invaluable. They helped us with the difficult task of converting the chapters submitted by the speakers into a uniform electronic format. With help from May West, Tonya also verified and corrected many of their references, and entered our seemingly endless editorial corrections. Terry Anderson, Kevin Johnston, and Sylvia MacBride provided quality graphic support.

We also benefited from strong financial support. A grant from the Alfred P. Sloan Foundation permitted us to bring invited speakers from all over the world to Stanford, while the National Science Foundation financed our work in editing the chapters that appear within this volume. A grant from the American Institute of Physics Center for the History of Physics allowed us to make audio recordings of the sessions. Finally, the U.S. Department of Energy supported this Symposium through the generous contributions – both financial and in-kind – of its national laboratories, Fermilab and SLAC, as well as from its contractor, the Universities Research Association.

To all these individuals and organizations, without whose contributions we could not have held this Symposium, we extend our deepest thanks.

Photographs of the Symposium on the following pages are by Joe Faust, and provided courtesy of SLAC.

Burton Richter helps Steven Weinberg with the microphone before his opening talk.

Peter Higgs and Donald Perkins.

David Gross, Alexander Polyakov, Gerhard 't Hooft, and Lenny Susskind chat during a coffee break.

Gross, Sam Schweber, and Harry Lipkin.

Michael Riordan and Sidney Bludman discussing the sequence of speakers.

John Heilbron and Laurie Brown.

Maurice Goldhaber, Pief Panofsky, and Bill Wallenmeyer during the panel session on Science Policy and the Sociology of Big Laboratories.

John Krige speaking during the panel session, while Mark Bodnarczuk and Robert Seidel listen.

Fermilab's first Director, Robert R. Wilson, talks with Norman Ramsey, the first President of Universities Research Association.

Jim Cronin, Nick Samios, Maurice Goldhaber, and Alan Wattenberg.

Maurice Goldhaber greets Murray Gell-Mann at the Symposium banquet, while
Lillian Hoddeson and Matt Sands look on.

Gell-Mann, Gosta Ekspong, and Nina Adelman Stolar.

Abbreviations and Acronyms

Acta Phys. Pol	*Acta Physica Polonica*
Am. J. Phys.	*American Journal of Physics*
Ann. Phys.	*Annals of Physics*
Ann. Rev. Nucl. Part. Sci.	*Annual Reviews of Nuclear and Particle Science*
Comm. Math. Phys.	*Communications in Mathematical Physics*
Dokl. Akad. Nauk SSR	*Doklady Akademii Nauk SSR*
Helv. Phys. Acta	*Helvetica Physica Acta*
Hist. Stud. Phys. Biol. Sci.	*Historical Studies in the Physical and Biological Sciences*
Hist. Stud. Phys. Sci.	*Historical Studies in the Physical Sciences*
JETP Lett.	*Journal of Experimental and Theoretical Physics Letters*
J. Math. Phys.	*Journal of Mathematical Physics*
J. Phys.	*Journal de Physique*
Lett. Nuovo Ciimento	*Lettres al Nuovo Cimento*
Natl. Acad. Sci. USA	*Proceedings of the National Academy of Sciences (USA)*
Nucl. Phys.	*Nuclear Physics*
Phys. Lett.	*Physics Letters*
Phys. Rep.	*Physics Reports*
Phys. Rev.	*Physical Review*

Phys. Rev. Lett.	Physical Review Letters
Proc. Roy. Soc.	Proceedings of the Royal Society (London)
Prog. Theor. Phys.	Progress of Theoretical Physics
Prog. Theor. Phys. Supp.	Progress of Theoretical Physics Supplement
Rep. Prog. Phys.	Reports on Progress in Physics
Rev. Mod. Phys.	Reviews of Modern Physics
Sov. J. Part. Nucl.	Soviet Journal of Particles and Nuclei
Sov. Phys. Dokl.	Soviet Physics – Doklady
Sov. Phys. JETP	Soviet Physics – Journal of Experimental and Theoretical Physics
Stud. Hist. Phil. Sci.	Studies in the History and Philosophy of Science
Supp. Nuovo Cimento	Supplemento al Nuovo Cimento
Z. Phys.	Zeitschrift fur Phyzik
AdA	Anello di Accumulatione (i.e., storage ring)
ADONE	"big AdA"
AEC	Atomic Energy Commission
AGS	Alternating Gradient Synchrotron (at Brookhaven)
ARGUS	third-generation particle detector built at DORIS; also the collaboration
BCS	Bardeen–Cooper–Schrieffer (theory)
BEPC	Beijing Electron–Positron Collider
BES	Beijing Spectrometer at BEPC
BNL	Brookhaven National Laboratory
BRS	Becci–Rouet–Stora (invariance or transformation)
CDF	Collider Detector at Fermilab; also the experimental collaboration
CEA	Cambridge Electron Accelerator
CELLO	solenoidal particle detector at PETRA; also the collaboration
CERN	Centre European pour la Recherche Nucleaire (now known as the European Center for Particle Physics)
CESR	Cornell Electron Storage Ring
CKM	Cabibbo–Kobayashi–Maskawa (matrix)

CLEO	solenoidal particle detector and experimental collaboration at CESR
CRT	cathode-ray tube
CUSB	Columbia University/Stony Brook detector and collaboration at CESR
CVC	conserved vector current (hypothesis)
DASP	Double-Arm Spectrometer (DESY); also the experimental collaboration
DELCO	Direct Electron Counter detector built at SPEAR
DESY	Deutsches Elektronen Synchrotron
DOE	Department of Energy
D0	large particle detector (at Fermilab); also the experimental collaboration
DORIS	Double-Ring Storage electron–positron collider at DESY
ERDA	Energy Research and Development Administration
FNAL	Fermi National Accelerator Laboratory (a.k.a. Fermilab)
GIM	Glashow–Iliopolous–Maiani (mechanism)
GWS	Glashow–Weinberg–Salam (model or theory)
HRS	large particle detector at PEP; also the experimental collaboration
ISABELLE	unfinished proton collider (at Brookhaven)
ISR	Intersecting Storage Rings (CERN)
JADE	solenoidal particle detector at PETRA; also the collaboration
LBL	Lawrence Berkeley Laboratory
LEP	Large Electron–Positron collider (CERN)
MAC	solenoidal particle detector and experimental collaboration at PEP
MARK I	The first SLAC–LBL particle detector at the SPEAR e^+e^- collider (SLAC)
MARK II	solenoidal particle detector built at SPEAR and moved to PEP (SLAC)

MARK III	third-generation detector built at SPEAR; also the collaboration
MARK-J	large particle detector at PETRA; also the collaboration
MIT	Massachusetts Institute of Technology
MURA	Midwestern Universities Research Association
NAL	National Accelerator Laboratory (renamed Fermilab)
PCAC	partially conserved axial current
PEP	Positron Electron Project (at SLAC)
PETRA	Positron–Electron Tandem Ring Accelerator (at DESY)
PLUTO	detector built at DORIS and moved to PETRA; also the collaboration
QCD	quantum chromodynamics
QED	quantum electrodynamics
SALT	Strategic Arms Limitations Talks
SLAC	Stanford Linear Accelerator Center
SPEAR	Stanford Positron–Electron Asymmetric Ring (at SLAC)
TASSO	Two-Armed Solenoidal Spectrometer at PETRA; also the collaboration
TPC	solenoidal particle detector at PEP; also the experimental collaboration
UA1	magnetic particle detector and collaboration at CERN $p\bar{p}$ collider
UA2	major particle detector and collaboration at CERN $p\bar{p}$ collider
URA	Universities Research Association

Mathematical Notation

A	axial-vector interaction (or current); also a scattering amplitude
$A\mu$	four-vector electromagnetic potential
$B(X \to Y)$	branching ratio for interaction $X \to Y$
B, B^0, B_u, B_d	B mesons – heavy mesons containing a bottom quark
B^+	baryon-matter field (in the Nagoya model); also used for B meson
BeV	billion electron volts (old usage, now written as GeV)
$b(\bar{b})$	bottom quark (antiquark); also the Yang–Mills field
C	charge-conjugation operator or quantum number
CP	charge-conjugation–parity operator or quantum number
CPT	charge-conjugation–parity–time-reversal operator; also a theorem
$c(\bar{c})$	charm quark (antiquark); also the speed of light in vacuum
D^0, D^+, D^-	charmed mesons – containing a charm quark and an up or down quark
D^{0*}, D^{+*}	charmed meson resonances
$d(\bar{d})$	down quark (antiquark)
E	energy
$E^-(E^+)$	hypothetical electron-like heavy lepton (antilepton)
E_{cm}	center-of-mass energy
e, e^-	electron; e is also the magnitude of the electron charge

e^+	positron – the antiparticle of the electron
e^+e^-	electron-positron (as in electron-positron collisions)
F, F^+, F^*	F mesons – composed of a charm quark and a strange antiquark
$F_{\mu\nu}$	electromagnetic tensor
G_F	Fermi coupling constant
G	G-parity operator or quantum number
GeV	billion electron volts, or gigaelectronvolts
g, g'	generalized coupling constants
I	isospin quantum number
I	isospin operator
I_1, I_2, I_3	isospin components
J	total angular momentum operator or quantum number
J	angular momentum operator
$J, J/\psi$	J or J-psi particle – a heavy neutral meson of mass 3.1 GeV
J^P	spin-parity quantum numbers (e.g., $0^+, 1^-, \ldots$) of a particle
J^{PC}	spin, parity and charge-conjugation quantum numbers of a particle
K, K^+, K^-	K mesons, or kaons – mesons that contain a strange quark
K^*	kaon resonance
K^0, K_L^0, K_S^0	neutral K mesons
K_1^0, K_2^0	neutral K mesons (now more commonly written as K_S^0, K_L^0)
$L^-(L^+)$	hypothetical heavy lepton (antilepton)
$\mathcal{L}$	Lagrangian (of a system)
$l, l^-(l^+)$	lepton (antilepton); l is also used to denote orbital angular momentum
M	generalized mass of a subatomic particle
$M^-(M^+)$	hypothetical muon-like heavy lepton (antilepton)
MeV	million electron volts, or megaelectron volts
m_x	mass of a subatomic particle x
N^*	nucleon resonance
$n(\bar{n})$	neutron (antineutron); also the neutron current
P	parity operation or quantum number
P	quantum state with orbital angular momentum $l = 1$
P_c	spin-1 particle (also known as χ) produced in ψ decays

P_μ, P_ν four-momentum components of a particle

$^3P_0, {}^3P_1, {}^3P_2$ P states with third component of angular momentum $m = 0, 1, 2$

p_T or P_t transverse momentum

$p(\bar p)$ proton (antiproton); also the proton current

p or P particle four-momentum; P also used for probability

$\vec p$ three-vector momentum of a particle

Q charge quantum number; also decay energy in K decays

$q(\bar q)$ generalized quark (antiquark); q also generalized charge on a particle

$R = \sigma_{had}/\sigma_{\mu\mu}$ ratio of hadron to muon-pair production in e^+e^- collisions

$R = \sigma_L/\sigma_T$ ratio of longitudinal to transverse photoabsorption in $e-N$ scattering

S strangeness operator or quantum number

$S, 1S, 2S, \ldots$ S states of a particle system – having orbital angular momentum $l = 0$

S-matrix   scattering matrix

SU(N)   special unitary group in N dimensions

$s(\bar s)$ strange quark (antiquark); also the center-of-mass energy squared

T time-reversal operator or quantum number

TeV   trillion electron volts, or teraelectronvolts

$t(\bar t)$ top quark (antiquark)

U "unknown" particle (e.g., the particle later called the tau lepton)

$u(\bar u)$ up quark (antiquark)

V vector interaction (or current)

V Cabibbo–Kobayashi–Maskawa matrix; V also used for V particle

$V-A$ vector minus axial-vector interaction (or current)

$V_{ij}(V_{ij}^*)$ ijth element (or its hermitian conjugate) of CKM matrix

W, W^+, W^- charged vector bosons

W_μ charged vector-boson field

Z, Z^0 neutral vector boson

Z_3, Z_S transverse, longitudinal components of the neutral vector boson field

α	fine-structure constant; also $g^2/4\pi$, where g = general coupling constant
β	beta particle – an electron or positron emitted in nuclear beta decay; also a parameter in Callan–Symanzik equation
β decay	nuclear decay by emission of a beta particle (plus a neutrino)
Γ	width of a resonance or interaction strength; also scattering amplitude
γ	photon, or gammy ray
γ_μ, γ_5	Dirac matrices
δ	*CP*-violating phase in the Cabibbo–Kobayashi–Maskawa matrix
ε	dielectric constant (of a medium)
$\varepsilon, \varepsilon'$	*CP*-violating parameters measured in neutral kaon decays
η, η^0	eta meson – neutral, composed of up and down quarks
η_{+-}, η_{00}	eta plus-minus, eta zero zero – neutral kaon decay parameters
θ, θ^0	theta mesons (now known as kaons); θ also used for polar angle
θ_c	Cabibbo angle
θ_W	weak mixing angle (or Weinberg angle)
$\theta_1, \theta_2, \theta_3$	mixing angles in the Cabibbo–Kobayashi–Maskawa matrix
$\Lambda(\overline{\Lambda})$	lambda baryon (antibaryon); also used for QCD scaling parameter
$\Lambda_c(\overline{\Lambda}_c)$	charmed lambda baryon (antibaryon)
$\mu, \mu^-(\mu^+)$	muon (antimuon); μ also the magnetic permeability
$\nu(\bar{\nu})$	neutrino (antineutrino), of any type
$\nu_e(\bar{\nu}_e)$	electron neutrino (antineutrino)
$\nu_l(\bar{\nu}_l)$	neutrino (antineutrino) of type or lepton "flavor" l
$\nu_\mu(\bar{\nu}_\mu)$	muon neutrino (antineutrino)
$\nu_\tau(\bar{\nu}_\tau)$	tau neutrino (antineutrino)
π, π^+, π^-, π^0	pi mesons, or pions – spin-0 mesons composed of up and down quarks
ρ	rho parameter (in deep-inelastic ν–N scattering)
$\rho, \rho^0, \rho^+, \rho^-$	rho mesons – neutral and charged, composed of up and down quarks
σ	generalized cross section; also used for sigma model

$\sigma(X \to Y)$	cross section for a scattering or decay process $X \to Y$
σ_{had}	cross section for hadron production (especially in e^+e^- collisions)
$\sigma_{\mu\mu}$	cross section for muon-pair production (especially in e^+e^- collisions)
σ_L	cross section for absorption of longitudinal virtual photons
σ_T	cross section for absorption of transverse virtual photons
$\tau, \tau^-(\tau^+)$	tau (antitau) lepton; also used for particle lifetimes, tau mesons
τ_x	mean lifetime of a subatomic particle x
$\Upsilon, \Upsilon', \Upsilon'' \ldots$	upsilon particles – composed of a bottom quark and its antiquark
ϕ	phi meson – composed of a strange quark and its antiquark
χ	chi particles – neutral spin-1 particles produced in decays of ψ particles
ψ, ψ'	wave function or spinor
ψ, ψ', ψ''	psi particles – composed of a charm quark and its antiquark
ψ_1, ψ_2	components of a wave function or spinor
Ω^-	omega-minus baryon – composed of three strange quarks
ω, ω^0	omega meson – neutral, composed of up and down quarks

Part one

Introduction

1

The Rise of the Standard Model: 1964–1979

LAURIE M. BROWN

Born 1923, New York City; Ph.D. (physics), Cornell University, 1951; theoretical physics and history of physics; Northwestern University.

MICHAEL RIORDAN

Born 1946, Springfield, Mass.; Ph.D. (physics), MIT, 1973; experimental physics, history of physics, and science writing; Stanford Linear Accelerator Center and University of California, Santa Cruz.

MAX DRESDEN

Born 1918, Amsterdam; Ph.D. (physics), University of Michigan, 1946; theoretical physics and history of physics; State University of New York at Stony Brook and Stanford Linear Accelerator Center.

LILLIAN HODDESON

Born 1940, New York City; Ph.D. (physics), Columbia University, 1966; history of science and technology; University of Illinois and Fermi National Accelerator Laboratory.

In the late 1970s elementary particle physicists began speaking of the "Standard Model" as *the* basic theory of matter. This theory is based on sets of fundamental spin-$\frac{1}{2}$ particles called "quarks" and "leptons," which interact by exchanging generalized quanta, particles of spin 1. The model is referred to as "standard," because it provides a theory of fundamental constituents – an ontological basis for describing the structure and behavior of all forms of matter (gravitation excepted), including atoms, nuclei, strange particles, and so on. In situations where appropriate mathematical techniques are available, it can be used to make quantitative predictions that are completely in accord with experiment. There are no well-established results in particle physics that clearly disagree with this theory.

This pleasing state of affairs is quite new in particle physics. It contrasts markedly with the theoretical situation in the early 1960s, when

3

there were a variety of different ideas about the subatomic realm. For example, in 1964 most particle physicists considered protons, neutrons, pions, kaons, and a host of other strongly interacting particles (i.e., hadrons) to be in a certain sense "elementary." By 1979 the consensus had emerged that the hadrons were not elementary after all but are composed of more basic building blocks called quarks, held together by the exchange of another kind of particle called the gluon. Or consider the particle interactions. In 1964 almost all physicists thought the strong, weak, and electromagnetic interactions were independent phenomena, perhaps requiring different types of theories for their description. But fifteen years later, all three interactions had been successfully described by quantum field theories, and the last two were considered to be different aspects of a single unified "electroweak" interaction. Whereas particle physicists of 1964 used many different tongues, those of 1979 spoke a common language.

In this same period, roughly between 1964 and 1979, the entire field of particle physics passed through a profound metamorphosis. All aspects of the discipline – including detection equipment, particle accelerators, and methods of experimental and theoretical analysis – underwent irreversible change. In a wider context, the role and status of particle physics in society, its political support and financial backing, and its demographic basis were also very much altered. It was a period of turbulence, marked by deep intellectual ferment, conflicts, confusion, and some misleading results. Balanced by many remarkable discoveries – most of them discussed in this book – the period of the rise of the Standard Model was an exciting and critical period in the evolution of physics.

The conference on which this book is based, – the Third International Symposium on the History of Particle Physics [held at the Stanford Linear Accelerator Center (SLAC), 24–27 June 1992), was convened to examine this period of particle physics, which we consider to be the third major period in particle physics.[1] Two earlier symposia in this series (held in 1980 and 1985 at Fermi National Accelerator Laboratory, or Fermilab) dealt with the birth of particle physics during the 1930s and 1940s and with the field's adolescence, the period of particle discoveries in the 1950s and early 1960s, which left physics with a veritable "particle explosion."[2]

Starting in 1964, physicists made a serious effort to reduce the number of elementary constituents to a set of three particles with strong interaction – the u, d, and s quarks – and two leptonic pairs consisting of the electron e, and its neutrino ν_e, plus the muon μ, and its neutrino ν_μ, together with the antiparticles of both quarks and leptons.[3] An alternative school of thought, the "bootstrap" approach, considered all hadrons to be equally elementary (partaking in "nuclear democracy").[4] However, by 1970 evidence from deep-inelastic electron-nucleon scattering experiments at SLAC began to mount in favor of the quark theory, while the 1974 discovery of the J/ψ particle at Brookhaven National Laboratory (BNL) and at SLAC added a fourth quark to the list (the c or "charm" quark). The quark model now proved convincing to many who had been skeptical of it.

In retrospect, the most influential work with regard to a theory of particle interactions was a 1954 paper by Chen Ning Yang and Robert Mills, which introduced the idea of non-Abelian gauge fields.[5] This work caught the imagination of some theorists, but more phenomenologically oriented physicists generally ignored it for years. However, in the 1960s gauge theory gradually brought about a revival of interest in quantum field theories, which, despite the successful renormalization of quantum electrodynamics (QED) in the late 1940s, had failed to provide a satisfactory theory of the nuclear interactions.

Before proceeding to a more detailed discussion of the period of the rise of the Standard Model, we will in the next section set the stage by characterizing with broad strokes all three periods covered by this series of history of physics conferences.

Three periods of elementary particle physics
The birth of particle physics

Modern particle physics began in the 1930s as an outgrowth of experimental studies of nuclear and cosmic-ray physics being carried out with much improved techniques: electronic counter-arrays, particle-triggered Wilson cloud chambers operated in strong magnetic fields, and new types of particle accelerators. These new instruments made it possible to study interactions between particles up to about 100 MeV (in the target rest frame) and thus to explore behavior at distances of the order of the nuclear radius ($\approx 10^{-15}$ m).

Efforts to understand the new higher-energy regime in terms of concepts extrapolated from below met with only partial success. There followed the usual attempts to develop ad hoc phenomenological models, as well as attempts to apply quantum fields, even though the most useful and best established prototype, QED, was known to have serious problems connected with its high-energy behavior – the problem of the so-called divergences. During the 1930s, the other main quantum field theories developed were Enrico Fermi's theory of β decay and Hideki Yukawa's meson theory. Although we now think of these theories as describing, respectively, the weak and strong nuclear forces, in the 1930s they existed as unified theories of both types of nuclear force.[6]

The 1930s actually *began* with a maximally unified picture: All matter, including nuclear matter, consisted of "positive and negative electricity" – protons and electrons, interacting electromagnetically through the exchange of photons. There was thus (to speak anachronistically) a trinity consisting of the first hadron, the first lepton, and the first gauge boson. Soon, however, another trio of particles was found necessary to complete the standard model of its day: the positron, neutron, and neutrino.[7] The last particle, whose existence was conjectured by Wolfgang Pauli, was not observed until 1956.

By 1937 physicists had established that there are new unstable charged particles of mass intermediate between those of the proton and the electron. Named *mesotrons* (as well as other names), these particles were soon conjectured to be the same as the *U particles* that had been the basis for the theory of nuclear forces proposed by Yukawa in 1935. Interrupted by World War II, the remainder of the first period of particle physics was largely concerned with establishing the mass, spin, mean lifetime, and interaction cross section of the mesotrons, and with trying to reconcile them with Yukawa's theory, while the latter was being adapted to the constraints placed upon it by the increasing knowledge of nuclear forces.

A decade after the mesotron discovery, the true Yukawa mesons, now called *pions*, were discovered. Charged pions were identified in 1947 by the University of Bristol group through the tracks pions left in photographic emulsions exposed to high-altitude cosmic rays; the neutral pion was isolated in 1950. The mesotron, renamed *muon*, was found to be a heavy version of the electron (and is now called a second-generation lepton). The renormalization of QED led to great successes in the understanding of electromagnetic interactions, and together with the conjectured "μ–e universality" of the weak interactions, physicists hoped that a consistent and closed (though not unified) "theory of everything" could well be within reach.

However, as is usually the case when the imagined "end of physics" is in sight, there were still a few clouds in the sky. One of these was the muon itself, which had no evident role in the structure of matter (e.g., I. I. Rabi's famous comment on the muon, "Who ordered that?"). And if the muon appeared to be an extra piece left over after the theory had been completed, what was one to make of the strange particles, the mysterious cloud-chamber V-tracks discovered in Manchester in 1947, whose variety began to increase at a disturbingly prolific rate?

During the first period of particle physics, research was carried out either by individuals or by groups of a few researchers, sometimes with the assistance of semiprofessional workers, usually women (who were inevitably underpaid and called "girls"). In spite of the terrible social, political, and material conditions prevailing in much of the world during this period, science itself remained rather international in scope and spirit. And for the most part the scientific community remained friendly and cooperative, so far as the war and its aftermath allowed.[8]

From pions to quarks

To explore the new world of the pions and the strange particles, it was not sufficient to rely on chance observations in cosmic rays, which had uncertain composition, diverse energy and momentum, and too low an intensity for systematic experimentation. Therefore, although almost all the new particle discoveries up to 1955 were made using cosmic rays as a source, from that year on particle accelerators took over the job. During the transition, as cosmic-ray physicists struggled to increase their statistics in order to compete with the new accelerators, they found it necessary to work in large groups. (In 1955 the G-stack nuclear emulsion collaboration involved 22 laboratories.) At the accelerator sites also, large groups became the rule, marking a significant change in the culture of high-energy physics.

Although they were used for a variety of studies (for example, to compare the properties of muons and electrons), accelerators of energy up to about 1 GeV were built mainly to study the strong nuclear force. It was argued that their relatively large cost to society was justified, indeed required, by the great practical relevance of nuclear weaponry and nuclear power. This argument carried special force during the period of the Cold War. The research field itself was referred to as *high-energy nuclear physics*, and the adjective *nuclear* was officially dropped only in 1958.[9]

The first sighting of the neutral pion came at the 184 inch, 380 MeV Berkeley cyclotron and was confirmed later that year in cosmic ray experiments. The first pion–nucleon resonance (the spin $J = \frac{3}{2}$ and isospin $I = \frac{3}{2}$ resonance, or so-called 3-3 resonance, now called N*) was found at the 450 MeV Chicago cyclotron. The interaction properties of pions were extensively explored at cyclotrons capable of producing them (energy $E > 300$ MeV), as well as with the electron synchrotrons that operated in a similar energy range. One important result of the comparison of positive and negative pion-scattering amplitudes from nucleons was the establishment of the charge independence of the strong nuclear force.

Electron-scattering experiments, carried out at Stanford and Cornell, revealed the true size and shape of the nucleons, verifying that they are extended objects with radii $\approx 10^{-15}$ m. They were, of course, already known to have anomalous (that is, non-Dirac) magnetic moments and were assumed to differ from leptons also in other ways, for example by the presence of an extended meson "cloud." This was a good indication that hadrons would have to be treated as complex systems, reinforcing the view of many theorists who rejected the idea that strong interactions could be described in terms of quantum fields.

The real particle explosion began with the accelerators that operated in the multi-GeV range: the Cosmotron at BNL and the Bevatron at the Berkeley Radiation Laboratory (now the Lawrence Berkeley Laboratory, or LBL). At the same time, more sensitive visual detectors, the high-pressure diffusion cloud chamber and the bubble chamber, whose medium could serve also as a target, permitted the convenient viewing of the resulting long high-energy tracks. Now detailed studies could be made of the production and decay of the strange particles. These studies verified the idea of *associated production* (i.e., strong production in pairs, but weak decay) to explain the puzzling behavior that had earned them the name "strange." The behavior was then attributed in 1955 by Kazuhiko Nishijima and Murray Gell-Mann (independently) to an additive quantum number, called "strangeness" by Gell-Mann. This strangeness quantum number, conserved by strong and electromagnetic forces, could be violated by the weak interaction.

The study of strange particles, especially those called K mesons of mass about 500 MeV, led in the mid-1950s to a major conundrum. Certain decay modes of K mesons (now generally called *kaons*) were shown, especially by Richard Dalitz, to have opposite parity. Since this feature appeared to violate the left-right symmetry principle taken as universally valid in quantum mechanics, physicists assumed these were decay modes of different particles of very nearly equal mass. But as data accumulated, this viewpoint became hard to maintain, and C. N. Yang and T. D. Lee proposed several experiments that could be performed to determine whether parity *violation* might be a general feature of weak interactions, including pion and muon decay and β decay. The experimental proof that this unexpected behavior was present in nature constituted the *parity revolution*.

The discovery that parity is not conserved in weak interactions, far from detracting from the significance of symmetry, actually increased interest in the discrete symmetry operators, which besides parity P included charge-conjugation C and time-reversal T. Theorists showed that relativistic locality required invariance of the Lagrangian of any system under the combined operation CPT. Then if invariance under time-reversal were assumed, the laws of physics would be invariant under the newly defined operation of *combined inversion CP*. This assumption had far-reaching consequences; for example, the neutrino turned out to have a left-handed *chirality* (i.e., handedness), while the antineutrino is right-handed.

A consequence of the neutrino's chirality was the necessary inclusion of vector (V) and axial vector (A) interactions from the five relativistic forms available for the weak interactions. This was a major step toward a truly universal weak interaction, since all weak interactions could be formulated as a mixture of V and A, usually written as $V–A$. To complete the story required several additional steps: The interaction was formulated as that of charged *currents* (e.g., $n–p$ or $\nu–e^+$). The V current was taken to be a generalized electric current (i.e., its neutral component *was* the electric current itself) and was therefore conserved; on the contrary, the A current was conserved only in the limit of zero pion mass (known as partially conserved axial current, or PCAC). Finally, each current was the sum of a strangeness-conserving and a strangeness-changing part, with coefficients proportional to $\cos \theta_c$ and $\sin \theta_c$, respectively, where θ_c is the empirically determined Cabibbo angle.

During this period, the nature of the lepton sector was greatly clarified. The actual detection of the neutrino as a particle was accomplished in 1956 after years of effort by Frederick Reines and Clyde L. Cowan, Jr. Two further important results bearing on the weak interactions occurred near the end of the pions-to-quarks period. The electron and the muon were found in 1962 to be associated with distinct neutrinos, and the concept of a conserved lepton number was thus found to be valid (Schwartz, Chapter 24). And in 1964 James Cronin and Val Fitch discovered that CP invariance was violated in certain rare kaon decay processes (Cronin, Chapter 7).

As noted, the detailed study of pion interactions and of the associated production and interactions of strange particles depended on the construction of multi-GeV accelerators: Brookhaven's Cosmotron (1952, 3 GeV), Berkeley's Bevatron (1954, 6 GeV), the Synchrophasotron at Dubna (1957, 10 GeV), and others.[10] In 1952 a new principle called "strong focusing" made possible the design of even more powerful accelerators. The first strong-focusing machines were electron synchrotrons (e.g., at Cornell in 1954 at 1 GeV), but the principle was soon applied also to proton accelerators, resulting in the European Center for Nuclear Research (CERN) Proton Synchrotron (1959, 28 GeV) and the BNL Alternating Gradient Synchrotron, or AGS (1960, 32 GeV). Used with the new liquid-hydrogen bubble chambers, these machines allowed a number of new resonant states to be isolated in 1960 and 1961. Bubble chambers at Brookhaven and Berkeley were used to discover strange particle resonances, both of baryonic and mesonic type. Three non-strange mesons, the $\rho(770)$, the $\omega(783)$, and the $\eta(549)$, were also observed.[11]

With these discoveries of groups of particles of given character (e.g., several isospin multiplets of the same spin and parity) emerged a new spectroscopy that various researchers then tried to analyze from the standpoint of group theory. Another approach was to develop composite models for all the hadrons. Already in 1949, Fermi and Yang had suggested that pions might be bound pairs of a nucleon and an antinucleon, and Shoichi Sakata in 1956 generalized this idea to include strangeness by taking the lambda hyperon to be a third fundamental constituent. Sakata's Nagoya associates developed this idea, pointing out that the group SU(3) was the appropriate generalization of the isospin group SU(2), which was the basis of the Fermi–Yang model.

The ultimately successful group characterization due to Gell-Mann and to Yuval Ne'eman (independently) was called the "Eightfold Way." This was also an SU(3) group characterization, but it did not require a set of observed particles to form the fundamental three-fold representation of the group. Instead, the lower-lying spin-$\frac{1}{2}$ baryons (including the nucleons) simply formed an octet representation of SU(3), while a similar octet representation was formed by the pions, the kaons, and the eta meson (hence the name Eightfold Way). This model, octet-broken SU(3), so called because the operator giving the mass differences between the isospin multiplets was assumed to transform like a member of an octet, was spectacularly confirmed by the 1964 discovery of the omega-minus or Ω^- particle, predicted in 1962, having three units of strangeness (Samios, Chapter 29).

The rise of the Standard Model

As this third period will be the subject of the remainder of this essay, we will mention here only the most general themes and discoveries, reserving further detail for later sections. We shall for the most part refer to appropriate chapters in this book, rather than to original sources.

In the year of the Ω^- discovery, Gell-Mann and George Zweig independently proposed that hadrons could be made of three elementary fermions; not the p, n, Λ of Sakata, but new objects having baryon number $1/3$ and electric charges $e/3$ and $-2e/3$. Gell-Mann called these previously unthinkable fractionally charged objects "quarks," and that is the name that survived. That we observe only integral multiples of the electron charge e is attributed to a conjectured property of the quarks called *confinement*.

In the mid-1960s there were two rather different attitudes toward Gell-Mann's quarks (or "aces," as they were referred to by Zweig). They were regarded, on the one hand, as useful "mathematical" constructs lacking any physical reality and, on the other hand, as real physical pointlike objects, no less real than electrons or other fermions.[12] (Gross, Chapter 11; Lipkin, Chapter 30; Morpurgo, Chapter 31; Gell-Mann, Chapter 35). These two attitudes were accompanied by successes of quark phenomenology and by unsuccessful searches for free quarks. The successes culminated in the "scaling" of structure functions observed in 1968–69 in electron–nucleon scattering at SLAC, which was interpreted as evidence for the presence within nucleons of pointlike constituents (Friedman, Chapter 32; Bjorken, Chapter 33).

It was disappointing that the formally appealing Yang–Mills theory could not be used in a straightforward manner to describe either strong or weak interactions, since that would have demanded a number of massless vector bosons that were not to be found. Even if the gauge symmetry were spontaneously broken, a theorem of Goldstone showed that the theory would always require some massless particle.[13]

However, in 1964 a method of evading Goldstone's theorem (the Higgs mechanism) made use of mixing between scalar and vector particles. It solved the mass problem for particles of spin-1 at the cost of introducing a new kind of massive particle, the spin-0 Higgs boson (Brown, Chapter 28). Steven Weinberg and Abdus Salam independently proposed a theory in 1967, now called *electroweak* theory, that used the Higgs mechanism to make a unified theory of electromagnetism and weak nuclear interactions. For the next four years this theory was essentially ignored, in part because it predicted processes involving neutral weak currents at about the same level as charged weak currents, which appeared to be ruled out by experiment.[14]

The situation changed remarkably, however, in the early 1970s because of several developments, both theoretical and experimental. For one thing, a method was found to make the contribution of strangeness-changing neutral weak currents negligible, at the expense of adding a fourth (*charm*) quark (Iliopoulos, Chapter 26). A second theoretical breakthrough was the proof that massive Yang–Mills theory is renormalizable when the mass is produced by the Higgs mechanism (Veltman, Chapter 9; 't Hooft, Chapter 10). Finally, in 1973 and 1974 the existence of strangeness-conserving neutral currents of magnitude comparable to the corresponding charged currents was established in neutrino-scattering experiments at CERN and at Fermilab (Perkins, Chapter 25).

But there was a period of confusion for several years, due to apparent anomalies in neutrino–nucleus scattering and measurements of atomic parity violation. In 1978 the Weinberg–Salam theory was directly confirmed by the observation of interference between weak and electromagnetic interactions in the scattering of polarized electrons on deuterium at SLAC (Prescott, Chapter 27). At that point, the electroweak theory became widely accepted as an integral part of today's Standard Model.

The other part of the model's interactions, the color SU(3) sector, can be traced back to 1965, when an unbroken SU(3) Yang–Mills gauge theory was proposed by Yoichiro Nambu, in part to solve an outstanding problem concerning the statistics of quarks. Unless the spin-$\frac{1}{2}$ quarks possess a new degree of freedom now called "color," they cannot satisfy Fermi statistics, as they must do. This threefold color "charge" acts as the source of an SU(3) quantum field, which was later found to have a surprising (and gratifying) behavior at short distances. Unlike the usual quantum field theories, such as QED, non-Abelian Yang–Mills theories become *asymptotically free*, that is, noninteractive (Gross, Chapter 11; Susskind, Chapter 12; Polyakov, Chapter 13). This means that well-developed perturbation techniques can be used to deal with high-energy processes. The successes of the other Yang–Mills theory, the electroweak, undoubtedly caused a revival of interest in this color theory, now called quantum chromodynamics, or QCD.

Observation of the charm quark in the mid-1970s played a key role in the acceptance of both the electroweak theory and QCD. Indeed this quark appeared in such a spectacular way, and had such an immediately riveting effect, that physicists speak of its 1974 discovery and the aftermath as the "November Revolution" (Goldhaber, Chapter 4). The charm quark together with its antiparticle appeared first in a hidden form as the constituents of a meson given the name J/ψ, an extremely narrow (hence long-lived) resonance of mass about 3.1 GeV.

With the rise of the Standard Model, and specifically its electroweak sector (see below), came the growing realization that leptons and quarks came in groupings called "generations" or "families," each bearing a pair of quarks and a pair of leptons. Ordinary, garden-variety matter is composed of particles from the first generation – the up and down quarks u and d plus the electron and its associated neutrino, the electron neutrino. The second generation includes the charm and strange quarks c and s, plus the muon and muon neutrino. In this picture the tau lepton discovered at SLAC in 1976 (Perl, Chapter 5) had to belong to a *third* generation of quarks and leptons. Indeed, such an additional generation had already been predicted in 1973 by Makoto Kobayashi and Toshihide Maskawa (Kobayashi, Chapter 8) to account for the occurrence of CP violation, which had been discovered a decade earlier and was well established by the middle of the 1960s.

The occurrence of a third generation meant that two more heavy quarks had to exist, called "bottom" and "top," b and t (or "beauty" and "truth" if one worked in Europe). The bottom or b-quark was found at Fermilab in 1977 (Lederman, Chapter 6) and direct evidence for the top or t-quark has finally shown up in experiments at that laboratory.

By 1979 then, particle physicists had a fairly compact description of subatomic processes in terms of a table of "fundamental" entities consisting of six quarks and six leptons that interacted with one another through the agency of gauge bosons. These interactions could all be described by Yang–Mills gauge theories; electromagnetism and the weak interaction had been successfully unified by the electroweak theory, to which we now turn.

The electroweak theory

The concept of intermediate vector bosons was crucial to the development of the electroweak theory. Yukawa made this a part of his meson theory, which he conceived as a unified theory of strong and weak interactions. His meson was used as an intermediate boson in β decay, connecting hadronic and leptonic currents. The Fermi constant G was given by

$$G = gg'/4\pi m^2,$$

where m is the meson mass and g and g' are the coupling constants of the meson with, respectively, nucleons and leptons. However, this constraint on the couplings proved unable to account simultaneously for β decay and meson decay (or so it was thought, due in part to the confusion of the muon with the Yukawa meson).

Abandoned at the beginning of the 1940s, the unified theory was taken up in a new form in 1957 by Julian Schwinger, who pointed out that a single coupling strength α would suffice for both QED and weak interactions, provided the intermediate bosons were sufficiently massive. Schwinger proposed electroweak unification, with two vector bosons and the photon forming a representation of the group SU(2), which amounted to the introduction of a kind of (badly broken) weak isospin.[15] No explanation was given for the large mass splitting in the multiplet.

In 1958 Sidney Bludman suggested a theory that implied the existence of some neutral current weak interactions but ruled out the case of neutrino scattering on leptons.[16] The paper was mainly concerned with a new derivation of the $V-A$ form of weak interaction, assuming invariance under a continuous global transformation.[17] This would imply its being "universal," except that, according to Bludman: "We are inclined to believe that the Fermi interactions do not allow changes of strangeness."[18] When the author generalized to a *local symmetry*, he obtained a Yang–Mills field b as the possible carrier of the weak interaction. But he pointed out: "At least two difficulties argue against a realistic interpretation of the b field."[19] In addition, the massiveness of the b field violated the gauge invariance. To explain the existence of strangeness-changing decays, another and different "weak Yukawa" interaction was required.

Schwinger's idea of electroweak unification was picked up in 1960 by Sheldon Glashow, who pointed out that an additional neutral vector boson would be required, as well as some principle of symmetry between weak and electromagnetic interactions.[20] For the latter he invoked the notion of partial symmetry (as in the idea of PCAC, which had been proposed earlier that year). The two neutral bosons, called Z_3 and Z_S (the latter by analogy with strangeness), were combined with a mixing angle into orthogonal linear combinations identified, respectively, as neutral quanta of the electromagnetic and weak interaction fields. Regarding the mass problem, Glashow said in his introduction to the paper: "It is a stumbling block we must overlook." He was also silent on the absence of observable weak neutral currents in strangeness-changing decays, and indeed suggested that there were some indirect experimental indications that they were present (as an explanation of the so-called $\Delta I = \frac{1}{2}$ rule). A theory similar to Glashow's was worked out also by Abdus Salam and John Ward.[21]

As mentioned above, the year 1964 brought important developments: the completion of the tenfold representation of SU(3) (the decuplet or decimet) by the discovery of Ω^-,[22] the proposal of quarks by Gell-Mann and Zweig,[23] the discovery of the Higgs mechanism for evading the Goldstone theorem,[24] and the proposal of a fourth quark.[25] The idea of spontaneous symmetry breaking, on which the Higgs mechanism is based, goes back a century or more to the Weiss theory of magnetism (see panel discussion of this topic in Chapter 28). It was first used in the context of producing the mass of an "elementary" particle by Nambu and Giovanni Jona-Lasinio in 1961.[26] The electroweak sector of the Standard Model uses non-Abelian gauge theory, with three of the four gauge bosons acquiring mass by the Higgs mechanism, and it needs a minimum of four quarks to avoid strangeness-changing neutral currents.

Thus, by the end of 1964 the main ingredients of the Standard Model were available. Yet the ultimately successful Weinberg–Salam model was proposed only in 1967, and it received remarkably little attention until 1971. There were a number of reasons for this lack of interest. In the first place, there was substantial distrust of quantum field theory from the 1950s when, notwithstanding the successes of renormalized QED, no other quantum field theory had been usefully applied in high-energy strong interactions. S-matrix theory, dispersion theory, or Regge-pole methods were used instead. Neither did quantum field theory work in weak interactions, except in lowest (tree) approximation. The elegant idea of spontaneous symmetry breaking was first applied in trying to explain the breaking of global SU(3) symmetry (which we now call "flavor" symmetry, to distinguish it from color). That is, it was applied to the mass splitting between the isospin multiplets within a given SU(3) representation, and also to the muon–electron mass difference.[27] For these purposes it is actually inappropriate – and there was the additional problem of the unwanted massless Goldstone bosons.

Theories of gauge vector bosons continued to attract attention, and there were conjectures that gauge theories might be renormalizable, provided the mass problem was solved, but their main application in the early 1960s came in the strong interactions.[28] As for unification of forces, it was thought preferable to try to unify the *strong* and electromagnetic forces, as in the theory called "vector meson dominance." This theory proposed "that the entire hadronic electric current operator is identical with a linear combination of the local field operators of the known neutral vector mesons."[29] Thus, a high energy photon was effectively coupled to a linear combination of the three vector mesons ρ, ω, and ϕ.

In 1967, Weinberg proposed an electroweak theory of leptonic interactions, based upon the gauge symmetry $SU(2) \times SU(1)$, with a triplet plus a singlet of spin-1 bosons, interacting with a complex scalar $SU(2)$ doublet field. This formulation gave mass to three of the vector bosons by the Higgs mechanism.[30] Weinberg restricted the theory to leptons, so that he would not have to deal with the problem of the apparent absence of strangeness-changing neutral currents (e.g., a neutral kaon decaying into two muons was ruled out experimentally).

Weinberg's and Salam's papers received no more than a handful of citations until 1971, when the young Dutch theorist Gerard 't Hooft showed that the theory was renormalizable.[31] A student of Martinus Veltman, 't Hooft used a set of mathematical techniques largely due to Veltman to prove renormalizability – first for massless gauge fields, and then for gauge fields whose mass arose through spontaneous symmetry breaking. It was 't Hooft's proof of the renormalizability of gauge theories that revived theoretical interest in them. The number of citations of Weinberg's electroweak *Physical Review Letter* rose from one in 1970 to 64 in 1972. However, there were other important experimental and theoretical developments, which were responsible for the acceptance of the Weinberg–Salam model by 1975, and made it possible for those authors – together with Glashow – to be awarded the Nobel Prize in 1979 (even though the predicted vector gauge bosons W and Z had not yet been directly observed). These crucial developments were connected with the discovery of neutral weak currents and quarks, including the charm quark.

By 1970 it was becoming increasingly common to think of the fundamental hadronic weak and electromagnetic currents as quark currents. By then, the ideas of conserved vector currents and PCAC had been extended to a complete algebra of currents, and these were taken as quark currents.[32] The idea of quark–lepton symmetry had been used to predict the existence of a fourth quark, and since quarks are supposedly as "elementary" as leptons, it was natural to extend the Weinberg–Salam model to include the quark currents.[33]

This extension immediately raised the question of neutral currents. All weak interaction theories predicted them to occur, at least as secondary processes, which would be extremely weak and difficult to observe. While the theories with a neutral vector intermediary predicted them to be of the same order of magnitude as the charged-current processes, neutral currents seemed to be embarrassingly absent from experiments. Since strangeness-*conserving* neutral currents would compete with strong and electromagnetic processes, they would be overwhelmed by the latter, except for processes where neutrinos were present and detected. However, for strangeness-*changing* effects like K-decay, which cannot take place electromagnetically, the neutral weak currents should have been prominent. Instead, they were ruled out experimentally in K-decay by six to eight orders of magnitude.[34]

The way out of this predicament was found by making use of the charm quark in constructing the quark currents. Cabibbo had shown, already in 1963, that the effective hadronic currents involved a mixture having unequal coefficients of strangeness-changing and strangeness-conserving currents. This was necessary in order to explain the relative weakness of the former type of decay with respect to the latter. Now Glashow and his collaborators, John Iliopoulos and Luciano Maiani, found that they could use an SU(4) multiplet of four quarks to construct weak currents whose neutral strangeness-changing effects vanish.[35] (This became known as the GIM mechanism, based on the initials of the authors.) The same set of currents, however, predicted that neutral and charged weak interaction effects of the strangeness-conserving sort were of the same order of magnitude, and that they should also be seen in the purely leptonic cases.

In 1973 the predicted neutral currents were indeed found in neutrino-scattering experiments at CERN, in the heavy-liquid bubble chamber Gargamelle. The same experiments also confirmed the charged-current scattering predictions of the Weinberg–Salam theory.[36] These results were all confirmed by neutrino-scattering experiments at Fermilab, using electronic detectors.[37] In both cases, the "signature" of the desired neutral-current events was the production of a group of hadrons by neutrino interaction, without the production of a charged lepton, that is, an electron or muon. Because similar events can be produced just as well by an invisible neutron as by an invisible neutrino, the most difficult part of the analysis consisted of distinguishing the "true" (neutrino) from the "background" (neutron) events. There have been extensive analyses of these experiments from the point of view of "social construction of science."[38]

The next crucial discovery toward establishing the correctness of the electroweak theory was the J/ψ particle in November 1974 (discussed below). Finally, before the theory's ultimate prediction of the W and Z particles was verified in 1983, the most significant confirmation was the observation in 1978 at SLAC, using a polarized electron beam, of a predicted interference between weak and electromagnetic electron scattering from nucleons.[39] The signature was an asymmetry, brought about by the parity-nonconserving weak interaction amplitude. According to one historian of physics, this experiment "contributed more than any other to the establishment of the Weinberg–Salam model as the Standard Model of electroweak interactions."[40]

Quantum chromodynamics

As already stated, the sector of the Standard Model that deals with the strong interactions of hadrons is called quantum chromodynamics, or QCD. Like the theory of quantum electrodynamics or QED, it is an unbroken renormalizable gauge theory, but unlike QED it is non-Abelian; that is, the result of successive gauge transformations on the field in question depends upon the order in which they occur. In QED the sources of the field are electric charges and magnetic dipoles, while in QCD they are the three generations of colored quarks – fermion doublets of fractional charge and baryon number.

The roots of the theory of color, a completely new property of matter, can be traced to the problem of statistics encountered by those who wished to consider the quarks Gell-Mann and Zweig proposed in 1964 to be "real" constituents of hadrons. The problem was that two identical fermions (e.g., two up quarks in a proton) cannot occupy the exact same quantum state, according to Pauli's exclusion principle. To circumvent this problem, quarks had to possess some kind of additional property that could *differ* from one quark to another. Several proposals were made in the mid-1960s to solve this statistics problem. The most fruitful of these was a proposal by Nambu to treat the required new quantum number as the source of a new quantum field, such that the forces between quarks would be the result of exchanges of eight massless vector particles with an unbroken SU(3) gauge symmetry.[41] This was a new symmetry, different from the SU(3) symmetry (now called flavor symmetry) of the Eightfold Way. The quarks in this theory, as further elaborated by M. Y. Han and Nambu, came in nine varieties and carried integer electric charges.[42]

The need for such a new quantum number and its accompanying color field did not become major concerns of the particle physics community, however, until it could be demonstrated that quarks indeed existed as the constituents of hadrons. A long series of phenomenological successes of the nonrelativistic quark model in reproducing hadron magnetic moments and transition probabilities (Lipkin, Chapter 30; Morpurgo, Chapter 31) was dogged for years by the lack of direct experimental evidence for the existence of quarks. But in 1968 came the first results from the MIT–SLAC experiments in deep-inelastic scattering, suggesting that high-energy electrons were rebounding from pointlike entities inside protons.[43] As these experiments were extended to larger angles and to include electron–neutron scattering, the quantum numbers of these objects, generically dubbed "partons" by Richard Feynman, were found to be consistent with those expected for quarks.

But analysis of the initial round of experiments in terms of the quark–parton model (Bjorken, Chapter 33) indicated that only about half the proton's momentum could be carried by fractionally charged quarks (Friedman, Chapter 32), with the other half due to uncharged entities. Early converts to the quark–parton viewpoint generally believed that these neutral partons had to be the "gluons" – a term that began to see wide use in the early 1970s – needed to bind quarks together inside hadrons.

In a pedestrian picture of this color idea, individual quarks come in three different colors – red, green, and blue, for example – while observable baryons and mesons such as the proton and pion are colorless. Just as red, green, and blue light combine to yield white light, so would a "red" quark, a "green" quark, and a "blue" quark combine to produce a "colorless" baryon. Most theories required that only colorless, or SU(3) singlet, states could be observed in Nature, but others (including Han and Nambus) suggested that colored hadrons might actually turn up as excited states at higher energies.

Colored quarks also helped to resolve the so-called triangle anomaly and correctly predict the rate of neutral pion decay. And they readily explained the unexpected profusion of hadrons being produced in electron–positron annihilation experiments at the Italian storage ring ADONE (see below).[44] By the beginning of the 1970s, therefore, there were at least three pieces of experimental evidence for the existence of this new quantum number, encouraging theorists to explore its further ramifications.

One of the major mysteries at that time was how the force between quarks, which seemed to be relatively weak at short distances (allowing use of the impulse approximation in calculations of deep-inelastic scattering), could become strong at separations approaching the size of a nucleon and somehow prevent their escape. Physicists were accustomed to forces such as gravity and electromagnetism that *decrease* with increasing separation between objects. A natural explanation of this counterintuitive behavior – and of the approximate scaling of nucleon structure functions in deep-inelastic scattering – came from applications of renormalization-group techniques and Yang–Mills gauge theories to the strong force (Shirkov, Chapter 14; Gross, Chapter 11). In the spring of 1973, David Gross and Frank Wilczek at Princeton and David Politzer at Harvard realized independently that non-Abelian gauge theories exhibited the unique property of "asymptotic freedom," whereby the force between two quarks would fall to zero as they approached one another closely at high energies.[45] This theoretical breakthrough opened the door to quantitative calculations of high-energy processes involving the strong interaction; it also made plausible the notion of "infrared slavery," or permanent confinement of color – the nonemergence of single quarks and gluons (Susskind, Chapter 12; Polyakov, Chapter 13). Quarks and gluons are thought to be permanently confined within hadrons because at large separations the effective confining potential (due to gluon exchanges) grows linearly, corresponding to a constant attractive force. Therefore separation into component quarks and gluons requires an infinite amount of work.

The specific gauge theory to use for the gluon field was still very much up in the air as late as the summer of 1973, when Gell-Mann began to focus the attention of the theoretical community on vector gluons with the SU(3) symmetry originally suggested by Nambu.[46] Possessed of the naming gift, he who had dubbed the new property "color" now named the theory of the interquark force "quantum chromodynamics" (Gell-Mann, Chapter 35), a clever choice that aided its rapid acceptance by the rest of the particle physics community. (Curiously, Nambu had used the term "charm" for his new property of quarks, perhaps unaware that Bjorken and Glashow had already appropriated the very same name for their fourth quark; "quantum charmdynamics" somehow doesn't have quite the same ring.)

Convincing experimental proof for color and QCD was, however, hard to find. To test QCD's predicted logarithmic deviations from exact structure-function scaling (Gross, Chapter 11) required extremely precise measurements of lepton–nucleon scattering cross sections over wide ranges of energy and momentum transfer. Even when such deviations appeared, they could be interpreted instead as the results of nonleading terms in a perturbation expansion or due to other, nonperturbative effects. For a brief time the dramatic 1974 appearance of the J and ψ particles at Brookhaven and SLAC was interpreted by some theorists as the appearance of colored hadrons, but that idea soon fell by the wayside as these measurements were improved and extended.

At the end of the decade the new storage ring PETRA (Positron–Electron Tandem Ring Accelerator) came on line at the Deutsches Elektronen Synchrotron (DESY) laboratory in Hamburg, colliding beams of electrons and positrons at combined energies in excess of 10 GeV. As the total energy increased in successive runs during 1979, a distinctly new phenomenon began to emerge (Wu, Chapter 34) that had not been witnessed at earlier machines. Whereas events with two back-to-back "jets" of hadrons had been clearly observed on the Stanford Positron-Electron Asymmetric Ring (SPEAR) storage ring at SLAC and interpreted as the remains of a quark and an antiquark produced in electron–positron annihilation, the four PETRA detectors yielded growing evidence for events with *three* distinct lobes.[47] The third jet in these "three-jet events" is now recognized to be due to the emergence of an energetic gluon in a process analogous to *bremsstrahlung* in QED. Although it took a few more years to make an absolutely convincing case, this *visual* evidence for gluons was perhaps the most influential factor in the acceptance of QCD as the correct theory of the strong interactions. The gluon, like the graviton, actually experiences the very force it carries.

While at the beginning of 1964 hadrons had been largely perceived as strongly interacting particles, possibly fundamental, with dimensions of about 10^{-15} m, by the end of 1979, hadrons were almost universally believed to be composite particles made up of two or three quarks (and an indefinite number of virtual quark–antiquark pairs) bound together by colored gluons. A major upheaval in physicists' understanding of hadrons had taken place. The strong force of 1964 was found to be merely the uncancelled residue of the far stronger long-range QCD forces between individual quarks and gluons within baryons and mesons – a kind of Van der Waals force of the hadrons.

Heavy leptons and quarks

The mid-1970s was marked by a spate of new particle discovery reminiscent of that which occurred during the late 1950s and early 1960s (Samios, Chapter 29). The experimental tools that fostered this later particle explosion were the storage ring colliders and their associated multielement electronic detectors (see below), although important discoveries also came in traditional fixed-target experiments. The hadrons discovered through 1964 had all been organized into a coherent framework by incorporating them into the singlets, octets, and decimets of Gell-Mann and Ne'eman's original SU(3) scheme.[48] In contrast, all the newfound particles of the mid-1970s could be readily accounted for by adding a few new entries – heavy leptons and quarks, with masses in excess of 1 GeV – to the physicists' standard table of fundamental entities.

The first of these new particles was the charm quark, which appeared in late 1974 in the form of the J and ψ particles, neutral mesons with masses of 3–4 GeV that were soon interpreted as combinations of a charm quark and its antiquark.[49] But these hadrons carried only "hidden" charm; other hadrons that are composed of a charm quark plus another quark and explicitly exhibit the new property, or "naked charm," were discovered during an intensive period of experimentation that occurred over the next two years (Goldhaber, Chapter 4; Samios, Chapter 29). The detailed spectroscopy of the J and ψ particles, which is highly reminiscent of the spectroscopy of the hydrogen atom, was an important factor in convincing the rest of the physics community that quarks are indeed real, fundamental particles rather than some kind of hypothetical mathematical entities.

During the series of experiments on the SPEAR storage ring (see below) that resulted in the ψ discoveries, a subgroup of physicists led by Martin Perl unearthed a handful of "anomalous e–μ events," in which electron–positron collisions had resulted in an electron (or positron) and a muon, plus other particles (Perl, Chapter 5). After this handful had grown to about a hundred in 1975, the SLAC–LBL collaboration announced the discovery of a new heavy lepton, identical to the much lighter electron and muon except for its mass of 1.78 GeV, called the τ particle, or tau lepton.[50] Pair-produced in electron–positron collisions, the tau and its antiparticle occasionally both decay semileptonically – one to an electron and the other to a muon – yielding the e–μ events that led to Perl's discovery.

The discovery of the bottom quark came in 1977, soon after the tau discovery indicated there should be additional quarks. After one false start, a group of physicists led by Leon Lederman reported the discovery at Fermilab of a series of three neutral mesons called the "upsilon" or Y particles (Lederman, Chapter 6);[51] similar to the ψ particles, these mesons are composed of a heavy quark, in this case the bottom quark, and its antiquark. The discovery of particles explicitly carrying this new quantum number had to await the early-1980s start-up of the Cornell Electron Storage Ring (CESR), which had the right energy to produce a fourth upsilon particle and allow its disintegration into a pair of B mesons.[52]

A major goal of the PETRA storage ring, and of a similar, competing collider at SLAC named PEP (Positron Electron Project), was to discover the top quark required by the emerging Standard Model. At the time, phenomenological models suggested that the top quark's mass might fall in the 10–20 GeV range accessible at these machines, but early experiments failed to find any evidence for its existence.[53] So firm was the belief in the Standard Model by 1980, however, that this failure did not lead to serious doubt about the theory's validity. Physicists widely supposed that the top quark had to be more massive than could be produced at PETRA and PEP, and that it would eventually be found at higher energies. After a global search lasting over a decade, the 1994 announcement at Fermilab of evidence for the top quark appears finally to have borne out this faith.[54]

Thus the 1964–79 period opened with a successful classification scheme for hadrons based on SU(3) symmetry, but their connection with leptons was largely a mystery. By the time the period closed, a deeper rationale for this scheme in terms of fundamental quarks was widely accepted, and the relationship between hadrons and leptons was obvious. Newly discovered particles were easily accommodated by adding a third generation of quarks and leptons to the fundamental particle table of the Standard Model.

Accelerators, detectors, and laboratories

While European physicists were concentrating their research efforts at CERN and DESY during the 1960s and 1970s, two major new laboratories for particle physics, SLAC and Fermilab, appeared on the American landscape. The first high-energy physics laboratories in America to cost hundreds of millions of dollars to build, they were also the first to quote the dimensions of their particle accelerators in kilometers or miles rather than meters or feet. They afforded experimenters roughly a factor of 20 increase in particle energies over what had been otherwise available in the early 1960s. The construction of both Fermilab and SLAC benefited from huge increases in U.S. science funding in the nervous decade following the Soviet launch of Sputnik – a rise that had leveled off by the early 1970s, and actually fell slightly during the rest of the decade.[55]

At first all of these laboratories focused on so-called fixed-target methodology of experimentation, dating back to Ernest Rutherford's scattering experiments in the 1910s. Here energetic particles are passed through a stationary target, or allowed to decay in flight. The epitome of this experimental style is the bubble-chamber experiment, developed in the 1950s and perfected in the 1960s.[56] Some of the key discoveries of the 1964–79 period, such as the omega-minus and neutral currents, continued to be made with bubble chambers. But by the period's close, these workhorse detectors were being rapidly superseded by multielement electronic detectors patterned after the SLAC–LBL detector used on the SPEAR storage ring (Schwitters, Chapter 17; Galison, Chapter 18).

The main problem with fixed-target experiments was that the center-of-mass energy of a collision rises only as the square root of the beam energy. Consequently, almost all the resulting secondary particles are pitched forward into an increasingly narrow bundle or jet, thereby making it difficult for experimenters to observe all the details of an event. Several of the pivotal experiments of the period were those such as the MIT–SLAC deep-inelastic scattering experiments that ignored forward scattering and recorded only those extremely rare events in which particles emerged from collisions at relatively large angles to the beam direction. (In fact, the MIT–SLAC experiments began as a traditional study of resonance electroproduction at mostly forward angles, where the initial evidence was discovered in 1967–68 for pointlike nucleon constituents; only then did the emphasis swing to high momentum-transfer events at large angles.) After some confusion, such "hard-scattering," high momentum-transfer events were determined to be the results of close encounters between two fundamental fermions.

Physicists had recognized these limitations of fixed-target experiments as early as the 1950s. During that decade, colliding-beam machines, in which two beams of high energy particles are made to clash with each other, graduated from the status of interesting speculations and began to emerge as promising new tools for particle physics research. In a geometry in which the two beams clash head-on, or nearly head-on, almost all the energy carried by two colliding particles (one in each beam) is available for producing massive new particles. Although initial estimates of the event rates in such colliding-beam machines were pessimistic, the collision cross sections proved to be orders of magnitude larger than expected (due largely to the quark substructure of hadrons), and a statistically meaningful number of events could be recorded fairly rapidly. By the end of the 1970s, particle colliders (as these machines had by then become known) were firmly established as the only kind of machine to use in searches for massive new particles.

The first colliding-beam approach to receive extensive design attention was based on the proposal for the fixed-field alternating-gradient accelerator pioneered in the 1950s by the Midwestern Universities Research Association (MURA).[57] Such a machine was thought to be capable of stacking the intense beams then considered necessary to compensate for the anticipated small collision rates. But its construction was never approved.

Princeton's Gerard O'Neill suggested a different approach using two small-aperture storage rings intersecting at a single point. Beginning in 1958, this concept was implemented at Stanford's High Energy Physics Laboratory by a small group that included O'Neill and Burton Richter (Richter, Chapter 15). Although several instabilities hindered accumulation and collision of two high-intensity electron beams in this first colliding-beam machine, these problems had largely been solved by 1963; the physicists involved were able to use this 500 MeV (per beam) facility to make important tests of quantum electrodynamics.

Meanwhile, Italian physicists led by Bruno Touschek succeeded in storing and colliding two beams – one of electrons and the other of positrons – in a single ring named Anello di Accumulatione, or AdA, built at their Frascati laboratory in 1961.[58] It was succeeded by a much larger storage ring called ADONE (meaning "big AdA"), which in the early 1970s began colliding electrons and positrons at center-of-mass energies up to 3 GeV. With the lion's share of high-energy physics funding then going to build the $250 million National Accelerator Laboratory in Illinois (Wilson, Chapter 19), the only U.S. facility able to compete with ADONE until 1972 was a small "bypass" section added to the 6 GeV Cambridge Electron Accelerator (CEA) at Harvard.

Europeans also led the way in building proton–proton colliding-beam machines. The first proton collider was the Intersecting Storage Rings (ISR) completed at CERN in 1971 (Johnson, Chapter 16), in which 28

GeV protons circulated in two separate rings that crossed at six points. The Fermilab machine, a cascade of four accelerators of increasing energy, was originally designed in 1968 as a fixed-target machine. Although work began in the late 1970s to convert this machine into a colliding-beam facility, not until the 1980s did it start to function as a collider.

When SLAC's electron–positron collider SPEAR began operating in 1972, the U.S. high-energy physics community had its first colliding-beam facility that could compete with the European machines. And compete it did. SPEAR had the twin advantages of a much higher collision rate and a large multielement electronic detector surrounding the collision point.[59] After confirming the large cross sections for hadron production observed at ADONE and the CEA Bypass, physicists from SLAC and LBL made a long series of major discoveries in 1974–77 that rocked the worldwide community of particle physics. So effective were this machine and detector that almost all the important follow-up discoveries about charmed particles were made at SPEAR, even though the initial discovery of the J particle came in a fixed-target experiment.[60]

With the rise of particle colliders came a gradual shift in the "balance of power" among laboratories and research groups in particle physics. The laboratories that had already embraced colliders, such as CERN, DESY, and SLAC, emerged as the new leaders of the field, with physicists flocking there to join the large collaborations needed to do research on these power-packed machines. By the late 1970s laboratories such as Brookhaven and Fermilab that had clung to their fixed-target ways (even while planning to build new colliders) found that their output had steadily declining value in the international commerce of particle physics. Heavily committed to its Tevatron project, which was designed to boost the energy of the Main Ring to 1 TeV by the addition of superconducting magnets, Fermilab slowly made the conversion to colliding beams using this machine. And a program began to form in 1976 that eventually gave birth to CDF and D0, two mammoth particle detectors that are now conducting the world's highest-energy experiments in particle physics. Brookhaven began a struggle to redefine itself that led not only to the abortive ISABELLE project but to Brookhaven's Relativistic Heavy Ion Collider. Another outgrowth of the expanding collider effort was the recently discontinued Superconducting Super Collider project.[61]

In a similar shift, those physicists who had labored for years in what had previously been considered to be the "backwaters" of electron-based physics finally achieved parity with their colleagues at the proton labs – many of whom began doing research at electron–positron colliders.

DESY's PETRA collider, which received steady construction funding from the West German government during the late 1970s, attracted four large collaborations of physicists (each about a hundred physicists strong) to build its four major detectors. In the United States, where government funding was no longer quite as reliable, SLAC's electron-positron collider PEP took an extra year to build and was completed in 1980; its five large detector collaborations had to be content with confirming the DESY discovery of gluon jets.

The mid-1970s squeeze on U.S. funding for high-energy physics (West-fall, Chapter 20), attributable to the need to repay Vietnam War debts, was probably a major factor in allowing European high-energy physicists to pull even with their American counterparts in the early 1980s. Carlo Rubbia's 1976 proposal to convert Fermilab's Main Ring into a proton–antiproton collider was turned down by that laboratory, partly because of limited funding, as Fermilab had to choose between implementing that idea and completing its superconducting ring.[62] But when Rubbia took the idea instead to CERN's management, they welcomed the plan as a way to leapfrog U.S. competition. The discovery in 1982–83 of the massive W and Z particles, the long-sought intermediate vector bosons of the Standard Model (see above), brought CERN its first Nobel prize and universal recognition as the world's leading laboratory for high-energy physics.

With the inexorable growth of accelerator energies and costs during the 1964–79 period came an inevitable concentration of particle physics research activities into a handful of laboratories worldwide. In the United States, smaller accelerators such as LBL's Bevatron, CEA's electron synchrotron, the Penn-Princeton Accelerator, and the Zero Gradient Synchrotron at Argonne National Laboratory had been shut down by the close of this period or converted to nuclear physics research; Brookhaven's Alternating Gradient Synchrotron and Cornell's electron synchrotron managed to survive the cut by finding specialized research niches at comparatively low energies. Particle accelerators in Britain, France, and Italy were converted to do other research or closed down entirely, as CERN and DESY began to dominate European work in the field.

The lifestyle of high-energy experimenters changed drastically over this period. To continue doing world-class science, it became necessary for university-based physicists to include a substantial amount of travel funds in their budgets and spend a significant portion of their lives aboard airplanes. Graduate students and postdoctoral research associates were increasingly based at the major laboratories, with group leaders commuting to and fro between these labs and their universities, where they typically had teaching responsibilities.

To cope with the organizational problems of such large research operations, often involving thousands of employees and hundreds of physicist "users," the various laboratories evolved different management syles and laboratory cultures.[63] For the most part, Brookhaven and SLAC continued to operate in the style of the 1950s, in which the home institution or a group of institutions maintained absolute control over the major decisions, and staff physicists exercised much more power over the use of laboratory resources than did their university-based colleagues. Committees were appointed from among these users to advise the laboratory directors on important decisions – particularly the program advisory committees of the individual labs that helped determine which experiments to pursue (Westfall, Chapter 20). But the final decision-making authority always rested with the laboratory directors, who often handpicked the members of the more influential committees.

In the building of Fermilab a conscious attempt was made to create a more democratic management structure and culture (Wilson, Chapter 19). As first advocated by Lederman, it was intended to be a "truly national laboratory" in which the users would have a major role in making important decisions and allocating scarce laboratory resources.[64] The Universities Research Association (URA), incorporated in Washington, D.C., was set up to be the "management and operations contractor" responsible to the Atomic Energy Commission (and its successor agencies, the Energy Research and Development Administration and the Deparment of Energy) for managing Fermilab. Membership in URA was open to all universities involved in high-energy physics research; it appointed an independent board of overseers to meet periodically with laboratory management and ensure that appropriate policies were being pursued. Whether these and other democratic provisions resulted in a more effective and productive laboratory than might otherwise have occurred is the subject of current historical study.[65]

Concentration also occurred within the laboratories themselves, as the scale of experiments grew with the particle energies and the size of collaborations of physicists needed to build and operate the equipment increased accordingly (Bodnarczuk, Chapter 21). In 1964 a large collaboration might have included 25 physicists;[66] by 1979 this would have been considered a fairly small group. Two other factors besides increasing energy drove the inexorable trend toward concentration of physicists and available resources into fewer and fewer experiments involving ever larger collaborations: the rise of colliders as the favorite experimental tool and the rise of the Standard Model as the dominant theory of particle physics. At a collider there were at most a few special places – namely where the two beams clashed – where one could even hope to do an experiment, as opposed to the multitudes of opportunities in fixed-target work. And the complex, multielement electronic detectors covering as much as possible of the solid angle surrounding these interaction points often required the combined efforts and talents of a hundred (or more) physicists and engineers to build and operate. The rise of the Standard Model meant that the legitimate targets of frontier research had become relatively few and were difficult to pin down; large collaborations built increasingly sophisticated detectors in attempts to isolate these rare birds.[67]

Effective management of these collaborations, which were spending millions of dollars, francs, or deutschmarks to build their detectors, required ponderous, hierarchical decision-making structures that reflected the laboratory managements themselves. Gone were the heady days of the 1950s and 1960s when a small, elite commando unit of experimenters could take over an unused beamline with their detector and record a little data while the competition was nursing balky equipment back into operation. The coordination of the many different inputs and specialized talents that were needed to design, test, build, and integrate the various components of a large, sophisticated particle detector required a spokesman (it was always a man) with strong technical and managerial expertise plus a loyal staff of physicists and engineers whose judgments he could trust. Few members of a big collaboration could claim to understand all the aspects of their detector and the physics research it was intended to perform.

Summary

As the 1980s began, elementary particle physics entered a period of "normal science," a phrase coined by Thomas Kuhn in *The Structure of Scientific Revolutions*. All experimental results were henceforth to be compared with the Standard Model – the new "paradigm" of the field. Where anomalies cropped up, they were usually resolved by modest elaborations of the theory (or by uncovering errors made by the experimenters). No truly fundamental changes have occurred in the Standard Model during the ensuing 16-year period, which will probably be examined in detail by physicists, historians, and philosophers in the fourth Symposium in this series. The W and Z particles were discovered at CERN in 1982–83 near their expected masses; the production of Z bosons in electron–positron collisions indicated in 1989 that there could be only three conventional quark–lepton families in the standard particle table; and in 1995, after years of searching, physicists at Fermilab at long last reported the discovery of the top quark in proton–antiproton collisions. These and many other experimental results found a ready home within the framework of the Standard Model as it had been recognized in 1979.

To characterize the 1964–79 period as a "scientific revolution," however, strikes us as inexact and misleading. It was indeed a time of great upheaval in particle physics, but the field lacked a crucial ingredient at the outset of this period – a single, dominant theory of the subatomic world agreed upon by the entire community of practitioners. Rather, particle physics in the early 1960s could be broken down into a number of fiefdoms, none of which could claim the unswerving allegiance of every single knight. There was a surfeit of nobles ready and eager to fight amongst themselves, but no all-powerful king to overthrow.

The rise of the Standard Model changed this chaotic situation dramatically. A cacophony of competing ideas was replaced by a single theory upon which essentially all practicing particle physicists agree – a single reference point against which all work is now compared. In a telling phrase, perhaps reflecting the influence of ideas imported from condensed matter physics, particle physicists sometimes refer to this convulsive period not as a revolution but as a "phase change." As if the field itself had somehow become supercooled by the late 1960s, and the addition of a few critical "seeds" – the MIT–SLAC experiments, say, and the renormalization of Yang–Mills gauge theories – led to a period of rapid crystallization during the 1970s. The many had become one.

Whatever the ultimate metaphor, there can be no doubt that this was a crucial period in the history of physics. The rise of the Standard Model completely redefined what it meant to be an elementary particle physicist. The research agenda had changed dramatically from its pre-1970s anarchy. Whereas only a small minority of practitioners in 1970 paid much heed to gauge field theories and hard-scattering experiments, a decade later these topics occupied the center of attention. While large contingents busied themselves with completing the standard particle table and working out all the ramifications of the new theory, others – not content with a supposedly "fundamental" theory that had so many arbitrary parameters – began the search for whatever might lie beyond it, a search that continues today. In all instances, the common reference point has been the Standard Model, which even when it is finally superseded will surely remain as one of the crowning achievements of twentieth-century physics.

Notes

1 The existing treatments of this history include: Andrew Pickering, *Constructing Quarks* (Edinburgh: Edinburgh University Press, 1984); and popular accounts by Michael Riordan, *The Hunting of the Quark* (New York: Simon & Schuster, Inc., 1987), and Robert P. Crease and Charles C. Mann, *The Second Creation* (New York: Macmillan Publishing Company, 1986).

2 See L. M. Brown and L. Hoddeson, eds., *The Birth of Particle Physics* (New York: Cambridge University Press, 1983); L. M. Brown, M. Dresden, and L. Hoddeson, eds., *Pions to Quarks* (New York: Cambridge University Press, 1989).

3 The quark proposal was made in 1964 by Murray Gell-Mann and by George Zweig. The idea of composite hadrons was suggested in 1956 by Shoichi Sakata, who tried to use the known hadrons p, n, and Λ as a basis.

4 See, for example, G. F. Chew, "Particles as S-matrix poles: hadron democracy," in L. M. Brown, M. Dresden, and L. Hoddeson, eds., *Pions to Quarks*, pp. 600–7.

5 C. N. Yang and R. Mills, "Isotopic spin conservation and a generalized gauge invariance," *Phys. Rev. 95* (1954), p. 631; "Conservation of isotopic spin and isotopic gauge invariance," *Phys. Rev. 96* (1954), pp. 191–5.

6 Fermi's 1934 theory did not consider the strong nuclear force, but it was extended to do so by Heisenberg and others. On the other hand, Yukawa's meson was explicitly designed from the start to be an intermediate boson for both strong and weak forces. (Yukawa was actually the first person to assign separate coupling constants to these two types of force.)

7 For other historical standard models, see John Heilbron's chapter in this volume.

8 See H. A. Bethe, "The Happy Thirties," in R. Stuewer, ed., *Nuclear Physics in Retrospect; Proceedings of a Symposium on the 1930s* (Minneapolis: University of Minnesota Press, 1979).

9 The annual "Rochester" conferences up to 1958 had the word *nuclear* in their title, but the CERN "Rochester" conference of 1958 omitted it.

10 This list is taken from A. Pais, *Inward Bound* (New York: Oxford University Press, 1986), p. 476.

11 There had been some theoretical grounds for anticipating the ρ and the ω from attempts to explain the electromagnetic form factors of the nucleon in terms of vector-meson exchange.

12 A "sociological history" of this period, Pickering's *Constructing Quarks* considers the quarks to be "social constructs."

13 J. Goldstone, "Field theories with 'superconductor' solutions," *Nuovo Cimento 19* (1961), pp. 154–64.

14 P. Galison, "How the first neutral current experiment ended," *Rev. Mod. Phys. B55* (1983), pp. 477–509.

15 Julian Schwinger, "A theory of the fundamental interactions," *Ann. of Phys. 2* (1957), pp. 407–34. In 1980 Schwinger said that he had already suggested this idea privately to Robert Oppenheimer in 1941. See his "Renormalization theory of quantum electrodynamics," in L. Brown and L. Hoddeson, eds., *The Birth of Particle Physics*, pp. 329–53, especially p. 349.

16 S. A. Bludman, "On the universal Fermi interaction," *Nuovo Cimento 9* (1958), pp. 433–45. Bludman's theory did not include the idea of electroweak unification.

17 A global transformation is one with constant group parameters, as opposed to a local transformation, whose parameters are space–time dependent. It is the local type that gives rise to gauge fields.

18 S. A. Bludman, "On the universal Fermi interaction," p. 435.

19 S. A. Bludman, "On the universal Fermi interaction," p. 441. The two objections mentioned are the value of the muon's decay ρ value and the nonrenormalizable electromagnetic interaction of the vector meson.

20 Sheldon L. Glashow, "Partial-symmetries of weak interactions," *Nucl. Phys. 22* (1960), pp. 579–88.

21 A. Salam and J. C. Ward, "Weak and electromagnetic interactions," *Nuovo Cimento 11* (1959), pp. 568–77; "On a gauge theory of elementary interactions," *Nuovo Cimento 19* (1961), pp. 165–70; "Electromagnetic and weak interactions," *Phys. Lett. 13* (1964), pp. 168–71.

22 V. E. Barnes, et al., "Observation of a hyperon with strangeness minus 3," *Phys. Rev. Lett. 12* (1964), pp. 204–6.

23 M. Gell-Mann, "A schematic model of baryons and mesons," *Phys. Lett. 8* (1964), pp. 214–15; G. Zweig, "An SU(3) model for strong interaction symmetry and its breaking," CERN preprint 8182/TH401 (17 Jan. 1964).

24 P. W. Higgs, "Broken symmetries, massless particles and gauge fields," *Phys. Rev. Lett. 12* (1964), pp. 132–3; P. W. Higgs, "Broken symmetries and the mass of gauge vector bosons," *Phys. Rev. Lett. 13* (1964), pp. 508–9; F. Englert and R. Brout, "Broken symmetry an the mass of gauge vector bosons," *Phys. Rev. Lett. 13* (1964), pp. 321–3.

25 Z. Maki, "The 'fourth' baryon, Sakata model, and modified B-L symmetry, I," *Prog. Theor. Phys. 31* (1964), pp. 331–4.; Z. Maki and Y. Ohnuki, "Quartet scheme for elementary particles," *Prog. Theor. Phys. 31* (1964), pp. 144–58; J. D. Bjorken and S. L. Glashow, "Elementary particles and SU(4)," *Phys. Lett. 11*, (1964), pp. 255–7.

26 Y. Nambu and G. Jona-Lasinio, "Dynamical model of elementary particles base upon an analogy with superconductivity, I," *Phys. Rev. 122* (1961), pp. 345–58: "Dynamical model of elementary particles based upon an analogy with superconductivity, II," *Phys. Rev. 124* (1961), pp. 246–54.

27 E.g., M. Baker and S. L. Glashow, "Spontaneous breakdown of elementary particle symmetries," *Phys. Rev. 128* (1962), pp. 2462–71.

28 J. J. Sakurai, "Theory of strong interactions," *Ann. Phys. (N.Y.) 11* (1960), pp. 1–48.

29 T. D. Lee and B. Zumino, "Field-current identities and algebra of fields,"

Phys. Rev. 163 (1967), pp. 1667–81, especially p. 1667. This article was preceded by N. Kroll, T. D. Lee, and B. Zumino, "Neutral vector mesons and the hadronic electromagnetic current," *Phys. Rev. 157* (1967), pp. 1376–99.

30 S. Weinberg, "A model of leptons," *Phys. Rev. Lett. 19* (1967), pp. 1264–6. An equivalent model was independently proposed in A. Salam, "Weak and electromagnetic interactions," in N. Svartholm, ed., *Elementary Particle Theory: Relativistic Groups and Analyticity* (Stockholm: Almqvist and Wiksell, 1968).

31 G. 't Hooft, "Renormalization of massless Yang–Mills fields," *Nucl. Phys. B33* (1971), pp. 173–99; "Renormalizable Lagrangians for massive Yang–Mills fields," *Nucl. Phys. B35* (1971), pp. 167–88.

32 M. Gell-Mann, "Symmetries of baryons and mesons," *Phys. Rev. 125* (1962), pp. 1067–84.

33 S. Weinberg, "Effects of a neutral intermediate boson in semileptonic processes," *Phys. Rev. D5* (1972), 1412–17.

34 J. H. Kleins, R. H. Hildebrand, and R. Stiening, "Limits on the $K^+ \rightarrow \pi^+ + \nu + \bar{\nu}$ decay rate," *Phys. Rev. Lett. 24* (1970), pp. 1086–90. Strangeness-conserving neutrino scattering was also reported to be at least thirty times weaker than charge-exchange neutrino scattering, but this conclusion was erroneous. M. M. Block, et al., "Neutrino interactions in the CERN heavy liquid bubble chamber," *Phys. Lett. 12* (1964), pp. 281–5.

35 S. L. Glashow, J. Iliopolis, and L. Maiani, "Weak interactions with hadron–lepton symmetry," *Phys. Rev. D2* (1970), pp. 1285–92.

36 F. J. Hasert, et al., "Observation of neutrino-like interactions without muon or electron in the Gargamelle neutrino experiment," *Phys. Lett. 46B* (1973), pp. 138–40, and *Nucl. Phys. B73* (1974), pp. 1–22.

37 A. Benvenuti, et al., "Observation of muonless neutrino-induced inelastic interactions," *Phys. Rev. Letters 34* (1974), pp. 800–3.

38 P. Galison, "How the first neutral current experiment ended"; A. Pickering, "Against putting the phenomena first," *Stud. Hist. Phil. Sci. 15* (1984), pp. 85–117. See also P. Galison, *How Experiments End* (Chicago: University of Chicago Press, 1987).

39 C. Y. Prescott, et al., "Parity non-conservation in inelastic electron scattering," *Phys. Lett. 77B* (1978), pp. 347–53.

40 A. Pickering, *Constructing Quarks*, p. 298.

41 Y. Nambu, "A systematics of hadrons in subnuclear physics," in A. De-Shalit, H. Feshbach and L. Van Hove, eds., *Preludes in Theoretical Physics* (Amsterdam: North Holland, 1966), pp. 133-42.

42 M. Y. Han and Y. Nambu, "Three-triplet model with double SU(3) symmetry," *Phys. Rev. B139* (1965), pp. 1006–10.

43 Initial results from the MIT–SLAC experiments were presented in a summary talk by W. K. H. Panofsky, "Low q^2 electrodynamics, elastic and inelastic electron (and muon) scattering," in J. Prentki and J. Steinberger, eds., *Proceedings of the 14th International Conference on High Energy Physics*, Vienna, 1968 (Geneva: CERN, 1968), pp. 23–39. For details of the experiments, consult R. E. Taylor, "Deep inelastic scattering: the early years," *Rev. Mod. Phys. 63* (1991), pp. 573–595; H. W. Kendall, "Deep inelastic scattering: experiments on the proton and the observation of scaling," ibid., pp. 597–614; and J. I. Friedman, "Deep

inelastic scattering: comparisons with the quark model," ibid., pp. 615–627. For a historical summary of these experiments, see M. Riordan, "The discovery of quarks," *Science 256* (1992), pp. 1287–93.

44 V. Silvestrini, "Electron–Positron Interactions," in J. D. Jackson and A. Roberts, eds., *Proceedings of the XVI International Conference on High Energy Physics*, National Accelerator Laboratory, Batavia, Illinois, 6–13 Sept. 1972 (Batavia: National Accelerator Laboratory, 1972), Vol. 4, pp. 1–40.

45 D. J. Gross and F. Wilczek, "Ultra-violet behavior of non-Abelian gauge theories," *Phys. Rev. Lett. 30* (1973), pp. 1343–46; H. D. Politzer, "Reliable perturbative results for strong interactions?," ibid., pp. 1346–49.

46 H. Fritsch, M. Gell-Mann, and H. Leutwyler, "Advantages of the color octet gluon picture," *Phys. Lett. 47B* (1973), pp. 365–8.

47 R. Brandelik, et al., "Evidence for planar events in e^+e^- annihilation at high energies," *Phys. Lett. 86B* (1979), pp. 243–9; D. P. Barber, et al., "Discovery of three-jet events and a test of quantum chromodynamics," *Phys. Rev. Lett. 43* (1979), pp. 830–33. See also J. Ellis, M. K. Gaillard, and G. Ross, "Search for gluons in e^+e^- annihilation," *Nucl. Phys. B111* (1976), pp. 253–71 for the theoretical underpinnings of the three-jet analysis; and R. Brandelik, "Evidence for a spin one gluon in three-jet events," *Phys. Lett. 97B* (1980), pp. 453–8.

48 M. Gell-Mann, "The eightfold way: a theory of strong interaction symmetry," Caltech Report No. CTSL-20 (1961), unpublished; and Gell-Mann, "Symmetries of baryons and mesons"; Y. Ne'eman, "Derivation of strong interactions from a gauge invariance," *Nucl. Phys. 26* (1961), pp. 222–29; see also M. Gell-Mann and Y. Ne'eman, *The Eightfold Way* (New York: W. A. Benjamin, 1964), pp. 7–9.

49 J. J. Aubert, et al., "Experimental observation of a heavy particle J," *Phys. Rev. Lett. 33* (1974), pp. 1404–5; J.-E. Augustin, et al., "Discovery of a narrow resonance in e^+e^- annihilation," *Phys. Rev. Lett. 33* (1974), pp. 1406–8; G. S. Abrams, et al., "Discovery of a second narrow resonance $\psi'(3684)$ in e^+e^- annihilation," *Phys. Rev. Lett. 33* (1974), pp. 1433–36. For details of these experiments, consult Samuel C. C. Ting, "The discovery of the J particle: a personal recollection," *Rev. of Mod. Phys. 49* (1977), pp. 235–249; and Burton Richter, "From the psi to charm: the experiments of 1975 and 1976," ibid., pp. 251–66.

50 M. L. Perl, et al., "Evidence for anomalous lepton production in e^+e^- annihilation," *Phys. Rev. Lett. 35* (1975), pp. 1489–92; M. L. Perl, et al., "Properties of anomalous $e\mu$ events produced in e^+e^- annihilation," *Phys. Lett. 63B* (1976), pp. 466–70.

51 S. W. Herb, et al., "Observation of a dimuon resonance at 9.5 GeV in 400-GeV proton–nucleus collisions," *Phys. Rev. Lett. 39* (1977), pp. 252–55.

52 D. Andrews, et al., "Observation of a fourth upsilon state in e^+e^- annihilations," *Phys. Rev. Lett. 45* (1980), pp. 219–21.

53 See, for example, R. Brandelik, et al., "Energy scan for narrow states in e^+e^- annihilation at C. M. energies between 29.90 and 31.46 GeV," *Phys. Lett. 88B* (1979), pp. 199–202; and B. Adeva, et al., "Search for top quark and a test of models without top quark up to 38.54 GeV at PETRA," *Phys. Rev. Lett. 50* (1983), pp. 799–802.

54 F. Abe, et al., "Evidence for top quark production in $\bar{p}p$ collisions at $\sqrt{s}$ = 1.8 TeV," *Phys. Rev. D50* (1994), pp. 2966–3026.

55 For further discussion of the impact of Sputnik on funding for physics research, consult D. Kevles, *The Physicists* (New York: Alfred A. Knopf, 1978), pp. 384–6.

56 See for example, P. Galison, "Bubbles, sparks and the postwar laboratory," in *Pions to Quarks*, pp. 213–251, and L. Alvarez, "The hydrogen bubble chamber and the strange resonances," ibid., pp. 299–306.

57 D. Kerst, "Accelerators and the Midwestern Universities Research Association in the 1950s," in *Pions to Quarks*, pp. 202–12; F. T. Cole, "Oh, Camelot! A memoir of the MURA years," unpublished, Fermilab Archives.

58 C. Bernardini, "AdA: the smallest e^+e^- ring," in M. de Maria, M. Grilli, and F. Sebastiani, eds., *The Restructuring of Physical Sciences in Europe and the United States, 1945–1960* (Singapore: World Scientific, 1989), pp. 444–8; F. Amman, "The early times of electron colliders," ibid., pp. 449–76.

59 B. Richter, "From the psi to charm."

60 S. C. C. Ting, "The discovery of the J particle."

61 L. Hoddeson, "The first large-scale application of superconductivity: the Fermilab energy doubler, 1972–1983," *Hist. Stud. Phys. Sci. 18:1* (1987), pp. 25–54; A. Kolb and L. Hoddeson, "The mirage of the 'world accelerator for world peace' and the origins of the SSC, 1953–1983," *Hist. Stud. Phys. Bio. Sci. 24:1* (1993).

62 L. Hoddeson, "The first large-scale applications of superconductivity."

63 An interesting study of laboratory cultures (with emphasis on SLAC) from an anthropologist's perspective is S. Traweek, *Beamtimes and Lifetimes: The World of High Energy Physicists* (Cambridge: Harvard University Press, 1988).

64 L. Lederman, "The truly national laboratory (TNL)," paper presented at the Brookhaven Super-High-Energy Summer Study, 25 June 1963, available as Brookhaven Report No. BNL-AADD-6.

65 For a study of the history of Fermilab, consult L. Hoddeson, C. Westfall, M. Bodnarczuk, and A. Kolb, *The Final Frontier: The History of Fermilab* (in preparation).

66 For example, there were 33 names on the 1964 paper reporting the discovery of the omega minus; much was made at the time about the fact that there were 33 physicists involved and only a single event being reported.

67 A similar point has been made by Andrew Pickering in *Constructing Quarks*, where he notes that after 1974 experimental physicists rapidly abandoned their prior focus on soft-scattering, low momentum-transfer events in favor of much more difficult searches for the rare hard-scattering, high momentum-transfer events indicative of encounters between the fundamental Standard Model entities.

2

Changing Attitudes
and the Standard Model*

STEVEN WEINBERG

Born New York City, 1933; Ph.D., 1957 (physics), Princeton University;
Professor of Physics and Astronomy at the University of Texas at Austin;
Nobel Prize in Physics, 1979; high-energy physics (theory) and cosmology.

The history of science is usually told in terms of experiments and theories and their interaction. But there is a deeper level to the story – a slow change in the attitudes that define what we take as plausible and implausible in scientific theories. Just as our theories are the product of experience with many experiments, our attitudes are the product of experience with many theories. It is these attitudes that one usually finds at the root of the explanation for the curious delays that often occur in the history of science, as for instance, the interval of 15 years between the theoretical work of Alpher and Herman and the experimental search for the cosmic microwave radiation background. The history of science in general and this conference in particular naturally deal with things that happened, with successful theories and experiments, but I think that the most interesting part of the history of science deals with things that did not happen, or at least not when they might have happened. To understand this sort of history, one must understand the slow changes in the attitudes by which we are governed. But it is not easy. Experimental discoveries are reported in *The New York Times*, and new theories are at least reported in physics journals, but the change in our attitudes goes on quietly and anonymously, somewhere behind the blackboard.

The rise of the Standard Model was accompanied by profound changes in our attitudes toward symmetries and toward field theory. In this chapter I will try to outline these changes, relying chiefly on my own memories of the period, and emphasizing theory because that is what I know best. In choosing certain developments to discuss, I do not mean

* Research supported in part by the Robert A. Welch Foundation and NSF Grant PHY 9009850.

to imply that these were the only important developments in theoretical particle physics during this period, but they were important and represented changes in our deepest attitudes.

As I recall the atmosphere of particle physicists toward symmetry principles in the 1950s and 1960s, symmetries were regarded as important largely for want of anything else to think about. We knew that the strong interactions were too strong to allow the use of perturbation theory, and even if perturbation theory could be used, there was no plausible dynamical theory to which to apply it. The discovery of large numbers of hadronic states had pretty well discredited the old meson field theories (such as the theory of pions and nucleons with pseudoscalar coupling); most of us were skeptical about the quark model; and although we had not forgotten the idea of Yang and Mills about a theory of strong interactions based on a local non-Abelian gauge symmetry, the masslessness of the vector bosons in such a theory seemed like an insuperable barrier to any physical application. Only analyticity, unitarity, and symmetry were generally regarded as reliable inputs for studies of strong interactions, but by the early 1960s the S-matrix theory program of using just these inputs to calculate the strong interactions had run out of steam.

In the theory of weak interactions, the nonrenormalizablity of the Fermi interaction kept us from going beyond the lowest order of perturbation theory. We could use first-order perturbation theory, but wherever hadrons were involved we faced the same problems as in the strong interactions; almost the only property of the nonleptonic weak Hamiltonian or of the hadronic currents in semileptonic interactions that could usefully be studied were their symmetry properties.

It was generally supposed that these symmetry principles were somehow very fundamental, a reflection of the simplicity of Nature at its deepest levels. But there was an obvious difficulty with this attitude: a fair number of these symmetries were not exact. How were we supposed to regard an approximate symmetry – as a reflection of the approximate simplicity of Nature? There was some idea of a hierarchy of symmetries: the exact symmetries such as Lorentz invariance, electromagnetic gauge invariance, and (supposedly) baryon and lepton conservation were universally valid; parity, charge conjugation, and strangeness were exactly valid in the strong and electromagnetic but not the weak interactions; isospin conservation was exactly valid only in the strong interactions; and SU(3) symmetry was approximate even in the strong interactions. No one knew why.

When particle theorists in the early 1960s began to think seriously about broken symmetries, at least part of their excitement was spurred by the notion that the approximate symmetries might be approximate because they were spontaneously broken. I recall that for a while I suffered myself from this misconception, and so apparently did Salam and Ward, who in 1960 tried to explain strangeness nonconservation as a consequence of a vacuum expectation value of the K_1^0 field. You can find this view expressed in the first sentence of a 1966 paper of Higgs: "The idea that the apparently approximate nature of the internal symmetries of elementary-particle physics is the result of asymmetries in the stable solutions of exactly symmetric dynamical equations, rather than an indication of asymmetry in the equations themselves, is an attractive one."[1]

It was just beginning to be realized in the mid-1960s that the question of whether a symmetry is spontaneously broken or not is orthogonal to the question of whether it is approximate or not, and that a spontaneously broken exact symmetry does not look at all like an approximate symmetry. However, confusion persisted in some quarters. As late as 1975, Heisenberg included isotopic spin conservation in a list of the fundamental symmetries of Nature, and interpreted the violations of isospin symmetry in the weak and electromagnetic interactions as due to "an asymmetric, degenerate ground state."[2]

I want to go a little further here into the growth of our understanding of broken symmetry in the 1960s, because my own work on the electroweak theory flowed directly from it, so much so that I can hardly separate the two in my mind.[3]

It was because we hoped in the early 1960s to use broken symmetries to understand approximate symmetries that it came to us as such a disappointment to learn from Goldstone that spontaneous symmetry breaking entails the existence of unobserved massless particles. In the 1962 paper by Goldstone, Salam, and myself, we referred to this as an "intractable difficulty."[4] Remember also that the 1964–66 work of Higgs; Guralnik, Hagen, and Kibble; and Brout and Englert on the spontaneous breakdown of local symmetries was motivated, not by a hope to explain massive vector bosons, but by the wish to get rid of massless scalar bosons, as illustrated by the title of the paper of Higgs that I quoted earlier: "Spontaneous Symmetry Breakdown without Massless Bosons." Indeed, the work of Higgs et al. was not followed up at first because around 1966 it became generally accepted that the pion was the Goldstone boson of a spontaneously broken approximate symmetry, and it no longer seemed so important to find exceptions to the Goldstone theorem.

This early attitude toward spontaneously broken symmetries may seem a bit odd. After all, spontaneous symmetry breaking had first appeared in particle physics in the late 1950s in connection with an effort to account for the success of the Goldberger–Treiman relation for the pion decay rate, which today we understand as a simple consequence of the fact that the pion is the Goldstone boson of broken chiral symmetry. Why then were physicists trying so hard from 1960 to 1966 to find exceptions to the Goldstone theorem? I think the reason may be that the theorists who were trying to understand the Goldberger–Treiman relation were concentrating, not on symmetries, but on currents. The Goldberger–Treiman relation was first explained by Nambu and others in terms of a partial conservation of the axial vector current of beta decay, and when theorists asked in what sort of theory the axial vector current would have this property, the examples they found were of theories that exhibited spontaneous symmetry breaking, but the symmetry itself was not at the center of attention. The 1960 papers of Gell-Mann and Levy and of Nambu and Jona-Lasinio gave examples of theories with partially conserved axial vector currents, and these theories did exhibit spontaneous symmetry breaking, but these papers barely mentioned spontaneous symmetry breaking as a fact of interest in its own right.

It is ironic that some of these developments were inspired by work on superconductivity, for much of the literature on superconductivity shows the same dismissive attitude toward broken symmetry. From a modern perspective (or at any rate, my perspective) a superconductor is nothing more or less than a piece of matter within which electromagnetic gauge invariance is spontaneously broken, but although workers on superconductivity knew very early that electromagnetic gauge invariance is violated in the London equation, there was hardly any explicit mention of spontaneous symmetry breaking in the classic papers on superconductivity (including the 1957 paper of Bardeen, Cooper, and Schrieffer, but with some of Anderson's work as an exception) and very little even now in textbook treatments of superconductivity.

This attitude toward broken symmetry persisted in the early 1960s in the work of Gell-Mann and others on current algebra. As its name implies, the important thing about current algebra was supposed to be the currents; the fact that the axial vector current is partially conserved was seen as an incidental though convenient assumption that made it possible to estimate matrix elements of this current that had not been measured in the weak interactions. The great change came in the mid-1960s with the derivation of the Adler–Weisberger sum rule. I would say that the Adler–Weisberger sum rule was one of three theoretical breakthroughs during our period whose importance was recognized at once, and changed the direction of physics research immediately. (The others were the discovery of asymptotic freedom by Gross and Wilczek and by Politzer, and 't Hooft's demonstration of the renormalizabilty of spontaneously broken gauge theories.) Although the initial derivations of the Adler–Weisberger sum rule were very much in the current-algebra style, based on current commutation relations, it soon became clear that these sum rules are nothing but dispersion relations for a low-energy pion–nucleon scattering amplitude, whose value follows directly from the role of the pion as the Goldstone boson of a broken symmetry. But unlike earlier work of Nambu and his collaborators on amplitudes involving a single soft pion, the derivation of the Adler–Weisberger sum rule for the first time required a commitment to a particular symmetry of the strong interactions: to get the right answer the symmetry has to be $SU(2) \times SU(2)$, rather than $SO(3,1)$ or their contraction $IO(3)$. Broken symmetries had now become central features of physical theories, not just incidental aspects of models of weak currents.

It was quantum chromodynamics that provided the solution to the problem posed by approximate symmetries. I remember how exciting it was when we first realized that the most general renormalizable Yang–Mills theory of quarks and gluons automatically conserves charge conjugation, strangeness, and (aside from later problems with anomalies) parity, and that for relatively small quark masses it also has an $SU(3) \times SU(3)$ symmetry, of which the chiral part is spontaneously broken. In combination with the electroweak theory, quantum chromodynamics also dictates that the currents of this $SU(3) \times SU(3)$ symmetry furnish the hadronic currents that enter into semileptonic weak interactions. We learned that the weak interactions do not conserve parity or charge conjugation or strangeness because there is no reason why they should; these symmetries never were anything but accidents. This marked a permanent change in our attitude toward approximate symmetries: from then on we would be suspicious of any approximate symmetry that could not be explained as an accidental consequence of the constraints imposed by renormalizability and the various exact symmetries.

This stunning success was largely responsible for the rapid acceptance of quantum chromodynamics after the discovery of asymptotic freedom by Gross and Wilczek and by Politzer in 1973, even before the J/ψ was found. Indeed, for a brief period after the discovery of asymptotic freedom it was widely assumed that gluons had not been discovered because they were heavy, getting their mass from some sort of Higgs mechanism. In order to account for the spontaneous breakdown of color SU(3), some theorists proposed adding colored scalars to the theory. This idea did not get very far, however, because introducing strongly interacting scalars would have undone the great success of quantum chromodynamics in explaining the approximate symmetries. One of the reasons we were so willing to believe in massless gluons and color trapping is that it saved us from having to reopen the problem of the approximate symmetries.

In parallel with the changes it brought in our attitude toward symmetries, the birth of the Standard Model marked changes also in our attitude toward quantum field theory. After 't Hooft's breakthrough in 1971, it became clear that the old problem of infinities in the weak interactions had been solved by the use of spontaneous symmetry breaking to give masses to the W and Z particles. Then the asymptotic freedom of quantum chromodynamics gave us a framework in which we could actually calculate something about the strong interactions – not everything, but at least something. But in scoring these victories, quantum field theory was preparing the way for a further change in our attitude, in which quantum field theory would lose its central position.

For all its success, the Standard Model was obviously far from a final theory. In addition to relying on a number of apparently arbitrary elements, the Standard Model did not unify the strong interactions with the electroweak interactions, and it left out gravity altogether. But it pointed the way to a better theory, perhaps a final theory. The slow decrease of the strong coupling constant with energy makes the strong coupling comparable with the electroweak couplings at an energy of order $10^{15} - 10^{16}$ GeV, and gravity becomes comparable in strength to the electroweak interactions at 10^{18} GeV. These energies are so much larger than the W and Z masses that the Standard Model can only be understood as a low-energy approximation to a more fundamental theory, one that may not involve fields for quarks and gluons and gauge fields, one perhaps that is not even a field theory, but a string theory or something like it.

According to this view, the Standard Model is a mere effective field theory; it takes the specially simple form of a field theory only because any relativistic quantum theory looks like a quantum field theory at sufficiently low energy, and it looks renormalizable only because any nonrenormalizable terms in the effective Lagrangian are suppressed by negative powers of a mass of order $10^{16} - 10^{18}$ GeV. Effective field theories are an old story in physics, going back to the 1936 Euler–Heisenberg nonlinear Lagrangian for photon–photon scattering, and they had been used to derive soft pion theorems since 1967, though it does not seem to have been realized until the late 1970s that effective field theories could be regarded as full-fledged dynamical theories, useful beyond the tree approximation. This led to the chastening reflection that the Standard Model of which we are so proud may itself be nothing but an effective field theory, involving quark and lepton and gauge boson fields that perhaps do not even appear in the deeper underlying theory, any more than the pion field appears in the Lagrangian of quantum chromodynamics. The justification of any particular effective field theory is that it is simply the most general possible theory that satisfies the axioms of analyticity, unitarity, and cluster decomposition along with the relevant symmetry principles, so in a way our use today of effective field theories is the ultimate revenge of S-matrix theory; an effective quantum field theory like the Standard Model is just our way of implementing symmetry principles and the axioms of S-matrix theory. Quantum field theory is nothing but S-matrix theory made practical.

The changes in our attitudes that I have described cannot be explained in the classic terms of deduction or induction, but rather as the result of something more like natural selection. This may give the impression that our theories are not much more than social constructions, as supposed by some radical commentators on science, such as Pickering, the author of a book entitled *Constructing Quarks*.[5] None of us who have lived through these changes thinks this way. We know of course that science is a social activity. As Latour and Woolgar commented after observing research in biochemistry, "The negotiations as to what counts as a proof or what constitutes a good assay are no more or less disorderly than any argument between lawyers and politicians."[6] But the same could be said about mountain climbing. Mountain climbers, like biochemists or lawyers, may argue over the best path to the peak, and of course these arguments will be influenced by the traditions of mountain climbing and the history and social structure of the expedition. But in the end the climbers will either get to the peak or they will not, and if they do get there they will know it. No mountaineer would write a book about mountain climbing with a title like "Constructing Everest."

Nor do I see in the last 30 years of particle physics anything like the incommensurability that Kuhn describes between the standards we use to judge theory now and in the past. It may be just that there has been no real revolution in the past 30 years of physics – some changes of government or palace coups or assassinations, perhaps, but no revolution comparable to the advent of Newtonian mechanics or quantum mechanics. Our attitudes today are different from those of 1960 – some things now seem more important, and others less. But in reading our old papers we see little change in what we have been trying to achieve – a fundamental theory that would be entirely satisfying in its completeness and simplicity.

Unfortunately progress toward this goal seems to have come nearly to a stop. We are paying the price of our own success; the Standard Model has done so well that we cannot easily see how to go beyond it. Our great hope for progress has been that the Superconducting Super Collider would settle the question of the mechanism for spontaneous symmetry breaking in the Standard Model, and in doing so give us some clue as to what to do next.

Recently we learned that this hope may not be fulfilled. I gather that the members of the House of Representatives after rejecting a balanced budget amendment were eager to demonstrate their concern about the budget in an election year, and also that the Republican representatives from California turned against the Super Collider to punish a Texas congressman who had challenged one of them for a leadership position. The funds taken from the Super Collider were not given to other scientific projects, and Fermilab was cut at the same time. It would be bad enough if after serious discussion Congress had decided that the funds for the Super Collider should be better spent in other areas of research, but for Congress mindlessly to discard years of work and a billion dollars already spent for the pettiest political motives is simply disgusting. We can only hope that if the Super Collider project is really dead, then our friends in Europe and Asia will carry on with the historic task of exploring the frontiers of physics that is now being abandoned by the United States.

Notes

1 P. W. Higgs, "Spontaneous Symmetry Breakdown without Massless Bosons," *Phys. Rev. 145* (1966), p. 1156.
2 W. Heisenberg, "Cosmic Radiation and Fundamental Problems in Physics," lecture to the 14th International Cosmic Ray Conference in Munich, August 18, 1975, reprinted in *Encounters with Einstein* (Princeton: Princeton University Press, 1983).
3 See, e.g., S. Weinberg, "Unified Theory of Weak and Electromagnetic Interactions," *Rev. Mod. Phys. 52* (1980), p. 515.

4 J. Goldstone, A. Salam, and S. Weinberg, "Broken Symmetries," *Phys. Rev. 127* (1962), p. 965.
5 A. Pickering, *Constructing Quarks: A Sociological History of Particle Physics* (Chicago: University of Chicago Press, 1984).
6 Bruno Latour and Steve Woolgar, *Laboratory Life: The Social Construction of Scientific Facts* (Beverly Hills and London: Sage Publications, 1979), p. 237.

3

Two Previous Standard Models

J. L. HEILBRON

Born San Francisco, California, 1934; Ph.D., 1964 (history of science),
University of California at Berkeley; Professor of History and the History
of Science, and Vice-Chancellor, University of California at Berkeley.

Having slipped so far down the chain of being – from physicist to historian to administrator – I was very much flattered by the invitation to contribute this chapter. I shall not abuse the invitation by discussing the Standard Model itself, for you all know much more about it than I do. Instead I shall discuss two earlier physical theories (or, rather, sets of theories) that may be considered standard models of their times. My purpose is not to place the modern version in perspective – for what larger setting is possible for a theory that covers all time and all space? My purpose is to remind you that others have had the same intellectual impulses that drive contemporary particle physicists and cosmologists, and that they could point to persuasive evidence in support of their own standard models.

To qualify as a discarded standard model, a theory must have been deemed fundamental and universal; also, it must have enjoyed a wide consensus among physicists and produced quantitative results testable by experiment. These criteria are satisfied by two, and perhaps only two, previous models, which I'll call the Napoleonic and the Victorian.

The Napoleonic standard model

I call the standard model of the years around 1800 Napoleonic, not because he had anything to do with creating it, but because he patronized its principal architects, because it rose and fell coincidentally with his own career, and because it operated with the same mixture of the aristocratic and the democratic, the chauvinistic and the cosmopolitan, that characterized his regime. As in war and politics, law and culture,

France took the lead in standard modeling – in particular, French mathematicians, or, to use the term they sometimes applied to themselves, *physiciens géomètres*, or mathematical physicists. Napoleon recognized the intellectual power of this set of savants and worried that they were not all gentlemen. "It's dangerous to give people who have no money too wide an acquaintance with mathematics," he once said, thinking, perhaps, of the talented impecunious students of the École Polytechnique, who had won their places by competitive examinations and were inclined toward democracy.[1] Some of these men, of whom Augustin Fresnel, the architect of the wave theory of light, was perhaps the most effective, were to subvert both Napoleon and the standard model of his time.

Chief among the modelers and, likewise, a great favorite of Napoleon, was the Marquis de Laplace, Senator of France and prince of the world's *physiciens géomètres*. Napoleon showered blessings on Laplace and once, on the theory that mathematicians with money can do everything, appointed him Minister of the Interior. Laplace lasted six weeks. Later Napoleon explained why: "Laplace did not look at any question from the proper point of view; he looked for subtleties everywhere, had only problematic ideas, and carried into administration the spirit of the infinitely small."[2] Having set down the burdens of office, Laplace could devote himself entirely to standard modeling, or, as he put it, to making physics as perfect as astronomy by importing into it the mathematics and the method of the theory of gravitation.[3]

The fruitfulness of this approach had been demonstrated in 1785 by Charles A. Coulomb, whose famous measurement of electric and magnetic forces rested on several results of the gravitational theory. Perhaps the most important of these was the theorem that a uniform gravitating shell acts outside itself as if its matter were concentrated at its center. Coulomb's invocation of this theorem assumed what his measurement was intended to show: that elements of the supposed electric fluid, and of the magnetic, attract and repel one another in accordance with the law of inverse squares. These fluids were a capital feature of the Napoleonic standard model: every distinct force had at least one such fluid as its carrier. The polar forces had two each: a positive and a negative fluid for electricity, an austral and a boreal for magnetism. Since no one could detect a change of weight attributable to electrification or magnetization, physicists characterized the associated fluids as "imponderables."

In chronological as well as logical parallel with the fluids of electricity and magnetism, the substance of heat, or "caloric," became a Napoleonic imponderable fluid. At the phenomenological level, strong analogies existed between tangible and latent heat, and heat capacity, on the one hand, and electric charge, the uncharged normal state, and electrical capacity, on the other; and, on the modeling level, between the expansivity of the several hypothetical fluids. It was therefore obvious to suppose that the particles of caloric interacted among themselves and with the molecules of matter according to a force whose dependence on distance experiment might seek.

We now have five imponderables – two each for electricity and magnetism, and one for heat – and ponderable matter, all interacting directly by forces dependent only upon the distance between the interacting elements. We need only two more to complete an eightfold way. One of these was light, then still conceived in Newton's terms as streams of particles, regulated in their commerce with ponderable matter by a direct distance force that caused reflection, refraction, and diffraction. The capstone, or omega-minus, of the system, the seventh imponderable and the eighth constituent of the natural universe, was discovered around 1800, when physicists found radiant heat beyond the red end of the visible spectrum. Radiant heat made a fine middle term between light and heat; heat tied all three to electricity and magnetism, via the analogies I have mentioned; and these seven imponderables, or leptons, were associated through a mathematical parallel with baryonic matter that carried the forces of cohesion, affinity, and gravity. As the Napoleonic standard modelers liked to say, they had tied together astronomy and microphysics; a great strength of their system, according to them, was the mutual reinforcement of their theories of the very large and the very small.

The system had its successes: Poisson's work on the distribution of electricity on conductors, and on the behavior of magnetized shells; Biot and Ampère on electromagnetic forces between wires; Laplace and Poisson on adiabatic processes; Dulong and Petit on specific heats; Malus on light; Laplace on refraction and capillarity; and Gay-Lussac on the combination of gases. Around 1810, the Napoleonic model, like Napoleon himself, seemed capable of absorbing everything. In both cases, the result was a juxtaposition of disparate elements tied together by overarching laws and institutions constructed on the same pattern. The physicists recognized explicitly that the union was formal rather than substantial, the functional equivalent of the requirement that all descriptions be renormalizable, gauge-invariant quantum field theories; and they conceded that they could not affirm, in the prototypical case of the theory, whether there existed two, one, or zero electrical fluids. As the standard textbook of the Napoleonic standard model put the point, the objects of physical theory had their place in the heads of mathematicians, not in the course of Nature. The system of distance forces and their specialized carriers had the merits of intelligibility, mathematical convenience, universality, and, sometimes, fit with experiment; it would be asking too much to require that it also reproduce God's blueprints.

The system aroused some opposition among physicists, particularly in Germany and England, and among the proletariat of the educated, for its rigidity of form and its mathematical demands. In order to retain the model's formal coherence, the modelers had to introduce some very complicated formulas, particularly for the interaction of polarized light with birefringent crystals. In Biot's version of this complication, the optical force depended on the angle between the direction of the light ray and the optic axis of the crystal; and also on the shape, orientation, and rate of rotation of the light particles.[4] As for the off-putting demand on mathematics, many natural philosophers of the time complained that they could no longer follow fundamental physics because standard modelers insisted on writing it in the language of exact astronomy. And previously interested laymen ignored altogether a line of thought that they considered stifled by the oppressive requirements of mathematical analysis.

Further afield, the scientific respectability of imponderables appeared to some optimists to open a place for the human soul among the substances constituting the world. I leave the last word on this matter to the first man of his age. In his retirement on Saint Helena, Napoleon declared his faith as follows: "I believe that man is the product of the [imponderable] fluids ... that the brain pumps them around and gives life, [and] that they constitute the soul."[5]

The Victorian standard model

The Napoleonic model did not long survive Napoleon. During the 100 days between the Emperor's return from exile on Elba and his definitive abdication after Waterloo, Fresnel was busy perfecting the wave theory of light. At about the same time Fourier was putting the final touches on his quantitative theory of heat, which accomplished wonders without requiring the concept of caloric. At the midcentury, the work of Joule, Mayer, Clausius, and William Thomson destroyed caloric altogether, and made heat a mode of motion of ponderable molecules. Then came James Clerk Maxwell, who reduced the remaining six leptons to one, the electromagnetic aether, whose motions brought forth the phenomena of electricity, magnetism, light, and radiant heat. By 1880, the basis of a new standard model had been set. I call it Victorian because it was British in inspiration and execution, and because it perfectly fit the materialism, clutter, and complacency supposed to have characterized Victoria's reign. It may also count that she elevated its chief spokesman, William Thomson, to the peerage (as Lord Kelvin) just as Napoleon had distinguished Laplace.

The Victorian standard model had as main ingredients an omnipresent aetherial continuum and discrete material particles. The interactions of the continuum and the particles were a prime subject of discussion. The most radical scheme tried to dissolve matter itself into an aether. In the best elaborated form of this grand unifying theory, atoms became vortex rings, or, to give sufficient variety, chains of linked vortex rings, in a universal frictionless fluid; they moved about like smoke rings in air, or superstrings in vacuum, collided and parted like the particles of a gas, or stuck and moved together like atoms combined into molecules. All that was required to make a world was an incompressible aether possessing mass and independently mobile parts; and, also, mathematical physicists clever and patient enough to extract from long hydrodynamical calculations some analogies between the behavior of vortex atoms and laboratory results. One encouraging finding was that no more than six vortex atoms could move together as a vortex molecule, which seemed relevant to the periodicity of Mendeleev's ouija board, or table of the elements. As Maxwell observed of this form of the Victorian standard model, "[its] difficulties ... are enormous, but the glory of surmounting them would be unique."[6]

Another elaborate form descended from Maxwell's machinery of the electromagnetic aether and J. J. Thomson's discovery that a body has greater inertia when charged than when uncharged. The most refined exercise in this form was Joseph Larmor's theory of aether knots, permanent, mobile centers of strain in a universal plenum having no properties but rotational elasticity and a capacity for internal motions. Just as the kinetic theory had made heat a mode of motion of matter particles and the electromagnetic synthesis had made electricity, magnetism, and light modes of motion of the aether, now inertia arose from the movement of Larmor's aether knots. He found that he could do without the concept of mechanical, or rest mass, and aetherialize inertia; moreover, since he could find no reason that all his knots should be twisted in the same sense, he allowed two types, differing chirally, that is, with opposed he-

licities, which he called positive and negative electrons. The rotation of
an electron of one type around one of the other, or a circulating ring
or rings made of positives and negatives symmetrically placed, constituted an atom. These ideas date from the 1890s, that is, from before the
discovery of the electron, and long before the invention of the nuclear
atom.[7]

These parsimonious theories, which invoked no more than the principle of least action and a substrate with a very few general mechanical
proprieties, were the ideals to which, in theory though rarely in practice,
the various mechanical analogies fancied by Victorian physicists might
be reduced. The analogies between the machinery of the aether and
the workings of springs, flywheels, universal joints, and rubber bands,
which Kelvin and his colleagues deemed essential to their comprehension
of physical theory, gave their enterprise the air of a Victorian factory;
or so its primary critic, Pierre Duhem, liked to say in deprecation of the
system that had superseded his ancestors' scheme of imponderable matters and distance forces. No more, however, than Laplace's school did
Kelvin's insist that their standard model was true of Nature. Rather,
they rested the priority of mechanical reduction on its visualizability,
on its conceptual, if not mathematical, convenience, and on their conviction, and experience, that the human mind can reason exactly only
about mechanical processes occurring in absolute space and time.[8]

This epistemology was nicely caught at the International Congress of
Physics held in Paris in 1900. The French, as hosts, gave the opening
speeches. The president of the French Physical Society, Alfred Cornu,
adroitly wedded Gallic foresight with British achievement; the mechanical reductions that had succeeded across the channel, he said, and especially the attempts at a grand unified theory of aetherial vortices,
promised nothing more, or less, than the imminent realization of the
dream of Descartes, to explain the entire physical world with a little
geometry and lots of brain power. Then Henri Poincaré put this project
in its proper epistemological place. The making of models and the reduction of all physical phenomena to matter in motion might be, and
even do, very well, he said, but that did not signify that theoretical
physicists were approaching the truth. According to Poincaré, a theorist properly regarded is no more than a librarian, who arranges the
collections of science – that is, confirmed experimental results – in the
way most convenient for overview and retrieval. It makes sense to ask if
a library catalogue is useful and reliable, but not whether it is true. The

physicists of 1900 approved Poincaré's talk as "one of the most perfect expressions of the state of mind of the masters of modern science."[9]

It turned out that the relations of aether and matter could not be catalogued conveniently by the physicist-librarians of 1900. One day Lord Kelvin, looking up from the springs and rubber bands wherewith he was wont to fashion aethers, spied two clouds over the dawning twentieth century. He reminded physicists that, despite the labors of a dozen Larmors, they had no acceptable account of the interaction of aether and charged bodies in motion and that an inescapable consequence of the kinetic gas model, the equipartition of energy, failed before the facts. These were profound observations. As we know, it took relativity to dissipate the first cloud and quantum theory to knock down the second.

Like its Napoleonic predecessor, the Victorian standard model inspired or supported a general philosophy that outran it and darkened its name. From the goal, or even definition, of physics as the expression of all natural phenomena in terms of mechanical quantities, the troublemaker could infer that modern science taught that those quantities, and the physical and chemical properties built up from them, were the only things that truly and objectively exist. Champions of yesteryear also found it convenient to foist on physics the doctrines that the brain secretes thought as the kidneys do urine, that the soul is a mistake, that religion and poetry are bunk, that man descended from monkeys, and so on. "Bring up a woman in the positivist school [they said], and you make of her a monster, the very type of ruthless cynicism, of all engrossing selfishness, of unbridled passion."[10] In an article entitled, "The scientific spirit of the age," published in the *Contemporary Review* for 1888, you can read that a man bred to science "will view his mother's tears not as expressions of her sorrow [about his career choice] but as solutions of muriates and carbonates of soda ... and he will reflect that they were caused not by his selfishness, but [by] cerebral pressure on her lachrymal glands."[11]

Very likely these attacks helped to recommend that meekness that, however uncharacteristic of physicists, Poincaré expounded to applause before his peers as the orthodox epistemology of the standard model of 1900. That was the way several perceptive critics of the time regarded the matter. The philosopher Eduard von Hartmann can speak for them all. "The more physics keeps its completely hypothetical character in mind [he wrote], the better will be its scientific reputation in public opinion."[12]

The Atlantic model

A few symmetries may be discerned in the decay processes of Napoleonic and Victorian physics. For one, the demands of unalterable characteristics of the theories came into flat contradiction with experiment. In the earlier case, the conception of heat as a conserved fluid could not be reconciled with Carnot's analysis of the workings of an ideal engine and Joule's measurements of the behavior of real ones. In the later case, the conception of electrons as charged billiard balls subject to the rules of Victorian mechanics and electrodynamics prevented physicists from combining them into stable atoms or accounting via a visualizable model for spectroscopic and other data. Second, the resolution of these and associated difficulties was accomplished through new and far-reaching discoveries: in the examples just mentioned, the second law of thermodynamics and the elementary quantum of action. Third, the overthrow of the standard models resulted at first in important savings in mathematical effort: the wave theory of light arrived at conclusions about the optics of birefringent crystals more directly than the complex theories of Laplace's school; and the old quantum theory, for all its faults, allowed computations of spectral series much simpler than the classical theory of radiation did.

A fourth point worth dilating is that neither model ever had the allegiance of all the good physicists of its time. Although the opponents could not duplicate the full range of exact descriptions available to the consensus physicists, they could make important discoveries by looking where the standard model did not point. Thus the so-called romantic physicists of Napoleon's time set in motion the discovery of the magnetism of current-carrying wires by Oersted and the detection of thermal electricity by Seebeck. Opponents of the Victorian standard model, the so-called energeticists, played an important part in the pre-history of relativity.

A fifth similarity concerns wider applications. Napoleon fancied, perhaps rightly, that his soul was an imponderable fluid, and the materialists of the nineteenth century seized on mechanistic reduction to prove that no one had a soul. This similarity seems to extend to the current Standard Model and its fellow traveler, the Big Bang, which have prompted declarations from physicists that sound theological. In his Nobel prize lecture, one of the chief architects of the Standard Model, Abdus Salam, acknowledged his privilege in being chosen to reveal the portion of God's plan that related to weak interactions; a professor of theoretical physics at Oxford, John Polkinghorne, quit his post to become pastor of theological physics for the Anglican church; and the contributors to *Physics, Philosophy, and Theology*, a volume published a few years ago by the Vatican, show, among much else, how to square the Big Bang with St. Augustine and *creatio ex nihilo*.[13] The recent discovery of large-scale inhomogeneities in the background radiation was announced in phrases that might have come from the pulpit. One astrophysicist called the discovery "the holy grail of cosmology." The discoverer himself, George Smoot, who works for the spiritual University of California at Berkeley, observed that, "if you're religious, its like looking at God."[14]

A sixth point of symmetry, very necessary to complete the scheme, might be dragged in from epistemology. The spokesmen for the previous standard models explicitly disclaimed that they had found the truth. No doubt they were right. And how is it with you? Do you regard the Standard Model as the foundation – or as an ingredient – of a theory of everything? Or as a hodgepodge of mathematics and phenomenology requiring the insertion of ad hoc constants that will almost certainly never be deducible from first principles? If this skepticism represents the dominant view, the parallel holds; if not, if physicists truly believe that they have found the truth or some of it, we have a broken symmetry of great interest to historians.

I have decorated the standard models of 1800 and 1900 with the names of Napoleon and Victoria; that of 2000 remains unattributed, awaiting, as we say in the fundraising game, the naming gift. Unfortunately, no single individual now gives the tone to an age. In view of this sad decline in the human race, and because the current Standard Model has been purchased by the taxpayers of Europe and the United States, I propose to call it the Atlantic model. This choice may have the further recommendations that the ancient island of Atlantis was ruled by a decuplet of five pairs of twin brothers all from the same parents;[15] that, although it did not exist, it had its place on medieval maps of the world; and that, in the hope of going beyond it, Columbus discovered America.

Notes

1 Letter of 23 March 1805, quoted by Joachim Fischer, *Napoleon und die Naturwissenschaften* (Stuttgart: Steiner, 1988), p. 198.

2 Notes on conversation on St. Helena, quoted ibid., p. 110; cf. Maurice Crosland, *The Society of Arcueil* (Cambridge, Mass.: Harvard University Press, 1967), pp. 63–4.

3 Details about the Napoleonic standard model can be found in J. L. Heilbron, *Weighing Imponderables and other Quantitative Science Around 1800* (Berkeley: University of California Press, 1992).

4 J. B. Biot, "Sur de nouveaux rapports entre la réflexion et la polarisation de la lumière," Société Philomatique, *Bulletin des Sciences 3* (1812), pp. 209–16, 226–9, on 229.

5 Quoted by Fischer, *Napoleon*, p. 283.

6 J. C. Maxwell, *Encyclopedia Britannica*, 9th edition (1875), s.v. "Atom."

7 Joseph Larmor, *Aether and Matter* (Cambridge: Cambridge University Press, 1900), pp. 26–8.

8 More on the Victorian standard model can be found in J. L. Heilbron, "Fin-de-siècle physics," in C. G. Bernhard, E. Crawford, and P. Sorböm, eds., *Science, Technology and Society in the Time of Alfred Nobel* (Oxford, 1982), pp. 51–73.

9 C. E. Guillaume, "The International Physics Congress," *Nature 62* (1900), pp. 425–8.

10 W. S. Lilly, "Materialism and morality," *Fortnightly Review 46* (1885), pp. 575–94, on 589.

11 E. P. Cobbe, "The scientific spirit of the age," *Contemporary Review 54* (1888), pp. 126–39, on 130.

12 E. von Hartmann, *Die Weltanschauung der modernen Physik* (Leipzig, 1902), p. 219.

13 Polkinghorne, *Science and Creation* (Boston: 1989), pp. 84–98; Robert John Russell, et al., eds., *Physics, Philosophy, and Theology* (Vatican City: Vatican Observatory, 1988), pp. 283, 375–8, 405.

14 *Los Angeles Times*, 24 April 1992, 25 May 1992, p. B.7; *San Francisco Chronicle*, 25 April 1992, p. A16.

15 Plato, *Critias*, 113–14, in Edith Hamilton and Huntington Cairns, eds., *The Collected Dialogues*, (Princeton: Princeton University Press, 1961), pp. 1218–19.

Part two

Quarks and Leptons

4

From the Psi to Charmed Mesons: Three Years With the SLAC–LBL Detector at SPEAR

GERSON GOLDHABER

Born Chemnitz, Germany, 1924; Ph.D., 1950 (physics), University of Wisconsin; Professor in the Graduate School, Physics Department and at the Lawrence Berkeley Laboratory and Center for Particle Astrophysics, University of California at Berkeley; high-energy physics (experimental).

As I look back at the first three years or so at SPEAR, I consider this one of the most revolutionary, or perhaps *the* most revolutionary, experiment in the history of particle physics. It was certainly the most exciting time – in a laboratory, that is – that I have ever experienced. In this chapter I discuss the period 1973–76, which saw the discoveries of the ψ and ψ' resonances, the χ states and most of the psion spectroscopy, the D^0, D^+ charmed meson doublet, and the D^{*0} and D^{*+} doublet. I will also refer briefly to some more recent results.

Most of these discoveries were made with the SLAC–LBL Magnetic Detector – or, as it later became known, the MARK I – that we operated at SPEAR from 1973 to 1976.[1] The groups involved in this work were led by Burton Richter and Martin Perl of SLAC and by William Chinowsky, Gerson Goldhaber, and George Trilling of LBL.

The discovery of the ψ

Some of my personal reminiscences regarding the weekend of the ψ discovery have already been published and I will only allude to them briefly here.[2]

Our first task was to learn how our detector behaved in the SPEAR environment. For this purpose we developed two independent analysis systems, one at LBL and the other at SLAC. The overall data acquisition was due to Martin Breidenbach. The SLAC system for data analysis was under Adam Boyarski, while Gerald Abrams was largely responsible for the LBL system. Having recently emerged from bubble-chamber experiments, our group had a tendency to produce visual displays of the data. We thus used track reconstruction based on bubble-chamber programs. This work was largely done by Chinowsky and his students, Robert Hollebeek and John Zipse, as well as by Fatim Bulos, Harvey Lynch, and Roy Schwitters of SLAC. At a later stage the track fitting routines were revised and improved by George Trilling with the help of David Johnson at LBL.

I worked on the Berkeley version of the displays, which we recorded on microfiche and then scanned manually. We soon learned how to distinguish cosmic-ray events from Bhabha scatterings, muon-pair production, beam–gas collisions, and electron–positron annihilation into hadrons. At each step Abrams, John Kadyk, and I, as we developed scanning and filtering programs, compared our results with those of Charles Morehouse, who had developed similar independent programs at SLAC. With time we were able to incorporate our results into the triggering procedures for the detector. These details and procedures were important later in following the on-line data acquisition with a one-event display on a CRT screen.

Following an engineering run in the spring of 1973, we started our experiment at SPEAR later that year with an energy scan. At that time we had not expected narrow structures, and thus we decided to measure the cross section in steps of 100 MeV in beam energy, that is, 200 MeV steps in E_{cm}.[3] Fig. 4.1 shows our first cross-section data and results for the ratio $R = \sigma_{had}/\sigma_{\mu\mu}$ as presented by Richter at the London Conference in July 1974. The data were in good agreement with the earlier results from the Cambridge Electron Accelerator and Frascati, and, contrary to expectations, showed a roughly constant cross section from 2.5 to 4.8 GeV.[4]

And yet the data were not completely flat; we were sufficiently intrigued with the high points at 3.2 and 4.2 GeV that we decided to take additional intermediate points in June 1974 at 3.1 and 3.3 GeV as well as around 4.2 GeV. It was an irregularity in the new 3.1 GeV data point – as reanalyzed by Roy Schwitters with the help of my student Scott Whitaker, in October 1974 – that convinced us in early November 1974 that we had to remeasure this region before we could publish our cross-section data. One speculation some of us had was that the SPEAR energy could have drifted upward into a region of higher cross section, yielding two anomalous runs at the nominal energy of 3.1 GeV.

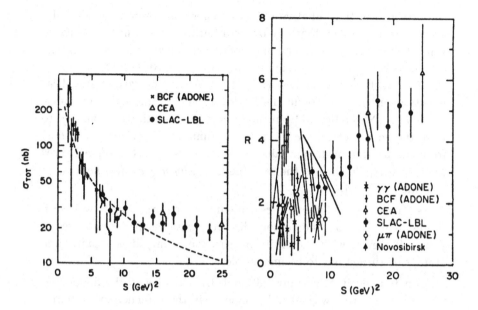

Fig. 4.1. The first cross section and $R = \sigma_{had}/\sigma_{\mu\mu}$ measurements with the SLAC–LBL detector taken in 200 MeV steps plotted here versus the square of the center-of-mass energy $S = E_{cm}^2$. Earlier data are also shown.

In the SPEAR control room on Saturday, November 9, we realized already that we were onto something momentous. When I came to SLAC that afternoon, we were scanning in small energy steps of 10 MeV across the 3.1 GeV region. In particular, Rudolph Larsen was watching the one-event display on the CRT monitor and recording and classifying the events as they came in for a background run at $E_{cm} = 3.0$ GeV. The important categories were e^+e^- or Bhabhas and hadronic events called "> 2P" or "$\geq$ 3P" for 3 or more prongs. As I took over the CRT watch from Larsen, the next energy was set at $E_{cm} = 3.12$ GeV. Now the most amazing thing happened before my eyes. The ratio hadronic/Bhabha events went from 10/61, in the background region, to 55/170 – an increase in this ratio, which is related to the hadronic cross section, by a factor of 2. Little did I know that on the next day this ratio would go up by nearly two orders of magnitude!

Fig. 4.3 shows the ψ signal found on November 10, 1974, by scanning in very small steps. We thus realized that the increase in cross section first noted at 3.2 GeV and the anomalies at 3.1 GeV were the result of the presence of the radiative tail of an enormous resonance. Fig. 4.4 shows a picture taken that day by Vera Lüth, who caught us discussing what the possible quantum numbers of this new resonance could be. The next day we learned from Samuel Ting about the MIT–BNL results on the J – clearly the same effect (Fig. 4.5).[5] As the messages about these results reverberated around the world, we got a rapid confirmation of the J/ψ on November 15 from the groups at Frascati, who managed to push the energy of their e^+e^- ring (by running all their magnets hot) from the maximum design value of 3.0 GeV up to 3.1 GeV. All three papers were published in the same issue of *Physical Review Letters*, on December 2, 1974.[6]

As good citizens of the physics community, we were going to wait with a press release on our momentous discovery until our paper appeared in print. However, with the entire physics community in a superexcited state, this turned out to be impossible.

As it happened, it was my talk at LBL that opened the floodgates. The gist of my talk was given by someone in the audience to a reporter from *The Daily Californian*, the Berkeley student newspaper. The reporter called me up, and I made a valiant attempt to have him wait until our paper was published. But all he wanted to know was had I given a talk, and had we really discovered a new particle? This started a chain reaction. We next heard from *The New York Times*. Why did we give this news to the *Daily Californian* and not to them as well?

A mad scramble followed on this same day, November 15, to coordinate a joint press release for the J and the ψ discoveries. Wolfgang Panofsky and Richter at SLAC, Martin Deutch and Ting at MIT as well as many others worked on this far into the night. Fortunately, the editors of *Physical Review Letters* "concurred that the news justified early public release," and agreed to publish the papers in spite of the press coverage.[7]

As we found out after the discovery of the ψ, our energy scale at SPEAR was off by 10 MeV. Thus the ψ, which was at first measured as 3105 MeV, was actually at 3095 MeV; this is shown in Fig. 4.3, which gives both the old and the new energy scales. Had we known the correct energy scale when the measurements were made by Breidenbach in June 1974 at 3.100 GeV, we would have seen all eight runs at about six times the normal cross section instead of just two anomalous runs with cross sections three and five times "normal"! This would certainly have led to the ψ discovery right then and there.

Jan. 1974	•*John Kadyk* (LBL) noted high σ by ~ 30% at E_{cm} = 3.2 GeV.
	•Confirmed by LBL and SLAC Collaborators.
June 1974	•*Martin Breidenbach* (SLAC) carried out measurements at 3.1, 3.2 and 3.3 GeV to check high point at 3.2 GeV.
July 1974	•Presentation of "flat" σ by *Burton Richter* at the London Conference.
	•However: anomalous point at 3.2 GeV.
Oct. 1974	•*Roy Schwitters* (SLAC) looked at all σ data to prepare for a paper. Found inconsistencies at 3.1 GeV.

$$
\begin{array}{lll}
6\ \text{runs} & \sigma\ \text{"normal"} & \sigma = \sigma_0 \\
1\ \text{run} & & \sigma = 3\,\sigma_0 \\
1\ \text{run} & & \sigma = 5\,\sigma_0
\end{array}
$$

•Finding confirmed by *Gerald Abrams* (LBL).

•Events in the 3.1 GeV data checked in detail independently by *Scott Whitaker* and *Gerson Goldhaber* (LBL).

•All looked normal except for an apparent increase of $K^{\pm}$ and K^0 in high σ runs (partly statistical fluctuation).

Week of Nov. 4, 1974	•Discussions with *Burton Richter* leading to decision to study 3.1 GeV data in detail to explain inconsistencies.
Nov. 9-10, 1974	•The weekend discovery. Paper written "on-line".
Nov. 11, 1974	•Reports on our result to our respective laboratories: *Roy Schwitters* at SLAC and *Gerson Goldhaber* at LBL.
	•*Samuel Ting* reports on discovery of the J.
Nov. 15, 1974	•The J/ψ confirmed at Frascati.

Fig. 4.2. Chronology of ψ discovery.

The discovery of the ψ'

Encouraged by our remarkable result, we decided to look for more sharp peaks. Richter together with Ewan Paterson and Robert Melen was able to modify the SPEAR operation so as to run in a mode in which the energy was stepped up by 1 MeV every 3 minutes, while Breidenbach was able to modify our analysis system so that the resulting cross-section points could be calculated *on-line*.[8] Fig. 4.6 illustrates a real-time test of this new setup by scanning the ψ mass region,[9] which shows clearly that such a resonance can be readily discovered in this mode of operation.

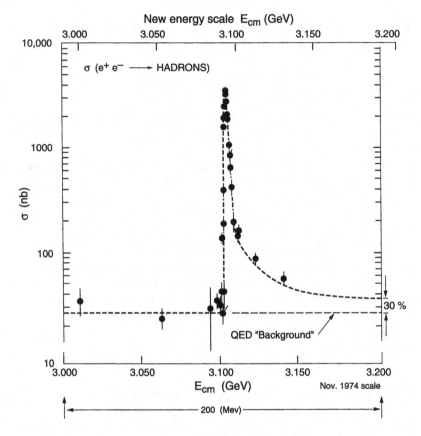

Fig. 4.3. The discovery of the ψ as observed on November 10, 1974. The lower energy scale was the one used at the time of discovery. The upper "new" energy scale is based on a recalibration of the orbits at SPEAR a few months later. The original 200 MeV energy step as well as the ~30% high value of σ at 3.2 GeV is also shown.

Indeed, ten days later, during the early morning hours of November 21, we discovered a second narrow resonance, the ψ', at 3.695 GeV (later renormalized to 3.685 GeV).

The properties of the J/ψ and ψ', and psion spectroscopy

Emboldened by this success, after taking a day or two off to write the ψ' paper, we continued our scan and scanned on and on and on. No other narrow resonance showed up (since SPEAR was not designed to

Fig. 4.4. November 10, 1974. SPEAR Control Room. William Chinowsky, Martin Perl, François Vannucci, and Gerson Goldhaber. What can the quantum numbers of this new resonance be? (Photo by Vera Lüth.)

reach 10 GeV!), but we did find a broad resonance at 4.4 GeV and considerable structure near 4.03 GeV. In Fig. 4.7 I show a later plot (1977) that indicates this structure as well as the $\psi''(3770)$ discovered by an extended collaboration using an upgraded MARK I detector.[10]

During the period November 1974 to May 1976 enormous progress was made in understanding the properties of the ψ and ψ' and in unraveling the entire psion spectroscopy. For the ψ and ψ' we measured the spin, parity, and charge conjugation in interference experiments with Bhabha scattering at the leading edge of the resonances. We found that $J^{PC} = 1^{--}$ – the quantum numbers of the photon and vector mesons. From final state studies in ψ decay we determined that $G = (-)$ from a predominance of an odd number of pions, and that $I = 0$ from the decay $\psi \to \Lambda\overline{\Lambda}$ among others. We observed the transitions $\psi' \to \psi\pi^+\pi^-$, the major decay mode of the ψ', and $\psi' \to \psi\eta$, which showed that if the ψ consisted of a combination of two quarks $q\overline{q}$, these quarks passed on to the final state. Following the DASP discovery of a P state intermediate between ψ and ψ', we observed the intermediate χ states $^3P_0$, $^3P_1$, and

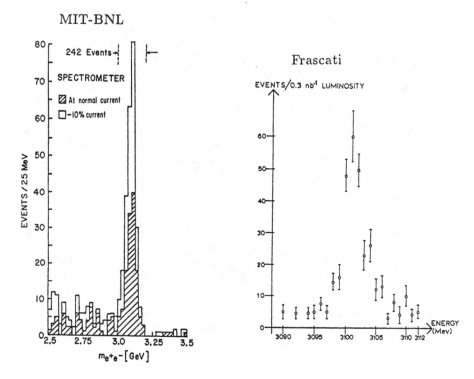

Fig. 4.5. The discovery of the J by the MIT–BNL group and the confirmation of the J/ψ discovery by the Frascati groups. See note 6.

$^3P_2$ obtained from

$$\psi' \to \chi\gamma$$
$$\hookrightarrow \gamma\psi$$

and also from $\psi' \to \chi\gamma$ followed by direct hadronic χ decays.[11] The detailed studies of the transitions between these states came later from work by the MPPSSD collaboration and the Crystal Ball collaboration.[12]

During this period also, Perl discovered the τ lepton, Gail Hanson demonstrated that jets are produced in the e^+e^- annihilation process, and Schwitters observed transverse beam polarization and demonstrated that the final-state particles followed the distribution expected for the spin-$\frac{1}{2}$ partons.[13]

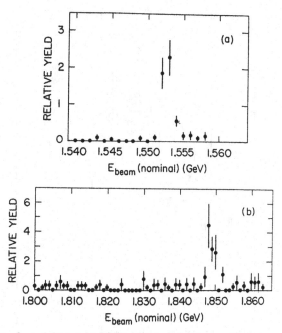

Fig. 4.6. Data for relative hadron yield taken (a) in a calibration run over the ψ, and (b) during the run in which the ψ' was discovered on November 21, 1974.

Where does the name psi come from?

We called the first resonance $SP(3105)$ for about one day where SP stood for SPEAR; however, we soon realized that a two-letter name was unsuitable.[14] The name ψ came from a cursory look I made through the LBL Particle Data Group booklet for an unused Greek letter – while on the phone to Trilling and then to Richter. In addition "PS" in "PSI" is "SP" spelled backward. Little did we know that the resonance would end up with two letters – J/ψ – anyhow! All the same, we "got a sign" (see Fig. 4.8) later, from the reaction

$$\psi' \rightarrow \psi\pi^+\pi^-$$
$$\hookrightarrow e^+e^-,$$

that our choice of the Greek letter ψ was an auspicious one!

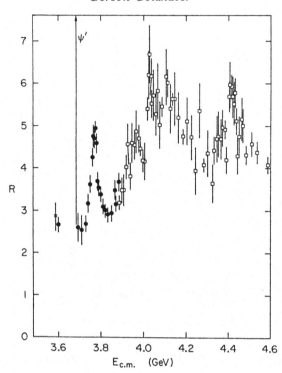

Fig. 4.7. Open squares represent R measurements in the SLAC–LBL detector at SPEAR. Closed circles represent measurements made after the detector was upgraded in 1976 with a lead glass wall for improved photon and electron detection.

What does all this have to do with charm?

Though our work on the ψ and ψ' was not influenced by theoretical predictions, the work on the psion spectroscopy was. In particular, there now came a groundswell of theoretical papers interpreting the effects we were observing. The front runner among these theories was the one suggesting that the J/ψ contained "hidden charm," namely, that it was a bound state of $c\bar{c}$ quarks, which had been predicted earlier, while the narrowness of the ψ was explained by the Okubo–Zweig–Iizuka rule.[15] If this was so, one expected to see particles with "naked charm," that is, a charm quark (or antiquark) bound with other quarks or antiquarks.[16] Yet it took us from November 1974 to May 1976 to find the expected

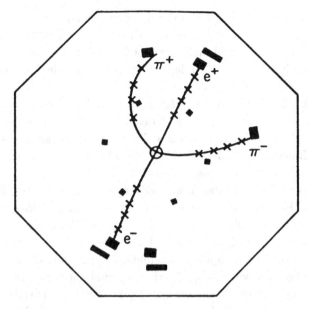

Fig. 4.8. Example of the decay $\psi' \to \pi^+\pi^-\psi$, followed by $\psi \to e^+e^-$.

signal – a clear peak in the $K^-\pi^+$ and $K^-\pi^+\pi^-\pi^+$ mass distributions, at a mass of 1865 MeV/c^2.[17]

Our difficulty in finding this signal was a major theme of a meeting held April 22–24, 1976, at the University of Wisconsin, my old alma mater. I talked on various aspects of the ψ, ψ', and χ studies involving baryonic final states.[18] Harvey Lynch talked on cross-section measurements and in particular about our inability to observe naked charm signals.[19]

I decided at the meeting that when I got back to Berkeley I would spend a month carefully sifting through our data and try to find charmed particles or to find out why charmed particles did not occur in our experiment. At the airport, on the way home, I met Sheldon Glashow, and we shared a plane ride to Chicago. He was particularly persistent that charmed particles just had to be there. Why were we incapable of finding them? I told him that I had resolved to spend at least a month reanalyzing the data to find the answer.

When I got back to Berkeley, I spent the rest of the week and the weekend at the lab. We had just taken some more data at SPEAR in the 3.9–4.6 GeV region. Previously, in various analyses I had carried out, I had always been careful to use strict criteria for K and π identification, using time-of-flight and momentum measurements to determine the masses of the particles produced in the e^+e^- annihilation. I decided to change my strategy and make very loose cuts. The thought was that I would not have a pure sample, but rather a sample enriched in K mesons.

I thus studied $K^\mp\pi^\pm$ mass distributions correlated with recoil mass distributions. To my surprise and delight, I did not have to wait a month. By Sunday, I had obtained a clear signal – a peak in the $K^\mp\pi^\pm$ mass distribution at about 1870 MeV (see Fig. 4.9) associated with an equal or larger recoil mass peak. (See Fig. 4.10.)

On Monday and Tuesday I was looking for my colleague François Pierre, a visitor with our group at LBL from Saclay, France, to show him my result. Finally I met up with him on Wednesday for lunch. As I found out, he had also observed a $K\pi$ as well as a $K\pi\pi$ signal. Right after lunch we compared distributions and realized we had each independently and with different criteria found the same mass peaks. We spent the next two hours writing a joint note to our collaboration showing our data. I called Schwitters, our spokesman at that time, to tell him about our results. There was much excitement both at LBL and SLAC. After our colleagues had a chance to check our results and convince themselves that we were right, we sent a paper off to *Physical Review Letters*. One question came up. How could we prove that we had really identified kaons? Jonathan Dorfan, who had just recently joined our collaboration, came up with the suggestion that we weight each track according to the probability that it be a K or a π and then plot the weighted $K\pi$ mass distribution, which is shown in our paper.[20]

On May 8, I called Glashow to tell him about our finding. He was extremely excited and happy. His long-standing predictions had finally come true. Shelly of course "knew" it all along. But now the rest of the world knew it as well! We were duly cautious in our paper and used the word "charm" only as the last word in the last sentence:[21]

> ... the narrow width of this state, its production in association with systems of even greater mass, and the fact that the decays we observe involve kaons form a pattern of observation that would be expected for a state possessing the proposed new quantum number charm.

When I told my friend Goesta Ekspong about our findings, he challenged me to prove that these data indeed represented charmed mesons and not just another $K^* \to K^-\pi^+$ decay. In the course of the next two years this proof became definitive.

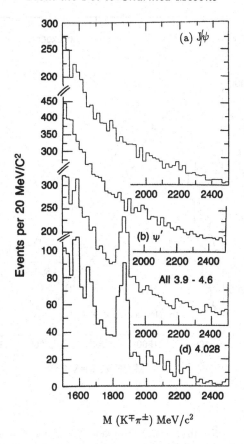

Fig. 4.9. A composite of the $K\pi$ mass distribution for the J/ψ region, the ψ' region, and the $E_{cm} = 3.9$–4.6 GeV region as well as the $E_{cm} = 4.028$ GeV data separately.

The case for charmed mesons

1. *Threshold.* For a new $K^*(1865)$ we also expect a threshold. But that is expected at $\sim$2.360 GeV $[K^*(1865) + K]$ or even $\sim$2.755 GeV $[K^*(1865)+K^*(890)]$. However, the experimental threshold lies above 3.7 GeV (see Fig. 4.9). In the charm theory, a threshold is expected at $E_{cm} = 2E_D \simeq 3.7$ GeV, corresponding to $e^+e^- \to D^0\overline{D}^0$. In fact, the $\psi''(3770)$, discovered later, is a resonance just above threshold that decays predominantly into $D^0\overline{D}^0$ and D^+D^-.[22]

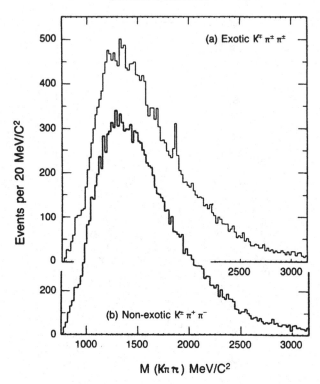

Fig. 4.10. (a) Mass distribution for the exotic $K\pi\pi$ combination showing the D^+ peak. (b) Mass distribution for the nonexotic $K\pi\pi$ combination.

2. *Associated production.* For a new $K^*(1865)$ we expect associated production with K or perhaps with $K^*(890)$, but there is no known reason to expect $K^*(1865) + \overline{K}^*(1865)$ associated production. Experimentally, we find that all observed events corresponding to the 1865 MeV peak occur in associated production with either equal or higher-mass objects.

3. *The charged decay mode.* For a K^* with $I = \frac{1}{2}$ we also expect a charged decay mode. For three-body decays this would have to be the nonexotic mode $K^{\mp}\pi^+\pi^-$. Experimentally we observe the exotic decay mode $K^{\mp}\pi^{\pm}\pi^{\pm}$ but do not observe the nonexotic decay mode (see Fig. 4.10); neither do we observe the $I = 5/2$ triply-charged $K^{\mp}\pi^{\mp}\pi^{\mp}$ decay mode (not shown here). Thus if the peak corresponds to a K^*,

it must have $I = 3/2$; that is, an exotic K^*, which (incidentally) would be the first clear case of an exotic meson state. If we adopt the point of view that we are dealing with an exotic K^*, we would still have to invent an explanation for the peculiar fact that the $I_z = \pm\frac{1}{2}$ states (the nonexotic combinations $K^\mp\pi^+\pi^-$) are suppressed.

On the other hand, our observations are in good agreement with charm theory in which Cabibbo-enhanced hadronic weak decays obey a $\Delta C = \Delta S$ rule, that is, the charm quark c decays weakly to $s\bar{d}u$. Thus in $D^+(C = 1, S = 0)$ decay, for example, the final state has $C = 0, S = -1$ together with $Q = +1$; that is, the charged final state is predicted to be exotic. This point holds explicitly for the charm model and would not necessarily be true for other new types of mesons M composed of $\bar{q}Q$.

4. *Experimental width.* For a K^* of mass 1865 MeV, we might expect a width $\Gamma \approx 50 - -200$ MeV, although for an exotic K^* we have no clear prediction. Experimentally, we find $\Gamma < 40$ MeV from the mass spectrum; however, by making use of the information from the recoil spectrum as well, this limit becomes $\Gamma < 2$ MeV. Charm theory predicts that the decays we are dealing with are weak decays and estimates are: $\tau \sim 10^{-13}$sec or roughly $\Gamma \sim 10^{-2}$ eV.

5. *Evidence for parity nonconservation, or the "τ–θ puzzle" revisited.* For a K^* we expect parity conservation in the decay; this should hold even for an exotic K^*. Experimentally, we find evidence for parity nonconservation. This is based on a study of the Dalitz plot for $K^\mp\pi^\pm\pi^\pm$ decay and the assumption that the charged and neutral states are an isospin multiplet. If parity is conserved in the $K^\mp\pi^\pm$ decay we must have the natural spin parity series $J^P = 0^+, 1^-, 2^+$, and so on. For the $K^\mp\pi^\pm\pi^\pm$ decay mode: $J^P = 0^+$ is ruled out for three pseudoscalars in the final state by angular momentum and parity conservation. $J^P = 1^-, 2^+$, and so on, give Dalitz plot distributions that vanish on the boundary. Our data rule this out clearly.[23] Thus we have strong evidence for parity nonconservation and hence a weak decay, consistent with the charm theory predictions.

6. *Higher mass states.* For a $K^*(1865)$ there is no specific prediction for a next higher mass state. Experimentally, we find from the recoil mass spectrum a next higher mass state at 2006 GeV. From charm theory a state D^* is predicted with mass $m_{D^*} \sim 2$ GeV. If, without prejudicing the case, we use the nomenclature of charm theory, the observed three peaks in the recoil spectrum can be interpreted as:

$$e^+e^- \rightarrow D^0\overline{D}^0 \qquad (4.1)$$

$$\rightarrow D\overline{D}^* \text{ and } \overline{D}D^* \qquad (4.2)$$

$$\rightarrow D^*\overline{D}^* \qquad (4.3)$$

although the detailed structure is complicated. The identity of a possible fourth peak in the recoil mass spectrum near 2.43 GeV was only recently established by the ARGUS experiment at DESY.

Furthermore, the decay modes

$$D^{*0} \rightarrow D^0\pi^0 \qquad (4.4)$$

$$\rightarrow D^0\gamma \qquad (4.5)$$

have been identified and proceed with comparable rates.[24] These two are the only important D^{*0} decay modes. The fact that D^{*0} has a large radiative decay indicates that it must be narrow and chooses to decay into a D^0 rather than directly into a $K^-\pi^+$ as might be expected for $K^*(2006)$. We must conclude that a special quantum number (presumably charm) is conserved in D^{*0} decay to the D^0.

Similar arguments can also be given for the decays[25]

$$D^{*+} \rightarrow D^0\pi^+ \qquad (4.6)$$

$$\rightarrow D^+\pi^0 \qquad (4.7)$$

$$\rightarrow D^+\gamma \qquad (4.8)$$

7. *Spin.* For a $K^*(1865)$ one might expect spin values of $J = 3$ or 4, although, again, for an exotic K^* all bets are off. An analysis of the events represented by reaction (4.2) given above can rule out simultaneous spin assignments for the states at 1865 and 2006, respectively, of 0 and 0 as well as 1 and 0, while the assignments 0 and 1 are consistent with the data.[26] Charm theory predicts $J^P = 0^-$ and 1^- for the D and D^*, respectively. These values have been confirmed in more recent measurements.[27]

8. *Lifetime.* For a K^* the lifetime is that typical of strong interaction or about 10^{-23}–10^{-24} sec. Charm theory predicts weak decay lifetimes in the 10^{-13} sec region. Emulsion measurements in cosmic rays and in neutrino beams had observed neutral and charged decays occurring $\sim$10–200 microns from the parent interaction.[28] Recently the lifetimes of the D^0 as well as the D^+ have been directly measured for identified

decays in emulsions, high-resolution bubble chambers, and electronic detectors with vertex chambers. The present best average values are[29]

$$\tau_{D^0} = 4.20 \pm 0.08 \times 10^{-13} \text{ sec}$$
$$\tau_{D^+} = 10.66 \pm 0.23 \times 10^{-13} \text{ sec}$$

9. *Semileptonic decays.* The DASP experiment at DESY has identified electrons in multiprong events ($N > 3$) with a maximum signal observed in the $E_{cm} = 4.0 - -4.2$ GeV region. They have also observed K^+-e correlations that peak in the same E_{cm} region. Furthermore, the PLUTO group at DESY has observed K_s^0-e correlations also peaked in the $E_{cm} = 4.05$ GeV region. More recently the decay modes

$$D^0 \rightarrow K^- e^+ \nu$$
$$\rightarrow K^{*-} e^+ \nu$$

have been identified and the decay spectrum measured in the lead-glass wall and DELCO experiments at SPEAR as well as in the DESY experiments.[30] The existence of semileptonic decays is further proof for the weak interaction being responsible for D decays as predicted for charmed quarks.

10. *The Cabibbo-suppressed decay modes.* The charm model also predicts a specific ratio between Cabibbo-enhanced and -suppressed decay modes. For example,

$$(D^0 \rightarrow \pi^- \pi^+)/(D^0 \rightarrow K^- \pi^+) = \tan^2 \theta_c$$

where θ_c is the Cabibbo angle. The decay modes

$$D^0 \rightarrow \pi^+ \pi^-$$

and

$$D^0 \rightarrow K^- \pi^+$$

were later observed in the SLAC–LBL MARK II detector.[31] The average value for the two decay modes is indeed consistent with the above relation.

Establishment of the Cabibbo-suppressed decay modes is another characteristic requirement of charmed quarks. The MARK III detector at SPEAR had in the 1980s identified many more Cabibbo-suppressed decay modes.[32]

11. *The F-meson.* In addition to the D^0 and D^+, the isodoublet of the charm model, which corresponds to $\bar{u}c$ and $\bar{d}c$, the singlet $\bar{s}c$ is also

predicted. This object was expected to have decay modes into two strange particles, $F^+ \to K^+ K^- \pi^+$, for example. This state was hard to find, at first. Early indications were observed at a mass of 2040 MeV, but later the clear observation has been made in the CLEO experiment at CESR, the ARGUS experiment at DORIS, and the TASSO experiment at PETRA.[33] These experiments observe the decay $F^+ \to \phi \pi^+$ at a mass of $M_F = 1970$ MeV. These observations, together with evidence for an F^* from ARGUS and the TPC at PEP, complete the picture and give us an unambiguous identification of the charmed mesons.

The SLAC–LBL collaboration at SPEAR

Not only were we lucky in that we were sitting on a "gold mine" at SPEAR, we also had a very congenial group of people. Since we had so much new data, a new discovery came up every few weeks, and there was very little infighting. There were plenty of data to go around, so that anyone who had something to report could give talks at conferences.

I am very proud of our record. I do not believe that any of the data we published had a serious flaw or were outright wrong. A lot of the credit for this must go to Trilling, my co-group leader at LBL who is a very liberal person but is very conservative when it comes to physics claims. He personally went through every word we published with a fine-toothed comb and checked the validity of every standard deviation we claimed.

There is of course another side to this coin. To never publish a wrong result, we had to set our threshold for the acceptance of any given result very high. Thus occasionally we decided not to publish a claim that actually turned out to be correct. I will give three examples:

When we published our paper on the $K\pi$ peak at 1865 MeV, the recoil spectrum appeared to have structure – later identified as $D\overline{D}^*$ and $D^*\overline{D}^*$ production; however, we decided not to claim this structure.

When we published this same recoil spectrum (with considerably higher statistics from the MARK II), there was a clear fourth peak at about 2.43 GeV. We never claimed the observation of a D^{**0}, which was later clearly identified by ARGUS at DESY.

Finally we had an isolated peak of about a dozen events in the $\phi\pi$ distribution at 1960 MeV. The data were, however, not completely understood. Thus we never claimed the discovery of the F, which was later clearly established by the CLEO group at Cornell.

Acknowledgments

I wish to thank Ms. Lonnette Robinson for help in assembling this manuscript. This work was supported by the U.S. Department of Energy under contract number DE-AC76SF00098.

Notes

1 The physicists involved in the initial experiment were: J.-E. Augustin, A. M. Boyarski, M. Breidenbach, F. Bulos, J. T. Dakin, G. J. Feldman, G. E. Fischer, D. Fryberger, G. Hanson, B. Jean-Marie, R. R. Larsen, V. Lüth, H. L. Lynch, D. Lyon, C. C. Morehouse, J. M. Paterson, M. L. Perl, B. Richter, P. Rapidis, R. F. Schwitters, W. M. Tanenbaum, and F. Vannucci of SLAC; and G. S. Abrams, D. Briggs, W. Chinowsky, C. W. Friedberg, G. Goldhaber, R. J. Hollebeek, J. A. Kadyk, B. Lulu, F. Pierre, G. H. Trilling, J. S. Whitaker, J. Wiss, and J. E. Zipse of LBL.

2 Gerson Goldhaber, "Discovery of Massive Neutral Vector Mesons: One Researcher's Personal Account," in Bogdan Maglich, ed., *Adventures in Experimental Physics* (Princeton, NJ: World Science Education, 1976), vol. 5, p. 131. It is interesting to note here that the earlier observation of J. H. Christenson, et al., "Observation of Massive Muon Pairs in Hadron Collisions," *Phys. Rev. Lett. 25* (1970), pp. 1523–26, showing a broad shoulder in the μ pair mass between 3 and 4 GeV, was finally understood (Leon Lederman, unpublished note, December 16, 1974; see also Chapter 6) when the J/ψ discoveries were made.

3 A paper by C. E. Carlson and P. G. O. Freund, "The Case for a Quartet Model of Hadrons," *Phys. Lett. 39B* (1972), pp. 349–52, talks about a vector meson with "hidden charm" but was not known to us. We found out later that Thomas Appelquist and David Politzer had a paper in preparation suggesting such mesons when our discovery was made, T. Appelquist and D. Politzer, "Heavy Quarks and e^+e^- Annihilation," *Phys. Rev. Lett. 34* (1975), pp. 43–9.

4 A. Litke, et al., "Hadron Production by Electron–Positron Annihilation at 4 GeV Center-of-Mass Energy," *Phys. Rev. Lett. 30* (1973), pp. 1189–92; G. Tarnopolsky, et al., "Hadron Production by Electron–Positron Annihilation at 5 GeV Center-of-Mass Energy," *Phys. Rev. Lett. 32* (1974), pp. 432–5; F. Ceradini, et al., "Multihadron Production in e^+e^- Collisions up to 3 GeV Total C.M. Energy," *Phys. Lett. 47B* (1973), pp. 80–4; C. Bacci, et al., "Multihadronic Cross Sections from e^+e^- Annihilation up to 3 GeV c.m. Energy," *Phys. Lett. 44B* (1973), pp. 533–6.

5 Samuel Ting, "Discovery of Massive Neutral Vector Mesons: One Researcher's Personal Account," in Bogdan Maglich, ed., *Adventures in Experimental Physics*, vol. 5, p. 115.

6 J. Aubert, et al., "Experimental Observation of a Heavy Particle J," *Phys. Rev. Lett. 33* (1974), pp. 1404–6; J.-E. Augustin, et al., "Discovery of a Narrow Resonance in e^+e^- Annihilation," *Phys. Rev. Lett. 33* (1974), pp. 1406–8; C. Bacci, et al., "Preliminary Result of Frascati (ADONE) on the Nature of a New 3.1-GeV Particle Produced in e^+e^- Annihilation," *Phys. Rev. Lett. 33* (1974), pp. 1408–10.

7 J. A. Krumhansl and George L. Trigg, "Editorial: Publication of a New Discovery," *Phys. Rev. Lett. 33* (1974), p. 1363.

8 Burton Richter, "Discovery of Massive Neutral Vector Mesons: One Researcher's Personal Account," in Bogdan Maglich, ed., *Adventures in Experimental Physics*, vol. 5, p. 143.

9 G. S. Abrams, et al., "Discovery of a Second Narrow Resonance in e^+e^- Annihilation," *Phys. Rev. Lett. 33* (1974), pp. 1453–5.

10 P. A. Rapidis, et al., "Observation of a Resonance in e^+e^- Annihilation Just above Charm Threshold," *Phys. Rev. Lett. 39* (1977), pp. 526–9.

11 W. Braunschweig, et al., "Observation of the Two Photon Cascade $3.7 \rightarrow 3.1 + \gamma\gamma$ via an Intermediate State P_c," *Phys. Lett. 57B* (1975), pp. 407–12; J. S. Whitaker, et al., "Radiative Decays of $\psi(3095)$ and $\psi(3684)$," *Phys. Rev. Lett. 37* (1976), pp. 1596–9; W. Tanenbaum, et al., "Radiative Decays of the $\psi(3684)$ into High-mass States," *Phys. Rev. 17D* (1978), pp. 1731–49.

12 C. J. Biddick, et al., "Inclusive γ-Ray Spectra from $\psi(3095)$ and $\psi'(3684)$ Decays," *Phys. Rev. Lett. 38* (1977), pp. 1324–7; E. D. Bloom and C. W. Peck, "Physics with the Crystal Ball Detector," *Ann. Rev. Nucl. Part. Sci. 33* (1983), pp. 143–97.

13 M. Perl, et al., "Evidence for Anomalous Lepton Production in e^+e^- Annihilation," *Phys. Rev. Lett. 35* (1975), pp. 1489–92; G. Hanson, et al., "Evidence for Jet Structure in Hadron Production by e^+e^- Annihilation," *Phys. Rev. Lett. 35* (1975), pp. 1609–12; R. F. Schwitters, et al., "Azimuthal Asymmetry in Inclusive Hadron Production by e^+e^- Annihilation," *Phys. Rev. Lett. 35* (1975), pp. 1320–2.

14 G. Goldhaber, "Discovery of Massive Neutral Vector Mesons," p. 131.

15 J. D. Bjorken and S. Glashow, "Elementary Particles and SU(4)," *Phys. Lett. 11* (1964), pp. 255–7; S. L. Glashow, J. Iliopoulo and L. Maiani, "Weak Interactions with Lepton–Hadron Symmetry," *Phys. Rev. D2* (1970), pp. 1285–92.

16 G. A. Snow, "If 'Charm' Particles Exist, Can They Be Detected?," *Nucl. Phys. B55* (1973), pp. 445–54; S. L. Glashow, "Charm: An Invention Awaits Discovery," in D. A. Garelick, ed., *Experimental Spectroscopy – 1974, AIP Conference Proceedings No. 21* (New York: American Institute of Physics, 1974), pp. 387–92; M. K. Gaillard, B. W. Lee, and J. L. Rosner, "Search for Charm," *Rev. Mod. Phys. 47* (1975), pp. 277–310; A. De Rújula, H. Georgi, and S. L. Glashow, "Hadron Masses in Gauge Theory," *Phys. Rev. D12* (1975), pp. 147–62; M. B. Einhorn and C. Quigg, "Nonleptonic Decays of Charmed Mesons: Implications for e^+e^- Annihilation," *Phys. Rev. D12* (1975), pp. 2015–30.

17 Earlier indication of possible charmed particle production have come from cosmic-ray emulsion studies, K. Niu, E. Mikumo, Y. Maeda, "A Possible Decay in Flight of a New Type Particle," *Prog. Theor. Phys. 46* (1971), pp. 1644–6, as well as from experiments involving neutrino interactions. See, for example: A. Benvenuti, et al., "Observation of New-Particle Production by High-Energy Neutrinos and Antineutrinos," *Phys. Rev. Lett. 34* (1975), pp. 419–22; E. G. Cazzoli, et al., "Evidence for $\delta S = -\delta Q$ Currents or Charmed-Baryon Production by Neutrinos," *Phys. Rev. Lett. 34* (1975), pp. 1125–8; J. Blietschau, et al., "Observation of Muon–Neutrino Reactions Producing a Positron and a Strange Particle," *Phys. Lett. 60B* (1976), pp. 207–10; J. von Krogh, et al., "Observation of

$\mu^- e^+ K S^0$ Events Produced by a Neutrino Beam," *Phys. Rev. Lett. 36* (1976), pp. 710–13; B. C. Barish, et al., "Investigations of Neutrino Interactions with Two Muons in the Final State," *Phys. Rev. Lett. 36* (1976), pp. 939–42. Subsequent to our work, evidence for a charmed baryon was also presented from a photoproduction experiment at Fermilab, B. Knapp, et al., "Observation of a Narrow Antibaryon State at 2.26 GeV/c^2," *Phys. Rev. Lett. 37* (1976), pp. 882–5, and in later work at CERN and with the MARK II detector. References to more recent papers can be found in G. Goldhaber and J. E. Wiss, "Charmed Mesons Produced in $e^+ e^-$ Annihilation," *Ann. Rev. Nucl. Part. Sci 30* (1980), pp. 337-81. See G. Goldhaber, et. al., "Observation in $e^+ e^-$ Annihilation of a Narrow State at 1865 MeV/c^2 Decaying to $K\pi$ and $K\pi\pi$," *Phys. Rev. Lett. 37* (1976), p. 255, and I. Peruzzi, et al., "Observation of a Narrow Charged State at 1876 MeV/c^2 Decaying to an Exotic Combination of $K\pi\pi$," *Phys. Rev. Lett. 37* (1976), p. 569.

18 G. Goldhaber, "Recent Results on the New Particle States below 3.7 GeV Produced in $e^+ e^-$ Annihilations," in D. B. Cline and J. J. Kolonko, eds., *Proceedings of the International Conference on Production of Particles with New Quantum Numbers*, University of Wisconsin, 22–24 April 1976, p. 40.

19 Harvey L. Lynch, "The Total Cross Section for $e^+ e^- \to$ Hadrons and its Associated Spectroscopy at SPEAR," in D. B. Cline and J. J. Kolonko, eds., *Proceedings of the International Conference on Production of Particles with New Quantum Numbers*, p. 20.

20 G. Goldhaber, et al., "Observation in $e^+ e^-$ Annihilation ...," pp. 255–9.

21 G. Goldhaber, et al., "Observation in $e^+ e^-$ Annihilation," p. 259.

22 P. A. Rapidis, et al., "Observation of a Resonance," pp. 526–9.

23 J. E. Wiss, et al., "Evidence for Parity Nonconservation in the Decays of the Narrow States near 1.87 GeV/c^2," *Phys. Rev. Lett. 37* (1976), pp. 1531–4.

24 G. Goldhaber, et al., "D and D^* Meson Production near 4 GeV in $e^+ e^-$ Annihilation," *Phys. Lett. 69B* (1977), pp. 503–7.

25 G. J. Feldman, et al., "Observation of the Decay $D^{*+} \to D^0 \pi^+$," *Phys. Rev. Lett. 38* (1977), pp. 1313–15.

26 H. K. Nguyen, et al., "Spin Analysis of Charmed Mesons Produced in $e^+ e^-$ Annihilation," *Phys. Rev. Lett. 39* (1977), pp. 262–5.

27 G. Goldhaber and J. E. Wiss, "Charmed Mesons," pp. 337–81.

28 K. Niu, E. Mikumo, and Y. Maeda, "A Possible Decay in Flight," pp. 1644–6; A. Benvenuti, et al., "Observation of New-Particle Production," pp. 419–22; E. G. Cazzoli, et al., "Evidence for $\delta S = -\delta Q$ Currents," pp. 1125–8; J. Blietschau, et al., "Observation of Muon–Neutrino Reactions," pp. 207–10; J. von Krogh, et al., "Observation of $\mu^- e^+ K_s^0$ Events," pp. 710–13; B. C. Barish, et al., "Investigations of Neutrino Interactions," pp. 939–42; B. Knapp, et al., "Observation of a Narrow Antibaryon State," pp. 882–5; G. Goldhaber and J. E. Wiss, "Charmed Mesons," pp. 337–81.

29 R. A. Sidwell, N. W. Reay, and N. R. Stanton, "Measurement of Charmed Particle Lifetimes," *Ann. Rev. Nucl. Part. Sci. 33* (1983), pp. 539–68.

30 G. Goldhaber and J. E. Wiss, "Charmed Mesons," pp. 337–81.

31 Ibid.

32 For very recent references see: Particle Data Group, *Phys. Rev. D 54*, part I (1996).
33 Ibid.

5

The Discovery of the Tau Lepton

MARTIN PERL

Born Brooklyn, New York, 1927; Ph.D., 1955 (physics), Columbia University; Professor of Physics at Stanford University; Wolf Prize, 1982; Nobel Prize in Physics, 1995; Stanford Linear Accelerator Center; high-energy physics (experimental).

I begin this chapter with the period 1965–1974 when my colleagues and I worked experimentally on the e–μ problem and I became immersed in the then hypothetical world of heavy leptons. I go on to describe the discovery, in the period 1974–1976, of the tau lepton by myself and my colleagues using the SLAC–LBL I detector at the SPEAR e^+e^- storage ring. I then recount the verification of our discovery by ourselves and others, research that occupied the years 1976 through 1978. In the final section I describe the period 1978–1985, in which the transition was made in experiment and theory to the modern phase of tau research. I have told much of this history in a paper given at the first Workshop on Tau Lepton Physics and so I have repeated here quite a bit of material from that paper.[1] A beautiful description of the discovery of the tau was given recently by Gary Feldman.[2] The discovery of the tau was the subject of a doctoral thesis by Jonathan Treitel at Stanford University.[3]

Before the tau: 1965–1974

The e–μ problem

The history of the discovery of the tau lepton begins in the late 1960s, when my colleagues and I and other experimenters worked on the problem, "How does the muon differ from the electron?" In fact, that was the title of a paper I wrote for *Physics Today* in 1971.[4] At SLAC my colleagues and I had been measuring the differential cross sections for inelastic scattering of muons on protons, and then comparing the μ–p cross sections with the corresponding e–p cross sections.[5] We hoped to find e–μ differences, particularly as we studied the differential cross

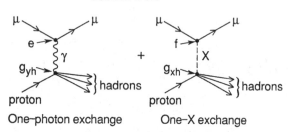

Fig. 5.1. The interaction of a muon with hadrons through exchange of a particle X, an example of the speculation that the muon has a special interaction with hadrons that is not possessed by the electron. See note 4.

sections at large momentum transfers. Some of our hopes – or at least of my hopes – were certainly naive by today's standards of knowledge of particle physics. For example, in my 1971 paper I speculated (Fig. 5.1), that the muon might have a special interaction with hadrons not possessed by the electron.

Other experimenters studied the differential cross section for μ–p elastic scattering and compared it with e–p elastic scattering.[6] But statistically significant differences between μ–p and e–p cross sections could not be found in either the elastic or inelastic case. Furthermore, systematic errors of the order of 5–10% were involved in comparing μ–p and e–p cross sections because the techniques were so different.

Thus it became clear that this was not a fruitful direction. I began to speculate that if we could not find the origin of the e–μ difference, perhaps we could find another charged lepton, and that this new lepton might lead us back into understanding the origin of lepton differences.

Varieties of leptons

In those naive days before the rise of the Standard Model, particle physicists thought about a large variety of types of leptons, and used or thought about using a variety of search methods. In 1972 I presented a paper in Moscow at the Seminar on the μ–e Problem, and I revised it in 1974 with Petros Rapidis.[7] In these papers we discussed many possible types of leptons:

- sequential leptons
- excited leptons
- paraleptons
- ortholeptons

- long-lived leptons
- stable leptons

In these papers we introduced the term "sequential lepton" to mean the sequence of pairs: (e^-, ν_e), (μ^-, ν_μ), (μ'^-, ν'_μ), (μ''^-, ν''_μ), and so forth. The list of search methods included:

- searches in particle beams
- production of new leptons by e^+e^- annihilation
- photoproduction of new leptons
- production of new leptons in p–p collisions
- searches in lepton *bremsstrahlung*
- searches in charged lepton–proton inelastic scattering
- searches in neutrino–nucleon inelastic scattering

Of all these methods the search for new charged leptons using e^+e^- annihilation was most appealing to me. The idea was to look for

$$e^+ + e^- \rightarrow \ell^+ + \ell^- \tag{5.1}$$

with

$$\ell^+ \rightarrow e^+ + \text{undetected neutrinos carrying off energy}$$
$$\ell^- \rightarrow \mu^- + \text{undetected neutrinos carrying off energy}$$

or

$$\ell^+ \rightarrow \mu^+ + \text{undetected neutrinos carrying off energy}$$
$$\ell^- \rightarrow e^- + \text{undetected neutrinos carrying off energy.}$$

This search method had many attractive features:

- If ℓ was a point particle, I could search up to lepton mass (m_ℓ) almost equal to the beam energy, given enough luminosity.
- The appearance of an $e^+\mu^-$ or $e^-\mu^+$ event with missing energy would be dramatic.
- The apparatus I proposed to use to detect these reactions was very poor in identifying types of charged particles (at least by today's standards), but the easiest particles to identify were the e and the μ.
- There was little theoretical ambiguity involved in predicting that the ℓ would have the weak decays

$$\ell^- \rightarrow \nu_\ell + e^- + \bar{\nu}_e$$
$$\ell^- \rightarrow \nu_\ell + \mu^- + \bar{\nu}_\mu, \tag{5.2}$$

with corresponding decays for the ℓ^+. One could simply argue by analogy with the well-known muon decay

$$\mu^- \rightarrow \nu_\mu + e^- + \bar{\nu}_e. \tag{5.3}$$

I incorporated the e^+e^- search method summarized by Eq. 5.1 in our 1971 proposal to use the not-yet-completed SPEAR e^+e^- storage ring.[8]

Sequential lepton theory

My thinking about sequential leptons and the use of e^+e^- annihilation to search for them was greatly helped and influenced by the seminal work of my long-time friend and colleague, Paul Tsai. His 1971 paper provided the theory for my work in sequential lepton searches from the beginning.[9] Table II from Tsai's paper gives the decay modes and their branching ratios for various lepton masses, branching ratios that we are still trying to measure precisely today. Tsai's work was incorporated in the heavy lepton search part of the detector proposal for SPEAR.[10]

In 1971 Thacker and Sakurai also published a paper on the theory of sequential lepton decays,[11] but it is not as comprehensive as the work of Tsai. His paper was the bible for my work on sequential heavy leptons, and in many ways it still is my bible in heavy lepton physics. A more general paper by Bjorken and Llewellyn Smith was also very important in my thinking.[12]

The detector proposal

My thoughts about heavy lepton searches using e^+e^- annihilation coincided with the beginning of the building of the SPEAR e^+e^- storage ring by a group led by Burton Richter and John Rees. Gary Feldman and I, and our Group E, joined with their Group C and a Lawrence Berkeley Laboratory group led by William Chinowsky, Gerson Goldhaber, and George Trilling. At that time, our group was also working with physicists from the Massachusetts Institute of Technology on a SLAC experiment on the electroproduction of hadrons.[13] I had given up on investigating the μ–e puzzle through studying inelastic charged lepton scattering, but I still hoped that anomalous lepton properties could be found in the interaction of leptons and hadrons.

We submitted the detector proposal in 1971;[14] its contents consisted of five sections and a supplement. The heavy lepton search was left for the last section and allotted just three pages because it seemed to be a remote dream. But the three pages contained the essential idea of searching for heavy leptons using e–μ events. I wanted to include a lot more about heavy leptons and the e–μ problem, but my colleagues thought that would unbalance the proposal. We compromised on a ten-page supplement, which began as follows:

> While the detector is being used to study hadronic production processes it is possible to simultaneously collect data relevant to the following questions:
>
> (1) Are there charged leptons with masses greater than that of the muon?
>
> We normally think of the charged heavy leptons as having spin $\frac{1}{2}$ but the search method is not sensitive to the spin of the particle. This search for charged heavy leptons automatically includes a search for the intermediate vector boson which has been postulated to explain the weak interactions.
>
> (2) Are there anomalous interactions between the charged leptons and the hadrons?
>
> In this part of the proposal we show that using the detector we can gather definitive information on the first question within the available mass range. We can obtain preliminary information on the second question – information which will be very valuable in designing further experiments relative to that question. We can gather all this information while the detector is being used to study hadronic production processes. Additional running will be requested if the existence of a heavy lepton, found in this search, needs to be confirmed.

My heart was in heavy lepton searches, but I continued to investigate the idea that an unknown e–μ difference could be revealed by an anomalous interaction of the e or μ with hadrons – a carryover from our old comparisons of e–p and μ–p inelastic scattering.

Heavy lepton searches at ADONE

While SPEAR and the SLAC–LBL detector were under construction, heavy lepton searches were being carried out at the ADONE e^+e^- storage ring in Frascati, Italy, by two groups of pioneer experimenters in electron–positron annihilation physics:[15] One group led by Antonino Zichichi reported its results in 1970 and 1973.[16] Fig. 5.2 is taken from their 1973 paper, from which I quote the first two paragraphs:

> Great interest in heavy leptons has recently been revived by the gauge theories of weak interactions. These theoretically wanted heavy leptons would be of two types, electronlike and muonlike, and would have the

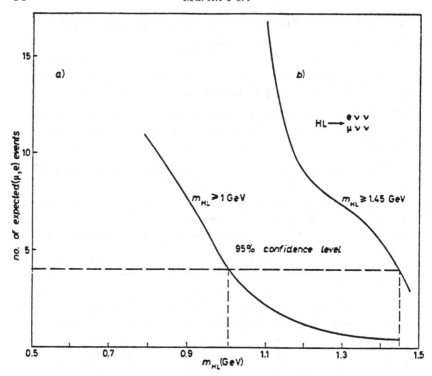

Fig. 5.2. Limits on heavy lepton masses from Bernadini, et al. Its caption reads: "The expected number of $(\mu^{\pm}e^{\mp})$ pairs *vs.* m_{HL} for two types of universal weak couplings of the heavy leptons. The dashed lines indicate the 95% confidence levels for m_{HL}. (a) HL universally coupled with ordinary leptons and hadrons, (b) HL universally coupled with ordinary leptons."[17]

leptonic number opposite to that of the same charge state of the ordinary lepton. So (E^+, ν_e, e^-) would be a triplet with leptonic number of the ordinary electron, and (M^+, ν_μ, μ^-) another triplet with leptonic number of the ordinary muon. Heavy leptons of a different type, each one with its own associated neutrino – and therefore with a leptonic number different from that of the ordinary leptons – were advocated a long time ago in the hope of understanding why the chain of leptons is short with respect to the long chain of hadrons.

All these types of heavy leptons can be produced via timelike photons in the reaction

$$e^+ + e^- \begin{cases} \to & E^+ + E^-, \\ \to & M^+ + M^-, \\ \to & L^+ + L^-. \end{cases}$$

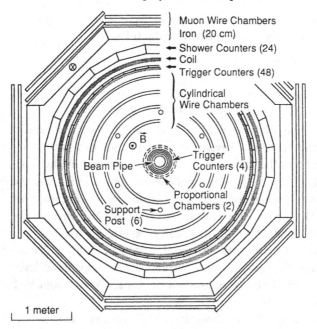

} Muon Wire Chambers
} Iron (20 cm)
← Shower Counters (24)
← Coil
← Trigger Counters (48)

Cylindrical
Wire Chambers

$\vec{B}$

Beam Pipe

Trigger
Counters (4)

Proportional
Chambers (2)

Support
Post (6)

1 meter

Fig. 5.3. Cross-sectional diagram of the SLAC–LBL detector, in its initial form.

The experiment covered two search regions, the mass reach depending upon the leptonic decay assumptions.

The other group of pioneer experimenters in electron–positron annihilation physics was led by Shuji Orito and Marcello Conversi. Their search region also extended to masses of about 1 GeV.[18]

Discovery of the tau in the SLAC–LBL experiment

The SPEAR e^+e^- collider began operation in 1973. Eventually it attained a total energy of about 8 GeV, but in the first few years the maximum energy with useful luminosity was 4.8 GeV.

We also began operating the SLAC–LBL detector in 1973 in the form shown in Fig. 5.3. It was one of the first large-solid-angle, general-purpose detectors built for colliding beams. The use of large solid-angle particle tracking and particle identification systems is obvious now, but it was not obvious 20 years ago.[19] The electron detection system used lead–scintillator sandwich counters built by our Berkeley colleagues. The muon-detection system was also crude, using the iron flux return, which was only 1.7 hadron absorption lengths thick.

Discovery of the $e\mu$ events

Both detection systems worked well enough, however, and in 1974 I began to find $e\mu$ events – events with an electron or positron, an opposite sign muon, no other charged particles, and no visible photons.

By early 1975 we had found dozens of $e\mu$ events, but those of us who believed we had found a heavy lepton faced two problems: how to convince the rest of our colleagues, and how to convince the physics world. The main focus of this early skepticism was the γ, e, and μ identification systems. Had we underestimated hadron misidentification into leptons? Since our γ and e system covered only about half of the 4π solid angle, what about undetected photons? What about inefficiencies and cracks in these systems?

I worked through this skepticism by gradually expanding the geographic range of the talks I presented, in which I answered objections if I could. If new objections were raised, I simply said that I had no answer. I then worked on the new objections before the next talk.

In June 1975 I gave my first international talk on the $e\mu$ events at the 1975 Summer School of the Canadian Institute for Particle Physics.[20] The talk had two purposes, to discuss possible sources of $e\mu$ events – heavy leptons, heavy mesons, and intermediate bosons – and to demonstrate that we had some good evidence for $e\mu$ events. The largest single-energy data sample, Table 5.1, was at 4.8 GeV, the highest energy at which we could then run SPEAR. The 24 $e\mu$ events in the first column was our strongest claim. One of the cornerstones of this claim was an informal analysis carried out by Jasper Kirkby. He showed me that just using the numbers in the first column of Table 5.1, we could calculate the probabilities for hadron misidentification in this class of events. There were not enough eh, μh, and hh events to explain away the 24 $e\mu$ events.

The misidentification probabilities determined from the three-or-more prong hadronic events and other considerations are given in Table 5.2. Compared to present experimental standards, the misidentification probabilities $P_{h\to e}$ and $P_{h\to\mu}$ of about 0.2 are enormous, but I could still show that the 24 $e\mu$ events could not be explained away.

Table 5.1. *A table of two-charged-particle events collected at 4.8 GeV in the SLAC–LBL detector. The table, containing 24 eμ events with zero total charge and no photons, was the strongest evidence at that time for the* τ.

	Total Charge = 0			Total Charge = ±2		
Number photons =	0	1	> 1	0	1	1
ee	40	111	55	0	1	0
eμ	24	8	8	0	0	3
μμ	16	15	6	0	0	0
eh	18	23	32	2	3	3
μh	15	16	31	4	0	5
hh	13	11	30	10	4	6
Sum	126	184	162	16	8	7

Table 5.2. *Misidentification probabilities for the 4.8 GeV sample given in Table 5.1.*

Momentum range (GeV/c)	$P_{h \to e}$	$P_{h \to \mu}$	$P_{h \to h}$
0.6–0.9	.130 ± .005	.161 ± .006	.709 ± .012
0.9–1.2	.160 ± .009	.213 ± .011	.627 ± .020
1.2–1.6	.206 ± .016	.216 ± .017	.578 ± .029
1.6–2.4	.269 ± .031	.211 ± .027	.520 ± .043
weighted average using hh, μh, and eμ events	.183 ± .007	.198 ± .007	.619 ± .012

The Montreal paper ended with these conclusions:[21]

1. No conventional explanation for the signature eμ events has been found.
2. The hypothesis that the signature eμ events come from the production of a pair of new particles – each of mass about 2 GeV – fits almost all the data. Only the θ_{coll} distribution is somewhat puzzling.
3. The assumption that we are also detecting ee and μμ events coming from these new particles is still being tested.

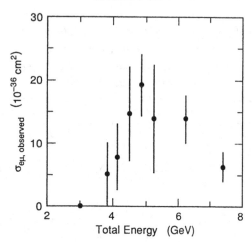

Fig. 5.4. The observed cross section for the signature $e\mu$ events from the SLAC–LBL experiment at SPEAR (reprinted from Perl, 1975). This observed cross section is not corrected for acceptance. There are 86 events with a calculated background of 22 events.[22]

I was still not able to specify the source of the $e\mu$ events: leptons, mesons, or bosons. But I remember that I felt strongly that the source was heavy leptons. It would take almost two more years to prove that.

A remark on the collision angle (θ_{coll}) distribution (Fig. 5 of Perl, 1975)[23] is in order. I was worried that there were no events with $\theta_{coll} >$ 80°. I knew that in small data sets it is unlikely for all distributions to fit predictions, but I was worried.

First publication

As 1974 passed we acquired $e^{+}e^{-}$ annihilation data at more and more energies, and at each of these energies there was an anomalous $e\mu$ event signal (see Fig. 5.4). Thus, my colleagues and I became more and more convinced of the reality of the $e\mu$ events and the absence of a conventional explanation.

An important factor in this growing conviction was the addition of a special muon-detection system to the detector (Fig. 5.5), called the muon tower. This addition was conceived and built by Feldman. Although we did not use events such as that in Fig. 5.6 in our first publication, seeing a few events like this was enormously comforting.

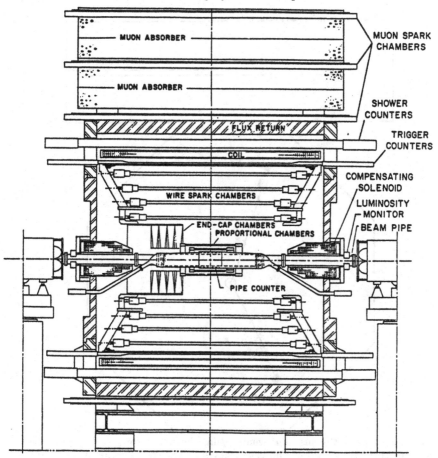

Fig. 5.5. The SLAC–LBL detector with the muon tower added.

Finally, in December 1975, we published a paper titled "Evidence for Anomalous Lepton Production in e^+e^- Annihilation."[24] The final paragraph read:

> We conclude that the signature $e\mu$ events cannot be explained either by the production and decay of any presently known particles or as coming from any of the well-understood interactions which can conventionally lead to an e and a μ in the final state. A possible explanation for these events is the production and decay of a pair of new particles, each having a mass in the range of 1.6 to 2.0 GeV/c^2.

We were not yet prepared to claim that we had found a new charged lepton, but we were ready to claim that we had found something new. To

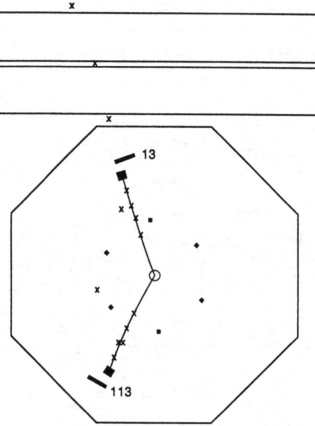

Fig. 5.6. One of the first $e\mu$ events using the muon tower. The μ moves upward through the tower and the e moves downward. The numbers 13 and 113 give the relative amounts of electromagnetic shower energy deposited by the μ and e.

accentuate our uncertainty I denoted the new particle by U for unknown in some of our 1975–1977 papers. The name τ came later. This name was suggested by Rapidis, who was then a graduate student and had worked with me in the early 1970s on the e–μ problem.[25] The letter τ is from the Greek $\tau\rho\iota\tau o\nu$ for "third" – the third charged lepton.

Is it a lepton? From uncertainty and controversy to confirmation: 1976–1978

Our first publication was followed by several years of confusion and uncertainty about the validity of our data and its interpretation. It is hard to explain this confusion a decade later when we know that τ pair production is 20% of the e^+e^- annihilation cross section below the Z^0, and when τ pair events stand out so clearly at the Z^0.

There were several reasons for the uncertainties of that period. It was hard to believe that both a new quark, charm, and a new lepton, tau, would be found in the same narrow range of energies. And while the existence of a fourth quark was required by theory, there was no such requirement for a third charged lepton. So there were claims that the $e\mu$ events were the complicated result of the decays of charm quarks. There were claims that the other predicted decay modes of tau pairs such as e–hadron and μ–hadron events could not be found. Indeed, finding such events was just at the limit of the particle-identification capability of the detectors of the mid-1970s.

It was a difficult time. Rumors kept arriving of definitive evidence against the τ: $e\mu$ events *not* seen, the $\tau \to \pi\nu$ decay *not* seen, theoretical problems with momentum spectra or angular distributions. With colleagues such as Feldman, I kept going over our data again and again. Had we gone wrong somewhere in our data analysis?

An illustration of the confusion about the tau is provided by two editions of a popular book on particle physics by Nigel Calder entitled *The Key to the Universe*. In the first edition Calder wrote:[26]

Martin Perl and his colleagues detected peculiar events occurring in SPEAR. From the scene of collision an electron and a heavy electron (the well-known muon) carrying opposite electric charges were ejected at the same moment without any other detectable particles coming out. No conventional process, involving conventional particles, could account for such events.

The particle called U was grotesquely weighty for an electron. Theorists had been asking one another for many years why the muon, a heavy electron of 105 mass-energy units, was two hundred times heavier than the ordinary electron (0.5 units). They had no answer even to that. And the U particle was estimated to be 1800 mass-energy units, twice as heavy as a hydrogen atom!

Doubts also overtook Stanford's heavy lepton, the U particle. There were suggestions, notably from the DORIS experiments in Hamburg, that it was an illusion – perhaps a misinterpretation of the decay of the charmed particles. At the time of writing these doubts have not been resolved and they illustrate again the difficulties and tensions of high-energy physics. A fair summary of the situation may be that there is no very compelling evidence so far for nature deploying more than four types of quarks and four members of the electron family.

But in the second edition Calder wrote:[27]

> The particle U was grotesquely weighty for an electron. Theorists had
> been asking one another for many years why the muon, a heavy electron
> of 105 mass-energy units, was two hundred times heavier than the ordi-
> nary electron (0.5 units). They had no answer even to that. And the U
> particle was estimated to be 1800 mass-energy units, twice as heavy as
> a hydrogen atom! Experiments with DORIS in Hamburg confirmed the
> discovery.

Anomalous muon events

The first advance beyond the $e\mu$ events came with three different demon-
strations of the existence of anomalous μ–hadron events from

$$
\begin{aligned}
e^+ + e^- &\rightarrow \tau^+ + \tau^- \\
\tau^+ &\rightarrow \bar{\nu}_\tau + \mu^+ + \nu_\mu \\
\tau^- &\rightarrow \nu_\tau + \text{hadrons}
\end{aligned}
\tag{5.4}
$$

I have in my files a June 3, 1976, note by Feldman discussing μ
events using the muon identification tower of the SLAC–LBL detector
(Fig. 5.5). For data acquired above 5.8 GeV he found the following:

> Correcting for particle misidentification, this data sample contains 8 μe
> events and 17 μ–hadron events. Thus, if the acceptance for hadrons is
> about the same as the acceptance for electrons, and these two anomalous
> signals come from the same source, then with large errors, the branching
> ratio into one observed charged hadron is about twice the branching ratio
> into an electron. This is almost exactly what one would expect for the
> decay of a heavy lepton.

This conclusion was published in a paper entitled "Inclusive Anomalous Muon Production in e^+e^- Annihilation."[28] The first and very welcome outside confirmation of anomalous muon events came from another
SPEAR experiment by Cavalli–Sforza, et al.[29]

The most welcomed confirmation, because it came from an experiment
at the DORIS e^+e^- storage ring, was from the PLUTO experiment. In
1977 the PLUTO collaboration published a paper that presented evidence for anomalous muon production in e^+e^- annihilation.[30] Because
PLUTO was also a large-solid-angle detector, for the first time we could
fully discuss the art and technology of τ research with an independent
set of experimenters, led by Hinrich Meyer and Eric Lohrman.

With the discovery of μ–hadron events I was convinced I was right
about the existence of the τ as a sequential heavy lepton. Yet there was

much to disentangle. It was still difficult to demonstrate the existence of anomalous e–hadron events and there were still rumors that the $\tau \to \pi\nu$ decay mode could not be found.

Anomalous electron events

The demonstration of the existence of e–hadron events

$$
\begin{aligned}
e^+ + e^- &\to \tau^+ + \tau^- \\
\tau^+ &\to \bar{\nu}_\tau + e^+ + \nu_e \\
\tau^- &\to \nu_\tau + \text{hadrons}
\end{aligned}
\tag{5.5}
$$

required improved electron identification in the detectors. A substantial step forward was made by the new DELCO detector at SPEAR, which I will discuss in connection with the determination of the mass of the τ.[31] The SLAC–LBL detector was also upgraded by SLAC Group E and a Lawrence Berkeley Laboratory group led by Angela Barbaro–Galtieri; some of the original experimenters had gone off to begin building the MARK II detector. We installed a wall of lead-glass electromagnetic shower detectors in the original detector, renamed MARK I (see Fig. 5.7). This led to the important paper whose abstract read:[32]

> We observe anomalous $e\mu$ and e–hadron events in e^+e^- collisions at SPEAR in an experiment that uses a lead-glass counter system to identify electrons. The anomalous events are observed in the two-charged-prong topology. Their properties are consistent with the production of a pair of heavy leptons in the reaction $e^+e^- \to \tau^+\tau^-$ with subsequent decays of $\tau^\pm$ into leptons and hadrons. Under the assumption that they come only from this source, we measure the branching ratios $B(\tau \to e\nu_e\nu_\tau) = (22.4 \pm 5.5)\%$ and $B(\tau \to h + \text{neutrals}) = (45 \pm 19)\%$.

The 1977 Lepton–Photon Conference at Hamburg

At the 1977 International Symposium on Lepton and Photon Interactions at High Energies, there were three papers that portrayed the then current state of knowledge of the τ. A paper from the DASP experiment at DORIS presented by S. Yamada described measurements of μ–hadron and e–hadron events, which confirmed the sequential lepton nature of the τ.[33] But Yamada reported that DASP could not find the pion decay mode

$$
\tau^- \to \nu_\tau + \pi^- \quad ,
\tag{5.6}
$$

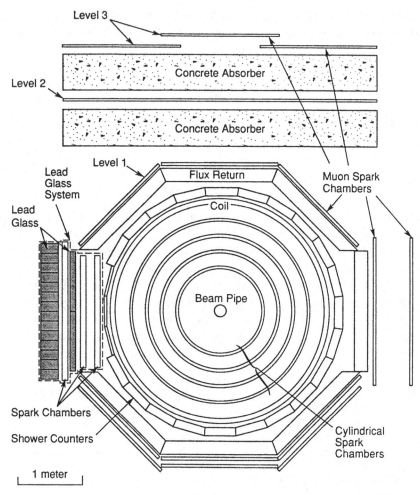

Fig. 5.7. The "lead-glass wall" modification of the SLAC–LBL detector used at SPEAR to find anomalous e–hadron events. See note 32.

setting an upper limit on the branching fraction of $(2. \pm 2.5)\%$ while theory predicted about 10%.

The second paper, entitled "Direct Electron Production Measurement by DELCO at SPEAR," was presented by Kirkby.[34] The abstract reads in part:

A comparison of the events having only two visible prongs (of which only one is an electron) with the heavy lepton hypothesis shows no disagreement. Alternative hypotheses have not yet been investigated.

Table 5.3. *The various branching fractions for* $\tau^- \to \pi^- \nu_\tau$, *as measured by late 1978.*

Experiment	Mode	Events	Background	$B(\tau \to \pi\mu)(\%)$
SLAC–LBL	$x\pi$	≈ 200	≈ 70	$9.3 \pm 1.0 \pm 3.8$
PLUTO	$x\pi$	32	9	$9.0 \pm 2.9 \pm 2.5$
DELCO	$e\pi$	18	7	$8.0 \pm 3.2 \pm 1.3$
MARK II	$x\pi$	142	46	$8.0 \pm 1.1 \pm 1.5$
	$e\pi$	27	10	$8.2 \pm 2.0 \pm 1.5$
Average				8.3 ± 1.4

Finally, in a paper entitled "Review of Heavy Lepton Production in e^+e^- Annihilation,"[35] I concluded that:

a. All data on anomalous $e\mu$, ex, ee and $\mu\mu$ events produced in e^+e^- annihilation is *consistent* with the existence of a mass 1.9 ± 0.1 GeV/c^2 charged lepton, the τ.
b. This data *cannot* be explained as coming from charmed particle decays.
c. Many of the expected decay modes of the τ have been seen. A very important problem is the existence of the $\tau^- \to \nu_\tau \pi^-$ decay mode.

The search for the decay $\tau^- \to \nu_\tau + \pi^-$

Thus in the summer of 1977 the major problem in fully establishing the nature of the τ was the uncertainty in the branching ratio, $B(\tau \to \pi\nu)$. This was a serious problem because from $B(\pi \to \mu\nu)$ and $B(\tau \to e\nu\nu)$ it follows directly that $B(\tau \to \pi\nu)$ should be about 10%. I cannot explain now why experimenters, including ourselves, had such difficulty with this mode, but we did have difficulty.

In the SLAC–LBL collaboration the first demonstration that $B(\tau \to \pi\nu)$ was substantial came from Gail Hanson in an internal note dated March 7, 1978. She looked at a sample of 2-prong, 0-photon events with one high-momentum prong and found an excess of events, particularly at large x, if $B(\tau \to \pi\nu)$ is taken as zero.[36] By mid-1978 there was no longer a problem with $\tau \to \pi\nu$; the clouds of confusion parted and the sun shone on a $B(\tau \to \pi\nu)$ close to the expected 10%. Table 5.3 shows the late-1978 measurements.[37]

Thus by the end of 1978 all confirmed measurements agreed with the hypothesis that the τ was a lepton that was produced by a known electromagnetic interaction and, at least in its main modes, decayed through the conventional weak interaction. I think of 1978 as the year when the first phase of research on the τ ended.

Nailing down the tau: 1978–1985

In the final section of this chapter, I sketch the history of τ research in the years 1978–1985, when that research made the transition from the verification of the existence of the tau to the present period of detailed studies of tau properties – studies that may lead to the discovery of new particle physics.

The tau mass

The initial history of measurements of the τ mass, m_τ, is brief. The first estimate $m_\tau = 1.6$–2.0 GeV$/c$ was made along with the initial evidence for the τ.[38] By the beginning of 1978 the DASP experiment showed $m_\tau = 1807 \pm 20$ MeV.[39] By the middle of 1978 the DELCO experiment made the best determination of $m_\tau = 1784^{+2}_{-7}$ MeV by measuring $\tau^+\tau^-$ production at threshold.[40]

It was only in 1992, fourteen years later, that there was an improvement in the measurement of m_τ. The BES Collaboration using the BEPC e^+e^- collider reported $m_\tau = 1776.9 \pm 0.5$ MeV$/c^2$.[41]

The tau at high energies – PETRA and PEP

As the 1970s ended, τ research began to be carried out at higher energies, first at the new PETRA e^+e^- collider at DESY, then at the new PEP collider at SLAC. Two of the earlier high-energy papers are from the TASSO and CELLO experiments.[42] This was the beginning of a tremendous amount of research in the 1980s on the tau by the CELLO, JADE, MARK-J, PLUTO, and TASSO experiments at PETRA; and by the DELCO, HRS, MAC, MARK II, and TPC experiments at PEP. The papers on the tau from these experiments number close to one hundred.

Table 5.4. *The status of τ lifetime measurements in 1982.*

Experiment	Number of Decays	Average Decay Length Error (mm)	$\tau_\tau(10^{-13}\text{sec})$
TASSO	599	10	0.8 ± 2.2
MARK II	126	4	4.6 ± 1.9
MAC	280	4	$4.1 \pm 1.2 \pm 1.1$
CELLO	78	6	$4.7^{+3.9}_{-2.9}$
MARK II Vertex Detector	71	0.9	$3.31 \pm .57 \pm .60$

The tau lifetime

Measurements of the τ lifetime, τ_τ, could not be made at the low energies at which SPEAR and DORIS usually operated; the first measurement of τ_τ required the higher energies of PETRA and PEP. The best measurements required, in addition, secondary-vertex detectors. Actually the first published measurement used a primitive secondary-vertex detector built by Walter Innes and myself to improve the triggering efficiency of the MARK II detector.[43] Led by Feldman and Trilling, we measured $\tau_\tau = (4.6 \pm 1.9) \times 10^{-13}$ sec. Another early measurement was from the MAC experiment with $\tau_\tau = (4.9 \pm 2.0) \times 10^{-13}$ sec.[44]

The modern era in τ lifetime measurements began with the pioneering work of John Jaros on precision vertex detectors.[45] Table 5.4 taken from that paper shows the status of τ lifetime measurements at the end of 1982. Today's average value of τ_τ is in the range of (2.95 to 3.04) $\times 10^{-13}s$ so these measurements were remarkably good for the detector technology of the early 1980s. Thus, by the beginning of 1984 a decade of τ research had ended with a value of the lifetime in agreement with conventional weak interaction decay theory and, although not discussed here, many measurements on decay modes and branching ratios. It seemed as though τ research was ready to settle down into a comfortable second decade.

Precise calculations on tau decays: 1984–1985

But comfort and ease did not appear. In 1984–1985 two papers appeared that carefully applied accepted decay theory to the many measurements on τ branching.[46] These papers showed that there was something wrong in the theory or in the measurements of the one-charged particle decay modes of the τ. We did not understand the τ at the 5–10% level in 1985.

The future of the tau

Seven years have passed since 1985 but the question still remains:[47]

Is the tau simply a Standard Model lepton, or will the physics of the tau lead us outside of the Standard Model?

For me, the remark of Francis Bacon – "they are ill discoverers that think there is no land when they can see nothing but sea" – is my guide. The Standard Model is today's "sea" of particle physics, but what does it conceal that is new and more fundamental? In 1975 the discovery of the tau was new land in the sea of two generations. Perhaps the tau will lead us out of the sea of the Standard Model.

Notes

1 M. L. Perl, "The Discovery of the τ and its Major Properties: 1970–1985," in M. Davier and B. Jean-Marie, eds., *Workshop on Tau Lepton Physics* (Gif-sur-Yvette, France: Editions Frontières, 1991), pp. 3–27.

2 G. J. Feldman, "The Discovery of the τ 1975–1977: A Tale of Three Papers," in J. Hawthorne, ed., *Proc. 20th SLAC Summer Inst. Particle Physics* (Stanford University, 1992), pp. 631–46.

3 J. Treitel, "Confirmation with Technology: The Discovery of the Tau Lepton," *Centaurus 30* (1987), p. 140.

4 M. L. Perl, "How Does the Muon Differ from the Electron?" *Physics Today*, July (1971), p. 34.

5 W. T. Toner, et al., "Comparison of Muon-Proton and Electron–Proton Deep Inelastic Scattering," *Phys. Lett. 36B* (1972), p. 251; T. Braunstein, et al., "Comparison of Muon–Proton and Electron–Proton Inelastic Scattering," *Phys. Rev. D6* (1972), p. 106.

6 R. W. Ellsworth, et al., "Muon–Proton Elastic Scattering at High Momentum Transfers," *Phys. Rev. 165* (1968), p. 1449; L. Camilleri, et al., "High Energy Muon–Proton Scattering: Muon–Electron Universality," *Phys. Rev. Lett. 23* (1969), p. 153; I. Kostoulas, et al., "Muon–Proton Deep Elastic Scattering," *Phys. Rev. Lett. 32* (1974), p. 489.

7 M. L. Perl, "Searches for Heavy Leptons and Anomalous Leptonic Behavior – the Past and the Future," SLAC Report No. SLAC-PUB-1062 (1972), unpublished; M. L. Perl and P. Rapidis, "The Search for Heavy Leptons and Muon–Electron Differences," SLAC Report No. SLAC-PUB-1496 (1974), unpublished.

8 R. R. Larsen, et al., "An Experimental Survey of Positron–Electron Annihilation into Multiparticle Final States in the Center-of-Mass Energy Range 2 GeV to 5 GeV," SLAC Proposal No. SP-2 (1971), unpublished.

9 Y. S. Tsai, "Decay Correlations of Heavy Leptons in $e^+ + e^- \to \ell^+ + \ell^-$," *Phys. Rev. D4* (1971), p. 2821.

10 R. R. Larsen, et al., op. cit.

11 H. B. Thacker and J. J. Sakurai, "Lifetimes and Branching Ratios of Heavy Leptons," *Phys. Lett. 36B* (1971), p. 103.

12 J. D. Bjorken and C. H. Llewellyn Smith, "Spontaneously Broken Gauge Theories of Weak Interactions and Heavy Leptons," *Phys. Rev. D7* (1973), p. 887.

13 J. T. Dakin, et al., "Measurement of the Inclusive Electroproduction of Hadrons," *Phys. Rev. Lett. 29* (1972), p. 746; J. F. Martin, et al., "Particle Ratios in Inclusive Electroproduction from Hydrogen and Deuterium," *Phys. Lett. 65B* (1976), p. 483.

14 R. R. Larsen, et al., op. cit.

15 A. Zichichi, "Heavy Lepton Searches Before the Tau," in J. Hawthorne, ed., *Proc. 20th SLAC Summer Inst. Particle Physics* (Stanford University, 1992), pp. 603–22.

16 V. Alles–Borelli, et al., "Limits on the Electromagnetic Production of Heavy Leptons," *Lett. Nuovo Cimento IV* (1970), p. 1156; M. Bernardini, et al., "Limits on the Mass of Heavy Leptons," *Nuovo Cimento 17A* (1973), p. 383.

17 Bernadini, et al., ibid.

18 S. Orito, et al., "A Search for Heavy Leptons with e^+e^- Colliding Beams," *Phys. Lett. 48B* (1974), p. 165.

19 See chapters 17 and 18 by R. Schwitters and P. Galison, this volume.

20 M. L. Perl, "Lectures on Electron–Positron Annihilation – Part II: Anomalous Lepton Production," in R. Heinzi and B. Margolis, eds., *Proc. Canadian Inst. Particle Physics Summer School* (Montreal: McGill Univ., 1975).

21 Ibid.

22 Ibid.

23 Ibid.

24 M. L. Perl, et al., "Evidence for Anomalous Lepton Production in e^+e^- Annihilation," *Phys. Rev. Lett. 35* (1975), p. 1489.

25 M. L. Perl and P. Rapidis, "The Search for Heavy Leptons."

26 N. Calder, *The Key to the Universe* (London: British Broadcasting Corporation, 1977).

27 N. Calder, *The Key to the Universe* (Harmondsworth: Penguin Books, 1978).

28 G. J. Feldman, et al., "Inclusive Anomalous Muon Production in e^+e^- Annihilation," *Phys. Rev. Lett. 38* (1977), p. 117.

29 M. Cavalli–Sforza, et al., "Anomalous Production of High-Energy Muons in e^+e^- Collisions at 4.8 GeV," *Phys. Rev. Lett. 36* (1976), p. 558.

30 J. Burmester, et al., "Anomalous Muon Production in e^+e^- Annihilation as Evidence for Heavy Leptons," *Phys. Lett. 68B* (1977), p. 297.

31 J. Kirkby, "Direct Electron Production Measurements by DELCO at SPEAR," in F. Gutbrod, ed., *Proc. Int. Symp. Lepton and Photon Interactions at High Energies* (Hamburg, 1977), p. 3.

32 A. Barbaro–Galtieri, et al., "Electron–Muon and Electron–Hadron Production in e^+e^- Collisions," *Phys. Rev. Lett. 39* (1977), p. 1058.

33 S. Yamada, "Recent Results from DASP," in F. Gutbrod, op. cit., p. 69.

34 J. Kirkby, "Direct Electron Production Measurements by DELCO at SPEAR," in F. Gutbrod, op. cit., p. 3.

35 M. L. Perl, "Review of Heavy Lepton Production in e^+e^- Annihilation," in F. Gutbrod, op. cit., p. 145.

36 M. L. Perl, "The Discovery of the τ and its Major Properties."

37 G. J. Feldman, "e^+e^- Annihilation," in S. Homma, M. Kawaguchi, and H. Miyazawa, eds., *Proc. XIX Int. Conf. on High Energy Physics* (Tokyo, 1978).

38 M. L. Perl, et al., "Evidence for Anomalous Lepton Production in e^+e^- Annihilation," *Phys. Rev. Lett. 35* (1975), p. 1489.

39 R. Brandelik, et al., "Measurements of Tau Decay Modes and a Precise Determination of the Mass," *Phys. Lett. 73B* (1978), p. 109.

40 W. Bacino, et al., "Measurement of the Threshold Behavior of $\tau^+\tau^-$ Production in e^+e^- Annihilation," *Phys. Rev. Lett. 41* (1978), p. 13.

41 J. Z. Bai, et al., "Measurement of the Mass of the Tau Lepton," *Phys. Rev. Lett. 69* (1992), p. 3021.

42 R. Brandelik, et al., "Production and Properties of the τ Lepton in e^+e^- Annihilation at C.M. Energies from 12 to 31.6 GeV," *Phys. Lett. 92B* (1980), p. 199. H. J. Behrend, et al., "Measurement of $e^+e^- \to \tau^+\tau^-$ at High Energies and Properties of the τ Lepton," *Phys. Lett. 114B* (1982), p. 282.

43 G. J. Feldman, et al., "Measurement of the τ Lifetime," *Phys. Rev. Lett. 48* (1982), p. 66.

44 W. T. Ford, et al., "Lifetime of the Tau Lepton," *Phys. Rev. Lett. 49* (1982), p. 106.

45 J. A. Jaros, "Measurement of the τ Lifetime," *Proc. 21st International Conference on High Energy Physics* (Paris, 1982), *J. Physique 43* Colloque C-3, p. 106.

46 T. N. Truong, "Hadronic τ Decay, Pion Radiative Decay, and Pion Polarizability," *Phys. Rev. D30* (1984), p. 1509; F. J. Gilman and S. H. Rie, "Calculation of Exclusive Decay Modes of the Tau," *Phys. Rev. D31* (1985), p. 1066.

47 M. L. Perl, "The Tau Lepton," *Rep. Prog. Phys. 55* (1992), p. 653. M. L. Perl, "Tau Physics," in J. Hawthorne, op. cit.

6

The Discovery of the Upsilon, Bottom Quark, and *B* Mesons

LEON M. LEDERMAN

Born New York City, 1922; Ph.D., 1951 (physics), Columbia University; Pritzker Professor of Physics at the Illinois Institute of Technology in Chicago; Director of Fermi National Accelerator Laboratory 1979–89; Wolf Prize, 1982; Nobel Prize in Physics, 1988; high-energy physics (experimental).

History conferences are designed to set the record straight or, depending on where you stand, make it as crooked as it can possibly be. In this case I intend to personalize the story and the complicated reason is that the discovery of the bottom quark, almost exactly fifteen years ago, was the culmination of a series of events in experimental physics which go back to the discovery of the muon neutrino just thirty years ago, in 1962.[1] I think it's important to emphasize that this story is one of missed opportunities, abysmal judgment, monumental blunders, stupid mistakes, and inoperative equipment. It was leavened only by the incredible luck and incandescent good fortune which you all know is an essential ingredient for any physics career. Lest you sneer that I am displaying false modesty, I beg you to hold your opinion until you've seen the data.

Preamble

In the period Haim Harari called "From the fourth lepton to the fifth quark,"[2] we found the muon neutrino but missed neutral currents. We discovered what became known as the Drell–Yan process but missed the J/ψ. We missed the J/ψ again at the ISR but stumbled on high-transverse-momentum hadrons. We missed the J/ψ at Fermilab in 1973, chasing single-direct-lepton yields that were a red herring. Then we found a false upsilon. But finally Nature, terrified that she would be stuck with us forever, yielded up her secret, the true upsilon (Υ), hoping this would make us go away.

101

A somewhat ameliorating factor was that several full-employment industries came out of this sequence. High-energy neutrino beams were
established in all the hadron labs. Drell–Yan became a bread-and-butter
technique, a kind of third avenue for studying structure functions and
a tool for probing the timelike domain. High-p_T hadrons became a cot-
tage industry providing grist for the QCD mill. Bottom physics has fully
occupied the relevant e^+e^- labs, provides a daunting challenge for fixed-
target physics, and is beginning to emerge rather powerfully from hadron
colliders. Someday soon the obsession with this physics will culminate
in the construction of a B factory. One or more of these experimental
approaches will sooner or later illuminate the origin-of-matter question
of CP violation.

The way to bottom

It is both historically and logically correct to begin with the two-neutrino
experiment. Establishing the muon neutrino in 1962 as the fourth lepton also established "flavor" and the structure of what grew, in the next
decade, to be the Standard Model. As soon as we finished this experiment at Brookhaven, we went on to the second neutrino experiment, a
much more elaborate and much less fruitful piece of research. We made
a new experimental area there, and among the many things we tried to
do was to look for neutrino production of intermediate bosons. Failing
to detect them, we established the limit that the mass of the intermediate boson was greater than 2 GeV. Finishing that run and realizing
that we had more neutrino collision events than the resolution of the
apparatus could justify, we went on to use primary protons to make a
more sensitive search for the W boson, assuming that if there was a W
and that if it were massive, then it would, upon decay, give rise to a
large-angle, high-energy muon. We had already instrumented the steel
shielding as a way of measuring the neutrino flux. Thus we decided to
look at single muons emitted at large angles to the proton beam in the
shielding of the neutrino apparatus. This experiment established that
the mass of the W was greater than 5 GeV.[3] At this rate we would have
gotten to the true mass of the W by the year 2022.

That experiment was criticized in a paper by Yoshio Yamaguchi,[4]
then at CERN, who pointed out that, had we found wide-angle muons
(which we had not) we couldn't prove they were generated by Ws. It
might be virtual photons that could yield wide-angle muons and, in
fact, the relation between these two processes was coupled by conserved

vector currents. I remember reading that paper and then asking my colleague Norman Kroll, "What's a virtual photon?"

His lecture intrigued me and led directly to the Brookhaven experiment of 1967–68. I was so excited by the properties of virtual photons that we decided to study them specifically and we set up this ingeniously stupid detector in order to measure muon pairs. We knew the cross section for massive virtual photons giving rise to muon pairs would be small and we wanted to produce these photons, using the highest energy primary protons. The way you measure the mass (well enough) is by summing up the transverse momentum of the two muons. This involved measuring the production angle with a scintillator counter and measuring the energy of the muon by noting in which of these blocks of absorber the muon would stop. Putting two muons together, we had a mass resolution of only 15%, but the experiment had remarkable sensitivity. Cross sections as low as 10^{-38}cm^2 could be seen. We got this famous curve used by many standup comics around the world (see Fig. 6.1a). The only thing in physics named after me is a shoulder.... Notice that the yield of dimuons goes down for seven decades and at the smallest level observed, it corresponds to a cross section of the order of 10^{-14} of the total cross section for pp collisions. Of course the price we paid was a mass resolution of about 15%. There's a lot of things one can say about Fig. 6.1 but I'll summarize it with my "IF" litany:

It is interesting to note, especially for those with historian inclinations, that IF our mass resolution had been 10%, and IF Drell and Bjorken had been professors on the East Coast, think of the consequences! First, I would have discovered the J/ψ! (I would have named it something else – probably after T.D. – the Lee-on!) Sid and Bj would have given the explanation of the data in terms of the dynamical existence of quarks. Ting[5] and Richter,[6] and Friedman, Kendall, and Taylor[7] wouldn't have gotten the Nobel prize, not, at least, for that for which they did receive their awards. The mind boggles at the consequences for physics.... This experiment, properly carried out, would have produced results that won five Nobel prizes!

Everyone has seen the contrast of these two, in principle, identical sets of data. I think that after the shock of seeing Ting's data (see Fig. 6.1b), I sent around a note saying that any apparatus that can convert this towering peak to this mound of rubble should be proscribed by SALT talks.

As you know our 1968 data did start a busy business, colloquially called the Drell–Yan process – or quark–antiquark annihilation. Now

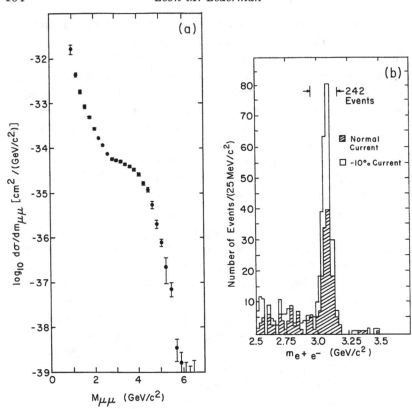

Fig. 6.1. (a) Dimuon yield from the 1968 Brookhaven experiment. (b) Dielectron yield from Ting's 1974 experiment.

the simple Brookhaven apparatus was so dumb that there was no way you could change it to improve the resolution even a little bit. Every element was designed to match the distortions generated by multiple scattering in ten feet of steel. This took a minimum of thought, so no matter *what* you did you were stuck with 15% resolution. We could have completely rebuilt the apparatus. At that time spark chambers were around (this was before wire chambers), but we felt that this kind of process would profit most from higher energy, so we decided to pursue this experiment at newer machines.

In 1971–72 I went to CERN where we set up a large lead-glass spectrometer at the ISR to look at electron pairs. We were specifically trying to find out what that shoulder was all about. Funny things were happening at the ISR, but in 1973, in our production of electron pairs, the mass distribution was found to be a smooth decreasing function, and we presented cross sections from 4 GeV all the way up to 20 GeV. We didn't go below 4 GeV because we couldn't. We were swamped by a strange electromagnetic background and yellowing lead glass due to radioactivity, except that we didn't know our glass was yellowing then. We decided in desperation to publish the background,[8] which in fact was π^0s of large transverse momentum. In the pre-ISR epoch, the data on transverse momentum of particles, which came from cosmic-ray experiments, varied as e^{-6p_T}, which was generally known and understood by many theorists. Our data indicated about seven orders of magnitude enhancement above this expectation, so we had to convince ourselves that it was real. It was. We were seeing high-transverse-momentum π^0s. We established an inverse eighth power law to replace the exponential out to a transverse momentum of 5 or 6 GeV/c.

The discovery of the upsilon

We then went to Fermilab to mess that lab up (why should they be any different?), and we stumbled on this so-called red herring of direct leptons. Many people at Fermilab, including Jim Cronin, were involved in this intriguing idea that the ratio of leptons to pions seemed to be an interesting constant, 10^{-4}, for p_T larger than something like 1 GeV. We spent a lot of time on this around 1973–74. By the time we decided that this wasn't a useful thing to continue, November 1974 was upon us. Just to show you how it was resolved, that ratio happens to be just a consequence of a pile-up of all kinds of things – the J/ψ, the Drell–Yan process, the upsilon, and so on. They all added up to give this mysterious constant that took a lot of time and effort but yielded very little.

Then we started learning how to observe pairs instead of single leptons, and in January 1976 we found an incredibly sharp peak that stood out like a healthy thumb. These electron pairs had a "clear" peak at 6 GeV. We were a little bit hedgy, but we said, "it looks like there's a particle" and we gave it the name "upsilon."[9] Then a few months later we found that when using muons there was no bump there at all.[10] We were helped by the inverse process at SPEAR, where they had also looked at 6 GeV and found nothing interesting. A statistical blooper. Oops, Leon.

Summarizing the runs from 1973 to 1975, for both electrons and muons, going back and forth, John Yoh noticed that there was another enhancement at 9.5 GeV. It was barely there, but it convinced him to put a bottle of Mumm's champagne in the refrigerator labeled 9.5. Everyone thought he was kidding.

Let me make some experimental comments on what was happening. Our program essentially stuck to primary protons because that was the best-quality beam and it was essentially of unlimited intensity. We never were able to use more than about 1% of the available protons just because the apparatus would start melting, one way or another. We learned from Brookhaven that mass resolution was essential and, in fact, in subsequent experiments (including the main one) we had achieved something like a mass resolution of 2% for muons. When compared to electrons, muons have the enormous virtue that you can screen out the hadrons and therefore raise the proton intensity by roughly four orders of magnitude. Electrons give you better resolution because you don't have the multiple scattering in the hadron filter. Later on we were able to achieve resolutions on the order of 0.2%. These experiments must be designed to see signals that are an enormously small fraction of the total cross section; therefore you have to know and control backgrounds with almost no uncertainty. For example, it's nice to be able to look at the same-sign muons because the mean of the same-sign muons (i.e., average of plus-plus and minus-minus), with small corrections for acceptance, is about the same as the background to the plus-minus ones you want. You have to take very, very high rates and you have to have redundant identifications because both electrons and muons can be simulated by rare processes.

The upsilon experiment, which collided 400-GeV protons with nuclei, was part of this program. A lot of the success came from the fact that, for some reason, we were sitting in the beam at Fermilab from 1973 to 1977 and acquired a great deal of experience. This learning process was very important. Another crucial thing was the presence of four super-postdocs: Chuck Brown, Steve Herb, Walter Innes, and John Yoh. We started running that experiment in 1977 after rearranging the apparatus and putting in all of our experience. Before that time we had a total dimuon yield of about 300 events above 4 GeV. In the total worldwide sample there were certainly less than 1000 events. But after the '77 run began we had 10,000 dimuon events with good resolution and 600 of them appeared in a peak near 10 GeV. A series of papers came out that summer.[11] We had a broad, somewhat lopsided peak near 10 GeV. We then found structure in the upsilon region[12] as more data came in. We

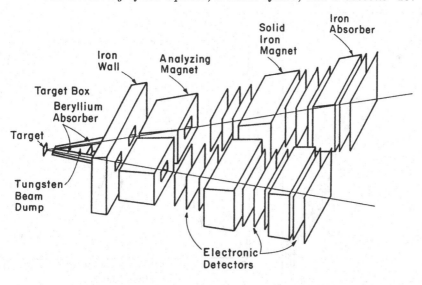

Fig. 6.2. Schematic diagram of Fermilab dimuon Experiment 288.

did nothing really different but modestly tune the experiment and just keep running.

The key to the experiment was the ability to correct mistakes on the basis of experience. Consider a drawing of the apparatus (see Fig. 6.2). Here's the target with the fierce proton intensity filtered by 20 feet of beryllium. We collected almost the entire world sample of beryllium – most of it from a weapons lab. We had a hard time because the metal came in classified shapes; we could have the beryllium but we couldn't have the shapes. That took a little bit of finagling. We had 20 feet of beryllium in the path of the muons, but using beryllium reduced the multiple scattering. We then had vertical bending so we could be pretty symmetric between same-sign dimuons and opposite-sign dimuons. We had lots of detectors by now, wire chambers essentially, and then in order to make sure it was a muon, we remeasured the momentum of the particle in a toroid magnet. Redundancy! In *every* place we would check. "Are you a muon? Are you *sure* you're a muon? Do you *swear* you're a muon?" We installed devices that would discourage as much as possible any muons having been derived from pions. Again, the same-sign pairs gave us a good measure of this background.

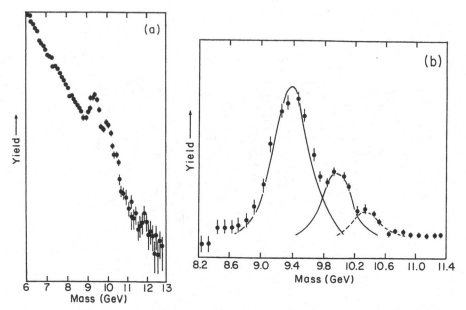

Fig. 6.3. (a) Yield of dimuons from the 1977 run of Fermilab E-288. (b) Dimuon yield after subtraction of the Drell–Yan continuum.

To reduce backgrounds, enormously ingenious attention was paid to the area around the target. It's interesting to see how the data differs from the Brookhaven J experiment, because we see the Drell–Yan continuum going all the way down to 14–15 GeV and then we find this enhancement sprawled out over the continuum. We made a very good fit to the continuum, and then subtracted it and found this set of enhancements (see Fig. 6.3). This was the data as of about June 1977, improving every few months as the data rolled in. You can see the resolution of the two main peaks and this very suggestive shoulder for the third peak.

The biggest problem we had after that was what to name it. Obviously we wanted to rescue our mistake so we decided on upsilon, or Υ. Because J/ψ was the pattern, we thought we might name it Υ/Υ.

Bottom quarks and B mesons

The interpretation of the upsilon peaks as bound states of a new quark and its antiquark came immediately. Theorists can learn too. If it was a new quark, and since the up–down and charm–strange quark pairs were now keenly understood, was this an up-charm type or a down-strange type? From our data, in particular, the ratio of the first (Υ) to second (Υ') peak indicated that we were dealing with a down-strange, or charge $-\frac{1}{3}$, type quark.

Let me make a very brief and subjective theoretical reprise. Why six quarks? I'll go through the theory as I understood it at the time ... or later. The 1962 two-neutrino experiment, or the fourth lepton experiment if you like, established this little button on which the Standard Model overcoat was eventually sewn. In 1964 Bjorken and Glashow (among others) said "Wouldn't it be *charming* if there were a fourth quark?"[13] so we could have a similar picture for the quarks. (Why should the leptons have such a nice picture?) Nothing happened until the GIM Mechanism[14] provided a compelling reason for charm. As soon as that happened, God put the charm mechanism in the SLAC data, and Gerson Goldhaber and his friends found it.[15] Now we had this prototype of the Standard Model, and in the period 1970–74 neutral currents were found,[16] so we were getting the forces right.

At that time there was something called a "high-y anomaly" that gave rise to ideas about right-handed quarks, and there were other reasons to speculate that there might be other quarks. Theorists were bored. Various people – Barnett, Gottfried, Eichten, Harari, Maki, Pakvasa, and especially noteworthy, Kobayashi and Maskawa[17] – proposed six-quark models. I think the three reasons were: (1) the *tau* (τ) had appeared in 1975 and as Martin Perl said "it was shaky for a while" (still it was possible there were going to be six leptons and it would be nice to have six quarks), (2) the high-y anomaly, and (3) the Kobayashi–Maskawa paper indicated, almost as an afterthought, that CP violation really required six quarks.

None of the above ideas motivated our research, incidentally, because we were obsessed with looking for massive particles using dileptons ever since Brookhaven. We were only vaguely aware that there were ideas around about six quarks, and I wasn't aware that we were trying to check up on the six-quark hypothesis.

Confirmation that upsilon was indeed a bound state of a charge $-\frac{1}{3}$ or b-quark and its antiquark came very quickly from DESY in a tour de force of accelerator work. They were able to jack up the machine's energy so that their e^+e^- collisions could produce Υs in order to check on our process and conclusions. We couldn't think of any other sensible alternative explanation but that it was a new quantum number and therefore a new quark, and it seemed sensible that if it were a new quark

Leon M. Lederman

Fig. 6.4. Composite data from dimuon searches at Fermilab.

in the pattern of the old ones, the charge would be $-\frac{1}{3}$. By measuring the rate of Υ production from e^+e^- collisions, they gave us the definitive proof of the quark interpretation and the identification of charge equal to $-\frac{1}{3}$. The three peaks seen in the dimuon mode at Fermilab in 1977 are the $1S$, $2S$, and $3S$ state of the $b\text{-}\bar{b}$ "atom," a 10 GeV atom that happily obeys the nonrelativistic Schrödinger equation.

Subsequent events

We of course continued to pursue this search for dimuon bumps, and here is a composite of the search at Fermilab (see Fig. 6.4). I think it finally went better than this but notice we're down to neutrino-type cross sections in this search. In fact, if I made the same graph in 1990, it would go out to 200 GeV with the dilepton data from CDF, including

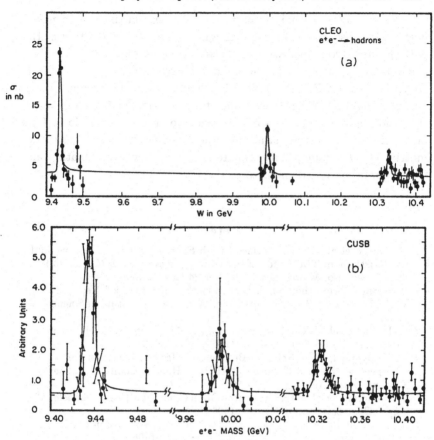

Fig. 6.5. Upsilon production in e^+e^- collisions at CESR: (a) CLEO experiment; (b) CUSB experiment.[18]

of course, a nice big peak at the Z^0, and then very few events because we really haven't looked very hard above the Z^0 yet.

I later helped to organize an experiment at the Cornell Electron Synchrotron (CESR). This was the CUSB (Columbia University with Stony Brook) detector, which was a lead-glass, and then later a bismuth-germanate, detector. I delayed my arrival at Fermilab for a year to make sure that experiment got off to a bad start. The same three peaks are seen in the CUSB work and by the CLEO detector with slightly better resolution (see Fig. 6.5.). That led of course to a very interesting and complex spectroscopy where eventually the $6S$, and probably the $7S$ resonance, is now seen, as well as P states. Also the agonizing delay between the J/ψ and naked charm didn't repeat in the case of naked bottom, because good evidence soon appeared for B mesons, that is, the bottom quark, combined with an up and down quark.

My final comment is on the great interest in B physics at CESR, DORIS, UA1, CDF and DZero, LEP, and many other experiments at Fermilab and CERN. These experiments are looking for B-mesons where the yield might be enormous but the backgrounds are horrendous. Lots of physics, spectroscopy, studies of the Cabbibo–Kobayashi–Maskawa matrix, QCD, mixing, and ultimately, presumably, CP violation and the origin of matter, remains to be done.

Notes

1 G. Danby, et al., "Observations of High-Energy Neutrino Reactions and the Existence of Two Kinds of Neutrinos," *Phys. Rev. Lett. 9* (1962), pp. 36–44. See also, M. Schwartz, Chapter 24 this volume.

2 H. Harari, "From the Fourth Lepton to the Fifth Quark," Report No. WIS-87/64/Sep-PH, (Rehovot, Israel: Weizmann Institute of Science, 1987).

3 R. Burns, et al., "Search for Intermediate Bosons in Proton–Nucleon Collisions," *Phys. Rev. Lett. 15* (1965), pp. 830–34.

4 Y. Yamaguchi, "The Relation Between W-Meson Production and Charged Lepton Pair Creation by Hadron–Hadron Collisions," *Nuovo Cimento A 43* (1966), pp. 193–99.

5 J. J. Aubert, et al., "Experimental Observation of a Heavy Particle J," *Phys. Rev. Lett. 33* (1974), pp. 1404–06.

6 J.-E. Augustin, et al., "Discovery of a Narrow Resonance in e^+e^- Annihilation," *Phys. Rev. Lett. 33* (1974), pp. 1406–08.

7 E. D. Bloom, et al., "High-Energy Inelastic e-p Scattering at 6° and 10°," *Phys. Rev. Lett. 23* (1969), pp. 930–34; M. Breidenbach, et al., "Observed Behavior of Highly Inelastic Electron–Proton Scattering," *Phys. Rev. Lett. 23* (1969), pp. 935–39; A. Bodek, et al., "Comparisons of Deep-Inelastic e-p and e-n Cross Sections," *Phys. Rev. Lett. 30* (1973), pp. 1087–91.

8 F. W. Busser, et al., "Observation of π^0 Mesons with Large Transverse Momentum in High-Energy Proton–Proton Collisions," *Phys. Lett. 46* (1973), pp. 471–76.

9 D. C. Hom, et al., "Observation of High-Mass Dilepton Pairs in Hadron Collisions at 400 GeV," *Phys. Rev. Lett. 36* (1976), pp. 1236–39.

10 D. C. Hom, et al., "Production of High-Mass Muon Pairs in Proton–Nucleus Collisions at 400 GeV," *Phys. Rev. Lett. 37* (1976), pp. 1374–77.

11 S. W. Herb, et al., "Observation of a Dimuon Resonance at 9.5 GeV in 400-GeV Proton–Nucleus Collisions," *Phys. Rev. Lett. 39* (1977), pp. 252–55.

12 W. R. Innes, et al., "Observation of the Υ Region," *Phys. Rev. Lett. 39* (1977), pp. 1240–42.

13 B. J. Bjorken and S. L. Glashow, "Elementary Particles and SU(4)," *Phys. Lett. 11* (1964), pp. 255–57.

14 S. L. Glashow, J. Iliopoulos, and L. Maiani, "Weak Interactions with Lepton–Hadron Symmetry," *Phys. Rev. D2* (1970), pp. 1285–92.

15 G. Goldhaber, et al., "Observation in e^+e^- Annihilation of a Narrow State at 1865 MeV/c^2 Decaying to $K\pi$ and $K\pi\pi\pi$," *Phys. Rev. Lett. 37* (1976), pp. 255–59.

16 F. J. Hasert, et al., "Observation of Neutrino-like Interactions without Muon or Electron in the Gargamelle Neutrino Experiment," *Phys. Lett. 46B* (1973), pp. 138–40.

17 M. Kobayashi and T. Maskawa, "CP-Violation in the Renormalizable Theory of Weak Interaction," *Prog. Theor. Phys. 49* (1973), pp. 652–57; see also Chapter 8, this volume.

18 P. Franzini and J. Lee-Franzini, "Upsilon Physics at CESR," *Phys. Rep. 81* (1982), pp. 241–91.

7

The Discovery of *CP* Violation

JAMES CRONIN

Born Chicago, Illinois, 1931; Ph.D., 1955 (physics), University of Chicago; Professor of Physics at the University of Chicago; Nobel Prize in Physics, 1980; high-energy physics (experimental).

This opportunity to discuss the discovery of CP violation has forced me to go back and look at old notebooks and records. It amazes me that they are rather sloppy and very rarely are there any dates on them. Perhaps this is because I was not in any sense aware that we were on the verge of an important discovery. In the first reference I list some of the literature on this subject, which provides different perspectives on the discovery.[1] I begin with a review of some of the important background that is necessary to place the discovery of CP violation in proper context.

Precursors

The story begins with the absolutely magnificent paper of Gell-Mann and Pais published in early 1955.[2] Each time I read it, it gives me goose bumps such as I experience while listening to the first movement of Beethoven's *Archduke* Trio. They gave the paper a very formal title, "Behavior of Neutral Particles under Charge Conjugation," but they knew in the end that this was something that concerned experiment. So the last paragraph reads:

> At any rate, the point to be emphasized is this: a neutral boson may exist which has a characteristic θ^0 mass but a lifetime $\neq \tau$ and which may find its natural place in the present picture as the second component of the θ^0 mixture.

> One of us, (M. G.-M.), wishes to thank Professor E. Fermi for a stimulating discussion.

The reference to Fermi acknowledges a comment he made to Gell-Mann in a class at the University of Chicago in the spring of 1954. It was the key remark that led Gell-Mann and Pais to write this paper.

It did not take Lederman long to get after this problem and with Lande, Booth, Impeduglia, and Chinowsky carry out a successful experiment at the Brookhaven Cosmotron. Their paper was published in 1956;[3] in the acknowledgments they write, "The authors are indebted to Professor A. Pais whose elucidation of the theory directly stimulated this research. The effectiveness of the Cosmotron staff is evidenced by the successful coincident operation of six magnets and the Cosmotron with the cloud chamber."

Figure 7.1 shows an event in the cloud chamber, which was located in a corn crib in the back yard of the Cosmotron. The event shows a three-body decay because both charged decay tracks emerge on the same side of the beam. A third, neutral particle (arrow labeled P_A) is required to balance the transverse momentum. By 1961 the combined world data showed that the upper limit for two-body decays was 0.3% of all decays.[4]

The paper of Gell-Mann and Pais used conservation of charge conjugation to argue for the necessity of a long-lived neutral K meson. With the discovery of parity violation, the conclusion was unaltered when the charge conjugation conservation was replaced by the combined conservation of charge conjugation and parity (CP).[5] The consequence was that the long-lived neutral K meson (K_2^0) was forbidden to decay to two pions.

There was another important consequence, the phenomenon of regeneration, which was described in a paper by Pais and Piccioni.[6] This paper deduced one of the beautiful aspects of the particle mixture theory. In passing through matter, neutral K mesons displayed a behavior very similar to light passing through a birefringent material. When a K_2^0 passes through matter, the positive and negative strangeness components are attenuated by different amounts. When the particle emerges from matter, the balance between the positive and negative components is altered so that there is a superposition of K_2^0 and short-lived K mesons (K_1^0). The K_1^0s decay to two pions immediately beyond the absorbing material.

Oreste Piccioni, with colleagues at the Berkeley Bevatron, demonstrated this phenomenon experimentally in a propane-filled bubble chamber.[7] The introduction to their paper pays tribute to the theory of Gell-Mann and Pais.

James Cronin

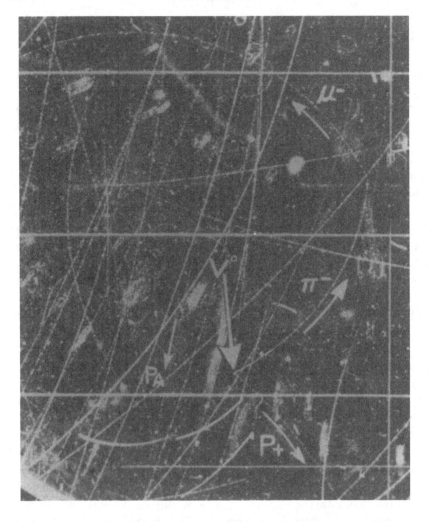

Fig. 7.1. Three-body decay of a long-lived K meson (at arrow labeled V^0), as observed in a cloud chamber at the Brookhaven Cosmotron (Source: Note 3).

It is by no means certain that, if the complex ensemble of phenomena concerning the neutral K mesons were known without the benefit of the Gell-Mann–Pais theory, we could, even today, correctly interpret the behavior of these particles. That their theory, published in 1955, actually preceded most of the experimental evidence known at present, is one of the most astonishing and gratifying successes in the history of the elementary particles.

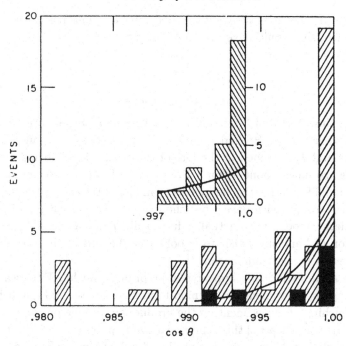

Fig. 7.2. Anomalous regeneration in hydrogen. "Angular distribution of events which have a 2π-decay Q-value consistent with K_1^0 decay, and a momentum consistent with the beam momentum. All events are plotted for which 180 MeV $\leq Q \leq$ 270 MeV, $p \geq$ MeV/c. The black histogram presents those events in front of the thin window. The solid curve represents the contribution from K_2^0 decays."[8]

After regeneration had been established, Adair, Chinowsky, and collaborators placed a hydrogen bubble chamber in a neutral beam at the Brookhaven Cosmotron to study the effect in hydrogen.[9] Figure 7.2 shows their result. In this experiment, as in subsequent ones, the vector momenta of the two charged tracks in the decay are measured. Assuming each track is a pion, the direction and mass of a parent particle is calculated. Two-body decays of K mesons will produce a peak in the forward direction at 498 MeV. The three-body decays will produce a background that can be estimated by Monte Carlo and extrapolation. The forward regenerated peak was found to be too large by a factor of 10 to 20.

Adair gave a very creative explanation; he postulated a fifth force that was very weak but had a long range and hence had a small total cross section with a large forward amplitude. It also differentiated between positive and negative strangeness producing a strong regeneration. If confirmed, this would have been a major discovery.

The experiment

At this time both Val Fitch and I were working at Brookhaven on separate experiments. He had spent much of his career working with K mesons and was steeped in the lore of these particles that had already revealed so much about Nature. He was one of the first to measure the individual lifetimes of the various decay modes of the charged K mesons. To avoid trouble with parity, one thought that the two-pion and three-pion decay modes were actually due to different particles.[10] On the occasion of Panofsky's visit to Brookhaven, he and Fitch detected the K_2^0 mesons by electronic means.[11]

The Adair experiment appeared in preprint form while Fitch was just finishing an experiment on the pion form factor at the AGS, and I was just finishing an experiment on the production of ρ mesons at the Cosmotron. At the heart of this experiment was a spark-chamber spectrometer designed to detect ρ mesons produced in hydrogen at low transverse momentum. At that time optical spark chambers were a new tool in which one could, by selective electronic trigger, record the trajectories of the desired events out of a very high-rate background. I was among the first to apply this technique to accelerator experiments.[12] Using the Chew–Low extrapolation, we could study pion–pion scattering. We could also look for ρ-ω interference. Though the paper we wrote on this experiment was not distinguished by a high rate of citations, the spectrometer was state-of-the-art at the time.[13]

Fitch, being tuned into K mesons, came to me and suggested that together we use our spectrometer to look for Adair's anomalous regeneration. Progress in physics thrives on good ideas. Fitch had a good one, it took me about one microsecond to agree to pursue his suggestion. Jim Christenson and René Turlay, who was visiting from France, joined us on the experiment. In addition to checking the Adair effect, it gave us opportunity to make other measurements on K_2^0 with greater precision.

The spectrometer I had built with Alan Clark, Christenson, and Turlay was ideally suited for the job. It was designed to look at pairs of particles with small transverse momentum. This was just the property needed to detect two-body decays in a neutral beam. We also had a 4-foot-long hydrogen target that would make a perfect regenerator.

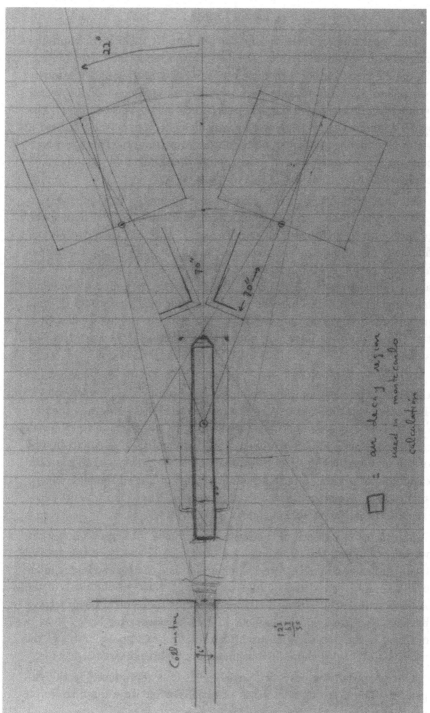

Fig. 7.3. Sketch of the planned spectrometer arrangement, from notebook of J. W. Cronin.

The spectrometer consisted of two normal 18 × 36 inch beam-line magnets turned on end so that the deflections were in the vertical plane. The angle between the two magnets was adjustable. Spark chambers before and after the magnet permitted the measurement of the vector momentum of a charged particle in each arm of the spectrometer. The spark chambers were triggered by a coincidence of scintillators and a water Čerenkov counter behind each spectrometer arm. This apparatus could accumulate data much more rapidly than the bubble chamber and had a mass resolution that was five times better.

Another fortunate fact was that we had an analysis system ready to measure the spark chamber photographs quickly. We had homemade projectors and measured, instead of points, only angles of tracks and fiducials. The angular measurement was made with a Datex encoder attached to an IBM Model 526 card punch. The least count of the angular encoder was 1.5 mrad. In addition we had bought a commercial high-precision bubble-chamber measuring machine that would become important in checking our results. Note that our support came from the Office of Naval Research; this was at a time before the Mansfield amendment!

We looked around for a neutral beam at both the Cosmotron and the AGS. The most suitable beam was one used by the Illinois group.[14] The beam was directed toward the inside of the AGS ring to a narrow, crowded area squeezed between the shielding of the machine and the wall of the experimental hall. Dubbed "Inner Mongolia" by Ken Green, one of the builders of the AGS, this area was mostly relegated to parasitic experiments working off the same target that produced the high-energy, small-angle beams for the major experiments. The beam was produced on an internal target at an angle of 30°.

Figure 7.3 is a sketch of the setup that I made in my notebook when we were planning the experiment. An angle of 22° between the neutral beam and each arm of the spectrometer matched the mean opening angle of $K_2^0 \rightarrow \pi^+ + \pi^-$ decays at 1.1 GeV/c which was at the peak of the K_2^0 spectrum at 30°. It also allowed room for the neutral beam to pass between the front spark chamber of each spectrometer arm. Heavily outlined is the decay region used for the Monte Carlo estimates of the rates. Fainter lines show the outline of the hydrogen target.

Our proposal was only two pages long. It is reproduced in the Appendix. The first page describes essentially what we wanted to do. It reads in part:

Fig. 7.4. The only existing photograph of the apparatus used in the experiment. The individual pictured is graduate student Wayne Vernon, who did his thesis on a subsequent experiment.

> It is the purpose of this experiment to check these results with a precision far transcending the previous experiment. Other results to be obtained will be a new and much better limit for the partial rate of $K_2^0 \rightarrow \pi^+ + \pi^-$.

One notes that we referred to a limit; we had no expectation that we would find a signal. We also proposed to measure a limit on neutral currents and study coherent regeneration. On the second page of the proposal one reads:

> We have made careful Monte Carlo calculations of the counting rates expected. For example, using the 30° beam with the detector 60-ft. from the A.G.S. target we could expect 0.6 decay events per 10^{11} circulating protons if the K_2 went entirely to two pions. This means that we can set a limit of about one in a thousand for the partial rate of $K_2 \rightarrow 2\pi$ in one hour of operation.

This estimate turned out to be somewhat optimistic.

We moved the spectrometer from the Cosmotron to the AGS in May 1963. It just barely fit inside the building. We began running in early

FLOOR PLAN OF EXPERIMENT

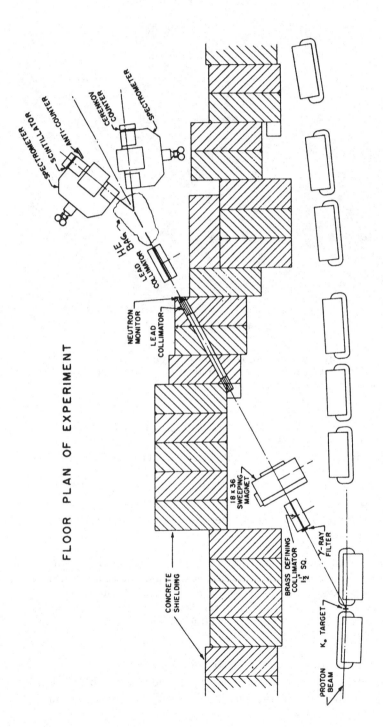

Fig. 7.5. Schematic view of the experimental arrangement for the *CP* invariance test.

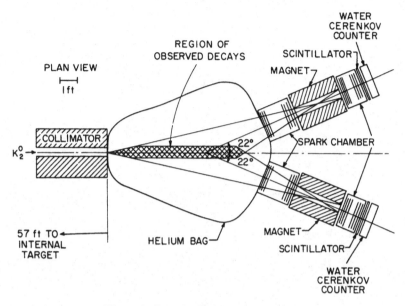

Fig. 7.6. Details of the spectrometer.

June. There was no air-conditioned trailer. The electronics, all home-made, was just out on the AGS floor in the summer heat. Figure 7.4 shows the only photograph that we have of the apparatus. Most prominent are the plywood enclosures that contained the optics for photographing the spark chambers. One can discern the two magnets set at 22° to the neutral beam. Also visible are the few racks of electronics.

Figures 7.5 and 7.6 show schematically the experimental arrangement for the CP invariance test. A large helium bag was placed in the decay region. By the time we were ready to begin the CP run on June 20, 1963, we had a better number for the flux of K_2^0s in the beam. The best monitor was a thin scintillation telescope placed in the neutral beam upstream of the decay region, which counted neutrons. In my notebook I estimated that there were $10^6 K_2^0$s per neutron count (in units of 10^5). The Monte Carlo efficiency to detect a $K \rightarrow 2\pi$ decay was 1.5×10^{-5}. Thus to set a limit of 10^{-4}, 666 neutron counts were needed; for safety I suggested 1200 neutron counts. The observed yield of K_2^0 in the beam turned out to be about one-third of the original estimate given in the proposal.

The page of our data book from the day that the CP run began is shown in Fig. 7.7. Only ten minutes into the run one finds the note:

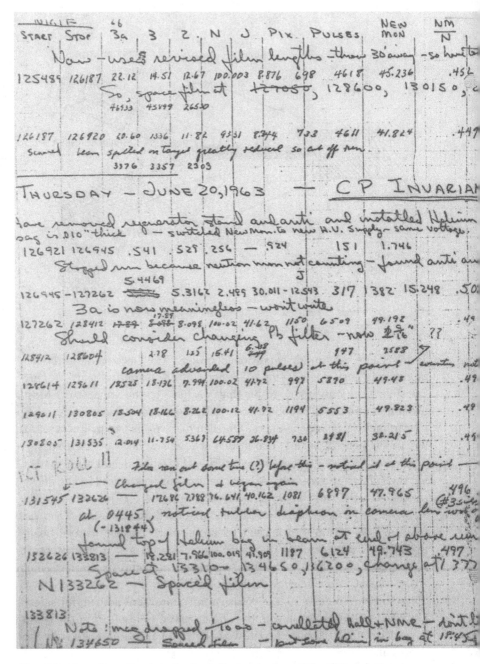

Fig. 7.7. Page from the data book at the beginning of the *CP* violation run.

$\frac{N}{PULSE}$	$67 \frac{2a}{N}$	$\frac{3}{N}$	$\frac{c}{N}$	$\frac{P_{ix}}{N}$	$\frac{N}{J}$	SWEEP HALL	

200 pictures — 1550 per "100' "

| 21.7 | .221 | .145 | .127 | 6.98 | 11.30 | | 63348 $b = 2\frac{5}{8}°$ |

film at 131700

$b = 0$ ($3"$ So
DIAL = 363

| 20.2 | .221 | .144 | .129 | 9.84 | 11.20 | 63335 alsso same |

II

...touches SF. windows and falls ~6" short of collimator

64437 63261

...transistors blown in coinc. circuit — replaced — A.O.K.

| 21.7 | .1816 | .1772 | .0830 | 10.6 | 2.390 | AT N127355→DIAL 36 |

| | .179 | .179 | .081 | 11.50 | 2.40 | ← Made adjustments Monitor Voltage this run |

| | .185 | .181 | .07954 | 9.97 | 2.40 | |

| | .185 | .182 | .0826 | 11.9 | 2.40 | 13 |
| | | | | | | 130151 10 pw |

| 16.3 | .186 | .1815 | .0832 | 11.3 | 2.41 | 63330 |

...before 131700 it would seem! Sorry.

| 14.0 | — | .183 | .0806 | 11.2 | 240 | |

...loops pictures taken till mag are no good?

more He in.

| 16.3 | — | .1831 | .080 | 11.9 | 2.39 | — | SAME DIAL - 36? |

ditch end closely!

64451 63.253
64530 63303 DIAL→?

at so what.

63.246

Stopped run because neutron monitor was not counting – found anti and collector transistors blown in coin. circuit – replaced – A. O. K.

This was not a smooth run – it was the real world! Other comments include:

- 3a is now meaningless – won't write
- Film ran out sometime (?) before this – noticed it at this point – sometime before 131700 it would seem! Sorry.
- at 0445, noticed rubber diaphram on camera lens (#3 side) was off – perhaps pictures taken till now are no good?
- found top of helium bag in beam at end of above run. – put more He in.

And so it went – not smoothly but working nevertheless. We ran for about 100 hours over five days. The AGS ran very well. We collected a total of more than 1800 neutron counts, more than the 1200 we thought we needed. During about a month of running, data were taken on many aspects of K_2^0s including a measurement of the K_1^0–K_2^0 mass difference and the density dependence of the coherent regeneration in copper. And of course a week of data was taken with the hydrogen target to search for anomalous regeneration.

Data analysis

We stopped running at the end of June and gave our first results at the Brookhaven Weak Interactions Conference in September. We reported on a new measurement of the mass difference. As I recall, we did not give high priority to the CP run in the early analysis, but it was Turlay who began to look at this part of the data in the fall. A quick look at the hydrogen regeneration did not reveal any anomaly.

All the events collected in the CP invariance run were measured with the angular encoder. This work was completed by early 1964, and Turlay produced the curves shown in Fig. 7.8. There were 5211 events that were measured and successfully reconstructed. The top curve shows the mass distribution of all the events assuming each charged track was a pion. Shown also is the Monte Carlo expectation for the distribution. The relative efficiency for all K_2^0 decay modes compared to the decay to two pions was found to be 0.23. The bottom curve shows the angular distribution of the events in the effective mass range of 490–510 MeV. The curve was plotted in $\cos\theta$ bins of 0.0001, presumably consistent with the angular resolution that could be obtained with the angular

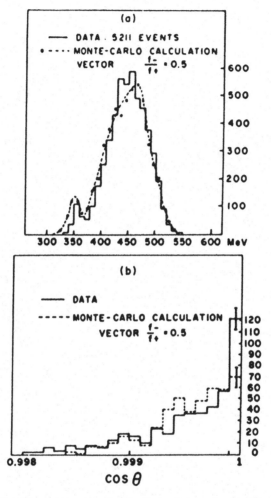

Fig. 7.8. (a) Experimental mass distribution compared with Monte Carlo calculation. The calculated distribution is normalized to the total number of observed events. (b) Angular distribution of those events in the mass range of 490–510 MeV. The calculated curve is normalized to the number of events in the complete sample.

measuring machines. There appeared to be an excess of about 50 events above what was expected. We then remeasured those events with $\cos\theta \geq$ 0.9995 on our precision bubble chamber measuring machine.

In looking over my old notebooks I found a page that is reproduced in Fig. 7.9. When the data measured with the angular encoder were plotted in finer $\cos\theta$ bins of 0.00001, the angular resolution was much

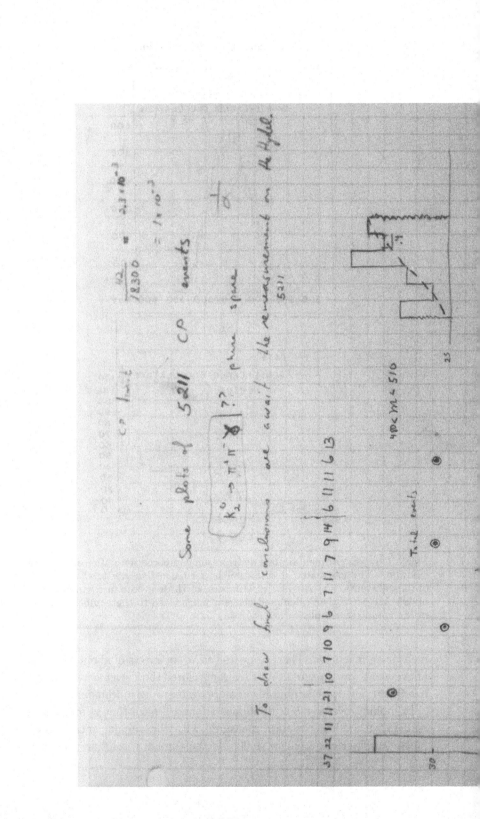

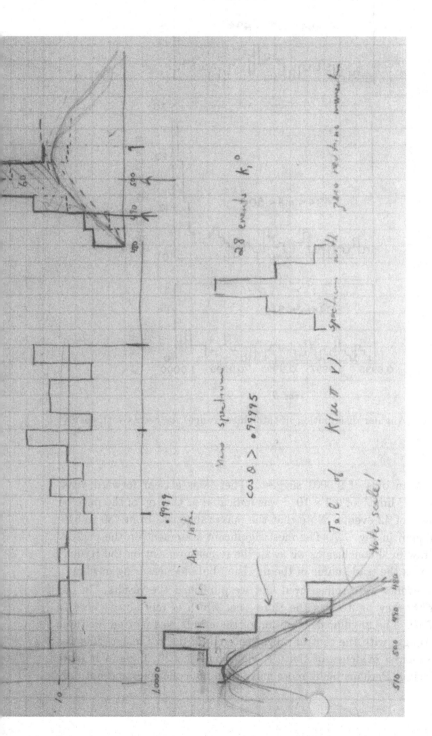

Fig. 7.9. Page from notebook of J. W. Cronin with comment on the first results of the analysis of the CP events measured with the angular encoder.

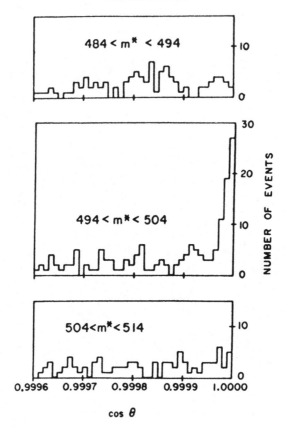

Fig. 7.10. Angular distribution in three mass ranges for events with $\cos\theta \geq 0.9995$.

better than bins of 0.0001 suggest. There was a clear forward peak, and a "*CP* limit" of 2.3×10^{-3} was indicated at the top of the page on the basis of 42 events. Note that the mass range was from 480 to 510, larger than in Fig. 7.7. The most significant statement on the page is: "To draw final conclusions we await the remeasurement on the Hydel," which was the trade name of the precision bubble-measuring machine.

The events were remeasured and we published the results. In our paper the key figure was the third one, which is reproduced here as Fig. 7.10. The angular distribution of the events was plotted for three mass ranges with the central range centered on the K_1^0 mass. This was our principal evidence of the *CP* violation effect.[15] I found it quite convincing. Perhaps being more naive than my colleagues and not fully appreciating the profound consequences of the result, I was not at all worried that the result might be wrong. We had done an important check to be sure of the calibration of the apparatus, placing a tungsten block at five positions along the decay region to simulate with two-pion decays of regenerated K_1^0s the distribution of the CP-violating events. We found the mass, angular resolution, and spatial distribution of the events observed with the helium bag to be identical with the regenerated K_1^0 events. From our own measurements of regeneration amplitudes, the regeneration in the helium was many orders of magnitude too small to explain the effect. We reported a branching ratio of $(2.0 \pm 0.4) \times 10^{-3}$, a result that within the error has not changed to this day. We also reported in this paper a value of the parameter that has come to be known as η_{+-}.[16]

Two weeks after our publication appeared, the Illinois group published a paper entitled, "Search for CP Nonconservation in K_2^0 Decays."[17] It reported some evidence for the two-pion decay of the K_2^0. The data were taken in the same AGS beam at an earlier date, during an experiment that was designed to study the form factor of the three-body decays. While their experiment was not optimized for CP studies, they reported some ten events in a mass range of 500–510 MeV that were consistent with two-body decays. One important aspect was the fact that in the Illinois apparatus the decay products passed through some material. Two of the events in the forward peak showed one of the decay products interacting in material, which identified them as strongly interacting particles.

At the time of the discovery all kinds of ideas were proposed to save the concept of CP violation. Among these theories were situations where the apparent pions in the CP violation would not be coherent with the pions of a K_1^0 decay. Thus it was important to first establish the coherence of the CP-violating decays. There was an experiment carried out by Fitch and his collaborators that has not received the proper attention of those who have reviewed the field.[18] In this experiment he showed explicitly that there was constructive interference between regenerated K_1^0 decays and the CP-violating decays. The idea was clever and grew out of our extensive experience with the regeneration phenomenon. A long, low-density regenerator, made of thin sheets of beryllium, was prepared with a regeneration amplitude that just matched the CP amplitude. The experiment showed definitively that there was maximum constructive interference, and strengthened the idea that in the constitution of the long-lived K there was a small admixture of CP-even state in what is a predominantly CP-odd state.

Concluding remarks

Since the discovery of CP violation there has been an enormous amount of work both on the neutral K meson system and on searches for time reversal violation in many systems. Technological improvements over the last 25 years have permitted very sensitive experiments on the CP-violating parameters of the K meson system. Event samples containing a million CP-violating decays have been obtained. In recent years great emphasis has been placed on the parameter η_{00}, which compares the CP-violating strength of charged pions to that of neutral pions.[19] An observed difference in η_{00} and η_{+-} would mean a second CP violating parameter would have been found. Sadly, the CP violation can still be characterized by only one parameter, which represents a small admixture of a CP-even state in the long-lived neutral K meson. To quote from a review written in 1981:[20]

> If we state that the mass matrix which couples K and $\bar{K}$ has an imaginary off-diagonal term given by:
> $$\mathrm{Im}M_{12} = -1.16 \times 10^{-8}\mathrm{eV}$$
> then all the experimental results related to CP violation can be accounted for.

With only one number there are any number of theories that can account for the effect. There is a need for at least a second parameter. Surely the most attractive "explanation" for CP violation lies in the innovative ideas of Cabbibo and of Kobayashi and Maskawa.[21] A phase in the so-called CKM matrix that generates relative imaginary amplitudes for the weak decays is compatible with the CP violation as seen in the neutral K system. The constraints on the parameters of the CKM matrix from the K system predicts large CP-violating effects in some of the rare decay modes in the neutral B meson system. Soon we will have B factories with sufficient luminosity to observe these effects.

I would like to conclude with a personal remark, though Nature is not going to pay any attention to what I think. I would be very disappointed if CP violation occurs only because there is a phase in the CKM matrix, which has as much or as little significance as the other constants that refer to the mixing of the quark states between the weak and the strong interactions. I would like to think that there is some more fundamental relationship between the manifest CP violation in the neutral K meson system and the significant fact that our galaxy and most likely our Universe is matter dominated. It may not be so. When parity violation was discovered, many thought that the fact that our biological molecules show a handedness was related to the manifest handedness of the weak interaction.[22] But subsequent experiments and theoretical considerations do not support this possibility.[23] Indeed it is almost certain that the CP violation observed in the K meson system is not directly responsible for the matter dominance of the universe, but one would wish that it is related to whatever was the mechanism that created the matter dominance. At present one number is adequate to describe our knowlege of CP violation. I hope that before I depart this Earth we will find the origin of CP violation and understand its significance.

Notes

1 R. K. Adair, "CP Non Conservation: The Early Experiments," in J. Tran Tranh Van, ed., *CP Violation in Particle Physics and Astrophysics*, (Gif-sur-Ynette, France: Editions Frontières, 1990), pp. 37–54; A. Pais, "CP Violation: – The First 25 Years," ibid., pp. 3–35; J. W. Cronin and Margaret Stautberg Greenwood, "CP Symmetry Violation," *Physics Today 35:7* (July 1982), pp. 38–44; Val L. Fitch, "A Personal View of the Discovery of CP Violation," in M. Doncel, ed., *Symmetries in Physics (1600–1980), Proceedings of the Conference at San Felice de Guixols, Spain, 1983* (University of Barcelona, 1987); A. Franklin, "The discovery and acceptance of CP violation," *Hist. Stud. Phys. Sci. 13* (1983), pp. 207–38.

2 M. Gell-Mann and A. Pais, "Behavior of Neutral Particles Under Charge Conjugation," *Phys. Rev. 97* (1955), pp. 1387–89.

3 K. Lande, et al., "Observation of Long-Lived Neutral V Particles," *Phys. Rev. 103* (1956), pp. 1901–04.

4 D. Neagu, E. O. Okonov, N. I. Okonov, N. I. Petrov, A. M. Rosanova, and V. A. Rusakov, "Decay Properties of K_2^0 Mesons," *Phys. Rev. Lett. 6* (1961), pp. 552–53.

5 T. D. Lee, R. Oehme, and C. N. Yang, "Remarks on Possible Noninvariance under Time Reversal and Charge Conjugation," *Phys. Rev. 106* (1957), pp. 340–45.

6 A. Pais and O. Piccioni, "Note on the Decay and Absorption of the θ^0," *Phys. Rev. 100* (1955), pp. 1487–90.

7 R. H. Good, et al., "Regeneration of Neutral K Mesons and Their Mass Difference," *Phys. Rev. 124* (1961), pp. 1223–39.

8 L. B. Leipuner, et al., "Anomalous Regeneration of K_1^0 Mesons from K_2^0 Mesons," *Phys. Rev. 132* (1963), pp. 2285–91.

9 L. B. Leipuner, et al., "Anomalous Regeneration."

10 R. Motley and V. Fitch, "Lifetime of the τ^+ Meson," *Phys. Rev. 105* (1957), pp. 265–67.

11 W. K. H. Panofsky, et al., "Measurement of the Total Absorption Coefficient of Long-Lived Neutral K Particles," *Phys. Rev 109* (1958), pp. 1353–57.

12 James W. Cronin and George Renninger, "Studies of a Neon-Filled Spark Chamber," in *Proceedings of an International Conference on Instrumentation for High Energy Physics* (New York: Interscience Publishers Inc., 1961), pp. 271–75.

13 A. R. Clark, et al., "Dipion Production at Low Momentum Transfer in π-p Collisions at 1.5 BeV/c," *Phys. Rev. 139B* (1965), pp. 1556–65; J. H. Christenson, A. R. Clark, and J. W. Cronin, "A Study of Spatial Accuracy in a System of Spark Chambers," *IEEE Trans. Nucl. Sci. 11* (June 1964), pp. 310–16.

14 L. Creegie, J. D. Fox, H. Frauenfelder, A. O. Hanson, G. Moscati, C. F. Perdrisat, and J. Todoroff, "Observation of the 2 Photon Decay of the K_2^0 Meson," *Phys. Rev. Lett. 17* (1966), pp. 150–153.

15 J. H. Christenson, J. W. Cronin, V. L. Fitch, and R. Turlay, "Evidence for the 2π decay of the K_2^0 meson," *Phys. Rev. Lett. 13* (1964), pp. 138–41.

16 T. T. Wu and C. N. Yang, "Phenomenological Analysis of Violation of CP Invariance in Decay of K^0 and $\bar{K}^0$," *Phys. Rev. Lett. 13* (1964), pp. 380–86.

17 A. Abashian, R. J. Abrams, D. W. Carpenter, G. P. Fisher, B. M. K. Nefkens, and J. H. Smith, "Search for CP Nonconservation in K_2^0 Decays," *Phys. Rev. Lett. 13* (1964), pp. 243–46.

18 V. L. Fitch, R. F. Roth, J. Russ, and W. Vernon, "Evidence for Constructive Interference Between Coherently Regenerated and CP Nonconserving Amplitudes," *Phys. Rev. Lett. 15* (1965), pp. 73–76.

19 J. R. Patterson, et al., "Determination of $R_e\left(\frac{\varepsilon'}{\varepsilon}\right)$ by the Simultaneous Detection of the Four $K_{L,S} \to \pi\pi$ Decay Modes," *Phys. Rev. Lett. 64* (1990), pp. 1491–95; H. Burkhardt, et al., "First Evidence for Direct CP Violation," *Phys. Lett. B206* (1988), pp. 169–76.

20 James W. Cronin, "CP Symmetry Violation – the Search for its Origin," *Rev. Mod. Phys. 53* (1981), pp. 373–84.

21 N. Cabbibo, "Unitary Symmetry and Leptonic Decays," *Phys. Rev. Lett. 10* (1963), pp. 531–33; M. Kobayashi, and T. Maskawa, "CP-Violation in the Renormalizable Theory of Weak Interaction," *Prog. Theor. Phys. 49* (1973), pp. 652–57; see also Chapter 8, this volume.

22 H. Krauch and F.Vester, *Naturwissenschaften 44* (1957), p. 49.

23 R. A. Hegstrom, A. Rich, and J. Van House, "New Estimates of Asymmetric Decomposition of Recessive Mixtures by Natural β Decay Sources," *Nature 313* (1985), pp. 391–93.

Appendix: Proposal for K_2^0 Decay and Interaction Experiment

PROPOSAL FOR K_2^0 DECAY AND INTERACTION EXPERIMENT

J. W. Cronin, V. L. Fitch, R. Turlay

(April 10, 1963)

I. INTRODUCTION

The present proposal was largely stimulated by the recent anomalous results of Adair et al., on the coherent regeneration of K_1^0 mesons. It is the purpose of this experiment to check these results with a precision far transcending that attained in the previous experiment. Other results to be obtained will be a new and much better limit for the partial rate of $K_2^0 \rightarrow \pi^+ + \pi^-$, a new limit for the presence (or absence) of neutral currents as observed through $K_2 \rightarrow \mu^+ + \mu^-$. In addition, if time permits, the coherent regeneration of K_1's in dense materials can be observed with good accuracy.

II. EXPERIMENTAL APPARATUS

Fortuitously the equipment of this experiment already exists in operating condition. We propose to use the present 30° neutral beam at the A.G.S. along with the di-pion detector and hydrogen target currently being used by Cronin, et al. at the Cosmotron. We further propose that this experiment be done during the forthcoming μ–p scattering experiment on a parasitic basis.

The di-pion apparatus appears ideal for the experiment. The energy resolution is better than 4 Mev in the m^* or the Q value measurement. The origin of the decay can be located to better than 0.1 inches. The 4 Mev resolution is to be compared with the 20 Mev in the Adair bubble chamber. Indeed it is through the greatly improved resolution (coupled with better statistics) that one can expect to get improved limits on the partial decay rates mentioned above.

III. COUNTING RATES

We have made careful Monte Carlo calculations of the counting rates expected. For example, using the 30° beam with the detector 60-ft. from the A.G.S. target we could expect 0.6 decay events per 10^{11} circulating protons if the K_2 went entirely to two pions. This means that one can set a limit of about one in a thousand for the partial rate of $K_2 \rightarrow 2\pi$ in one hour of operation. The actual limit is set, of course, by the number of three-body K_2 decays that look like two-body decays. We have not as yet made detailed calculations of this. However, it is certain that the excellent resolution of the apparatus will greatly assist in arriving at a much better limit.

If the experiment of Adair, et al. is correct the rate of coherently regenerated K_1's in hydrogen will be approximately 80/hour. This is to be compared with a total of 20 events in the original experiment. The apparatus has enough angular acceptance to detect incoherently produced K_1's with uniform efficiency to beyond 15°. We emphasize the advantage of being able to remove the regenerating material (e.g., hydrogen) from the neutral beam.

IV. POWER REQUIREMENTS

The power requirements for the experiment are extraordinarily modest. We must power one 18-in. x 36-in. magnet for sweeping the beam of charged particles. The two magnets in the di-pion spectrometer are operated in series and use a total of 20 kw.

8

Flavor Mixing and CP Violation

MAKOTO KOBAYASHI

Born Nagoya, Japan, 1944; Ph.D., 1972 (physics), Nagoya University; Professor at KEK National Laboratory for High Energy Physics, Tsukuba, Japan; high-energy physics (theory).

I begin with a few remarks about early studies on the heavy flavors done in Japan. In 1960, Ziro Maki, Masami Nakagawa, Yoshio Ohnuki, and Shoichi Sakata proposed a model, known as the Nagoya model, in which the symmetry between the baryons and the leptons in the weak interactions is explained by a possible composite structure of the baryons:

$$p = (\nu B^+), \quad n = (e^- B^+), \quad \Lambda = (\mu^- B^+),$$

where B^+ is what they called B-matter.[1]

In 1962, soon after the existence of two kinds of neutrinos was revealed by experiments, the Nagoya model was modified in the following manner:[2]

$$p = (\nu_1 B^+), \quad n = (e^- B^+), \quad \Lambda = (\mu^- B^+), \quad p' = (\nu_2 B^+),$$

$$\begin{aligned} \nu_1 &= \cos\theta \, \nu_e + \sin\theta \, \nu_\mu, \\ \nu_2 &= -\sin\theta \, \nu_e + \cos\theta \, \nu_\mu. \end{aligned}$$

In this scheme the weak current of the baryons induced by their leptonic components is nothing but what we now call the GIM current.[3] Although the fundamental particles are supposed to be ordinary baryons, flavor mixing and the possible existence of a fourth fundamental particle are clearly mentioned in these articles.

The modified Nagoya model was recalled by Shuzo Ogawa in 1971, when Kiyoshi Niu and his group discovered a new kind of event in emulsion chambers exposed to cosmic rays.[4] They found a few events of short-lived particles with a mass around 2 GeV. Immediately, Ogawa pointed out that they might be particles including the fourth element

p', and then several people started to study the quartet scheme.[5] Toshihide Maskawa and I were among them. At that time Maskawa was a research associate at Kyoto University and I was a graduate student at Nagoya University.

Meanwhile the renormalizabilty of the Glashow–Weinberg–Salam (GWS) model of the weak interaction began to attract attention.[6] We accepted its extension to the hadron based on the GIM scheme as a quite realistic possibility, because the fourth quark already existed for us in a sense. Sometimes it is said that our CP paper was written before the discovery of charm.[7] In this sense, however, our paper came after the charm.

Anyhow, we realized that the GWS model was a quite promising one to describe the weak interactions of both the leptons and the hadrons. What we thought then is that, for the model to be true, all interactions should be described in a renormalizable and therefore gauge-invariant way. In particular, to incorporate CP-violating interactions seemed most important. In spring of 1972, after receiving a Ph.D. from Nagoya, I moved to Kyoto University, and my collaboration with Maskawa on CP violation started.

The problem we tried to solve was the following. The GIM scheme implies that the left-handed components of the four quarks consist of two doublets of the following form:

$$\begin{pmatrix} u \\ d' \end{pmatrix}, \qquad \begin{pmatrix} c \\ s' \end{pmatrix},$$

$$\begin{pmatrix} d' \\ s' \end{pmatrix} = \begin{pmatrix} \cos\theta & \sin\theta \\ -\sin\theta & \cos\theta \end{pmatrix} \begin{pmatrix} d \\ s \end{pmatrix}$$

The minimal theory of the GWS type based on this assignment does not break CP invariance and therefore does not explain the CP-violating phenomena observed in the neutral K meson system.

At first sight, it looks easy to break CP invariance, because the mixing matrix between the d and s quarks can be a 2×2 unitary matrix and the complex coupling constant would violate CP invariance. However, the relative phase of the states with different flavors can be observed only through the interaction that causes transition between them. Therefore, insofar as no other flavor-changing interaction exists, the complex character of the 2×2 unitary matrix is absorbed into the phase convention of the quark fields and reduced to the original GIM scheme. This kind of argument on the phase convention is now familiar, but at that time it was seldom mentioned in textbooks. We spent a considerable amount of time trying to convince ourselves.

Confirming that no other possibilities of CP violation exist in the minimal theory, we came to a conclusion that in order to explain the observed CP-violating phenomena we needed some new particles, in addition to the charm quark. We felt excitement in the fact that we could conclude the inevitable existence of new particles from quite logical reasoning.

Once we admit the introduction of new particles, however, we are faced with a vast number of possibilities. This can be understood qualitatively as follows: As the number of particles increases, the number of their mutual interactions increases more rapidly than the number of particles itself and this implies a growing chance of complex coupling constants occurring whose phase cannot be absorbed into the phase convention of the fields. The difficulty of the CP-violation problem lies here. Even now we cannot pin down the origin of CP violation among many possibilities because of very limited experimental information.

Although the main purpose of our paper was not to propose a particular model of CP violation, a very attractive model came out naturally from the above observation. The previous mechanism, which reduces a unitary matrix to a real (orthogonal) one utilizing the phase convention of the quark fields, no longer works for three generations. For the latter case we are left with one phase factor in the mixing matrix that violates CP invariance. An example of the parametrization of the mixing matrix is the following:

$$\begin{pmatrix} u \\ d' \end{pmatrix}, \quad \begin{pmatrix} c \\ s' \end{pmatrix}, \quad \begin{pmatrix} t \\ b' \end{pmatrix},$$

$$\begin{pmatrix} d' \\ s' \\ b' \end{pmatrix} = \mathbf{V} \begin{pmatrix} d \\ s \\ b \end{pmatrix} = \begin{pmatrix} c_1 & -s_1 c_3 & -s_1 s_3 \\ s_1 c_2 & c_1 c_2 c_3 - s_2 s_3 e^{i\delta} & c_1 c_2 s_3 + s_2 c_3 e^{i\delta} \\ s_1 s_2 & c_1 s_2 c_3 + c_2 s_3 e^{i\delta} & c_1 s_2 s_3 - c_2 c_3 e^{i\delta} \end{pmatrix} \begin{pmatrix} d \\ s \\ b \end{pmatrix},$$

where $c_i = \cos\theta_i$ and $s_i = \sin\theta_i$. There are four observable mixing parameters, $\theta_1, \theta_2, \theta_3$, and δ. If δ is not 0 or π, CP invariance is violated.

This six-quark scheme began to attract attention after the τ lepton was discovered. In the renormalizable theory, the anomaly due to the τ lepton should be canceled by something new, and the six-quark scheme is the simplest possibility. At an early stage, the concrete analysis of the model was made by Sandip Pakvasa and Hirotaka Sugawara; Luciano

Maiani; and John Ellis, Mary K. Gaillard, and Dimitrius V. Nanopoulos right after the discovery of the τ lepton.[8]

One of the key issues of the analysis was its distinction from the superweak model, proposed by Lincoln Wolfenstein in 1964. Its major prediction is that ϵ', one of the measurable quantities describing CP violation in the neutral K meson system, is essentially zero.[9] In the six-quark model, CP violation comes from two types of diagrams. One is the box diagram that contributes to K-$\bar{K}$ transition and the other is the so-called penguin diagram that contributes to the decay amplitudes. The latter contribution gives nonvanishing ϵ' and is therefore crucial for the test of the model.[10] The theoretical estimate of ϵ', however, has been a function of time. The estimated value has been decreasing, as if it escapes from the experimental upper bound. The main reason for this decrease is that the top quark mass used for the estimate has increased gradually. According to recent estimates, ϵ' could even be zero for a certain value of the top quark mass, due to cancellation between QCD and electroweak penguin contributions.

Meanwhile, the Υ particle was found and the six-quark scheme became a standard picture, and the top quark has finally been discovered. Now direct measurements of V_{cb} and V_{ub} are available, and much effort has been devoted to the experimental determination of the four mixing parameters. The parameter θ_1 is essentially the Cabibbo angle and it is well measured already. Information on the rest of the three parameters is nicely described by the so-called unitarity triangle, which represents the following unitarity relation in the complex plane:

$$V_{ud}^* V_{ub} + V_{cd}^* V_{cb} + V_{td}^* V_{tb} = 0.$$

An interesting fact is that all the CP violation effects in the Standard Model are proportional to the area of the triangle.

The B meson system is actually a very good place to test the model of CP violation. For example, the asymmetry between $B_d \to J/\Psi + K_S^0$ and $\bar{B}_d \to J/\Psi + K_S^0$ is related to one of the angles of the unitarity triangle with the least theoretical ambiguity.[11] Various plans for future experiments on the B meson system, including a dedicated machine called the B factory, are under construction in Japan and the United States.

Finally, let me briefly mention a couple of issues related to CP violation. One is the so-called strong CP problem. In QCD theory, the topological degeneracy of the classical ground state is resolved due to the quantum tunneling known as the instanton effect, and the quantum vacuum is labeled by a new parameter, θ. When θ is not 0 or π, P and CP are not conserved. The parameter θ is severely constrained by the experimental upper bound of the neutron electric dipole moment and believed to be less than 10^{-9}. To cure the unnaturalness of such a small number, Roberto Peccei and Helen Quinn proposed a mechanism that predicts the existence of a new particle called the axion.[12] The strong CP problem is also still an open question.

Another important issue related to CP violation is the problem of cosmological baryon-number generation, which was first discussed by Andrei Sakharov in 1967 and revived by Motohiko Yoshimura and by Alexander Ignatiev, Nikolai Krasnikov, Vadim Kuzmin and Albert Tavkhelidze in connection with grand-unified theories.[13] As is well known, baryogenesis requires three fundamental conditions, that is, in-equilibrium, baryon number nonconservation, and CP violation. The inflation scenario, which is considered to be necessary to explain the uniformity of the Universe, amplifies the importance of baryogenesis, be-cause the large dilution factor makes it unlikely to attribute the baryon number excess of the present Universe to initial conditions. However, implementation of baryogenesis in the inflation scenario is not an obvious matter. There seems to be no evidence that indicates a direct connection between the origins of CP violation observed in the K meson system and that required for baryogenesis.

After more than 30 years since the discovery of CP violation, we are far from understanding its origin. Even if the standard six-quark model succeeds in explaining CP violation in the B meson system, as well as in the K meson system, it is clearly not the end of the story. CP violation will continue to be a challenging problem.

Notes

1 Z. Maki, M. Nakagawa, Y. Ohnuki and S. Sakata, "A Unified Model for Elementary Particles," *Prog. Theor. Phys. 23* (1960), pp. 1174–80. A. Gamba, R. E. Marshak and S. Okubo, "On a Symmetry in Weak Interactions," *Natl. Acad. Sci. USA 45* (1959), pp. 881–85; Y. Yamaguchi, "A Composite Theory of Elementary Particles," *Prog. Theor. Phys. Supp. 11* (1959), pp. 1–36.

2 Y. Katayama, K. Matumoto, S. Tanaka, and E. Yamada, "Possible Unified Models of Elementary Particles with Two Neutrinos," *Prog. Theor. Phys. 28* (1962), pp. 675–89; Z. Maki, M. Nakagawa, and S. Sakata, "Remarks on the Unified Model of Elementary Particles," *Prog. Theor. Phys. 28* (1962), pp. 870–80.

3 S. L. Glashow, J. Iliopoulos, and L. Maiani, "Weak Interactions with Lepton–Hadron Symmetry," *Phys. Rev. D2* (1970), pp. 1285–92.

4 K. Niu, E. Mikumo, and Y. Maeda, "A Possible Decay in Flight of a New Type Particle," *Prog. Theor. Phys. 46* (1971), pp. 1644–46.

5 See S. Ogawa, "The Sakata Model and Its Succeeding Development toward the Age of New Flavours," *Prog. Theor. Phys. Supp. No. 85* (1985), pp. 52–60.

6 S. L. Glashow, "Partial-Symmetry of Weak Interactions," *Nucl. Phys. 22* (1961), pp. 579–88; S. Weinberg, "A Model of Leptons," *Phys. Rev. Lett. 19* (1967), pp. 1264–66; A. Salam, "Weak and Electromagnetic Interactions," in N. Svartholm, ed., *Elementary Particle Theory, Proc. of the Eighth Nobel Symposium* (Stockholm: Almquist and Wiksell, 1968), pp. 367–77.

7 M. Kobayashi and T. Maskawa, "CP-Violation in the Renormalizable Theory of Weak Interaction," *Prog. Theor. Phys. 49* (1973), pp. 652–57.

8 S. Pakvasa and H. Sugawara, "CP Violation in the Six-Quark Model," *Phys. Rev. D14* (1976), pp. 305–08; L. Maiani, "CP Violation in Purely Lefthanded Weak Interactions," *Phys. Lett. 62B* (1976), pp. 183–86; J. Ellis, M. K. Gaillard, and D. V. Nanopoulos, "Left-Handed Currents and CP Violation," *Nucl. Phys. B109* (1976), pp. 213–43.

9 L. Wolfenstein, "Violation of CP Invariance and the Possibility of Very Weak Interactions," *Phys. Rev. Lett. 13* (1964), pp. 562–64.

10 F. J. Gilman and M. B. Wise "The $\Delta I = 1/2$ Rule and Violation of CP in the Six-Quark Model," *Phys. Lett. 83B* (1979), pp. 83-86.

11 I. I. Bigi and A. I. Sanda, "Note on the Observability of CP Violations in B Decays," *Nucl. Phys. B193* (1981), pp. 85–108.

12 R. D. Peccei and H. R. Quinn, "CP Conservation in the Presence of Pseudoparticles," *Phys. Rev. Lett. 38* (1977), pp. 1440–43; "Constraints Imposed by CP Conservation in the Presence of Pseudoparticles," *Phys. Rev. D16* (1977), pp. 1791–97.

13 A. D. Sakharov, "Violation of CP Invariance, C Asymmetry, and Baryon Asymmetry of the Universe," *JETP Lett. 5* (1967), pp. 24–27. M. Yoshimura, "Unified Gauge Theories and the Baryon Number of the Universe," *Phys. Rev. Lett. 41* (1978), pp. 281–84; *Phys. Rev. Lett. 42* (1979), p. 746(E). A. Yu. Ignatiev, N. V. Krasnikov, V. A. Kuzmin, and A. N. Tavkhelidze, "Universal CP-Noninvariant Superweak Interaction and Baryon Asymmetry of the Universe," *Phys. Lett. 76B* (1978), pp. 436–38.

Part three

Toward Gauge Theories

9

The Path to Renormalizability

MARTINUS VELTMAN

Born Waalwijk, The Netherlands, 1931; Ph.D., 1963 (physics), University of Utrecht; Professor of Physics at the University of Michigan; high-energy physics (theory).

This is the history of the proof of renormalizability of gauge theories as I perceive it. It is a personal account.

The importance of the proof of renormalizability is well known to all. Personally I have always felt that the proof was much more important than the actual construction of a model, the Standard Model. I felt that, once you knew the recipe, the road to a realistic description of Nature would be a matter of time and experiment. There some may disagree with me; I think, however, that a careful study of the recent history of high-energy physics will lead to this conclusion. Seldom has there been such a clear watershed. Old models, truly "dormant" (as Steven Weinberg put it), became credible and popular. Quantum chromodynamics came into being almost overnight. The proof of renormalizability also provided detailed technical methods such as, for example, suitable regularization methods, next to indispensable for any practical application of the theory. In longer perspective, the developments in supersymmetry and supergravity have been stimulated and enhanced by the renewed respectability of renormalizable field theory (including the absence of anomalies). If anything "turned the wheel," as SLAC people have put it, it is this proof of renormalizability. Of course, the theory needs experimental verification, and whether people were convinced after the discovery of neutral currents, or after the discovery of charm, or W and Z, is another matter.

Whatever one may argue, there can be little disagreement on the importance of renormalizability with respect to quantitative understanding. Radiative corrections such as, for example, the vector boson mass shifts can and have been computed and measured.[1] The stunning agreement between theory and experiment reinforces the belief in our theoretical insights and prepares the way for further progress. Thus very importantly, the detailed quantitative understanding of the theory has influenced the direction of present-day research; for example, the realization that radiative corrections to W-pair production may provide essential information on the Higgs sector has pushed the design energy of LEP to well over the W-pair threshold. The design and construction of very-high-energy hadron colliders owes much to the concept of a second threshold, related to the very existence of the Higgs sector. Progress comes from a full and detailed understanding, and for that renormalizability is essential.

Technical introduction

A very abbreviated technical introduction may be helpful, and if nothing else, serve as a background for the discussion to come.

In present-day field theory the starting point is a Lagrangian. This Lagrangian, somehow, defines an S-matrix. The square of the absolute values of the S-matrix elements are the link to physical reality: they are transition probabilities. A transition probability specifies the chances of observing a certain configuration at time plus infinity given a certain configuration at minus infinity. These configurations are systems of particles, supposedly so far removed from each other that they can be taken to be free particles. In other words, S-matrix elements, when squared, describe the transition probability of a system of free particles to collide and emerge as another system of free particles.

The key word here is "free." A free particle is one whose energy is kinetic only; that is, it can be computed if the momentum is known: $E^2 = \vec{p}^2 + m^2$. A particle for which energy and momentum satisfy this relationship is said to be "on the mass shell." The S-matrix refers to initial and final states whose particles are all on the mass shell.

Very roughly speaking, Green's functions are S-matrix elements extrapolated to off-mass-shell values of energy and momentum for the incoming and final particles. By themselves, Green's functions have no special physical significance. All physics is contained in the (on-mass shell) S-matrix. However, the renormalization procedure needs Green's functions with certain requirements of smoothness with respect to the extrapolation mentioned.

Because the S-matrix really contains the relevant physics, it must satisfy a number of basic requirements. It must be Lorentz invariant, and it must conserve probability, to name two important requirements.

In practice, and in any case in the context of this overview, the important property is unitarity, implying conservation of probability.

The relation between Lagrangian and S-matrix is somewhat nebulous, to say the least. There are two formalisms that span the bridge from Lagrangian to S-matrix. Certain desirable properties of the S-matrix may be guaranteed by the formalism. These two formalisms are the canonical operator formalism and the path-integral formalism.

The derivation using the operator formalism guarantees unitarity of the S-matrix, but there are difficulties with Lorentz invariance. The path-integral formalism has its own troubles, one of them being that the interpretation and relation with physics is somewhat mystical. The most important shortcoming is that the formalism has no guarantee with respect to unitarity. Therefore, within that formalism, the proof of unitarity must be specified separately, something that in any case was taken seriously by Feynman, the inventor of the path-integral method.

Both formalisms need completely ad hoc modifications in the case of gauge theories, that is, if the Lagrangian obeys a gauge invariance. The path-integral formalism can be modified in a rather elegant way, but the operator formalism remains rather ugly.

The result of these derivations is in terms of Feynman rules. These rules allow one to calculate S-matrix elements. As a matter of fact, these rules define Green's functions as well. Green's functions generally contain infinities, to be removed by means of some subtraction procedure (renormalization). This requires some scheme for handling those infinities (regularization), and today most physicists use the scheme of dimensional regularization. It is invoked at the very last stage of perturbation theory. Interestingly, that scheme cannot be formulated within the operator or path-integral formalisms. That makes the formalisms appear even more artificial.

As it happens, for a given Lagrangian, the S-matrix is generally unique (insisting on unitarity, etc.), but the extrapolation to Green's functions is not. In other words, for a given Lagrangian there may be different sets of Feynman rules, defining different Green's functions but always the same (unitary) S-matrix. Since the renormalization procedure needs Green's functions, one may be able to carry the renormalization procedure through in some sets of Feynman rules, but not in others. Unitarity remains an important constraint; today's prescription for gauge choosing is such that unitarity is satisfied.

If one is not aware of the possibility of different Green's functions for a given Lagrangian, then it is quite possible to "prove" that some theory is nonrenormalizable if one happens to work with an unfortunate set of Green's functions.

The "proof of renormalizability" amounts then to the following:

- Understanding fully the relationship between Lagrangian and Feynman rules, or rather possible sets of Feynman rules. These rules must be such that the resulting S-matrix is unitary.
- Identifying those Lagrangians that have at least one set of renormalizable Feynman rules.

Certain theories, seemingly renormalizable, turn out not to be, owing to anomalies. Such theories are to be avoided, as indeed Nature seems to do. The Standard Model has no anomalies, or so we think. Knowing about anomalies is important when constructing models. A crucial instrument here is the regulator method. The dimensional regularization method provides us with knowledge as to which theories have anomalies and where they might be. As such, one might consider the construction of a regulator method as part of the proof of renormalizability.

Lacking a good regulator method, one needs Ward identities and must try to renormalize such that the symmetries are not broken in the process, that is, such that the Ward identities remain valid. That is one reason why Ward identities played a much larger role in the beginning. Renormalization is then very complicated, especially if anomalies may occur.

Modern physicists are usually not overly concerned about the apparent lack of any complete and consistent formalism leading from Lagrangian to the S-matrix. As long as the Feynman rules are known, and the resulting S-matrix can be shown to be satisfactory, one tends to say, with Alfred E. Neuman: "What, me worry?"

Old views

It is necessary to recall some of the notions of field theory and renormalizability as understood in the fifties.

Field theory itself was understood in terms of canonical field theory involving operators, interaction Hamiltonians, and the like. Thus, unitarity was something that you understood from the formal expression for the S-matrix in terms of that interaction Hamiltonian. Studying field theory meant manipulating operators and applying subsidiary conditions to state vectors. Of course, no one understood fully the question of gauge invariance in quantum electrodynamics. The Gupta–Bleuler method is a very partial solution. The Stueckelberg formalism enjoyed some popularity in studying massive electrodynamics, and there were some attempts to generalize it to the non-Abelian case.[2] There existed some quaint concepts, such as the path-integral formalism, and there were things such as functional methods; these were known to some students of Schwinger and a very small set of mathematically oriented people. Gauge invariance was not understood within these frameworks either. Mostly physicists thought that gauge invariance meant that you could arbitrarily change the longitudinal part of the vector boson propagator.

The first thing is that people were generally not that well aware that renormalizability required Green's functions rather than the S-matrix, and that there is quite some arbitrariness in Green's functions for a given theory. That has nothing to do with gauge invariance; a simple canonical transformation of fields may turn a perfectly reasonable set of Feynman rules into an unrenormalizable mess. Let me emphasize: unrenormalizable. An example of that is a gauge theory in the physical (or unitary) gauge. That is an unrenormalizable theory. Even if you subtract the known (that is, known from the renormalizable version) infinities, you do not wind up with a finite theory. Green's functions have infinities all over the place. Only when you pass to the S-matrix do these infinities go away, assuming that your regularization method is quite perfect.

That arbitrariness was not generally understood. Canonical transformations were not part of the game yet. It was thought that once you demonstrated some bad infinity in some Green's function, then that constituted proof of nonrenormalizability. I will quote proof of this statement, by considering published work by Komar and Salam, and Glashow and Iliopoulos.[3] In fact, since 1948 the prevalent opinion was that charged massive vector-boson theories were hopelessly divergent.

Second, there was in people's minds a one-to-one relationship between infinities and physical quantities. Thus, a certain infinity related to the electron mass, another to the electric charge of the electron. That notion, in a subtle way, is not true in gauge theories. The relation is more abstract, and must be formulated differently. A theory has a certain number of free parameters, and an equal number of data points is needed to fix these parameters (to generally infinite values). The relation between infinities and parameters is more complex. A good example is the weak mixing angle. There is no clear-cut infinity related to that. The old concept of "bare" mixing angle versus "dressed" mixing angle is not appropriate. This point is not that relevant for the question of renormalizability, but I just take the opportunity to mention it. Infinities have lost their "physical" meaning. Not everybody has realized this. But you see, if you believe that infinities have something physical, then you need to find only one somewhere to demonstrate that a theory is bad. Certainly, you cannot gauge them away.

Unitarity

My first enterprise in field theory was the work for my predoctoral thesis (in Dutch, "scriptie"). The subject was the Lorentz condition, Gupta–Bleuler method, and so on. I do remember thinking that the choice of gauge seemed so much more limited in quantum theory as compared to classical theory. The work itself did not contain anything original, it was a review, as is generally the case with this type of thing.

After that, and military service, I started work for a doctoral thesis under the guidance of Leon Van Hove, then professor of theoretical physics in Utrecht. I remember that the question of renormalizability of weak interactions came to the forefront as a consequence of the $V-A$ theory. That theory revived the intermediate vector-boson hypothesis, and many theorists at that time tried to prove renormalizability, without success. A published example is an article by Glashow, claiming renormalizability.[4]

At a Scottish summer school in 1960,[5] without fully realizing it, I learned something very important from people such as Chew and Jackson: you do not have to know a Hamiltonian to do S-matrix theory. It is quite possible to study the S-matrix all by itself, and then the important thing is that you establish causality and unitarity (but how?). At that school, inspired by Chew's lecture, and after discussions with Derek Robinson, I decided that I wanted to investigate unitarity in a theory containing unstable particles. I remember, in fact, that I had the mistaken idea that something in Chew's vision was wrong there. Often, you start from something wrong. I discussed this with Van Hove, who thought that this could be an appropriate subject for a thesis.

In the end (1961), the work on unitarity of the S-matrix and unstable particles worked out very well.[6] There was already some work by Cutkosky, and that was usually quoted in this context.[7] My proof of unitarity was, I think, simple and elegant. I learned a lot from it. First, I did not really worry about interaction Hamiltonians, but started essentially directly from the S-matrix, defined in terms of diagrams. Then I proved unitarity, and also some form of causality (more or less following the formulation of Bogoliubov) for this S-matrix. Unitarity and causality of the S-matrix as functions of the Feynman rules became transparent to me. Bye-bye, operator formalism.

I want to pause for a moment here to emphasize these points, as I consider them crucial. One may consider a theory from the point of view of the canonical formalism, or from the point of view of path integrals, or, as I have done since then, just from the point of view of diagrams. For gauge theories it is very difficult to derive Feynman rules using the canonical formalism; in the path-integral formulation one is unsure about unitarity, not to speak of difficulties with fermions. To treat unstable particles correctly one must make a partial summation of perturbation theory, and that cannot be done easily in the canonical or path-integral formalism. In my world, looking directly at diagrams, these problems simply did not exist. Mind you, certain things are easier in the canonical formalism or with path integrals. To me, however, they are merely convenient heuristic tools to guess properties of diagrams.

Feynman, in his famous talk in Poland, also studied unitarity, trying to understand it directly from the Feynman rules.[8] He succeeded up to one loop. I did not see that paper till 1968; more about it later.

Going to CERN in 1961, I completed my thesis with a study of Coulomb corrections to W-production (a Dutch thesis tends to be a weighty affair).[9]

From that moment on till 1966 I involved myself mostly with phenomenological things. The only field-theoretical work that I would like to quote from that time are the works of Lee and Yang, and Lee on Feynman rules for vector bosons and the ξ-limiting formalism.[10] Anyone interested in understanding the state of affairs concerning Feynman rules for vector bosons should consult these papers. They were quite complicated. The first version of my algebraic program Schoonschip (December 1963) was actually aimed at extending the work of Lee on charged vector-boson interactions with photons. I computed symbolically the triangle diagram, with as free parameter the vector-boson magnetic moment, which is, unlike the electron magnetic moment, not fixed by the principle of minimal electromagnetic interactions (replacement of ∂_μ by $D_\mu = \partial_\mu - ieA_\mu$).

At the end, toying around with this, I established for which value of the magnetic moment the divergences would be minimal. The outcome was the value suggested by a Yang–Mills theory; that is, the W-photon vertex became the very familiar Yang–Mills vertex.[11] I did not know

Yang–Mills theory at the time, but I certainly remembered that vertex, and recognizing that was a factor in my later decision to study renormalization of Yang–Mills theories. Nothing of that work was published, but Schoonschip became an important tool.

Divergence conditions

In 1966 Gell-Mann's current algebra and Schwinger terms were the hot topic of the day, and I decided to try to understand the issue.[12] I succeeded in that, at least as far as Schwinger terms were concerned. The result of this work was a set of equations, simple extensions of the CVC (conserved vector current) and PCAC equations.[13] The equations were extended to include higher order weak and electromagnetic effects by means of the replacements ∂_μ by $\partial_\mu - W_\mu \times$ in those equations. The equations came about in a two-step process: first the minimal electromagnetic replacement, and then the extension to include vector bosons along the lines of the CVC hypothesis itself. I did not really understand what I was doing, but at least I could derive the Adler–Weisberger relation from the divergence equations without Schwinger-term ambiguities.[14] I felt quite happy about that.

Then I went to the 1966 high-energy physics conference in Berkeley and was told that Feynman had worked on the issue of Schwinger terms. Naturally I looked up Feynman and we spent an amusing hour or so discussing this point. He opened his notebook, and it turned out that he had done much the same thing as I had. My work was in fact already published, or at least on the way; Feynman never published his results. I think that he was busy working further toward Yang–Mills theories, but I am not sure, and I forgot to ask about that later.

In my view, my divergence equations resolved the issue of Schwinger terms. It was never much recognized as such; other people eventually did more or less the same and were more successful in advocating their views. I do not hold that to be very important at this point. The main issue, as far as I was concerned, is that these Schwinger terms were the result of fancy operator manipulations, not really of much physical interest. More importantly, the divergence equations, as I called the extensions of CVC and PCAC, were things that you could apply directly to diagrams. You did not have to know about operators.

As an amusing anecdote I may mention that some time later, in Utrecht, Ward visited me in my office. He sternly told me that my divergence equations were really Ward identities, and then he left again, leaving me bewildered in my office. He was right, of course, but I did not know what he was talking about.

At that time I left CERN for Utrecht. Preparing for a talk at the Royal Society in London, November 1966, I tried to make it transparent how divergence conditions could be used to derive interesting results.[15] In the process I discovered that neutral pion decay into photons was forbidden, and that was reported in London. I thought the proof entirely trivial and did not report details. I actually tried to change PCAC so as to repair this forbiddenness, and at first introduced the same term as known now, deducing its coefficient from the observed decay rate. I remember being astonished at it being a remarkable multiple of the fine-structure constant. I rejected the term, because it did not cure η-decay in three pions, also forbidden, as demonstrated before by Sutherland.[16] I cooked up some other term and reported that in London. I gave one seminar about the above, at SLAC, but most likely nobody remembers. I do remember discussing things with Treiman at that occasion.

John Bell, in the audience in London, picked up this remark on pion decay, and going back to CERN he tried to understand it. I will not go into details here; he thought that I was wrong, we had a correspondence and several telephone conversations, but somehow he did not accept my argument. Instead, with the help of Sutherland, a proof using current commutators came about.[17] As you may know, this led ultimately to the discovery of the anomaly.[18] Adler followed some other road, also discovering the anomaly.[19] My argument, incidentally, is the same as that found in Adler's paper.

Adler–Weisberger relations and gauge invariance

In 1966, as a reaction to my paper on divergence conditions, John Bell presented a more formal derivation of the divergence equations.[20] Gauge transformations were the starting point of his considerations. In a very transparent way this made it clear to me that the successes of current algebra, notably the Adler–Weisberger relation, must be considered a consequence of gauge invariance. In the spring of 1968, while spending a month at Rockefeller University, I tried to think it through. Because of the success of the Adler–Weisberger relation, I considered it an experimentally proven fact that currents satisfy divergence conditions (or current-commutation rules, if you prefer). Divergence conditions follow from gauge invariance. Why would Nature choose its currents so that they satisfied divergence conditions (or current-commutation relations)? Since I firmly believed in vector bosons, the question then was: why would Nature couple vector bosons according to the rules of a gauge symmetry? In light of the old vector-boson magnetic moment question, the (to me) obvious answer was: because that makes the theory renormalizable. In other words, I interpreted the Adler–Weisberger relation as experimental evidence for the renormalizability of a vector-boson theory of weak interactions. I have put this argument several times in print.[21]

Actually, Gell-Mann's commutation rules were, as far as I understand, constructed according to the recipe: start with a Yang–Mills theory of vector mesons, then take away the vector bosons. What is left are currents satisfying certain current-commutation rules. One could say that I thus reinstated the vector bosons, after experiment verified the correctness of the commutation rules.

This was my physics reason for entering the field of Yang–Mills theories. I repeated that argument to myself an untold number of times, with invariably the same conclusion: weak interactions must be some renormalizable Yang–Mills theory. It has to be.

Massive Yang–Mills

Thus I started to study vector bosons interacting with fermions, with in addition the vector boson self-couplings as specified by a Yang–Mills type of theory. More specifically, I started out with electrons and neutrinos. Things became very complicated very quickly and in no time I collected large amounts of diagrams, with an untransparent number of canceling divergences. Thus I started simplifying, and as a first step I threw away the fermions, considering just the vector-boson interactions. Moreover, I decided not to worry about which gauge symmetry would be appropriate, but just took the simplest one, namely SU(2). I kept, however, a finite vector boson mass, even if the mass term violated the Yang–Mills symmetry. Certainly, the vector bosons were not massless, but my main motivation was that massless vector bosons had other unrelated problems, namely infrared divergences.

At that time I became a specialist in shelving problems: shelve fermions, shelve the symmetry group, shelve the neutral-current problem, in fact shelve all attempts at phenomenology. Renormalizability is the problem. If that is understood, the rest will follow. Later on I also shelved the anomaly problem, which I took as a great danger with respect to renormalizability.

The theory, at least one-loop diagrams, became more or less manageable. There were a lot of cancellations of infinities. I noted that there were many more cancellations if the external lines were kept on the mass shell. Trying to understand the cancellations in general became again hopelessly confusing. I therefore decided that somehow I had to transform the theory, changing the Feynman rules in such a way that the new rules incorporated the cancellations. Clearly, gauge invariance had a lot to do with the cancellations. But how to translate gauge invariance? It occurred to me that a change of gauge would perhaps be translatable into a change of Feynman rules. But how? I just did not know how to do that. Nobody knew.

Then I had an idea. Introduce a free scalar field, not interacting with the vector bosons. Now replace the vector field with some combination of vector field and scalar field; at the same time add vertices such that the scalar field remains a free field. Surely then the physics remains the same. But the Feynman rules for the new theory were different: the propagator for the W-field was replaced by the propagator for the combination, and that combination could be chosen so as to lead to less divergent Feynman rules. The price to be paid were the new vertices, and the new particle entered as a ghost (remember that it was a free particle). That is how ghosts entered into my scheme. I called the technique the free-field technique, and the transformation was named the Bell–Treiman transformation. Neither Bell nor Treiman was responsible.

It was a very crude beginning. But the idea was there, and the technique worked. Let me emphasize what I think was the main idea. I changed the Feynman rules before attempting to prove renormalizability. That was really the new thing. Second, I found a technique for changing the rules, that is, for changing gauge. Needless to say, I was not fully aware of the fact that I had done something new. To be clear, that I had done something new was obvious, because of the results. But I did not care too much about what precisely happened. After all, the new S-matrix was unitary and causal, so who cares where it comes from?

In 1968 I presented these first attempts at a Danish summer school. These notes were not published.[22]

In the autumn of 1968, tickled by the French "revolution," I took a sabbatical at Orsay, returning to Utrecht in August 1969. In Orsay I worked the argument to some more detail, profiting from discussions with Bouchiat, Boulware, and Mandelstam, and published the results.[23] The results were quite astonishing: Yang–Mills theory with an explicit mass term turned out to be one-loop renormalizable.

Perhaps the most important point of this paper is that it destroyed a myth, namely that charged massive vector boson theory is hopelessly divergent. To me, and some others, that stimulated further work and it just made me feel very sure that I was on the right track. You see, if you make some sophisticated reasoning and then things fall into place, there is after that nothing that will stop you. This is a very personal feeling; it is the moment of discovery.

The technical virtues of this paper that made this discovery possible will be discussed further below. They were also essential for the subsequent development of the theory. A barrier had been overcome.

Technical discussion

The Bell–Treiman transformation can briefly be described as follows. A canonical transformation is a change of variables, and physics (i.e., the S-matrix) is invariant for such a transformation (provided the transformation is local). Now let there be given an invariance, or partial invariance. Then make a canonical transformation corresponding to that invariance. Physics (but not Green's functions) must remain the same. That statement allows then derivation of identities among diagrams, and these are of course precisely the identities related to the symmetry of the theory.

Again, the invariance of the S-matrix is a consequence of the fact that it is a canonical transformation. It is a change of variables. That has nothing to do with gauge invariance. Then make a smart choice for this change of variables, thus exploiting gauge invariance.

Thus, a Bell–Treiman transformation is a canonical transformation that looks like a gauge transformation with the gauge parameter replaced by a field. I used in the above-described paper the finite form of the gauge transformations. As is well known from group theory, everything follows from the infinitesimal, and that is what I used in subsequent work.

The transformation is thus also useful if there are symmetry-breaking terms in the Lagrangian. They may be the mass terms, as above, or the gauge-breaking term. In particular, to derive generalized Ward identities (as I called them) it is advantageous to include source terms in the Lagrangian. Such terms are not invariant under the gauge symmetry.

If the field used in the Bell–Treiman transformation was a free field, that is what it remains. That is the free-field technique.

Fradkin and Tyutin extended the formalism to nonlocal transformations (local means depending only on fields located at the same point in space-time as the field being transformed).[24] That is useful if one wants to directly connect different gauges without using Ward identities.

Bell–Treiman transformations of one kind or another are now tools to derive the Ward identities of the theory. Those may be used for various purposes. The BRS transformation is the ultimate sophisticated example, the usage by 't Hooft is on the most elementary level.[25] A direct and ingenious example is the use by Slavnov in deriving the Slavnov–Taylor identities for the massless Yang–Mills theory.[26] His field is not a free field, but satisfies an equation such that also the Faddeev–Popov part of the Lagrangian is invariant. That has some analogy to classical electrodynamics: you chose a subsidiary condition, such as the Lorentz condition, and after that there remains a gauge invariance, namely invariance under gauge transformations with a function whose d'Alembertian is zero.

No one uses the name Bell–Treiman transformations these days. It is one of those things: when you do not know them you are stuck (as demonstrated by many), but once you have them you say "of course" (as also demonstrated by many). People usually call them simply gauge transformations, obscuring the fact that they are really canonical transformations, changes of variables, involving fields. Fields are replaced by combinations of fields.

The meaning of the statement "one-loop renormalizable" needs further clarification. To properly renormalize a theory with a symmetry, one needs to regularize in accordance with that symmetry. Here I had several problems. First, the theory did not have, strictly speaking, a symmetry. The mass term broke the Yang–Mills symmetry. Second, I did not have a reasonable regularization scheme. That I found particularly troublesome; I was not sure of the infinities. Now, the infinities that I found were those of a renormalizable theory with respect to power counting – that is, up to quadratic for the self-energies, linear for the three-point vertex, logarithmic for the four-point function. However, the divergences did not, at least in the crude way that I found them, obey the Yang–Mills symmetry. In that somewhat stricter sense the theory was not renormalizable.

I did not really worry very much about that point. Much more troublesome was the fact that I could not obtain similar results for two-loop diagrams. You might think, why should that be possible? The reason is that I had, or so I thought, an excellent argument. Here I have to digress a moment, to massless Yang–Mills theory.

Massless theory

Rather early in this development someone (I truly do not remember who; it was a one-day visitor to Rockefeller University) mentioned to me that Feynman had done something in this field.[27] Eventually I found out about this work, that is, the Polish lecture, and in addition discovered other relevant work on the subject. In Feynman's case, the subject was formally gravitation, but in actual fact the article contained also a discussion on Yang–Mills theory. Feynman's paper is not understandable if you do not already know the answer, but at least he made a clear statement: ghost loops were needed. He could do it only for one-loop diagrams. That was partly due to his way of understanding unitarity, and partly, I think, because he really did not study the massless case but rather the massive case in the limit of zero mass. I find it hard to tell whether Feynman obtained the correct one-loop rules for either gravity or massless Yang–Mills fields. There are not enough details. But I think he obtained those of the massive case.

The main consequence of Feynman's article is that it inspired a few physicists to study the question more precisely. Bryce DeWitt made a monumental effort and established the correct rules for gravitation in the Feynman gauge (no momenta in the numerator of the propagator).[28] To be frank, I am not entirely sure, because his papers are very complicated and I have not really digested them. But the rules seem to be there, with the correct ghost. The correct ghost for this case is a ghost with an arrow. The arrow is not mentioned by Feynman.

Faddeev and Popov, starting with whatever they saw in Feynman's work, found the ghost rules using the now-familiar argument involving path integrals.[29] They found it for one particular gauge, the Landau gauge, and their argument leaves the question of unitarity unanswered. Since the ghost is really there for reasons of unitarity, that is an important question. The Faddeev–Popov procedure amounts to gauge breaking in a certain way, and it is nontrivial that the way chosen provides for a unitary result. Other methods are needed to establish that.

I was editor of *Physics Letters* (1966–1968) when the Faddeev–Popov article arrived. Of course, it was totally incomprehensible to me, being about path integrals. But I felt that Faddeev was man enough to be responsible for his own work, and I accepted it without further ado.

Another physicist working his way through the massless Yang–Mills theory was Mandelstam.[30] He used his own formalism, and his results agreed with those of DeWitt and Faddeev and Popov. In other words, at this time (June 1968) the rules for the massless case were known, at least for some specified choices of gauge. The rules were those of a renormalizable theory, at least by power counting. Whether the infinities obeyed the symmetry was not clear. On top of this there were of course the infrared problems.

Now here is the "excellent" argument mentioned above. The non-renormalizability of the massive Yang–Mills theory relative to the massless theory is a direct consequence of the form of the vector-boson propagator. In the massive case there is the extra term $p_\mu p_\nu / M^2$ in the numerator of the propagator. This term causes all the problems. It is simply absent in the massless case. Now, let us assume that the limit of zero mass exists for the massive case. After all, it is only one term in the Lagrangian, not appearing very menacing. If the limit of zero mass exists, then obviously the extra term in the propagator must behave reasonably. This means that the product of the two momenta must be equivalent to something that behaves as M^2.

Now why could I not get through for the two-loop diagrams? I could not understand that. There were terms blowing up in the limit of zero mass. But how could that be? I assumed that it might have something to do with the perturbation expansion, and I mentioned that in an appendix.

Reactions

There were two immediate reactions to my article. The first was from Salam, who stated that "somewhat similar work" had been done by Komar and himself.[31] I looked it up and found that they had proven that massive Yang–Mills theory was not renormalizable. The proof was a calculation of the three vertex, in the one-loop approximation, in the unitary gauge. Indeed, that is nonrenormalizable. Let me, however, mention that Salam was perhaps the only one who realized the fact that I first transformed the rules and only then considered renormalizability. He appreciated this progress, and he has stated that to me at some occasion. In fact, if you read his Nobel Symposium article, you see that he and Ward (and presumably before him Higgs and Kibble) were on the way.[32] As he put it, he did not have the dictionary to go from one gauge to the other. The missing part is the idea that Green's functions may change, as long as the S-matrix remains the same. The dictionary works only on mass shell. That is what he always told me: "Ah, Veltman, the mass shell." I usually felt hollow after that remark. I did not read this lecture until much later. But frankly, who had read this before 1971? The proceedings of the Nobel Symposium is not a particularly popular channel of communication.

The other reaction was a letter from David Boulware. He succeeded in summing the series that I mentioned in my appendix, with the result that it was still singular in the limit of zero mass. He added a comment stating that he did not know how serious this was. Neither did I.

Subsequently Boulware, working within the path-integral formalism, confirmed my results up to one loop.[33] His formulation was general, not restricted to one loop; his result for two or more loops was that the theory was not renormalizable. The essential technique is the path-integral version of what I called the free-field technique. I looked upon this with some suspicion, largely because of my unfamiliarity with the path-integral formalism, and also because within that formalism unitarity is not obvious; this suspicion was here unjustified. It was a good paper that very probably influenced Russian authors.

The zero-mass limit

I now set out to look for gaps in the argument. There was the question of the Feynman rules themselves, more precisely the relation between the Lagrangian and Feynman rules in the canonical formalism. This rather old problem concerns the handling of derivatives in the Lagrangian; vector boson Lagrangians have many derivatives. It was first tackled in connection with the case of pion–nucleon interactions, axial-vector coupling, by Matthews.[34] As cited above, Lee and Yang also considered this problem.[35] Actually, the difficulties encountered when following the usual canonical procedure are rather similar to Schwinger-term problems. We (this work was done with J. Reiff, then a graduate student) discovered a very simple way of settling this problem, leaving no doubt as to the correctness of the Feynman rules in the unitary gauge.[36]

Next, we started on the two-loop problem. We investigated the two-loop self-energy diagrams, using the rules that were valid at the one-loop level. We established that the result definitely violated unitarity. In other words, it really seemed that the limit of zero mass did not result in the massless theory. This was reported first at a conference at CERN, in January 1969.[37] It was also clear that an extra vertex had to be added at the two-loop level; this vertex was of a nonrenormalizable type. It was now definitely clear that the limit of zero mass was not the massless theory, as the extra term contained a factor $1/M$. It left me confused for quite some time. Clearly, I needed more understanding of the unitarity problem and the rules at the two-loop level.

In 1969 a visitor and old friend of mine, H. Van Dam from North Carolina, lectured on Schwinger's source formalism. In addition he imported a certain amount of enthusiasm for gravity. Both were crucial to the next development, as will become clear. Let me first, however, describe what happened with respect to the zero-mass limit problem.

By this time the Russians were entering the arena. I will enlarge upon that shortly, but for now I would like to mention the work of Slavnov and Faddeev, following up on the work of Boulware; their presentation of the essentials made the work more transparent, at least to me.[38] They refer to an unpublished article by Slavnov (of which I have a preprint) written in response to mine. It contains a treatment of the massive case, but the conclusion, namely that the massive Yang–Mills theory was renormalizable, is wrong.

Treating massive and massless theories on the one-loop level, Faddeev and Slavnov noted that there was a factor of 2 difference between the contribution of a ghost one-loop diagram in the massive case as compared with the massless case. I do remember discussions on that issue in Orsay, involving Boulware and Mandelstam. But I think none of us noted this factor. To me the observation came through a discussion with Bruno Zumino, quoting Faddeev and Slavnov.

Thus there was now also an explicit difference between the massive and massless case at the one-loop level. The factor of 2 relates to the fact that the ghost of the massive case has no arrow, while the massless ghost does. Proper symmetrization requires a factor of $\frac{1}{2}$ for the massive case.

For the experts: the rules for the massive case were much as we have them today. There are now two ghosts, the Faddeev–Popov ghost with a factor -1, and a Higgs ghost with a factor $+\frac{1}{2}$. They combine to what I had for the massive case, a diagram with a factor $-\frac{1}{2}$. To some extent that reasoning is in the Slavnov–Faddeev article.

At the one-loop level, by this time, things were transparent. It now was a matter of carefully analyzing the situation, and it became clear that indeed, the massless case is not the limit of the massive case, simply because the longitudinal polarization of the massive vector boson is not decoupling in the limit of zero mass. It sounds trivial at this time, but it really was a shocking thing. How shocking was clearly demonstrated in a paper that Van Dam and I published.[39] It so happens that a similar phenomenon occurs for gravitation, and we demonstrated that the limit of zero-mass gravity is not the same as zero-mass gravity. The difference is non-zero, already at the tree level. From the observed deflection of light by the sun and the perihelion movement of Mercury, one can then deduce that gravitation is strictly zero mass. No mass, however small, is allowed. Gravitation is not limited in range, not even at distances on the galactic scale or beyond. Many people refused to accept this result; a learned colleague in the Netherlands put it in Latin: *"Natura non facit saltum."*

This insight cheered me up. While there was clearly trouble at the two-loop level, at least the situation had become clear. And mind you, having already one-loop renormalizability is not bad. In fact, this is all you need to cover the experimental situation of today. But it was the principle that I was after.

Ward identities

The problem of unitarity of a Yang–Mills theory is more complex than in non-gauge theories. In a renormalizable gauge the Feynman rules are not manifestly unitary; there are ghosts and one must show that the contributions of the ghosts vanish. This requires the use of Ward identities. The problems that presented themselves at this stage were these:

- What are the precise rules for two or more loops?
- What are the Ward identities?

For quite some time I did not know how to handle that. Then I started to use sources, undoubtedly inspired by Van Dam's lectures on Schwinger's formalism, although I thought that I did it by myself. It often goes that way. This turned out to be a really big step forward. The derivation of Ward identities became easy; these Ward identities for the non-Abelian off-mass-shell case, later called Slavnov–Taylor identities, could then be used to work out two-loop diagrams.[40] (To avoid misunderstanding: Slavnov and Taylor derived these identities for the massless theory, which is not what I was dealing with at that time.) This removed all doubts concerning unitarity and renormalizability. The paper in which these results were published was actually written before

I understood the zero-mass limit problem.[41] In August 1970 I fully understood the massless and massive case. What now?

In the autumn of 1970 I was pondering these results. I developed the concept of a cancelable divergence: a divergence that can be canceled by a physical particle with legitimate interactions, as compared to a non-cancelable divergence that needs a particle of indefinite metric, or wrong statistics, and so on. Then I decided to somehow subtract the massless case (but with a mass inserted in the denominator of the propagator) from the massive case and see if the resulting divergences could be canceled by a physical interaction. This of course is not legitimate history, since I cannot prove that I was drifting along these lines. It is possible that I would have discovered the Higgs particle this way. It is also possible that I would have remembered Glashow's remark (see below) and tried spontaneous symmetry breakdown. All that is irrelevant. At this time 't Hooft entered in the field, and things resolved themselves in another way.

Other work

It is first necessary to paint a picture of mainstream physics in the period 1967–1971. The Adler–Weisberger relations marked the last successes of field theory; weak interactions appeared more nonrenormalizable than ever; and people turned to other methods. Most physicists considered the anomaly only as a modification of PCAC. Effective Lagrangians, low-energy theorems, Regge poles were at the center of interest. Popular opinion was that charged vector-boson theories were nonrenormalizable, and the very idea of working on such theories marked you as halfway to insanity. Thus working on Yang–Mills theories was considered far out, and many a remark in that sense came to me. Some persisted in field theory, and their contributions survived and are well known today, but they were few. Not the least among them were those that kept on doing calculations in QED, thereby fortifying the idea of renormalizability.

In the period mentioned I was a regular visitor to the Orsay group (now the theory group at the École Normale in Paris). Every summer they organized a summer institute; Bouchiat and Meyer, and later also Iliopoulos were the hosts. Regular visitors included Coleman and Glashow, and I attended every summer. Every year I reported the latest on the subject of Yang–Mills theories. After a few years Coleman once expressed his doubts: "Tini, you are just sweeping an odd corner of weak interactions."

However, not all reacted this way. I do remember positive comments from Glashow. First, he apparently read my Copenhagen lectures, notably the part involving Cabibbo matrices and neutral strangeness-changing currents; he said something like "I see that you have also been working on this problem." Also, somewhere in 1969 or 1970 he said to me, "You should try to get masses by means of spontaneous symmetry breaking," to which I answered, "I am not yet ready for that." While his remark kept on spooking through my head, I somehow never did it. Psychologists may see something here.

I firmly believe that my work of 1968 was the stimulus to the work of Glashow, Iliopoulos, and Maiani.[42] There is even a reference to this effect, although that same reference tries to weaken the case. In footnote 12, after referring to various papers, mine among them, they write: "Note however, that none of these references consider the far more difficult case of vector mesons coupled to non-conserved currents."

A few months after the GIM papers, Glashow and Iliopoulos decided to clear up the massive Yang–Mills case.[43] I quote here their footnote 4, referring to my papers (the second reference should have been to *Nucl. Phys. B21*):

> For an extensive discussion of the problem of divergences in massive Yang–Mills theory, see M. Veltman, *Nucl. Phys. B7*, 637 (1968); *B20*, 288 (1970). In the last reference a set of on-mass-shell Ward identities has been obtained which are used to analyze Feynman diagrams. The cancellations of divergences found by M. Veltman go beyond the theorem proven in this paper, but they only apply to on-mass-shell amplitudes and they depend on the assumption that the W's are coupled to conserved currents. For alternative discussions see D. Boulware, *Ann. Phys.* (N.Y.) *56*, 140 (1970), and S. K. Wong, this issue, *Phys. Rev. D3*, 945 (1971).

This footnote shows that they had not understood the fundamental point: you first change gauge, then consider renormalizablity. I quote this mainly to show that this was not a trivial thing, but that it truly prevented people from discovering renormalizability.

Let me comment on, again, this reference to nonconserved currents. That problem is not and never was any more serious than the nonvanishing of the W-mass. If they had applied my technique to the fermion–W coupling they would have seen that; it is quite trivial, and I knew it. They never asked me about it. Today we use the same Higgs to cure both problems. In the fermion sector, in the limit of large Higgs mass, at the one-loop level there is a term logarithmic in the Higgs mass. That is what they would have found had they used my technique: a logarithmic divergence. Precisely like the four-point W vertex. I have used that information later to locate effects that go to infinity as the Higgs mass becomes large, as such effects might be used to deduce the Higgs mass from experiment.[44] The "screening theorem" reflects the fact that all such effects are logarithmic in the Higgs mass.

This is perhaps the moment to cite Weinberg's unpublished attempts at renormalizing his model. This was in his "dormant" period. I quote from his Nobel lecture:[45]

> The next question now was renormalizability. The Feynman rules for Yang–Mills theories ... had been worked out by deWitt, Faddeev and Popov and others, and it was known that such theories are renormalizable. But in 1967 I did not know how to prove that this renormalizability was not spoiled by the spontaneous symmetry breaking. I worked on the problem on and off for several years, partly in collaboration with students, but I made little progress. With hindsight, my main difficulty was that in quantizing the vector fields I adopted a gauge now known as the unitary gauge: this gauge has several wonderful advantages, it exhibits the true particle spectrum of the theory, but it has the disadvantage of making renormalizability totally obscure.

Russian work

Russian authors, notably Fradkin and Tyutin, contributed in an important way to the subject. They, obviously well informed on the subject of path integrals, introduced and extended my techniques in that context. Eventually, Fradkin and Tyutin established a procedure by which one obtains the Feynman rules for a general gauge. The ghost part of the Lagrangian was still written in terms of a determinant, as in the work of Faddeev and Popov.[46] Initially Fradkin and Tyutin thought that massive Yang–Mills theories were renormalizable; I have not tried to trace the mistake.[47] I think that they bought my argument and Slavnov's work on the limit to zero mass. In 1970 Fradkin and Tyutin published a paper applying to massless Yang–Mills theories and gravitation (i.e., theories without any symmetry breaking).[48] Let me quote from their introduction: "The basic idea of the method proposed is to choose the Lagrange multiplier in such a way that one is led to free equations of motion for the additional field." With this work one could write Feynman rules for different choices of gauge. They did it for a number of gauges.

Utrecht

In 1966 I began teaching in Utrecht, and in the period until 1971 several students (J. Reiff, P. Van Nieuwenhuizen, G. 't Hooft, and B. de Wit) started graduate work under my supervision. High-energy physics, at that time, was not a popular subject in the Netherlands, traditionally strong in the field of statistical mechanics. The disadvantage was a certain isolation, the advantage a certain isolation. Starting on Yang–Mills theories in 1968, I found it extremely pleasant that I did not have to defend my aberrant views.

As a general rule I avoided dragging students into the field of Yang–Mills theories. It was too risky. Armed solely with knowledge on that subject, they were at a disadvantage in finding proper employment afterwards, or so I thought. For at least part of their thesis work, I insisted on more phenomenologically oriented work.

Another subject very popular in Utrecht was the sigma model. I always felt that this model, due to Schwinger and employed in the article of Gell-Mann and Levy in relation to PCAC, was of fundamental importance.[49]

In the beginning of 1969 a student, Gerhard 't Hooft, was assigned to me for his predoctoral thesis. For a good understanding, we (my colleagues and I) shared this type of task, and students were more or less distributed among the professors. A student could, however, express his preferences, and when expressing a preference for high-energy physics, as 't Hooft did, they tended to wind up with me. The work that I asked him to consider was the sigma model, axial currents, and anomalies. His predoctoral thesis was completed in 1969.[50] It contained a discussion about PCAC, the sigma model, the anomaly, and renormalization of the sigma model.

In October 1969 he was offered a position at the Institute, in order to enable him to complete a doctoral thesis. He expressed his interest to work under my supervision, in high-energy physics, rather than the alternative, statistical mechanics with his uncle, N. G. Van Kampen.

At that time I felt an urgent need to understand path integrals. The best way to learn is to lecture on the subject, and so I did, first in Orsay, in 1968–1969. Ben Lee was in the audience. I felt that I still needed further education and proposed a course in Utrecht in 1969–1970 on the same subject, in collaboration with Van Kampen, who agreed. We thus lectured on the use of path integrals in statistical mechanics and high-energy physics. 't Hooft was assigned the task of writing the lecture notes. I remember being quite happy about it; at times he improved the derivations considerably.

At some point I had to specify a thesis subject. I do not remember precisely at what time, but I assume it was in the autumn of 1969. As explained before, I was not particularly happy about students going into Yang–Mills theories; I therefore mentioned the "hot" topic of the day, the $A2$ resonance splitting.

Somewhat later he expressed his disdain, rightly so, for the subject of $A2$ splitting. Actually, I certainly had not much sympathy for the subject either. The point then was, what now? We discussed this together in the presence of Van Kampen, and I gave in: if he so wanted, let him have a try at Yang–Mills theory. More specifically, I suggested to him the problem of finding a good regulator method to be used with Yang–Mills theories. That was something for which I felt a real need, and which seemed just the type of thing for a proper initiation into the subject.

In the summer of 1970, like many other European students, 't Hooft went to a summer school. In his case that school was in Cargese. Ben Lee lectured there, on the renormalization of the sigma model, in particular focusing on what happens to infrared problems when spontaneous symmetry breaking occurs.[51] According to 't Hooft, he found there the inspiration to introduce spontaneous symmetry breaking into the pure Yang–Mills theory.[52] However, he first worked on the unbroken theory, and somewhere near the end of 1970 his first article came to my desk: "Renormalization of Massless Yang–Mills Fields."[53]

I do remember quite clearly a number of discussions, but I will not elaborate on them here. Mainly I remember the moment when I, alone in my office, pondered whether this should be published; that is, I tried to weigh what was really new in the paper. In my view these were the truly new elements: a new cutoff method that, however, worked up to one loop only, and also a quite elegant extension of the work by Fradkin and Tyutin on Feynman rules in an arbitrary gauge.[54] The ghost determinant was replaced by explicit ghost Feynman rules.

Actually, I did not know the work of Fradkin and Tyutin in detail at that time. I just knew that the rules were known (also to myself) for a few gauges in the massless case. I liked 't Hooft's derivation a lot. There could be no doubt that this was a nice article.

The cutoff method was a precursor to dimensional regularization. For one loop he introduced a fifth dimension. Dimensional regularization came from the idea that somehow this fifth dimension should be distributed over the loops. I once said to 't Hooft that once you collaborate you forget about who invented what. I am not going to violate my own rules here.

Somewhere in the autumn or winter of 1970–1971 we walked together from one building of the Institute to another. I complained about theories of charged vector bosons. I said something like "All this stuff about massless theories is very nice, but if we only had one renormalizable theory of massive charged vector bosons, no matter how far removed from reality. In any case all possible models exist already." He answered "I can do that." This moment is grafted in my brain, as I almost ran into a tree. I said "What?!" He repeated his statement. I said, "Write it down, we will see." And he did, and we saw.[55]

The moment his second article came under my eyes, I knew that this was it. In fact, I think that he was very surprised at my immediate acceptance. He expected, I think, a lot of arguments about the Higgs mechanism, and we did argue some about it. The fact is that I did, indeed, not like it very much, not then and not now, but at that time I was only interested in the result and could not care less how it came about. Actually, we did not know that it was the Higgs mechanism; to us it was the spontaneous symmetry breaking of the sigma model, as in the articles of Schwinger and Gell-Mann and Levy.[56] In my opinion spontaneous symmetry breaking, at least in this context, owes nothing to the work by Anderson in superconductivity and subsequent developments.[57] This is different for Weinberg's paper.[58]

As a testimony to our ignorance, I remember sitting with 't Hooft and musing that this probably had something to do with Goldstone's theorem. Since neither of us really knew precisely what that meant, and since the theory was obviously correct, we decided to forget about it. My unease with the Higgs mechanism remained throughout, and eventually I realized why I felt that way: it is the problem of the cosmological constant.[59]

At the end I said to 't Hooft: "Now the time has come to construct a realistic model." Then I took the article with me to CERN and verified the whole lot, in particular two-loop unitarity, with the help of Schoonschip. As Jacques Prentki put it: "If it is wrong, you will get the blame; if it is correct, he will get the credit." Furthermore, I asked Zumino to read it, and to provide me with references that he thought to be relevant. He was always my infallible guide to the literature. And so it came to pass that references to Kibble and Weinberg were included.[60] Well, this time Zumino was not so infallible, or else he should have mentioned the work of Englert and Brout, which slightly predates the work of Higgs.[61] These are evidently independent pieces of work. Actually Englert and Brout saw clearly the connection with renormalizability, and they advised Weinberg of that, at the Solvay Conference of 1967.[62] Students of the history of the Standard Model may want to check this little-known reference.

I am reasonably sure that 't Hooft deduced the Weinberg model by himself. The vector-boson part was already in his work when I took it to Geneva. We discussed the lepton part on the telephone. When I informed him of Weinberg's article, his first reaction was that it was wrong; a few days later he said that it was the same as his version. As I have stated before: once you know the rules it is easy. The number of models that were cranked out in the ensuing years testifies to this fact.

I then set out to promote 't Hooft's work, starting at the Amsterdam Conference of 1971.[63] I organized a session, inviting Salam and T. D. Lee to present their views on finiteness of field theory (these were nonpolynomial Lagrangians and unstable particles with negative metric, respectively). After that 't Hooft presented his work.

Admittedly, at that time this gave me some pleasure of a dubious kind. Being chairman, I was up on the podium. Salam was sitting in the first row, looking glassily. Coleman, about ten rows back on the left, looked at least as intelligent. I did not see Glashow; he probably was not interested in all this field-theory stuff. As far as I know, he has never mentioned his presence in Amsterdam. Ben Lee was perhaps the only one who understood what was going on; we talked about it afterwards.

Little did I realize the contributions of Glashow and Salam at that time! Students of the history of the Standard Model may want to check the references that Salam added to his own paper in the proceedings in relation to 't Hooft's talk.

If somebody would have told me then about the 1979 Nobel prize, I would have laughed. Later I got used to the idea. Such is life.

Mopping up

There were a number of loose ends. First, there still was no regularization scheme. Furthermore, renormalization, Ward identities, and the like had to be established on a rigorous base.

The idea of dimensional regularization was first hinted at publicly in the middle of 1971. In the autumn of 1971 I spent most of my time on this subject; 't Hooft, stimulated by Symanzik, became very interested in asymptotic freedom and the massless theory as a model for strong interactions. In fact, he established asymptotic freedom and reported that at the Marseille Conference in June 1972 – a full ten months before the publications of Politzer, Gross, and Wilczek.[64] This is actually quite well known although not published. The communication was in the form of a comment after a lecture by Symanzik. However, the connection with observed physics was not mentioned, and certainly I did not understand the relevance of the affair. Nor do I know the complete history; it appears that there are even earlier Russian papers containing all or parts of the calculation.

There is not much to say about the paper on dimensional regularization, ready and submitted by the end of February 1972.[65] It did take a lot of effort, but it was essentially straightforward. The effort concerns the formulation of the method beyond the one-loop level, and given that there was already a working one-loop method, it was felt that this was essential. It may perhaps be mentioned that dimensional regularization cannot be formulated within the path-integral formalism nor the conventional operator formalism. Therefore a general formulation of the method beyond one loop is not entirely trivial.

I would like to comment here on the subject of competing papers. A number of other papers came out containing the same idea. The idea of continuation in the number of dimensions was known to us since somewhere in the begining of 1971, and we made no particular secret of it.[66] It was most explicitly mentioned at the Orsay conference of January 1972. I did not recognize, and have not in general recognized, papers dated after February, in particular if they limited their treatment to one-loop diagrams. There is one exception: the paper by Bollini and Giambiagi, received 8 February 1972.[67] I received a letter from them (9 March 1972) while correcting some misprints in our article, and I added a reference to their work; they added similarly to theirs. In their article they referred to a preprint that I have never seen and that they did not send me; though showing a received date of October 1971, it was finally published almost a year later (August 1972).[68] Evidently they had a lot of trouble getting the paper accepted. That paper explicitly mentions the idea of dimensional regularization and a few one-loop diagrams are worked out. The motivation was certainly very different from ours, and the essential advantage of the method, that is, the respecting of non-Abelian gauge invariance, was not mentioned (QED gauge invariance was mentioned in their second paper).[69] They did not consider the extension to more than one loop. Their work is clearly independent, even if it is almost unbelievable that an outlandish idea such as dimensional regularization would happen simultaneously in two unrelated instances. Let me add, though, that they had worked in this field before: I believe that they are among the inventors of analytic regularization.[70] For more information see Speer.[71]

At the Orsay Conference 't Hooft and I presented various subjects such as a clear exhibition of general gauge fixing, ghost generation, and dimensional regularization. The matter of anomalies was thrashed out there; 't Hooft thought he had an argument showing that anomalies were harmless in the Weinberg model.[72] Bardeen argued against it, with success; the main argument centered around a diagram with two triangle anomalies, showing clearly that the Weinberg model contains anomalies and is as such nonrenormalizable. The paper of Bouchiat, Iliopoulos, and Meyer shows how to avoid anomalies, by including quarks.[73] It was, I believe, inspired by the arguments at this conference.

The final work, as far as I am concerned, is a paper by 't Hooft and myself containing a formal combinatorial derivation of the Ward identities and a proof of renormalizability.[74] A preliminary version was presented at the Marseille Conference; Bell pointed out that there was a difficulty with respect to external-line renormalization.[75] He referred to an important piece of work by Bialynicki-Birula, which I am happy to acknowledge.[76] It was quite serious criticism, and took some time to correct. At the conference we also presented a very explicit example [pure SU(2) case], with a two-parameter choice of gauge and illustrating the content of the more formal paper.[77] There is a delicate sentence in ref 74, just above section 3.

This, from my perspective, was the road to the proof of renormalizabilty. After this 't Hooft and I collaborated on a few more papers, among them an investigation of the divergencies in gravitation.[78] I believe that that paper as well as the related lectures in Les Houches have had their impact, but that is another history.[79] Later I interested myself in the Higgs mechanism and radiative corrections. After all, measuring and comparing radiative corrections with the predictions of the theory is in my view an indispensable part of the proof of renormalizability. But this is again another chapter of history.

Other authors, notably Ben Lee and Jean Zinn-Justin, have published work after July 1971 that differs from ours in the fact that heavy use is made of path-integral methods.[80] That may have helped acceptance of the formalism; apparently formal path-integral methods are more readily accepted than combinatorial arguments relating to diagrams. Our attempt at popularization is a CERN yellow report entitled "Diagrammar."[81]

The contribution of Becchi, Rouet, and Stora concerning Ward identities should perhaps be mentioned as the final part of the formalism.[82] What remained and partly remains after that are some technical questions concerning the handling of spinors and γ^5.

Assessment

At this moment, after 20 years of experimentation, the Standard Model has been verified, including radiative corrections. That is, effects needing renormalization for their calculation have been verified. The agreement so far is excellent.

An interesting point may now be raised. To what extent has the theory been tested in all its glory and renormalizability? My first paper established one-loop renormalizability; 't Hooft's paper specified fully renormalizable models including at least one extra particle, the Higgs particle. On the phenomenological level, Glashow's paper as compared with that of Weinberg did have masses put by hand rather than generated by the Higgs mechanism.[83] If full renormalizability has indeed been tested, then we ought to have a statement on the Higgs mass.

Well, there is no statement on the Higgs mass. We have no clue to its magnitude from experiment. That is, experiment has verified Glashow's model, using my one-loop renormalizability result. Quantum chromodynamics is a pure Yang–Mills theory. The final word on the renormalizability of that one was established with the advent of dimensional regularization, undoubtedly an indispensable tool in present-day theory. I do not know to how many loops this theory has been established with certainty, but I would say well beyond two loops. That is where we stand. To be complete, another fact that some might interpret as a tie to the Higgs system is that the ρ parameter turns out to be close to 1, experimentally.[84]

The problem of the cosmological constant has further aggravated the Higgs problem.[85] It has motivated me to investigate the theory without a Higgs – that is, essentially the same thing that I started with.[86] At this time there will be few physicists who would bank on actually finding the Higgs particle. Many of us feel that the world is more complicated than that. Even so, the fully renormalizable theories with a Higgs sector provide the framework for parametrizing the present-day situation.

So, all told, the word is still out. However, the psychological effect of a complete proof of renormalizability has been immense. This then is the important point. The proof of renormalizability gives certain theories a certain internal strength that makes them credible. People (at least most) did not go into Yang–Mills theories after Glashow's, Weinberg's, or my paper. The proof of renormalizability provided the necessary psychological impact.

Notes

The date on which the article was received is indicated here in square brackets.

1 F. Antonelli, M. Consoli, and G. Corbó, "One-Loop Correction to Vector Boson Masses in the Glashow–Weinberg–Salam Model of Electromagnetic and Weak Interactions," *Phys. Lett. B91* (1980), pp. 90–4 [11 Jan 1980]. M. Veltman, "Radiative Corrections to Vector Boson Masses," *Phys. Lett. B91* (1980), pp. 95–8 [11 Jan 1980].

2 Selected references to the Stueckelberg formalism: E. C. G. Stueckelberg, "Interaction Forces in Electrodynamics and in the Field Theory of Nuclear Forces," *Helv. Phys. Acta 11* (1938), pp. 299–328 (in German); W. Pauli, "Relativistic Field Theories of Elementary Particles," *Rev. Mod. Phys. 13* (1941), pp. 203–32; Y. Miyamoto, "On the Interaction of the Meson and Nucleon Field in the Super-Many-Time Theory," *Progr. Theor. Phys. 3* (1948), pp. 124–40; F. J. Belinfante, "Quantum Electrodynamics" *Phys. Rev. 75* (1949), p. 1321; F. Coester, "Quantum Electrodynamics with Nonvanishing Photon Mass," *Phys. Rev. 83* (1951), pp. 798–800; R. J. Glauber, "On the Gauge Invariance of the Neutral Vector Meson Theory," *Prog. Theor. Phys. 9* (1953), pp. 295–8; H. Umezawa, *Quantum Field Theory* (Amsterdam: North-Holland, 1956), pp. 113 and 204; E. C. G. Stueckelberg, "Theory of the Radiation of Photons of Small Arbitrary Mass," *Helv. Phys. Acta 30* (1957), pp. 209–35 (in French); A. Fujii, "On the Analogy Between Strong Interaction and Electromagnetic Interaction," *Prog. Theor. Phys. 21* (1959), pp. 232–40; H. Umezawa and S. Kamefuchi, "Equivalence Theorems and Renormalization Problem in Vector Field Theory (The Yang–Mills Field with Non-Vanishing Masses)," *Nucl. Phys. 23* (1961), pp. 399–429; D. Boulware and W. Gilbert, "Connection between Gauge Invariance and Mass," *Phys. Rev. 126* (1962), pp. 1563–7; S. Bonometto, "On Gauge Invariance for a Neutral Massive Vector Field," *Nuovo Cimento 28* (1963), pp. 309–19; J. A. Young and S. A. Bludman, "Electromagnetic Properties of a Charged Vector Meson," *Phys. Rev. 131* (1963), pp. 2326–34; A. Fujii and S. Kamefuchi, "A Generalization of the Stueckelberg Formalism of Vector Meson Fields," *Nuovo Cimento 33* (1964), pp. 1639–56; A. Slavnov, "Renormalization of Gauge Invariant Theories," *Sov. J. Part. and Nucl. 5* (1975), pp. 303–17.

3 A. Komar and A. Salam, "Renormalization Problem for Vector Meson Theories," *Nucl. Phys. 21* (1960), pp. 624–30 [22 Aug 1960]. A. Salam, "Renormalizability of Gauge Theories," *Phys. Rev. 127* (1962), pp. 331–4

[27 Nov 1961]; S. L. Glashow and J. Iliopoulos, "Divergences of Massive Yang–Mills Theories," *Phys. Rev. D3* (1971), pp. 1043–5 [15 Sep 1970].

4 S. L. Glashow, "The Renormalizability of Vector Meson Interactions," *Nucl. Phys. 10* (1959), pp. 107–17 [24 Nov 1958].

5 G. R. Screaton, ed., *Dispersion Relations: Scottish Universities Summer School, 1960* (New York: Interscience Publishers, 1961; and Edinburgh: Oliver and Boyd, 1961), pp. 186–205.

6 M. Veltman, "Unitarity and Causality in a Renormalizable Field Theory with Unstable Particles," *Physica 29* (1963), pp. 186–207 [5 Nov 1962].

7 R. E. Cutkosky, "Singularities and Discontinuities of Feynman Amplitudes," *J. Math. Phys. 1* (1960), pp. 429–33 [31 Mar 1960].

8 R. P. Feynman, "Quantum Theory of Gravitation," *Acta Phys. Pol. 24* (1963), pp. 697–722 [Talk July 1962, received 3 Jul 1963].

9 M. Veltman, "Higher Order Corrections to the Coherent Production of Vector Bosons in the Coulomb Field of a Nucleus," *Physica 29* (1963), pp. 161–85 [24 Oct 1962].

10 T. D. Lee and C. N. Yang, "Theory of Charged Vector Mesons Interacting with the Electromagnetic Field," *Phys. Rev. 128* (1962), pp. 885–98 [29 May 1962]; T. D. Lee, "Application of ξ-Limiting Process to Intermediate Bosons," *Phys. Rev. 128* (1962), pp. 899–910 [29 May 1962].

11 C. N. Yang and R. L. Mills, "Conservation of Isotopic Spin and Isotopic Gauge Invariance," *Phys. Rev. 96* (1954), pp. 191–5 [28 Jun 1954].

12 M. Gell-Mann, "The Symmetry Group of Vector and Axial Vector Currents," *Physics 1* (1964), pp. 63–75 [25 May 1964].

13 M. Veltman, "Divergence Conditions and Sum Rules," *Phys. Rev. Lett. 17* (1966), pp. 553–6 [29 Jul 1966].

14 S. L. Adler, "Calculation of the Axial-Vector Coupling Constant Renormalization of β Decay," *Phys. Rev. Lett. 14* (1965), pp. 1051–5 [17 May 1965]; W. I. Weisberger, "Renormalization of the Weak Axial-Vector Coupling Constant," *Phys. Rev. Lett. 14* (1965), pp. 1047–51 [26 May 1965].

15 M. Veltman, "Theoretical Aspects of High Energy Neutrino Interactions," *Proc. Roy. Soc. A301* (1967), pp. 107–12 [2 Nov 1966].

16 D. Sutherland, "Current Algebra and the Decay $\eta \rightarrow 3\pi$," *Phys. Lett. 23* (1966), pp. 384–5 [24 Oct 1966].

17 D. Sutherland, "Current Algebra and Some Non-Strong Mesonic Decays," *Nucl. Phys. B2* (1967), pp. 433–40 [30 May 1967].

18 J. S. Bell and R. Jackiw, "A PCAC Puzzle: $\pi^0 \rightarrow \gamma\gamma$ in the σ-Model," *Nuovo Cimento A60* (1969), pp. 47–60 [11 Sep 1968].

19 S. L. Adler, "Axial-Vector Vertex in Spinor Electrodynamics," *Phys. Rev. 177* (1969), pp. 2426–38 [24 Sep 1968].

20 J. S. Bell, "Current Algebra and Gauge Variance," *Nuovo Cimento 50A* (1967), pp. 129–34 [16 Dec 1966].

21 See, in particular, M. Veltman, ref. 23.

22 For a belated reprinting, see M. Veltman, "Relation Between the Practical Results of Current Algebra Techniques and the Originating Quark Model," in R. Akhoury, B. De Witt, P. Van Nieuwenhuizen, and H. Veltman, eds., *Gauge Theory–Past and Future* (Singapore: World Scientific, 1992), pp. 293–336.

23 M. Veltman, "Perturbation Theory of Massive Yang–Mills Fields," *Nucl. Phys. B7* (1968), pp. 637–50 [10 Sep 1968].

24 E. S. Fradkin and I. V. Tyutin, "Feynman Rules for the Massless Yang–Mills Field Renormalizability of the Theory of the Massive Yang–Mills Field," *Phys. Lett. 30B* (1969), pp. 562–3 [15 Oct 1969]; E. S. Fradkin, E. Esposito, and S. Termini, "Functional Techniques in Physics," *Rivista del Nuovo Cimento 2* (1970), pp. 498–560 [26 Sep 1970].

25 C. Becchi, A. Rouet, and R. Stora, "Renormalization of Gauge Theories," *Ann. Phys. (N.Y.) 98* (1976), pp. 287–321 [8 Dec 1975]; G. 't Hooft, "Renormalization of Massless Yang–Mills Fields," ref. 53.

26 A. A. Slavnov, "Ward Identities in Gauge Theories," *Theoretical and Mathematical Physics 10* (1972), pp. 153–61, (English translation pages 99–104) [23 Jun 1971]; J. C. Taylor, "Ward Identities and Charge Renormalization of the Yang–Mills Field," *Nucl. Phys. B33* (1971), pp. 436–44 [25 Jun 1971].

27 R. P. Feynman, "Quantum Theory of Gravitation," ref. 8.

28 B. S. DeWitt, "Theory of Radiative Corrections for Non-Abelian Gauge Fields," *Phys. Rev. Lett. 12* (1964), pp. 742–6 [12 May 1964]; "Quantum Theory of Gravity. I. The Canonical Theory," *Phys. Rev. 160* (1967), pp. 1113–48; "Quantum Theory of Gravity. II. The Manifestly Covariant Theory," *Phys. Rev. 162* (1967), pp. 1195–239.

29 L. D. Faddeev and V. N. Popov, "Feynman Diagrams for the Yang–Mills Field," *Phys. Lett. 25B* (1967), pp. 29–30 [1 Jun 1967].

30 S. Mandelstam, "Feynman Rules for Electromagnetic and Yang–Mills Fields from the Gauge-Independent Field-Theoretic Formalism," *Phys. Rev. 175* (1968), pp. 1580–1603 [17 Jun 1968].

31 A. Komar and A. Salam, "Renormalization Problem"; A. Salam, "Renormalizability," ref. 3.

32 A. Salam, "Weak and Electromagnetic Interactions," in Nils Svartholm, ed., *Elementary Particle Theory* (Stockholm: Almqvist & Wiksell, 1968), pp. 367–77.

33 D. Boulware, "Renormalizeability of Massive Non-Abelian Gauge Fields: A Functional Integral Approach," *Ann. Phys. 56* (1970), pp. 140–71 [14 May 1969].

34 P. T. Matthews, "The Application of the Tomonaga–Schwinger Theory to the Interaction of Nucleons with Neutral Scalar and Vector Mesons," *Phys. Rev. 76* (1949), pp. 1657–74 [28 Jun 1949].

35 T. D. Lee and C. N. Yang, "Theory of Charged Vector Mesons," ref. 10.

36 J. Reiff and M. Veltman, "Massive Yang–Mills Fields," *Nucl. Phys. B13* (1969), pp. 545–64 [11 Aug 1969].

37 M. Veltman, "Massive Yang–Mills Fields," in J. S. Bell, ed., *Proc. Topical Conf. on Weak Interactions* (CERN, Geneva, Switzerland, 14–17 Jan 1969), CERN yellow report no. 69–7, pp. 391–3.

38 A. A. Slavnov and L. D. Faddeev, "Massless and Massive Yang–Mills Fields," *Teoreticheskaya i Matematicheskaya Fizika* Vol. 3, No. 1 (April 1970), pp. 18–23 [4 Nov 1969]; D. Boulware, "Renormalizeability," ref. 33.

39 H. Van Dam and M. Veltman, "Massive and Massless Yang–Mills and Gravitational Fields," *Nucl. Phys. B22* (1970), pp. 397–411 [8 Jun 1970].

40 A. A. Slavnov, "Ward Identities in Gauge Theories"; J. C. Taylor, "Ward Identities and Charge Renormalization," ref. 26.

41 M. Veltman, "Generalized Ward Identities and Yang–Mills Fields," *Nucl. Phys. B21* (1970), pp. 288–302 [16 Apr 1970].

42 S. L. Glashow, J. Iliopoulos, and I. Maiani, "Weak Interactions with Lepton–Hadron Symmetry," *Phys. Rev. D2* (1970), pp. 1285–92 [5 Mar 1970]. See also, Yasuo Hara, "Unitary Triplets and the Eightfold Way," *Phys. Rev. B134* (1964), pp. 701–4 [23 Dec 1963]; and J. D. Bjorken and S. Glashow, "Elementary Particles and SU(4)," *Phys. Lett. 11* (1964), pp. 255–8 [19 Jun 1964].

43 S. L. Glashow and J. Iliopoulos, "Divergences of Massive Yang–Mills Theories," ref. 3.

44 M. Veltman, "Second Threshold in Weak Interactions," *Acta Phys. Pol. B8* (1977), pp. 475–92 [7 Jan 1977].

45 S. Weinberg, "Conceptual Foundations of the Unified Theory of Weak and Electromagnetic Interactions," *Rev. Mod. Phys. 52* (1980), pp. 515–23, on p. 518.

46 L. D. Faddeev and V. N. Popov, "Feynman Diagrams," ref. 29

47 E. S. Fradkin and I. V. Tyutin, "Feynman Rules"; E. S. Fradkin, E. Esposito, and S. Termini, "Functional Techniques in Physics," ref. 24.

48 E. S. Fradkin and I. V. Tyutin, "S Matrix for Yang–Mills and Gravitational Fields," *Phys. Rev. D2* (1970), pp. 2841–57 [19 Jan 1970].

49 J. Schwinger, "A Gauge Theory of Fundamental Interactions," *Ann. Phys. 2* (1957), pp. 407–35 [31 Jul 1957]. M. Gell-Mann and M. Lévy, "The Axial Vector Current in Beta Decay," *Nuovo Cimento 16* (1960), pp. 705–26 [19 Feb 1960].

50 G. 't Hooft, Utrecht scriptie, 1969 (unpublished); xerox copy in existence.

51 B. Lee, "Chiral Dynamics," *Cargese Lecture in Physics*, Vol. 5 (New York: Gordon and Breade, 1971), pp. 1–119; D. Bessis and Turchetti, "Renormalization of the σ model through Ward Identities," ibid., pp. 119–179.

52 G. 't Hooft, thesis, Utrecht, March 1972. This thesis contains essentially the papers of refs. 53 and 55, with an additional introduction, summary, and short curriculum vitae in Dutch. There is also, according to Dutch tradition, a sheet with 15 "stellingen" (propositions). They must be arguable.

53 G. 't Hooft, "Renormalization of Massless Yang–Mills Fields," *Nucl. Phys. B33* (1971), pp. 173–99 [12 Feb 1971].

54 E. S. Fradkin and I. V. Tyutin, "S Matrix for Yang–Mills," ref. 48.

55 G. 't Hooft, "Renormalizable Lagrangians for Massive Yang–Mills Fields," *Nucl. Phys. B35* (1971), pp. 167–88 [13 Jul 1971].

56 J. Schwinger, "A Gauge Theory of Fundamental Interactions"; M. Gell-Mann and M. Levy, "The Axial Vector Current," ref. 49.

57 P. W. Anderson, "Plasmas, Gauge Invariance and Mass," *Phys. Rev. 130* (1963), pp. 439–42 [8 Nov 1962].

58 S. Weinberg, "A Model of Leptons," *Phys. Rev. Lett. 19* (1967), pp. 1264–6 [17 Oct 1967].

59 M. Veltman, "Cosmology and the Higgs Mechanism," Rockefeller University preprint May 1974 (unpublished). M. Veltman, "Cosmology and the Higgs Mass," *Phys. Rev. 34* (1975), pp. 777–8 [5 Dec 1974].

60 T. W. B. Kibble, "Symmetry Breaking in Non-Abelian Gauge Theories," *Phys. Rev. 155* (1967), pp. 1554–61 [24 Oct 1966]; S. Weinberg, " A Model of Leptons," ref. 58.

61 F. Englert and R. Brout, "Broken Symmetry and the Mass of Gauge Vector Mesons," *Phys. Rev. Lett. 13* (1964), pp. 321–3 [26 Jun 1964]; P.

W. Higgs, "Broken Symmetries, Massless Particles and Gauge Fields," *Phys. Lett. 12* (1964), pp. 132–3 [27 Jul 1964].

62 *Fundamental Problems in Elementary Particle Physics*, Proceedings of the Fourteenth Conference on Physics, University of Brussels, 2–7 October 1967 (New York: John Wiley, 1968). See discussion after the lecture of H. P. Durr, page 18. Weinberg distributed there one or more copies of the handwritten manuscript of his 1967 article; the difference between that and the published version is minimal.

63 A. Tenner and M. Veltman, eds., *Proceedings of the Amsterdam International Conference on Elementary Particles*, June 30–July 6, 1971 (Amsterdam: North-Holland, 1972).

64 H. D. Politzer, "Reliable Perturbative Results for Strong Interactions," *Phys. Rev. Lett. 30* (1973), pp. 1346–8 [3 May 1973]. D. Gross and F. Wilczek, "Ultra-Violet Behavior of Non-Abelian Gauge Theories," *Phys. Rev. Lett. 30* (1973), pp. 1343–6 [27 Apr 1973].

65 G. 't Hooft and M. Veltman, "Regularization and Renormalization of Gauge Fields," *Nucl. Phys. B44* (1972), pp. 189–213 [21 Feb 1972].

66 It was alluded to in section 5 of G. 't Hooft, "Renormalization of Massless Yang-Mills Fields," ref. 53.

67 C. Bollini and J. Giambiagi, "Dimensional Renormalization: The Number of Dimensions as a Regularizing Parameter," *Nuovo Cimento 12B* (1972), pp. 20–6 [8 Feb 1972].

68 C. Bollini and J. Giambiagi, "Lowest Order Divergent Graphs," *Phys. Lett. 40B* (1972), pp. 566–70 [18 Oct 1971].

69 C. Bollini and J. Giambiagi, "Dimensional Renormalization," ref. 67.

70 C. G. Bollini, J. J. Giambiagi, and A. Gonzalez Dominguez, "Analytic Regularization and the Divergences of Quantum Field Theories," *Nuovo Cimento 31* (1964), pp. 550–61 [15 Jul 1963].

71 Eugene R. Speer, "Analytic Renormalization," *J. Math. Phys. 9* (1968), pp. 1404–10 [1 Dec 1967].

72 See comments in G. 't Hooft, "Prediction for Neutrino–Electron Cross Sections in Weinberg's Model," *Phys. Lett. 37B* (1971), pp. 195–9 [27 Oct 1971]. This paper contains in footnote 2 a statement suggesting the existence of an argument that the Weinberg model of leptons as such is renormalizable, i.e., that anomalies are harmless.

73 C. Bouchiat, J. Iliopoulos, and Ph. Meyer, "An Anomaly-Free Version of Weinberg's Model," *Phys. Lett. 38B* (1972), pp. 519–23 [11 Feb 1972].

74 G. 't Hooft and M. Veltman, "Combinatorics of Gauge Fields," *Nucl. Phys. B50* (1972), pp. 318–53 [31 Jul 1972].

75 G. 't Hooft and M. Veltman, "Example of a Gauge Field Theory," in C. P. Korthals-Altes, ed., *Renormalization of Yang-Mills Fields and Applications to Particle Physics*, Marseille Conference 19–23 June 1972 (Marseille: Centre de Physique Théorique, 1972), pp. 37–75. . Note: J. S. Bell is not mentioned in the list of participants; he was there.

76 I. Bialynicki-Birula, "Renormalization, Diagrams and Gauge Invariance," *Phys. Rev. D2* (1970), pp. 2877–86 [27 Aug 1970].

77 G. 't Hooft and M. Veltman, "Example of Gauge Field Theory," ref. 75. Actually typed at CERN in October 1972.

78 G. 't Hooft and M. Veltman, "One-loop divergencies in the theory of gravitation" *Annales de l' Institut Henri Poincaré 20* (1974), pp. 69–94 [4 Sep 1973].

79 M. Veltman, "Quantum Theory of Gravitation," in R. Balian and J. Zinn-Justin, eds., *Structural Analysis of Collision Amplitudes*, Proceedings of the Les Houches Summer School on Theoretical Physics, 2–27 June 1975 (Amsterdam: North-Holland, 1976), pp. 265–327.

80 B. Lee and J. Zinn-Justin, "Spontaneously Broken Gauge Symmetries," *Phys Rev. D5* (1972), pp. 3121–37, 3137–55, 3155–60 [10 Mar 1972]; B. Lee and J. Zinn-Justin, "Spontaneously Broken Gauge Symmetries," *Phys. Rev. D7* (1973), pp. 1049–56 [30 Oct 1972].

81 G. 't Hooft and M. Veltman, "Diagrammar," CERN yellow report No. 73–9 (Geneva, 1973).

82 C. Becchi, A. Rouet, and R. Stora, "Renormalization of Gauge Theories," ref. 25.

83 S. L. Glashow, "Partial-Symmetries of Weak Interactions," *Nucl. Phys. 22* (1961), pp. 579–88 [5 Sep 1960].

84 D. A. Ross and M. Veltman, "Neutral Currents and the Higgs Mechanism," *Nucl. Phys. B95* (1975), pp. 135–47 [11 Apr 1975].

85 M. Veltman, "Cosmology and the Higgs Mechanism" ref. 59.

86 M. Veltman, "Second Threshold in Weak Interactions," ref. 44.

10

Renormalization of Gauge Theories

GERARD 'T HOOFT

Born Den Helder, The Netherlands, 1946; Ph.D., 1972 (physics), University of Utrecht; Professor of Physics at the Institute for Theoretical Physics, University of Utrecht; high-energy physics (theory).

Like most other presentations by scientists in this Symposium, my account of the most important developments that led toward our present view of the fundamental interactions among elementary particles is a personal one, recounting discoveries I was just about to make when someone else beat me to it. But there is also something else I wish to emphasize. This is the dominant position reoccupied during the last two decades by theory, in its relation to experiment. In particular quantum field theory not only fully regained respectability but has become absolutely essential for understanding those basic facts now commonly known as the "Standard Model." So much happened here, so many discoveries were made, that the space allotted to theory in this volume runs far too short to cover it all. Therefore, I will limit myself only to the nicest goodies among the many interesting developments in theory, and of those I'll only pick the ones that were of direct importance to me.

Renormalization

Before the seventies there was only one renormalizable quantum field theory that seemed to give a reasonable and useful description of (parts of) the real world: quantum electrodynamics. Its remarkable successes in explaining, among others, the Lamb shift and the anomalous magnetic moment of the electron did not go unnoticed.[1] Yet the idea that other interactions should also be described in the context of renormalizable field theories became less and less popular. Indeed, the notion of renormalizability was quite controversial, and to some it still is.

The reason for this controversy is quite understandable: there are many misconceptions concerning the real meaning of renormalization in quantum field theories, and these are – partly – due to inaccurate pre-

179

sentations of the notion of renormalization, in particular the "infinite" renormalization apparently required in these constructs. A correct presentation would have to explain elaborately *why* theories are constructed the way they are, in a logically coherent way. Instead of that, however, it is often much more convenient to explain how renormalization works *in practice*.

In the latter case one is tempted to short-circuit the original delicate physical arguments. And then one gets some useful mathematical prescriptions, roughly to be summarized as follows:

> Start with the "naive" unrenormalized theory. You will see that it contains "infinities." Renormalization simply amounts to "subtracting" or "removing" the infinite terms.

Now this sounds like: "You hit upon difficulties; just ignore them, cover them up!" As if by miracle, the resulting prescriptions are now claimed to be completely unique and self-consistent. But of course the explanations as to why they work are then lacking, and many textbooks that contain only this version of the argument have added to the widespread mistrust and contempt for such an obviously shaky procedure, in spite of its experimental success, which, according to some, had to be accidental.[2]

Quite a few investigators tried to launch "infinity subtraction" as a first principle in renormalization. That renormalization turned theories with infinities into finite – hence useful – theories was used as a commercial that, in my opinion, did not betray deep insight concerning the real underlying physics.

To resolve this confusion one must realize that all known quantum field theories (in $3 + 1$ space–time dimensions) must be viewed as *models*. They do not pretend to describe any possible system of interacting particles with *infinite* accuracy, although some models allow us to make far more accurate predictions than others. The reason for this is that nothing in the model can be calculated with infinite precision.

All one has is some power expansion. An amplitude Γ will always be represented in terms of a series such as

$$\Gamma = a_o + a_1 g^2 + a_2 g^4 + \dots, \qquad (10.1)$$

where g is some coupling constant. In all known realistic theories this series will be an asymptotic series at best, which means that there is no value for g small enough for the series to converge completely, apart from $g = 0$ (in which case the particles do not interact at all).

In practice the convergence question is often of little importance, that is, when we have g so small that the first few terms suffice. But if it comes to mathematical rigor, we have to state that mathematically these models are well-defined only if the coupling strength(s) g is (are) *infinitesimally* small.

Since in reality the coupling strengths are non-zero, we must admit that our quantum field theories must be viewed upon as *effective* field theories having a very accurate, but not infinitely accurate, predictive power. One must terminate the series when the next term becomes bigger than the previous. The value of that next term then roughly represents the error bar. This is mathematically acceptable if we simply replace the *field* of (real or complex) *numbers* by the field of *asymptotic series expansions*.

In our effective field theory we must assume that at a very tiny length scale $a = 1/\Lambda$ the basic interactions are not understood, but can be approximated by a simple model with a cutoff, for instance defined by a lattice, or by assuming the presence of unphysical particles as described by Wolfgang Pauli and F. Villars.[3] Now at this point we must replace all numbers by power-series expansions in terms of some expansion parameter z (for instance the coupling strength g^2), which tends to zero when all interactions vanish.

As a next step we express all quantities that can be directly observed in an experiment at low energy, hence large distance scale, in terms of z, and then we also replace z itself by an expansion parameter that can be observed at large distances (such as the physically observed electric charge of an electron). For instance, we will not consider the "bare" (i.e., original) mass, but only the physically observed mass (i.e., "renormalized mass") of a particle. One then discovers that for a certain class of models the limit $a \Rightarrow 0$ $\Lambda \Rightarrow \infty$ exists, in the sense that all expansion coefficients of the asymptotic series remain finite. All artifacts due to the (lattice or Pauli–Villars) cutoff disappear in the limit. This class of models is called renormalizable.

The expansion coefficients for the *bare* mass and charge do not exist in the limit but may diverge, logarithmically in most cases. This means that for finite z one should not allow a to become much smaller than some exponential function like $\exp(-1/z)$, but in practice this is of little concern because it is far beyond the region where we expect the model to be physically reliable anyway. Thus the answer to many critical objections against the renormalization procedure is that the limits $z \Rightarrow 0$ and $a \Rightarrow 0$ *must* be taken in this order: z first, a last.

It is only when we streamline and short-circuit this long series of arguments in order to obtain a convenient manual for calculating the coefficients $a_o, a_1, \ldots$, that we find as a prescription that "infinities must be subtracted."

Actually one may consider five categories of sophistication for quantum field theories:

1. Nonrenormalizable field theories. If z is the expansion parameter representing the coupling strength, these theories allow us only to consider the lowest expansion term, for instance:

$$\Gamma = a_1 z + \mathcal{O}(z^2) . \qquad (10.2)$$

   Examples are the old (but still quite useful) Fermi theory for the weak interactions,[4] and quantum gravity with quantized matter fields.

2. One-loop renormalizable field theories. In some theories such as Yang–Mills theory with mass term,[5] and pure quantum gravity without matter,[6] the existing symmetry allows us to compute unambiguously the next term but not more:

$$\Gamma = a_1 z + a_2 z^2 + \mathcal{O}(z^3) , \qquad (10.3)$$

   where both a_1 and a_2 are unique and calculable.

3. Renormalizable theories. For these all expansion coefficients are uniquely defined and calculable, but the series are only asymptotic. Hence one has typically

$$\Gamma = \Sigma a_n z^n + \mathcal{O}(e^{-1/z}); \quad a_n = \mathcal{O}(n!). \qquad (10.4)$$

4. Asymptotically free theories. These theories are also renormalizable, but have as an additional bonus that if we scale to very small distances the expansion parameter z approaches to zero, so that there the expansion (10.4) becomes extremely accurate. Consequently these theories are very accurately defined even if at large distances z is large. However, we still do not know whether these theories allow for *infinitely* precise calculations, although this is generally conjectured. Examples are pure non-Abelian gauge theories coupled to a limited number of fermion species only, such as quantum chromodynamics.

5. Borel summable theories.[7] These are theories that allow a rigorous definition of all amplitudes, typically obtaining

$$\Gamma(z) = \int_o^\infty B(u) \, e^{-u/z} \, du, \qquad (10.5)$$

   where the power expansion of B in terms of u not only has a finite radius of convergence but also allows for an analytic extension toward the entire real axis. Theories of this sort are not known in $3 + 1$ dimensions, apart from some special limiting cases.[8]

The early days of Yang–Mills theory

Just a few classical papers in the older literature stand out as real jewels, and they were inspiring examples of theoretical reasoning to all of us for many years. First let me mention the marvelous paper by Chen Ning Yang and Robert Mills.[9] They pointed out that the only interparticle force that was well understood at that time, QED, can be seen as a construction built upon a fundamental principle: local gauge invariance. And this principle can be generalized if we have more than one type of fermionic fields $\psi(x, t)$, which we can arrange as isovectors:

$$\psi = \begin{pmatrix} \psi_1 \\ \psi_2 \end{pmatrix}. \tag{10.6}$$

Consider transformations of the type

$$\psi \Rightarrow \Omega(x, t)\, \psi\,, \tag{10.7}$$

where Ω is a 2×2 (or possibly larger) matrix. One can then construct the *covariant derivative* $D_\mu \psi$ as follows:

$$D_\mu \psi = \partial_\mu \psi + g b_\mu^a\, T^a \psi, \tag{10.8}$$

which transforms just as (10.7) if the new fields b_μ^a transform in a very special way. Here g is just some coupling constant, and the matrices T^a are the generators of infinitesimal rotations. One can formulate dynamical equations of motion for the new fields b_μ^a by first defining the covariant fields

$$F_{\mu\nu}^a = \partial_\mu b_\nu^a - \partial_\nu b_\mu^a + g f^{abc}\, b_\mu^b\, b_\nu^c, \tag{10.9}$$

where f^{abc} are the structure constants of the Lie group of matrices Ω. The field equations are generated by the Lagrangian

$$\mathcal{L}^{inv} = -\frac{1}{4}\, F_{\mu\nu}^a F_{\mu\nu}^a - \bar{\psi}(\gamma_\mu D_\mu + m)\psi. \tag{10.10}$$

It is invariant under local gauge transformations and as such a direct generalization of QED.

Since the rigid, space–time independent analog of the transformation group (henceforth called the *global* group) was known as isospin invariance for the strong interactions, Yang and Mills viewed their theory as

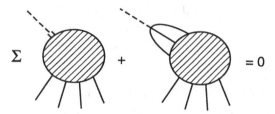

Fig. 10.1. Veltman–Ward identity among diagrams.

a scheme to turn isospin into a *local* symmetry, but they immediately recognized that then there was a problem: the Lagrangian describes a *massless* vector particle with three (or more) components, in general electrically charged as well as neutral ones. In spite of its beauty, this theory was therefore considered to be unrealistic. Besides, since these massless particles interact with each other, the theory showed horrible infrared divergences.

Proposals to cure this "disease" were made several times. Richard Feynman, who looked upon this model as a toy model for quantum gravity, proposed simply to add a small mass term just to avoid the infrared problem:[10]

$$\mathcal{L} = \mathcal{L}^{inv} - \frac{1}{2} M^2 (b_\mu^a)^2. \tag{10.11}$$

Sheldon Glashow and Martinus Veltman proposed to use the same Lagrangian as a model for the weak intermediate vector boson.[11] It was hoped that the mass term would not spoil the apparent renormalizability of the Lagrangian. Probably the philosophy here was that the mass term is only a mild symmetry-breaking correction of a kind we see more often in Nature: isospin invariance itself is also softly broken.

Indeed Veltman initially reported progress here: the theory (10.11) is renormalizable at the one-loop level.[12] He made use of field transformations that look like gauge transformations, even though the mass term in (10.11) is not gauge invariant:

$$b_\mu^{a'} = b_\mu^a + g f^{abc} \Lambda^b b_\mu^c - \partial_\mu \Lambda^a \quad ; \quad \psi' = \psi + g \Lambda^a T^a \psi. \tag{10.12}$$

Here Λ may be any function of some arbitrarily chosen field variable. Veltman called this a "Bell–Treiman trasformation." The identities among amplitudes corresponding to different Feynman diagrams obtained in this way (see Fig. 10.1) should have been called *Veltman–Ward* identities.

To me it came as a surprise that Veltman managed to renormalize his theory up to one loop with this method. The mass term renders the longitudinal part of the gauge field observable, in spite of the fact that the Lagrangian carries no kinetic term for it. This theory should self-destruct. This it does, as Veltman found out, but only if you try to renormalize diagrams with two or more loops. To render the "massive Yang–Mills theory" renormalizable, a better theory was needed.

The Gell-Mann–Lévy sigma model

It was one of those caprices of fate that brought me, as a young student of Veltman's, to the 1970 Cargèse Summer Institute. (I had first applied to Les Houches, where my application was turned down.) The champions of renormalization were gathered there to discuss the Gell-Mann–Lévy sigma model, which had been proposed by Murray Gell-Mann and Maurice Lévy in 1960 in another classic jewel.[13] In order to explain the existence of a partially conserved axial-vector current, they added a fourth component to the three pion fields, the sigma field, transforming together as a 2×2 representation of chiral $SU(2) \otimes SU(2)$. The Lagrangian was

$$
\begin{aligned}
\mathcal{L}\left(\vec{\pi}, \sigma, \psi, \bar{\psi}\right) = \\
-\frac{1}{2}\left[\partial_\mu \vec{\pi}^2 + \partial_\mu \sigma^2\right] - \frac{1}{2}\mu_o^2\left[\vec{\pi}^2 + \sigma^2\right] - \frac{1}{4}\lambda_o^2\left[\vec{\pi}^2 + \sigma^2\right]^2 \\
- \bar{\psi}\left[\gamma_\mu \partial_\mu + g_o\left(\sigma + i\gamma_5 \vec{\pi} \cdot \vec{\tau}\right)\right]\psi + c\sigma.
\end{aligned}
\tag{10.13}
$$

If we take μ_o^2 here to be negative, then the potential for the scalar fields has the by now familiar dumbbell shape. The sigma field gets a vacuum expectation value,

$$
\langle \sigma \rangle = F = |\mu_o|/\lambda_o,
\tag{10.14}
$$

so in a perturbative expansion we write $\sigma = F + s$, and expand in s. The nucleon fields ψ get a mass $g_o F$, the pions have a tiny mass-squared proportional to the small constant c, whereas the sigma field s becomes a heavy resonance.

Jean-Loup Gervais, Benjamin Lee, and Kurt Symanzik explained in their Cargèse lectures how this model could be renormalized, and that its beautiful features would not be seriously affected by renormalization.[14] It was clear to me at that time that one can produce mass terms for Yang–Mills fields in a way very similar to this sigma model. I did not ask many questions in this school, but I did ask one question to Lee and to Symanzik: "Do your methods also apply to the Yang–Mills case?" They both gave me the same answer: "If you are Veltman's student, you should ask him; I am not an expert in Yang–Mills theory."

Massless Yang–Mills

This I did, as soon as I was back in Utrecht. But Veltman replied that he found it difficult to believe in such a spontaneous symmetry breakdown in particle theory. His opinion was that if that happens the vacuum would have a tremendously large energy density, which would give the physical vacuum an enormously large cosmological constant.

But we know it happens in the sigma model, which describes strong interactions pretty nicely. And if it is not symmetry breaking, then at least all other vacuum fluctuation effects also contribute to the cosmological constant, not as much as in a weak interaction theory with Higgs mechanism, but still far more than the experimental upper bound. The cosmological constant problem should be postponed until we solve quantum gravity; we should not let it affect our theories at the GeV or TeV scale.

It was then that we decided what my research program would be. First I would try to really understand all details of the massless, unbroken Yang–Mills system, and then I would add the mass, by a "spontaneous local symmetry-breaking mechanism."[15]

The status of pure Yang–Mills theory was somewhat vague. Strong *formal* arguments existed that this theory had to be renormalizable. But there were competing and conflicting ideas as to what its Feynman rules were. One paper on this subject was my third classical gem: a short *Physics Letters* paper by Ludwig Faddeev and Victor Popov.[16] It was all I needed to understand what was going on. Faddeev and Popov argued that a gauge-invariant functional integral expression for the amplitudes had to have the form

$$\Gamma = \int e^{i \int \mathcal{L}^{\text{inv}}(B) d^4 x} \prod_x dB(x), \tag{10.15}$$

where $B(x)$ stands for all field components of the gauge and matter system. However, since the integrand is invariant under gauge transformations, one only needs to integrate over the inequivalent field configurations, each being constrained by some gauge condition. As a gauge condition one typically takes

$$\partial_\mu b_\mu^a = 0. \qquad (10.16)$$

If we impose this constraint on the integrand, however, we need a Jacobian factor. So if we keep track of the measure, this turns the integral into

$$\Gamma = C \int e^{i \int \mathcal{L}^{inv}(B) d^4 x} \prod_x \left(dB(x) \, \delta(\partial_\mu b_\mu^a) \right) \det \left(\frac{\delta \, \partial_\mu b_\mu^a}{\delta \Lambda} \right). \qquad (10.17)$$

The theory produces a transverse propagator:

$$\frac{\delta_{\mu\nu} - \frac{k_\mu k_\nu}{k^2 - i\epsilon}}{k^2 - i\epsilon}. \qquad (10.18)$$

Other theories led to a Feynman gauge propagator,

$$\frac{\delta_{\mu\nu}}{k^2 - i\epsilon}, \qquad (10.19)$$

and how this could be related to a functional integral was not clear.[17] More important, I thought, was that none of the existing papers provided for a precise prescription as to how the infinities should be subtracted. The *formal* arguments were there, but how does it work in practice?

This became the subject of my first publication.[18] Several things had to be done. First, the formalism to obtain the Feynman rules from the functional integrals could be simplified. The existing procedure to deduce the ghost Feynman rules from the determinant was not satisfactory. I observed that one can write

$$(\det \mathcal{M})^{-N} = C \int \mathcal{D}\vec{\phi} \mathcal{D}\vec{\phi}^* \, e^{-\vec{\phi}^* \mathcal{M} \vec{\phi}}, \qquad (10.20)$$

where $\vec{\phi}$ is a complex Lorentz-scalar field with N components. One now reads off directly the Feynman rules for closed loops of ϕ fields. A factor N goes with each closed loop. Since we want N to be -1, our closed loops will usually go with a factor -1, just like the rules for fermions. Indeed, one can also write

$$\det \mathcal{M} = C \int \mathcal{D}\eta \mathcal{D}\bar{\eta} \, e^{-\bar{\eta} \mathcal{M} \eta}, \qquad (10.21)$$

where η is an anticommuting (Grassmann) variable.

Next, I could also see how Faddeev and Popov's trick could produce the Feynman gauge. Just take an auxiliary field variable F and impose the gauge

$$\partial_\mu b_\mu^a = F^a. \qquad (10.22)$$

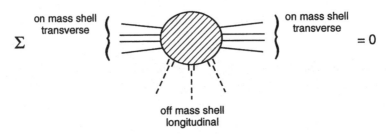

Fig. 10.2.

One then sees that

$$e^{-\frac{1}{2}\left(\partial_\mu b_\mu^a\right)^2} = \int \mathcal{D}F \, e^{-\frac{1}{2}F^2} \delta \left(\partial_\mu b_\mu^a - F^a\right). \qquad (10.23)$$

To see that the renormalization counterterms do not spoil gauge invariance, we needed Ward identities. It turned out to be sufficient to prove identities of the form of Fig. 10.2.

Since reducible and irreducible diagrams must all be added together, these identities are sufficient to restrict all counterterms completely up to gauge-invariant ones. This point was often not realized by later investigators. The proof of these Ward identities was much more complicated than the Veltman–Ward identities mentioned before, because we had to disentangle carefully the contributions of various ghost lines. See Fig. 10.3, which was an intermediate step.

I was annoyed that I could not use a simple symmetry argument for the proof as Veltman had done for his case. Only much later it was discovered how to do this. Becchi, Rouet, and Stora found that the underlying symmetry for this identity is an *anticommuting* one.[19] Their marvelous discovery was this. Take as an invariant Lagrangian, for instance

$$\mathcal{L}^{inv} = -\frac{1}{4}F_{\mu\nu}^a F_{\mu\nu}^a - D_\mu \phi^* D_\mu \phi - V\left(\phi, \phi^*\right) - \bar{\psi}\left(\gamma D + m\right)\psi + \dots \qquad (10.24)$$

and add as a gauge-fixing term

$$\mathcal{L}^{gauge} = i\frac{1}{2}\left(\ell^a\right)^2, \qquad (10.25)$$

where ℓ^a is anything like $\partial_\mu b_\mu^a$, b_4^a, and so on. Introduce the ghost fields η^a and $\bar{\eta}^a$, which must be anticommuting. Consider then the

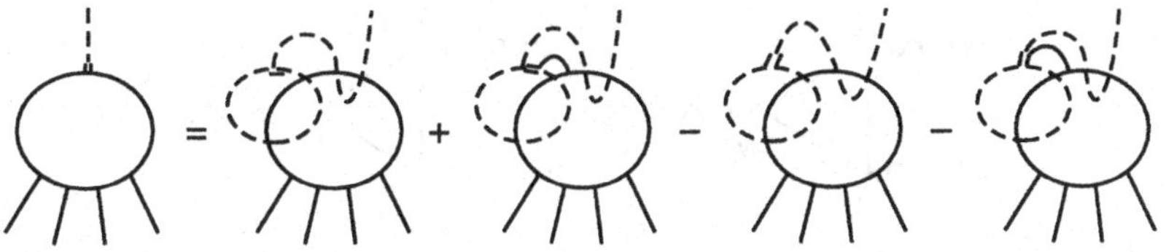

Fig. 10.3.

anticommuting variations:

$$
\begin{aligned}
\delta b_\mu^a &= D_\mu \eta^a; \\
\delta \phi &= -ig T^a \eta^a \phi; \\
\delta \eta^a &= \tfrac{1}{2} g f^{abc} \eta^b \eta^c; \\
\delta \bar\eta^a &= \ell^a(b, \phi, \ldots).
\end{aligned}
\tag{10.26}
$$

Here the first two equations are just gauge transformations. Then the total Lagrangian of the theory when taken to be

$$
\mathcal{L} = \mathcal{L}^{inv} + \mathcal{L}^{gauge} + \mathcal{L}^{ghost},
\tag{10.27}
$$

with

$$
\mathcal{L}^{ghost} = -\bar\eta^a \, \delta\ell^a(b, \phi, \ldots, \eta),
\tag{10.28}
$$

is invariant under this *global* transformation. The above identities are nothing but an expression of this invariance, now called BRS invariance.

I had to convince myself that the rules obtained produced a *unitary* theory. The new identities were sufficient to guarantee this. Just one problem remained: the identities *overdetermined* the renormalization counterterms. Would there never be a conflict? There was a well-known example of just such a conflict in the literature: the Adler–Bell–Jackiw anomaly. Steve Adler, and independently from him John Bell and Roman Jackiw, had discovered that diagrams of the kind depicted in Fig. 10.4 cannot be renormalized in such a way that both the vector current and the axial-vector current are conserved.[20] If something like this would happen in a gauge theory, there would be deep trouble. I could prove that if no gauge fields are coupled to the axial charge, clashes of this sort will not destroy renormalizability in diagrams with up to one loop. The trick was to use a fifth dimension for the internal lines inside the loop.

What if you have more than one loop? I tried to use six, seven, or more dimensions but this does not work. (Recently a book appeared in which a "proof" of renormalizability along these lines appeared. The

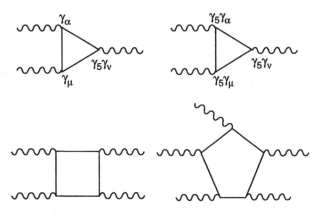

Fig. 10.4. Diagrams that contribute to the Adler–Bell–Jackiw anomaly.

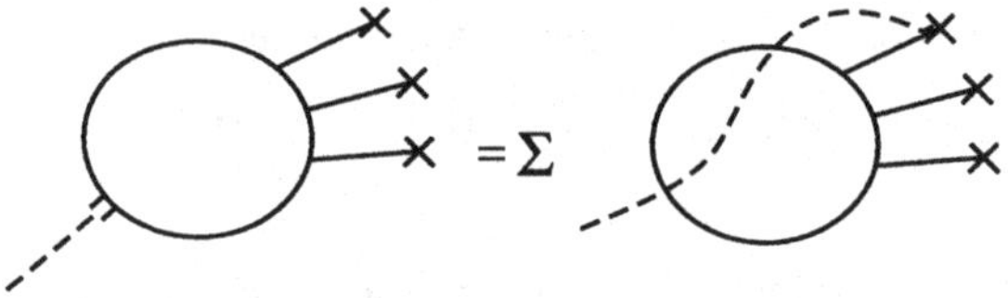

Fig. 10.5.

proof is incorrect.) I was confident the problem could be solved, but was unable to do it then.

Soon after my paper had come out, two other papers appeared, one by Andrei Slavnov and one by John Taylor.[21] Both observed that the identities I had written down could be generalized. If some of the external lines are neither longitudinal nor on mass shell, one gets extra contributions where the ghost line ends up at one of these lines. See Fig. 10.5.

The derivation went the same way as that for my own identities. The only reason why I had not written these identities in this new form before was that I thought the extra pieces would be cumbersome, requiring new renormalization counterterms of their own, and, furthermore, I didn't need them. It is clear now that these newer identities are more complete. And so it happened that they were to become known as the Slavnov–Taylor identities.

Massive Yang–Mills theories

For my advisor, Veltman, all this was just *spielerei*. Massless Yang–Mills fields seem not to occur in Nature. They are just there for exercises. The real thing is the massive case, and he thought that would be an entirely different piece of cake. Actually, however, the step remaining to be taken was a small one.[22] As I knew from Cargèse, the actual nature of the vacuum state has little effect upon renormalization counterterms. All that needed to be done was to add to the gauge-invariant Lagrangian the by-now familiar Higgs terms,

$$\mathcal{L}^{Higgs} = -\frac{1}{2}(D_\mu\phi)^2 - V(\phi), \qquad (10.29)$$

where $V(\phi)$ has the familiar dumbbell shape, just as in the Gell-Mann–Lévy sigma model (for simplicity I take the ϕ field here to be a real multiplet). Writing $\phi = F + \varphi$ we get a gauge-invariant Lagrangian for the b and φ fields, such that now the b fields get the required mass. To appease Veltman, I wrote the self-interaction as

$$V = \frac{1}{8}\lambda\left(2F\varphi + \varphi^2\right)^2, \qquad (10.30)$$

so that at least at lowest order the vacuum energy density vanishes. In terms of the φ fields, the gauge-transformation laws for these and the b field look very similar:

$$\begin{aligned}
b_\mu^{a'} &= b_\mu^a + f^{abc}\Lambda^b b_\mu^c - \frac{1}{g}\partial_\mu\Lambda^a; \\
\varphi' &= \varphi + T^a\Lambda^a\varphi + T^a\Lambda^a F.
\end{aligned} \qquad (10.31)$$

Everything else went exactly as in the previous paper. Because the local gauge invariance is still exact, we again have Slavnov–Taylor identities and BRS invariance, and from them one can prove unitarity and equivalence of the various gauge choices. A judicious gauge choice was found such that the propagators for the massive gauge fields and the other fields became as simple as possible.

The problem of regularizing and renormalizing diagrams with two or more loops was still there. Veltman and I discussed a lot about this problem and eventually agreed that the best strategy was continuous variation of the number of space–time dimensions.[23] As if it were a seed from outer space, this idea germinated simultaneously in various places as an answer to different problems. Kenneth Wilson and Michael Fisher were writing a paper proposing to calculate critical phenomena in statistical physics in $4 - \epsilon$ dimensions as an expansion in ϵ.[24] And independently of us C. Bollini and J. Giambiagi, and J. Ashmore, also suggested to use analyticity in space–time dimensions as a regulator.[25]

One may notice that by now I entirely address the problem of renormalization as a procedure for infinity subtraction. As explained at the beginning, this is not at all what renormalization really is from a physical point of view. It is preferable to talk about *regularization* first, and then *renormalization* afterwards. Regularization is the replacement of a theory by a slightly mutilated theory, using a cutoff. We must show that the effects of the cutoff become negligible at large distance scales, and then make the transition toward renormalized observables, after which it must be demonstrated that the limit where the cutoff goes away exists and is perturbatively finite. It does not matter much how crazy the mutilation was in the beginning, as long as the limit is well behaved. Going to $4 - \epsilon$ dimensions is just such a crazy regularization scheme. It turns out to be extremely elegant technically. Anyway, the important thing was that this method works fine at all orders of perturbation expansion and not just up to one loop, like the five-dimensional procedure found earlier.

We now had a general scheme for producing theories with interacting massive vector particles.[26] At first I was thinking about applying it to ρ mesons, as a nice generalization of the Gell-Mann–Lévy sigma model. But of course Veltman could convince me that the weak interactions were a much more promising application. I had practically reproduced Weinberg's model before I saw his 1967 paper. I also reproduced an error (the neutral cross section calculated by Weinberg was much too large because of a sign error in the Fierz transformation), but managed to correct it before my paper was published.[27] My own paper said in a footnote that the anomalies do not render the theory nonrenormalizable. Of course, this should be interpreted as saying that renormalizability can be restored by adding an appropriate amount of various kinds of fermions (quarks), but I admit that I also thought that perhaps this was not even necessary.[28] Now we know it certainly is necessary to have the anomalies cancel (see below).

More important to my mind was that we now had a large class of renormalizable theories with massive and massless vector mesons. A crucial argument was added to this by Chris Llewellyn-Smith and by John Cornwall, David Levin, and George Tiktopoulos; they showed that requiring unitarity implies that the *only* such theories are gauge theories.[29] So not only do we have a large class of new models, we have the *complete* class of renormalizable vector theories.

Asymptotic freedom

While searching for a decent regularization method for gauge theories, I had also studied scaling behavior; in 1971 I already knew that when you scale all momenta by a common factor upwards, then the gauge coupling constant decreases, whereas for QED the coupling strength increases. (I still had an error in the coefficient, now known as the β function. I found the right coefficient in 1972.)

I did dream about the possibility of pure gauge theories for quarks, but since I thought that strong interactions were infinitely complicated, I did not dare to ruin my reputation by launching such crackpot ideas. After the work with Veltman in 1972 in which we had carefully exhibited all detailed properties of the renormalization counterterms, I knew how to do the scaling calculation precisely.[30] Veltman convinced me, however, that our work on the counterterms for quantum gravity was much more important.[31]

In a Marseille conference in 1972 I met Symanzik again.[32] He explained his attempts to construct a theory with a negative β coefficient in order to explain Bjorken scaling.[33] I was delighted to announce after Symanzik's talk that what he was looking for was a non-Abelian gauge theory, and wrote its β function, in modern notation:

$$\beta\left(g^2\right) = \frac{1}{16\pi^2}\left(-\frac{11}{3}C_1 + \frac{1}{6}C_2\,N_{scalar} + \frac{2}{3}C_3 N_{fermion}\right), \quad (10.32)$$

so that, if you take SU(2), 11 fermion species would be needed to cancel the vector-boson contribution ($C_1 = 2, C_3 = 1$). Symanzik said to me that he believed I had made a sign error, but if it was correct I should publish it, because it was important, and if I did not, somebody else would.

I knew I had made no sign error (the origin of the sign differences was evident in the calculations). But there was still much work to do on quantum gravity, and also I would have to explain in detail my calculational procedure for which much time was needed. And so, my remarks remained largely unnoticed, by all except Symanzik. He first remained quiet (reportedly to allow me to correct my "mistake"), but then mentioned my result to Giorgio Parisi, who came to CERN and discussed the topic with me. Then, when the news about "asymptotic freedom" came, from the United States, Symanzik was the first to point out to everyone involved that the discovery was first made in Europe, and that an announcement made at a conference counts when matters of priority are concerned.[34]

Topological aspects of gauge theories

In weak interaction theories all phenomena related to the nonlinearity of the field equations are rare and weak. The perturbative expansion converges rapidly. And so, in the early days, it was natural to think that topologically nontrivial field configurations would never play a significant role whatsoever.

Yet the first interesting idea about topologically nontrivial structures in gauge theories was launched by Holger-Bech Nielsen and Poul Olesen, and independently by Bruno Zumino.[35] They considered stable magnetic flux vortices in the Abelian Higgs model, suspecting that these might have something to do with the dual string theory for mesons. Then, when you try to generalize this to the non-Abelian case, you hit upon a paradox, and this led to the discovery of the magnetic monopole.[36] This monopole was also found along a different route. Alexander Polyakov at the Landau Institute studied three-dimensional "hedgehog solutions" and found that these topologically stable objects have finite energy when they are coupled to a non-Abelian gauge theory.[37] According to a footnote in his publication, it was Lev Okun who remarked that his hedgehog must carry magnetic charge, which is why I believe that perhaps also Okun's name should be attached to the monopole.

The flux vortex is stable in two dimensions, the monopole in three; is there anything that is topologically stable in four dimensions? Sure there is! Alexander Belavin, Polyakov, Albert Schwarz, and Yuri Tyupkin were the first to point out that pure gauge theories allow for such a structure.[38] But what are the physical consequences of this idea? Naturally, the fields are localized not only in three-space, but also in time. Hence they describe an event. This is why we called this an "instanton."[39]

And instantons may be important events, in particular if the gauge theory is coupled to fermions. The solutions of the coupled Dirac equation near an instanton are so special that several kinds of fermionic conservation laws may be broken there. One of those laws was a chiral symmetry law that would prevent the η meson from having a mass. We now know that the η mass is entirely due to QCD instantons.[40] The electroweak instanton, on the other hand, gives rise to more drastic violations of conservation laws, but it is very rare. Three quarks of each generation and one lepton of each generation could be absorbed by one instanton:

$$u + u + d + c + c + s + t + t + b \Rightarrow e^+ + \mu^+ + \tau^+ \qquad (10.33)$$

Now this will probably never be seen. (There are claims that this process becomes detectable at energies in the 10–20 TeV region.[41] It is, however, practically certain that even at high accelerator energies, it remains exponentially damped.) However, one experimental consequence of the electroweak instanton is immediate. We see that (10.33) would be at odds with electric charge conservation if the number of baryonic and leptonic generations were not equal. These numbers *must* be equal for the theory to be self-consistent. We observe that experiment agrees with us here.

Further developments

It is a characteristic of successful theories that they provide further understanding in many different areas of the field, in elegant and unsuspected ways. As for the Standard Model, we now know that the roles of asymptotic freedom, monopoles, and instantons are crucial in our present picture of quark confinement, the hadron spectrum, the scaling phenomena, and jet physics. The renormalized theory allows us to reproduce the observed data on the Z and W bosons with unprecedented precision. The Standard Model, as a gauge theory with fermions and at most only one scalar, is indeed tremendously successful.

Of course, after two decades have passed, the deficiencies in our theory are also standing out clearly. A theory that explains why the local symmetry is as it is, where the fermion spectrum comes from, and how the values of some 20 constants of Nature are determined is still being sought, but it is difficult to believe that such a giant leap in particle theory as occurred in the 1970s will be repeated in the near future.

Notes

1 B. E. Lautrup, A. Peterman, and E. de Rafael, "Recent developments in the comparison between theory and experiment in QED," *Phys. Rep. 3C* (1972), pp. 196–259.

2 A nice account of these views is presented in T. Y. Cao and S. S. Schweber, "The Conceptual Foundations and the Philosophical Aspects of Renormalization Theory," *Synthese 97* (1993), pp. 33–108; S. Schweber, "A Historical Perspective on the Rise of the Standard Model, Traditions, Men," Chapter 38, this volume.

3 W. Pauli and F. Villars, "On Regularization in Quantum Electrodynamics," *Rev. Mod. Phys. 21* (1949), pp. 434–41.

4 R. P. Feynman and M. Gell-Mann, "Theory of the Fermi Interactions," *Phys. Rev. 109* (1958), pp. 193–8; E. C. G. Sudarshan and R. E. Marshak, "Charality Invariance and the Universal Fermi Interactions," *Phys. Rev. 109* (1958), pp. 1860–1.

5 M. Veltman, "Perturbation Theory of Massive Yang–Mills Fields," *Nucl. Phys. B7* (1968), pp. 637–50; J. Reiff and M. Veltman, "Massive Yang–Mills Fields," *Nucl. Phys. B13* (1969), pp. 545–64; M. Veltman, "Generalized Ward Identities and Yang–Mills Fields," *Nucl. Phys. B21* (1970), pp. 288–302.

6 G. 't Hooft and M. Veltman, "One-loop divergencies in the theory of gravitation," *Annales de l' Institut Henri Poincaré 20* (1974), pp. 69–94.

7 G. 't Hooft, in "The Whys of Subnuclear Physics," ed., A. Zichichi (New York, London: Plenum), p. 943; G. 't Hooft, "Borel Summability of a Four-Dimensional Field Theory," *Phys. Lett. 119B* (1982), pp. 369–71; G. Parisi, "On Infrared Divergences," *Nucl. Phys. B150* (1979), pp. 163–72.

8 G. 't Hooft, "On the Convergence of Planar Diagram Expressions," *Comm. Math. Phys. 86* (1982), pp. 449–63; G. 't Hooft, "Rigorous Construction of Planar-Diagram Field Theories in Four-Dimensional Euclidian Space," *Comm. Math. Phys. 88* (1983), pp. 1–25.

9 C. N. Yang and R. L. Mills, "Conservation of Isotopic Spin and Isotopic Gauge Invariance," *Phys. Rev. 96* (1954), pp. 191–5; see also R. Shaw, Cambridge University Ph.D. thesis (unpublished).

10 R. P. Feynman, "Quantum Theory of Gravitation," *Acta Phys. Pol. 24* (1963), pp. 697–722.

11 S. L. Glashow, "Partial Symmetries of Weak Interactions," *Nucl. Phys. 22* (1961), pp. 579–88. M. Veltman, "Perturbation Theory of Massive Yang–Mills Fields"; J. Reiff and M. Veltman, "Massive Yang–Mills Fields"; M. Veltman, "Generalized Ward Identities."

12 M. Veltman, "Perturbation Theory of Massive Yang–Mills Fields"; J. Reiff and M. Veltman, "Massive Yang–Mills Fields"; M. Veltman, "Generalized Ward Identities."

13 M. Gell-Mann and M. Lévy, "The Axial Vector Current in Beta Decay," *Nuovo Cimento 16* (1960), pp. 705–26.

14 B. W. Lee, "Renormalization of the $\sigma-$ Model," *Nucl. Phys. B9* (1969), pp. 649–72; J.-L. Gervais and B. W. Lee, "Renormalization of the $\sigma-$ model," *Nucl. Phys. B12* (1969), pp. 627–46; B. W. Lee, "Chiral Dynamics," in *Cargèse Lectures in Physics*, Vol. 5 (New York: Gordon and Breach 1972); K. Symanzik, "Renormalizable Models with Simple Symmetry Breaking," *Lett. Nuovo Cimento 2* (1969), p. 10 and *Comm. Math. Phys. 16* (1970), pp. 48–80.

15 Strictly speaking a local gauge symmetry is never spontaneously broken, since the vacuum is always completely symmetric. The Higgs mechanism is really something else: a rearrangement of the spectrum of states. But because of its analogy with spontaneous breakdown of a global symmetry one often uses this incorrect phrase. Hence the quotation marks.

16 L. D. Faddeev and V. N. Popov, "Feynman Diagrams for the Yang–Mills Field," *Phys. Lett. 25B* (1967), pp. 29–30; see also L. D. Faddeev, "The Feynman Integral for Singular Lagrangians," *Theoretical and*

Mathematical Physics 1 (1969), pp. 3–18 (in Russian), pp. 1–13 (Engl. transl).

17 S. Mandelstam, "Feynman Rules for Electromagnetic and Yang–Mills Fields From the Gauge-Independent Field-Theoretic Formalism," *Phys. Rev. 175* (1968), pp. 1580–1603.

18 G. 't Hooft, "Renormalization of Massless Yang–Mills Fields," *Nucl. Phys. B33* (1971), pp. 173–99.

19 C. Becchi, A. Rouet, and R. Stora, "Renormalization of Gauge Theories," *Ann. Phys. (N.Y.) 98* (1976), pp. 287–321; I. V. Tyutin, "Cargèse Lectures," *Lebedev Report No. FIAN39* (1975), unpublished; R. Stora, "Continuum Gauge Theories," in M. Lévy and P. Mitter, eds., *New Developments in Quantum Field Theory and Statistical Mechanics* (New York: Plenum Press, 1977), pp. 201–24; J. Thierry-Mieg, "Geometrical Reinterpretation of Faddeev–Popov Ghost Particles and BRS Transformations," *J. Math. Phys. 21* (1980), pp. 2834–8.

20 S. L. Adler, "Axial Vector Vortex in Spin Electrodynamics," *Phys. Rev. 177* (1969), pp. 2426–38; J. S. Bell and R. Jackiw, "A PCAC Puzzle $\pi^0 \to \gamma\gamma$ in the $\sigma-$ Model," *Nuovo Cimento A60* (1969), pp. 47–60.

21 A. Slavnov, "Ward Identities in Gauge Theories," *Theoretical and Mathematical Physics 10* (1972) (English translation), pp. 99–104; J. C. Taylor, "Ward Identities and Charge Renormalization of the Yang–Mills Fields," *Nucl. Phys. B33* (1971), pp. 436–44.

22 G. 't Hooft, "Renormalizable Lagrangians for Massive Yang–Mills Fields," *Nucl. Phys. B35* (1971), pp. 167–88.

23 G. 't Hooft and M. Veltman, "Regularization and Renormalization of Gauge Fields," *Nucl. Phys. B44* (1972), pp. 189–213.

24 Kenneth G. Wilson, "Renormlization Group and Strong Interactions," *Phys. Rev. D3* (1971), pp. 1818–46; Kenneth G. Wilson and Michael E. Fisher, "Critical Exponents in 3.99 Dimensions," *Phys. Rev. Lett. 28* (1972), pp. 240–3.

25 C. Bollini and J. Giambiagi, "Dimensional Renormalization," *Nuovo Cim. 12B* (1972), pp. 20–6; J. Ashmore, "A Method of Gauge-Invariant Regularization," *Lett. Nuovo Cim. 4* (1972), pp. 289–90.

26 G. 't Hooft and M. Veltman, "Combinatorics of Gauge Fields," *Nucl. Phys. B50* (1972), pp. 318–53.

27 G. 't Hooft, "Prediction for Neutrino–Electron Cross-Sections in Weinberg's Model of Weak Interactions," *Phys. Lett. 37B* (1971), pp. 195–6.

28 Stephen L. Adler and William A. Bardeen, "Absence of Higher-Order Corrections in the Anomalous Axial-Vector Divergence Equation," *Phys. Rev. 182* (1969), pp. 1517–36; William A. Bardeen, "Anomalous Ward Identities in Spinor Field Theories," *Phys. Rev. 184* (1969), pp. 1848–59; David G. Boulware, "Quantum Field Theory in Schwarzschild and Rindler Spaces," *Phys. Rev. D11* (1975), pp. 1404–23; David G. Boulware, "Hawking Radiation and Thin Shells," *Phys. Rev. D13* (1976), pp. 2169–87.

29 C. Llewellyn-Smith, "High Energy Behaviour and Gauge Symmetry," *Phys. Lett. B46* (1973), pp. 233–6. John M. Cornwall, David N. Levin, and George Tiktopoulos, "Uniqueness of Spontaneously Broken Gauge Theories," *Phys. Rev. Lett. 30*, (1973), pp. 1268–70.

30 G. 't Hooft and M. Veltman, "Combinatorics of Gauge Fields."

31 M. Veltman, "Perturbation Theory of Massive Yang–Mills Fields";
    J. Reiff and M. Veltman, "Massive Yang–Mills Fields"; M. Veltman,
    "Generalized Ward Identities."

32 C. P. Korthals-Altes, ed., *Renormalization of Yang–Mills Fields and
    Applications to Particle Physics*, Marseille, 19–23 June 1972 (Marseille:
    Centre de Physique Théorique, 1972).

33 K. Symanzik, "On Theories with Massless Particles," in
    C. P. Korthals-Altes, ed., *Renormalization of Yang–Mills Fields*;
    K. Symanzik, "A Field Theory with Computable Large-Momenta
    Behaviour," *Lett. Nuovo Cimento 6* (1973), pp. 77–80.

34 David J. Gross and Frank Wilczek, "Ultraviolet Behavior of Non-Abelian
    Gauge Theories," *Phys. Rev. Lett. 30* (1973), pp. 1343–6; H. David
    Politzer, "Reliable Perturbative Results for Strong Interactions," *Phys.
    Rev. Lett. 30* (1973), pp. 1346–9; H. David Politzer, "Asymptotic
    Freedom: An Approach to Strong Interactions," *Phys. Rep. 14C* (1974),
    pp. 129–80.

35 H. B. Nielsen and P. Olesen, "Vortex-Line Models for Dual Strings,"
    Nucl. Phys. B61 (1973), pp. 45–61. Bruno Zumino, "Relativistic Strings
    and Supergauges," pp. 367–81, and "Application of Gauge Theories to
    Weak and Electromagnetic Interactions," pp. 383–98, both in Eduardo
    R. Caianiello, ed., *Renormalization and Invariance in Quantum Field
    Theory*, Capri Summer Meeting, July 1973 (New York: Plenum Press,
    1974).

36 G. 't Hooft, "Magnetic Monopoles in Unified Gauge Theories," *Nucl.
    Phys. B79* (1974), pp. 276–84.

37 A. M. Polyakov, "Particle Spectrum in Quantum Field Theory," *JETP
    Lett. 20* (1974), pp. 194–5.

38 A. A. Belavin, A. M. Polyakov, A. S. Schwartz, and Yu. S. Tyupkin,
    "Pseudoparticle Solutions of the Yang–Mills Equations," *Phys. Lett. 59B*
    (1975), pp. 85–7.

39 G. 't Hooft, "Symmetry Breaking Through Bell–Jackiw Anomalies,"
    Phys. Rev. Lett. 37 (1976), pp. 8–11; "Computation of the Quantum
    Effects due to a Four-Dimensional Pseudoparticle," *Phys. Rev. D14*
    (1976), pp. 3432–50; R. Jackiw and C. Rebbi, "Vacuum Periodicity in a
    Yang–Mills Quantum Theory," *Phys. Rev. Lett. 37* (1976), pp. 172–5;
    C. G. Callan, Jr., R. F. Dashen, and D. J. Gross, "The Structure of the
    Gauge Theory Vacuum," *Phys. Lett. 63B* (1976), pp. 334–40; Curtis G.
    Callen, Jr., Roger Dashen, and David J. Gross, "Toward a Theory of the
    Strong Interactions," *Phys. Rev. D17* (1978), pp. 2717–63.

40 G. 't Hooft, "How Instantons Solve the U(1) Problem," *Phys. Rep. 142*
    (1986), pp. 357–87.

41 A. De Rujula, H. Georgi, S. L. Glashow, and H. R. Quinn, "Fact and
    Fancy in Neutrino Physics," *Rev. Mod. Phys. 46* (1974), pp. 391–407.

11

Asymptotic Freedom
and the Emergence of QCD*

DAVID GROSS

Born Washington, D.C., 1941; Ph.D., 1966 (physics), University of California at Berkeley; Professor of Physics at Princeton University; high-energy physics (theory).

The Standard Model is surely one of the major intellectual achievements of the twentieth century. In the late 1960s and early 1970s, decades of path-breaking experiments culminated in the emergence of a comprehensive theory of particle physics. This theory identifies the basic fundamental constituents of matter and describes all the forces of nature relevant at accessible energies – the strong, weak, and electromagnetic interactions.

Science progresses in a much more muddled fashion than is often pictured in history books. This is especially true of theoretical physics, partly because history is written by the victorious. Consequently, historians of science often ignore the many alternate paths that people wandered down, the many false clues they followed, the many misconceptions they had. These alternate points of view are less clearly developed than the final theories, harder to understand and easier to forget, especially as these are viewed years later, when it all really does make sense. Thus reading history one rarely gets the feeling of the true nature of scientific development, in which the element of farce is as great as the element of triumph.

The emergence of quantum chromodynamics, or QCD, is a wonderful example of the evolution from farce to triumph. During a very short period, a transition occurred from experimental discovery and theoretical confusion to theoretical triumph and experimental confirmation. We were lucky to have been young then, when we could stroll along the newly opened beaches and pick up the many beautiful shells that exper-

---

* Work supported by NSF Grant PHY90-21984.

iment had revealed. In trying to relate this story, one must be wary of the danger of the personal bias that occurs as one looks back in time. It is not totally possible to avoid this. Inevitably, one is fairer to oneself than to others, but one can try. In any case the purpose of this volume, I gather, is to provide raw material for the historians. One can take consolation from Emerson, who said that "There is properly no history; only biography."

The theoretical scene

I would like first to describe the scene in theoretical particle physics, as I saw it in the early 1960s at Berkeley, when I started as a graduate student. The state of particle physics was then almost the complete opposite of today. It was a period of experimental supremacy and theoretical impotence. The construction and utilization of major accelerators were proceeding at full steam. Experimental discoveries and surprises appeared every few months. There was hardly any theory to speak of. The emphasis was on phenomenology, and there were only small islands of theoretical advances here and there. Field theory was in disgrace; S-matrix theory was in full bloom. Symmetries were all the rage.

The field was divided into the study of the weak and the strong interactions. In the case of the weak interactions, there was a rather successful phenomenological theory but not much new data. The strong interactions were where the experimental and theoretical action was, particularly at Berkeley. They were regarded as especially unfathomable. The prevalent feeling was that it would take a very long time to understand the strong interactions, and that it would require revolutionary concepts. For a young graduate student this was clearly the major challenge.

The feeling at the time was well expressed by Lev Landau in his last paper, called "Fundamental Problems," which appeared in a memorial volume to Wolfgang Pauli in 1959.[1] In this paper he argued that quantum field theory had been nullified by the discovery of the zero-charge problem. He wrote:

> It is well known that theoretical physics is at present almost helpless in dealing with the problem of strong interactions.... By now the nulli-fication of the theory is tacitly accepted even by theoretical physicists who profess to dispute it. This is evident from the almost complete disappearance of papers on meson theory and particularly from Dyson's assertion that the correct theory will not be found in the next hundred years.

Let us explore the theoretical milieu at this time.

Quantum field theory

Quantum field theory was originally developed for the treatment of electrodynamics almost immediately after the completion of quantum mechanics and the emergence of the Dirac equation. It seemed to be the natural tool for describing the dynamics of elementary particles. The application of quantum field theory had important early success. Fermi formulated a powerful and accurate phenomenological theory of beta decay, which was to serve as a framework for exploring the weak interactions for three decades. Yukawa proposed a field theory to describe the nuclear force and predicted the existence of heavy mesons, which were soon discovered. On the other hand, the theory was confronted from the beginning with severe difficulties. These included the infinities that appeared as soon as one went beyond lowest-order perturbation theory, as well as the lack of any nonperturbative understanding of dynamics. By the 1950s the suspicion of field theory had deepened to the point that a powerful dogma emerged – that field theory was fundamentally wrong, especially in its application to the strong interactions.

The renormalization procedure, developed by Richard Feynman, Julian Schwinger, Sin-itiro Tomanaga, and Freeman Dyson, was spectacularly successful in quantum electrodynamics. However, the physical meaning of renormalization was not truly understood. The feeling of most was that renormalization was a trick. This was especially the case for the pioneering inventors of quantum field theory (for example Dirac and Wigner). They were prepared at the first drop of an infinity to renounce their belief in quantum field theory and to brace for the next revolution. However, it was also the feeling of the younger leaders of the field, who had laid the foundations of perturbative quantum field theory and renormalization in the late 1940s. The prevalent feeling was that renormalization simply swept the infinities under the rug, but that they were still there and rendered the notion of local fields meaningless. To quote Feynman, speaking at the 1961 Solvay conference,[2] "I still hold to this belief and do not subscribe to the philosophy of renormalization."

Field theory was almost totally perturbative at that time. The nonperturbative techniques that had been tried in the 1950s had all failed. The path integral, developed by Feynman in the late 1940s, which later proved so valuable for a nonperturbative formulation of quantum field theory as well as a tool for semiclassical expansions and numerical approximations, was almost completely forgotten. In a sense the Feynman rules were too successful. They were an immensely useful, picturesque,

and intuitive way of performing perturbation theory. However, these alluring qualities also convinced many that all that was needed from field
theory were these rules. They diverted attention from the nonperturbative dynamical issues facing field theory. In my first course on quantum
field theory at Berkeley in 1965, I was taught that *field theory equals
Feynman rules.*

No examples were known of four-dimensional field theories that one
could handle nonperturbatively. Indeed, except for free field theory and
a few examples of soluble two-dimensional field theories, there were no
models that could serve as practice grounds for developing field-theoretic
tools and intuition. The more mathematically inclined warned us that
most naive manipulations in quantum field theory were unjustified and
that the whole structure was potentially unsound. They initiated a
program of axiomatic analysis that promised to inform us in a decade
or two whether quantum field theory made sense.

In the United States, however, I think the main reason for the aban-
donment of field theory was simply that one could not calculate. American physicists are inveterate pragmatists. Quantum field theory had
not proved to be a useful tool with which to make contact with the
explosion of experimental discoveries. The early attempts in the 1950s
to construct field theories of the strong interactions were total failures.
In hindsight this was not surprising since a field theory of the strong
interactions faced two enormous problems. First, which fields to use?
Following Yukawa, the first attempts employed pion and nucleon fields.
Soon, with the rapid proliferation of particles, it became evident that
nothing was special about the nucleon or the pion. All the hadrons,
the strange baryons and mesons as well as the higher-spin recurrences
of these, appeared to be equally fundamental. The obvious conclusion
that all hadrons were composites of more fundamental constituents was
thwarted by the fact that no matter how hard one smashed hadrons at
each other, one had not been able to liberate these hypothetical constituents. This was not analogous to the paradigm of atoms made of
nucleons and electrons or of nuclei composed of nucleons. The idea of
permanently bound, confined constituents was unimaginable at the time.
Second, since the pion–nucleon coupling was so large, perturbative expansions were useless. All attempts at nonperturbative analysis were
unsuccessful.

In the case of the weak interactions, the situation was somewhat better. Here one had an adequate effective theory – the four-fermion Fermi
interaction, which could be usefully employed, using perturbation theory

to lowest order, to organize and understand the emerging experimental picture of the weak interactions. The fact that this theory was non-renormalizable meant that beyond the Born approximation it lost all predictive value. This disease increased the suspicion of field theory. Yang–Mills theory, which had appeared in the mid 1950s, was not taken seriously. Attempts to apply Yang–Mills theory to the strong interactions focused on elevating global flavor symmetries to local gauge symmetries. This was problematic since these symmetries were not exact. In addition, non-Abelian gauge theories apparently required massless vector mesons – clearly not a feature of the strong interactions.

In the Soviet Union field theory was under even heavier attack, for somewhat different reasons. Landau and collaborators, in the late 1950s, studied the high-energy behavior of quantum electrodynamics. They explored the relation between the physical electric charge and the bare electric charge (essentially the electric charge that controls the physics at energies of order of the ultraviolet cutoff). They concluded, on the basis of their approximations, that the physical charge vanishes, for any value of the bare charge, as we let the ultraviolet cutoff become infinite (this is of course necessary to achieve a Lorentz-invariant theory):[3]

> We reach the conclusion that within the limits of formal electrodynam-
> ics a point interaction is equivalent, for any intensity whatever, to no
> interaction at all.

This is the famous problem of *zero charge*, a startling result that implied for Landau that *"weak coupling electrodynamics is a theory which is, fundamentally, logically incomplete."*[4] This problem occurs in any theory that is not asymptotically free. Even today, many of us believe that any such theory – for example, QED – if taken by itself, is inconsistent at very high energies. In the case of QED this is only an academic problem, since the trouble shows up only at enormously high energy. However, in the case of the strong interactions, it was an immediate catastrophe. In the Soviet Union this was thought to be a compelling reason why field theory was wrong. Landau decreed:[5]

> We are driven to the conclusion that the Hamiltonian method for strong
> interaction is dead and must be buried, although of course with deserved
> honor.

Under the influence of Landau and Pomeranchuk, a generation of physicists was forbidden to work on field theory. One might wonder why the discovery of the zero-charge problem did not inspire a search for asymptotically free theories that would be free of this disease. The

answer, I think, is twofold. First, many other theories were explored –
in each case they behaved as QED. Second, Landau and Pomeranchuk
concluded, I think, that this problem was inherent in any quantum field
theory, that an asymptotically free theory could not exist. (As Kenneth
Johnson pointed out,[6] a calculation of the charge renormalization of
charged vector mesons was carried out by V. S. Vanyashin and M. V.
Terentév in 1964.[7] They got the magnitude wrong but did get the
correct sign and concluded that the result was absurd. They attributed
this wrong sign to the non-renormalizability of charged vector-meson
theory.)

The bootstrap

The bootstrap theory rested on two principles, both more philosophical
than scientific. First, local fields were not directly measurable. Thus
they were unphysical and meaningless. Instead, one should formulate the
theory using only observables. The basic observables are the S-matrix
elements measured in scattering experiments. Microscopic dynamics
was renounced. Field theory was to be replaced by S-matrix theory,
a theory based on general principles, such as unitarity and analyticity,
but with no fundamental microscopic Hamiltonian. The basic dynamical
idea was that there was a unique S-matrix that obeyed these principles.
It could be determined without the unphysical demand of fundamental
constituents or equations of motion that was inherent in field theory. To
quote Geoffrey Chew:[8]

> Let me say at once that I believe the conventional association of fields
> with strong interacting particles to be empty. It seems to me that no
> aspect of strong interactions has been clarified by the field concept.
> Whatever success theory has achieved in this area is based on the uni-
> tarity of the analytically continued S-matrix plus symmetry principles.
> I do not wish to assert (as does Landau) that conventional field theory
> is necessarily wrong, but only that it is sterile with respect to the strong
> interactions and that, like an old soldier, it is destined not to die but
> just to fade away.

In hindsight, it is clear that the bootstrap was born from the frustration of being unable to calculate anything using field theory. All models
and approximations produced conflicts with some dearly held principle.
If it was so difficult to construct an S-matrix that was consistent with
sacred principles, then maybe these general principles had a unique manifestation. The second principle of the bootstrap was that there were

no elementary particles. The way to deal with the increasing number of candidates for elementary status was to proclaim that all were equally fundamental, all were dynamical bound states of each other. This was called "nuclear democracy" and was a response to the proliferation of candidates for fundamental building blocks. As Chew stated,[9]

> The notion, inherent in conventional Lagrangian field theory, that certain particles are fundamental while others are complex, is becoming less and less palatable for baryons and mesons as the number of candidates for elementary status continues to increase.

The bootstrap idea was immensely popular in the early 1960s, for a variety of reasons. Superseding quantum field theory, it rested on the solid principles of causality and unitarity. It was real and physical. It promised to be very predictive, indeed to provide a unique value for all observables, satisfying a basic desire of particle physicists to believe that the world around us is not arbitrary and that, to quote Einstein,

> Nature is so constituted that it is possible logically to lay down such strongly determined laws that within these laws only rationally, completely determined constants occur, not ones whose numerical value could be changed without destroying the theory.

The bootstrap promised that this hope would be realized already in the theory of the strong interactions. This is of course false. We now know that there are an infinite number of consistent S-matrices that satisfy all the sacred principles. One can take any non-Abelian gauge theory, with any gauge group, and many sets of fermions (as long as there are not too many to destroy asymptotic freedom). The hope for uniqueness must wait for a higher level of unification.

In Berkeley, as in the Soviet Union, S-matrix theory was supreme, and a generation of young theorists was raised ignorant of field theory. Even on the calmer East Coast, S-matrix theory swept the field. For example, Marvin Goldberger wrote:[10]

> My own feeling is that we have learned a great deal from field theory ... that I am quite happy to discard it as an old, but rather friendly, mistress whom I would be willing to recognize on the street if I should encounter her again. From a philosophical point of view and certainly from a practical one the S-matrix approach at the moment seems to me by far the most attractive.

S-matrix theory had some notable successes: the early application of dispersion relations and the development of Regge pole theory. However, there were drawbacks to a theory that was based on the principle that there was no theory, at least not in the traditional sense. As Francis Low stated:[11]

> The distinction between S-Matrix theory and field theory is, on the one hand, between a set of equations that are not formulated, and on the other hand between a set of equations that are formulated if you knew what they were and for which you do not know whether there is a solution or not.

Nonetheless, until 1973 it was not thought proper to use field theory without apologies. For example, as late as the National Accelerator Laboratory (NAL) Conference of 1972, Murray Gell-Mann ended his talk on quarks with the summary:[12]

> Let us end by emphasizing our main point, that it may well be possible to construct an explicit theory of hadrons, based on quarks and some kind of glue, treated as fictitious, but with enough physical properties abstracted and applied to real hadrons to constitute a complete theory. Since the entities we start with are fictitious, there is no need for any conflict with the bootstrap or conventional dual parton point of view.

Symmetries

If dynamics was impossible, one could at least explore the symmetries of the strong interactions. The biggest advance of the early 1960s was the discovery of an approximate symmetry of hadrons, SU(3), by Gell-Mann and Yuval Ne'eman, and then the beginning of the understanding of spontaneously broken chiral symmetry. Since the relevant degrees of freedom, especially color, were totally hidden from view due to confinement, the emphasis was on flavor, which was directly observable. This emphasis was enhanced because of the success of SU(3). Nowadays we realize that SU(3) is an accidental symmetry, which arises simply because a few quarks (the up, down, and strange quarks) are relatively light compared to the scale of the strong interactions. At the time it was regarded as a deep symmetry of the strong interactions, and many attempts were made to generalize it and use it as a springboard for a theory of hadrons.

The most successful attempt was Gell-Mann's algebra of currents.[13] In an important and beautiful paper, he outlined a program for abstracting relations from a field theory, keeping the ones that might be generally true and then throwing the field theory away.[14]

> In order to obtain such relations that we conjecture to be true, we use the method of abstraction from a Lagrangian field theory model. In

other words, we construct a mathematical theory of the strongly interacting particles, which may or may not have anything to do with reality, find suitable algebraic relations that hold in the model, postulate their validity, and then throw away the model. We may compare this process to a method sometimes employed in French cuisine: a piece of pheasant meat is cooked between two slices of veal, which are then discarded.

This paper made quite an impression, especially on impoverished graduate students like me, who could only dream of eating such a meal. It was a marvelous approach. It gave one the freedom to play with the forbidden fruit of field theory, abstract what one wanted from it, all without having to believe in the theory. The only problem was that it was not clear what principle determined what to abstract.

The other problem with this approach was that it diverted attention from dynamical issues. The most dramatic example of this is Gell-Mann and George Zweig's hypothesis of quarks, the most important consequence of the discovery of SU(3).[15] The fact was that hadrons looked as if they were composed of (colored) quarks whose masses (either the current quark masses or the constituent quark masses) were quite small. Color had been introduced by Yoichiro Nambu, M. Y. Han and Nambu, and O. W. Greenberg.[16] Nambu's motivation for color was twofold, first to offer an explanation of why only (what we would now call) color singlet hadrons exist by postulating a strong force (but with no detailed specification as to what kind of force) coupled to color which was responsible for the fact that color-neutral states were lighter than colored states. His second motivation, which he explored with Han, was the desire to construct models in which the quarks had integer-valued electric charges. Greenberg's motivation was to explain the strange statistics of nonrelativistic quark model hadronic bound states (a concern of Nambu's as well). He introduced parastatistics for this purpose, which equally well solved the statistics problem, but clouded the dynamical significance of this quantum number. Yet quarks had not been seen, even when energies were achieved that were ten times the threshold for their production. This was not analogous to atoms made of nuclei and electrons or to nuclei made of nucleons. The nonrelativistic quark model simply did not make sense. The conclusion was that quarks were fictitious, mathematical devices. With this attitude one could ignore the apparently insoluble dynamical problems that arose if one tried to imagine that quarks were real.

It was a pity that particle theorists at that time, for the most part, totally ignored condensed matter physics. There were of course notable exceptions such as Nambu, and the last of the true universalists, Landau, who unfortunately was incapacitated at an early age. This attitude was largely a product of arrogance. Particle physics was much more fundamental and basic than the solid-state physics that studied collections of many atoms, whose basic laws of interaction were well understood. Thus particle physicists thought that they had little to learn from "dirt physics" (or "squalid-state physics"). This attitude was unfortunate. We would have profited much from a deeper study of superconductivity – the preeminent advance in condensed matter physics in this period: not only the insight it gave into broken gauge symmetry, stressed by Philip Anderson, but also of the possibility of confinement. The Meissner effect that occurs in the superconducting state is a very good, dual (under interchange of electric and magnetic fields) analog of confinement. Indeed if magnetic monopoles existed, they would form, in the superconducting state, magnetically neutral bound states that would be quite analogous to hadrons. This idea was not explored by condensed-matter physicists either, perhaps since monopoles had not been found. The situation might have been different if monopoles had existed to provide a live example of confinement.

This attitude toward quarks persisted until 1973 and beyond. Quarks clearly did not exist as real particles: therefore, they were fictitious devices (see Gell-Mann above). One might "abstract" properties of quarks from some model, but one was not allowed to believe in their reality or to take the models too seriously.

For many this smelled fishy. I remember very well Steven Weinberg's reaction to the sum rules Curtis Callan and I had derived using the quark–gluon model. I described my work on deep-inelastic scattering sum rules to Weinberg at a Junior Fellows dinner at Harvard. As I needed him to write a letter of recommendation to Princeton, I was a little nervous. I explained how the small longitudinal cross section observed at SLAC could be interpreted, on the basis of our sum rule, as evidence for quarks. Weinberg was emphatic that this was of no interest since he did not believe anything about quarks. I was somewhat shattered.

Experiment

This was a period of great experimental excitement. However, I would like to discuss an interesting phenomenon, in which theorists and experimentalists reinforced each other's conviction that the secret of the strong interactions lay in the high-energy behavior of scattering amplitudes at low momentum transfer. Early scattering experiments concentrated, for obvious reasons, on the events that had the largest rates. In the case of the strong interactions, this meant searching for resonant bumps or probing near forward scattering, where the cross section was largest. It was not at all realized by theorists that the secret of hadronic dynamics could be revealed by experiments at large momentum transfer that probed the short-distance structure of hadrons. Instead, prompted by the regularities that were discovered at low momentum transfer, theorists developed an explanation based on the theory of Regge poles. This was the only strong interaction dynamics that was understood, for which there was a real theory. Therefore, theorists concluded that Regge behavior must be very important and forward-scattering experiments were deemed to be the major tool of discovery. Regge theory was soon incorporated into the bootstrap program as a boundary condition. In response to this theoretical enthusiasm, the interest of experimentalists in forward scattering was enhanced. Opportunities to probe the less easily accessible domains of large momentum transfer were ignored. Only much later, after the impact of the deep-inelastic scattering experiments that had been ridiculed by many as unpromising, was it understood that the most informative experiments were those at large momentum transfers that probe short or light-like distances.

It used to be the case that when a new accelerator was initiated, one of the first and most important experiments to be performed was the measurement of the total p–p cross section. Nowadays, this experiment is regarded with little interest, even though the explanation of Regge behavior remains an interesting, unsolved, and complicated problem for QCD. Ironically, one of the principal justifications for this experiment today is simply to calibrate the luminosity of the machine.

My road to asymptotic freedom
From N/D to QCD

I was a graduate student at Berkeley at the height of the bootstrap and S-matrix theory. My Ph.D. thesis was written under the supervision of Geoff Chew, the main guru of the bootstrap, on multibody N/D equations. I can remember the precise moment at which I was disillusioned with the bootstrap program. This was at the 1966 Rochester meeting, held at Berkeley. Francis Low, in the session following his talk, remarked that the bootstrap was less of a theory than a tautology.[17]

> I believe that when you find that the particles that are there in S-matrix theory, with crossing matrices and all the formalism, satisfy all these conditions, all you are doing is showing that the S-matrix is consistent with the world the way it is; that is the particles have put themselves there in such a way that it works out, but you have not necessarily explained that they are there.

For example, the then-popular finite-energy sum rules (whereby one derived relations for measurable quantities by saturating dispersion relations with a finite number of resonance poles on the one hand and relating these to the assumed Regge asymptotic behavior on the other) were not so much predictive equations, but merely checks of axioms (analyticity, unitarity) using models and fits of experimental data.

I was very impressed with this remark and longed to find a more powerful dynamical scheme. This was the heyday of current algebra, and the air was buzzing with marvelous results. I was very impressed by the fact that one could assume a certain structure of current commutators and derive measurable results. The most dramatic of these was the Adler–Weisberger relation that had just appeared.[18] Clearly the properties of these currents placed strong restrictions on hadronic dynamics.

The most popular scheme then was current algebra. Gell-Mann and Roger Dashen were trying to use the commutators of certain components of the currents as a basis for strong-interaction dynamics.[19] After a while I concluded that this approach was also tautological – all it did was test the validity of the symmetries of the strong interactions. This was apparent for vector SU(3). However, it was also true of chiral SU(3), especially as the current-algebra sum rules were interpreted, by Weinberg and others, as low-energy theorems for Goldstone bosons. This scheme could not be a basis for a complete dynamical theory.

I studied the less understood properties of the algebra of local current densities. These were model dependent; but that was fine, they therefore might contain dynamical information that went beyond statements of global symmetry. Furthermore, as was soon realized, one could check one's assumptions about the structure of local current algebra by deriving sum rules that could be tested in deep-inelastic lepton–hadron scattering experiments. James Bjorken's 1967 paper, on the application of U(6)× U(6), particularly influenced me.[20]

In the spring of 1968, Curtis Callan and I proposed a sum rule to test the then-popular "Sugawara model," a dynamical model of local currents, in which the energy–momentum tensor was expressed as a product of currents.[21] The hope was that the algebraic properties of the currents and the expression for the Hamiltonian in terms of these would be enough to have a complete theory. (This idea actually works in the now very popular two-dimensional conformal field theories.) Our goal was slightly more modest – to test the hypothesis by exploiting the fact that in this theory the operator-product expansion of the currents contained the energy–momentum tensor with a known coefficient. Thus we could derive a sum rule for the structure functions that could be measured in deep-inelastic electron–proton scattering.[22]

In the fall of 1968, Bjorken noted that this sum rule, as well as dimensional arguments, would suggest the scaling of deep-inelastic scattering cross sections.[23] This prediction was shortly confirmed by the new experiments at SLAC, which were to play such an important role in elucidating the structure of hadrons.[24] Shortly thereafter Callan and I discovered that by measuring the ratio $R = \frac{\sigma_L}{\sigma_T}$ (where σ_L and σ_T are the cross sections for the scattering of longitudinal and transverse polarized virtual photons), one could determine the spin of the charged constituents of the nucleon.[25] We evaluated the moments of the deep-inelastic structure functions in terms of the equal-time commutators of the electromagnetic field using specific models for these – the *algebra of fields* in which the current was proportional to a spin-1 field on the one hand, and the quark–gluon model on the other. In this popular model quarks interacted through an Abelian gauge field (which could, of course, be massive) coupled to baryon number. The gauge dynamics of the gluon had never been explored, and I do not think that the model had been used to calculate anything until then. We discovered that R depended crucially on the spin of the charged constituents. If the constituents had spin-0 or 1, then $\sigma_T = 0$, but if they had spin-$\frac{1}{2}$, then $\sigma_L = 0$. This was a rather dramatic result. The experiments quickly showed that σ_L was very small.

These SLAC deep-inelastic scattering experiments had a profound impact on me. They clearly showed that the proton behaved, when observed over short times, as if it were made out of pointlike objects of spin one-half. In the spring of 1969, which I spent at CERN, Chris Llewelynn-Smith and I analyzed the sum rules that followed for deep-inelastic neutrino–nucleon scattering using similar methods.[26] We were clearly motivated by the experiments that were then being performed at CERN. We derived a sum rule that measured the baryon number of the charged constituents of the proton. The experiments soon indicated that the constituents of the proton had baryon number $\frac{1}{3}$ – in other words, they again looked like quarks. I was then totally convinced of the reality of quarks. They had to be more than just mnemonic devices for summarizing hadronic symmetries, as they were then universally regarded. They had to be physical pointlike constituents of the nucleon. But how could that be? Surely strong interactions must exist between the quarks that would smear out their pointlike behavior.

After the experiments at SLAC, Feynman came up with his "parton" picture of deep-inelastic scattering. This was a very picturesque and intuitive way of describing deep-inelastic scattering in terms of assumed pointlike constituents – partons.[27] It complemented the approach to deep-inelastic scattering based on the operator product of currents, and had the advantage of being extendible to other processes.[28] The parton model allowed one to make predictions with ease, ignoring the dynamical issues at hand. I felt more comfortable with the approach based on assuming properties of current products at short distances. I felt somewhat uneasy about the extensions of the parton model to processes that were not truly dominated by short-distance singularities.

At CERN I studied, with Julius Wess, the consequences of exact scale and conformal invariance.[29] However, I soon realized that in a field-theoretic context only a free, noninteracting theory could produce exact scaling. This became very clear to me in 1970, when I came to Princeton, where my colleague Curtis Callan (and Kurt Symanzik) had rediscovered the renormalization group equations, which they presented as a consequence of a scale-invariance *anomaly*.[30] Their work made it abundantly clear that once one introduced interactions into the theory, scaling, as well as my beloved sum rules, went down the tube. Yet the experiments indicated that scaling was in fine shape. But one could hardly turn off the interactions between the quarks, or make them very weak, since then one would expect hadrons to break up easily into their quark constituents. Why then had no one ever observed free quarks? This paradox and the search for an explanation of scaling were to preoccupy me for the following four years.

How to explain scaling

About the same time that all this was happening, string theory was invented, in one of the most bizarre turns of events in the history of physics. In 1968 Gabriele Veneziano came up with a remarkably simple formula that summarized many features of hadronic scattering. It had Regge asymptotic behavior in one channel and narrow resonance saturation in the other.[31] This formula was soon generalized to multi-particle S-matrix amplitudes and attracted much attention. The dual resonance model was born, the last serious attempt to implement the bootstrap. It was only truly understood as a theory of quantized strings in 1972. I worked on this theory for two years, first at CERN and then at Princeton with John Schwarz and Andre Neveu. At first I felt that this model, which captured many of the features of hadronic scattering, might provide the long-sought alternative to a field theory of the strong interactions. However, by 1971 I realized that there was no way that this model could explain scaling, and I felt strongly that scaling was the paramount feature of the strong interactions. In fact the dual resonance model led to incredibly soft behavior at large momentum transfer, quite the opposite of the hard scaling observed. Furthermore, it was clear that it required for consistency many features that were totally unrealistic for the strong interactions – massless vector and tensor particles. These features later became the motivation for the hope that string theory might provide a comprehensive and unified theory of all the forces of Nature. This hope remains strong. However, the relevant energy scale is not 1 GeV but rather 10^{19} GeV !

The data on deep-inelastic scattering were getting better. No violations of scaling were observed, and the free-field-theory sum rules worked. I remember well the 1970 Kiev conference on high-energy physics. There I met Sasha Polyakov and Sasha Migdal, uninvited, but already impressive participants at the meeting. Polyakov, Migdal, and I had long discussions about deep-inelastic scattering. Polyakov knew all about the renormalization group and explained to me that naive scaling cannot be right. Because of renormalization the dimensions of operators change with the scale of the physics being probed. Not only that, dimensionless couplings also change with scale. They approach at small distances fixed-point values that are generically those of a strongly coupled theory, resulting in large anomalous scaling behavior quite different from free-field-theory behavior. I retorted that the experiments showed otherwise. He responded that this behavior contradicts field theory. We departed; he convinced, as many were, that experiments at higher energies would change, I that the theory would have to be changed.

The view that the scaling observed at SLAC was not a truly asymptotic phenomenon was rather widespread. The fact that scaling set in at rather low momentum transfers, "precocious scaling," reinforced this view. Thus the cognoscenti of the renormalization group (Wilson,

Polyakov, and others) believed that the noncanonical scaling indicative of a nontrivial fixed point of the renormalization group would appear at higher energies.

Much happened during the next two years. Gerard 't Hooft's spectacular work on the renormalizability of Yang–Mills theory reintroduced non-Abelian gauge theories to the community.[32] The electroweak theory of Sheldon Glashow, Weinberg, and Abdus Salam was revived. Field theory became popular again, at least in application to the weak interactions. The path integral reemerged from obscurity.

Kenneth Wilson's development of the operator-product expansion provided a tool that could be applied to the analysis of deep-inelastic scattering. Most important from my point of view was the revival of the renormalization group by Wilson.[33] The renormalization group stems from the fundamental work of Gell-Mann and Low, E. Stueckelberg and A. Petermann, and Bogoliubov and Shirkov.[34] This work was neglected for many years, partly because it seemed to provide only information about physics for large spacelike momenta, which are of no direct physical interest. Also, before the discovery of asymptotic freedom, the ultraviolet behavior was not calculable using perturbative methods, and there were no others. Thus it appeared that the renormalization group provided a framework in which one could discuss, but not calculate, the asymptotic behavior of amplitudes in a physically uninteresting region. Wilson's development of the operator-product expansion provided a new tool that could be applied to the analysis of deep-inelastic scattering. The Callan–Symanzik equations simplified the renormalization group analysis, which was then applied to the Wilson expansion.[35] The operator-product analysis was extended to the light cone, the relevant region for deep-inelastic scattering.[36] Most influential was Wilson's deep understanding of renormalization, which he was then applying to critical behavior. Wilson gave a series of lectures at Princeton in the spring of 1972.[37] These had a great impact on many of the participants, certainly on me.

So by the end of 1972, I had learned enough field theory, especially renormalization group methods from Wilson, to tackle the problem of scaling head on. I decided, quite deliberately, to prove that local field theory could not explain the experimental fact of scaling and thus was not an appropriate framework for the description of the strong interactions. Thus, deep-inelastic scattering would finally settle the issue as to the validity of quantum field theory.

The plan of the attack was twofold. First, I would prove that "ultraviolet stability," the vanishing of the effective coupling at short distances, later called asymptotic freedom, was necessary to explain scaling. Second, I would show that there existed no asymptotically free field theories. The latter was to be expected. After all, the paradigm of quantum field theory – quantum electrodynamics (QED) – was *infrared stable*; in other words, the effective charge grew larger at short distances and no one had ever constructed a theory in which the opposite occurred.

Charge renormalization is nothing more (certainly in the case of QED) than vacuum polarization. The vacuum or the ground state of a relativistic quantum-mechanical system can be thought of as a medium of virtual particles. In QED the vacuum contains virtual electron–positron pairs. A charge, e_0, put in this medium polarizes it. Such a medium with virtual electric dipoles will screen the charge and the actual, observable, charge e, will differ from e_0 as e_0/ϵ, where ϵ is the dielectric constant. Now ϵ is frequency dependent (or energy or distance dependent). To deal with this one can introduce the notion of an effective coupling $e(r)$, which governs the force at a distance r. As r increases, there is more medium that screens, thus $e(r)$ decreases with increasing r, and correspondingly increases with decreasing r! The β-function, which is simply minus the derivative of $\log[e(r)]$ with respect to $\log(r)$, is therefore positive.

If the effective coupling were, contrary to QED, to decrease at short distances, one might explain how the strong interactions turn off in this regime and produce scaling. Indeed, one might suspect that this is the only way to get pointlike behavior at short distances. It was well understood, due to Wilson's work and its application to deep-inelastic scattering, that one might expect to get scaling in a quantum field theory at a fixed point of the renormalization group. However, this scaling would not have canonical, free-field-theory-like behavior. Such behavior would mean that the scaling dimensions of the operators that appear in the product of electromagnetic currents at lightlike distances had canonical, free-field dimensions. This seemed unlikely. I knew that if the fields themselves had canonical dimensions, then for many theories this implied that the theory was trivial, that is, free. Surely this was also true if the composite operators that dominated the amplitudes for deep-inelastic scattering had canonical dimensions.

By the spring of 1973, Callan and I had completed a proof of this argument, extending an idea of Giorgio Parisi to all renormalizable field theories, with the exception of non-Abelian gauge theories.[38] The essential idea was to prove that the vanishing anomalous dimensions of the composite operators, at an assumed fixed point of the renormalization group, implied the vanishing anomalous dimensions of the fields. This then implied that the theory was free at this fixed point. The conclusion was that naive scaling could be explained only if the assumed fixed point of the renormalization group was at the origin of coupling space – that is, the theory must be asymptotically free.[39] Non-Abelian gauge theories were not included in the argument since both arguments broke down for these theories. The discovery of asymptotic freedom made this omission irrelevant.

The second part of the argument was to show that there were no asymptotically free theories at all. I had set up the formalism to analyze the most general renormalizable field theory of fermions and scalars – again excluding non-Abelian gauge theories. This was not difficult, since to investigate asymptotic freedom it suffices to study the behavior of the β-functions in the vicinity of the origin of coupling-constant space, that is, in lowest-order perturbation theory (one-loop approximation). I almost had a complete proof but was stuck on my inability to prove a necessary inequality. I discussed the issue with Sidney Coleman, who was spending the spring semester in Princeton. He came up with the missing ingredient, and added some other crucial points – and we had a proof that no renormalizable field theory that consisted of theories with arbitrary Yukawa, scalar, or Abelian gauge interactions could be asymptotically free.[40] Tony Zee had also been studying this. He too was well aware of the advantages of an asymptotically free theory and was searching for one. He derived at the same time a partial result, indicating the lack of asymptotic freedom in theories with SU(N) invariant Yukawa couplings.[41]

The discovery of asymptotic freedom

Frank Wilczek started work with me in the fall of 1972. He had come to Princeton as a mathematics student, but soon discovered that he was really interested in particle physics. He switched to the physics department, after taking my field-theory course in 1971, and started to work with me. My way of dealing with students, then and now, was to involve them closely with my current work and very often to work with them directly. This was certainly the case with Frank, who functioned more as a collaborator than a student from the beginning. I told him about my program to determine whether quantum field theory could account for scaling. We decided that we would calculate the β-function for Yang–Mills theory. This was the one hole in the line of argument I was pursuing. It had not been filled largely because Yang–Mills theory still seemed strange and difficult. Few calculations beyond the Born approximation had ever been done. Frank was interested in this calculation for other reasons as well. Yang–Mills theory was already in use for the electroweak interactions, and he was interested in understanding how these behaved at high energy.

Coleman, who was visiting in Princeton, asked me at one point whether anyone had ever calculated the β-function for Yang–Mills theory. I told him that we were working on this. He expressed interest because he had asked his student, H. David Politzer, to generalize the mechanism he had explored with Eric Weinberg – that of dynamical symmetry breaking of an Abelian gauge theory – to the non-Abelian case. An important ingredient was the knowledge of the renormalization flow, to decide whether lowest-order perturbation theory could be a reliable guide to the behavior of the energy functional. Indeed, Politzer went ahead with his own calculation of the β-function for Yang–Mills theory.

Our calculation proceeded slowly. I was involved in the other parts of my program and there were some tough issues to resolve. We first tried to prove on general grounds, using spectral representations and unitarity, that the theory could not be asymptotically free, generalizing the arguments of Coleman and me to this case. This did not work, so we proceeded to calculate the β-function for a Yang–Mills theory. Today this calculation is regarded as quite simple and even assigned as a homework problem in quantum field-theory courses. At the time it was not so easy. This change in attitude is the analogue, in theoretical physics, of the familiar phenomenon in experimental physics whereby yesterday's great discovery becomes today's background. It is always easier to do a calculation when you know what the result is and you are sure that the methods make sense.

One problem we had to face was that of gauge invariance. Unlike QED, where the charge renormalization was trivially gauge invariant (because the photon is neutral), the renormalization constants in QCD were all gauge dependent. However, the physics could not depend on the gauge. Another issue was the choice of regularization. Dimensional regularization had not really been developed yet, and we had to convince ourselves that the one-loop β-function was insensitive to the regularization used. We did the calculation in an arbitrary gauge. Since we knew that the answer had to be gauge invariant, we could use gauge invariance as a check on our arithmetic. This was good since we both kept on making mistakes. In February the pace picked up, and we completed the calculation in a spurt of activity. At one point a sign error in one term convinced us that the theory was, as expected, not asymptotically free. As I sat down to put it all together and to write up our results, I caught the error. At almost the same time Politzer finished his calculation and we compared, through Sidney, our results. The agreement was satisfying.

A month or two after this Symanzik passed through Princeton and told us that 't Hooft had made a remark in a question session during a meeting at Marseilles the previous fall to the effect that non-Abelian gauge theories worked in the same way as an asymptotically free scalar theory he had been playing with. (This scalar theory was ruled out, as Coleman and I argued, since one could prove it had no ground state and therefore was unstable.[42]) He did not publish and apparently did not realize the significance for scaling and for the strong interactions.

Why are non-Abelian gauge theories asymptotically free? Today we can understand this in a very physical fashion, although it was certainly not so clear in 1973. It is instructive to interrupt the historical narrative and explain, in modern terms, why QCD is asymptotically free.

The easiest way to understand this is by considering the magnetic screening properties of the vacuum.[43] In a relativistic theory one can calculate the dielectric constant, ϵ, in terms of the magnetic permeability, μ, since $\epsilon\mu = 1$ (in units where c = velocity of light = 1). In classical physics all media are diamagnetic. This is because, classically, all magnets arise from electric currents and the response of a system to an applied magnetic field is to set up currents that act to decrease the field (Lenz's law). Thus $\mu < 1$, a situation that corresponds to electric screening or $\epsilon > 1$. However, in quantum-mechanical systems paramagnetism is possible. This is the case in non-Abelian gauge theories, where the gluons are charged particles of spin one. They behave as permanent color magnetic dipoles that align themselves parallel to an applied external field increasing its magnitude and producing $\mu > 1$. We can therefore regard the antiscreening of the Yang–Mills vacuum as paramagnetism!

QCD is asymptotically free because the antiscreening of the gluons overcomes the screening due to the quarks. The arithmetic works as follows. The contribution to ϵ (in some units) from a particle of charge

q is $-\frac{q^2}{3}$, arising from ordinary dielectric (or diamagnetic) screening. If the particle has spin s (and thus a permanent dipole moment γs), it contributes $(\gamma s)^2$ to μ. Thus a spin-1 gluon (with $\gamma = 2$, as in Yang–Mills theory) gives a contribution to μ of

$$\delta\mu = (-\frac{1}{3} + 2^2)q^2 = \frac{11}{3}q^2;$$

whereas a spin-$\frac{1}{2}$ quark contributes

$$\delta\mu = -(-\frac{1}{3} + (2 \times \frac{1}{2})^2)q^2 = -\frac{2}{3}q^2$$

(the extra minus arises because quarks are fermions). In any case, the upshot is that as long as there are not too many quarks, the anti-screening of the gluons wins out over the screening of the quarks.

The formula for the β-function of a non-Abelian gauge theory is given by

$$\beta(\alpha) \equiv \mu\frac{d}{d\mu}\alpha(\mu)|_{\alpha_{\text{bare fixed}}} = \frac{\alpha^2}{\pi}b_1 + (\frac{\alpha^2}{\pi})^2 b_2 + \ldots; \qquad \alpha = \frac{g^2}{4\pi}. \quad (11.1)$$

Our result was that

$$b_1 = -[\frac{11}{6}C_A - \frac{2}{3}\sum_R n_R T_R]. \quad (11.2)$$

Here C_R is the eigenvalue of the quadratic Casimir operator in the representation R of SU(N) (for the adjoint representation $C_A = N$, for the fundamental $C_F = \frac{N^2-1}{N}$), T_R is trace of the square of the generators for the representation R of SU(N) ($T_A = N$ and $T_F = \frac{1}{2}$), and n_R is the number of fermions in the representation R. In the case of a SU(3) gauge group such as QCD, $C_A = 3$, $T_F=2$, and thus $b_1 = -[\frac{11}{2} - \frac{n}{3}]$. Thus one can tolerate as many as 16 triplets of quarks before losing asymptotic freedom.

Non-Abelian gauge theories of the strong interactions

For me the discovery of asymptotic freedom was totally unexpected. Like an atheist who has just received a message from a burning bush, I became an immediate true believer. Field theory wasn't wrong; instead scaling must be explained by an asymptotically free gauge theory of the strong interactions. Our first paper contained, in addition to the report of the asymptotic freedom of Yang–Mills theory, the hypothesis that this could offer an explanation for scaling, a remark that there would be logarithmic violations of scaling and most important of all the suggestion that the strong interactions must be based on a color gauge theory. The first paragraph reads:[44]

> Non-Abelian gauge theories have received much attention recently as a means of constructing unified and renormalizable theories of the weak and electromagnetic interactions. In this note we report on an investigation of the ultraviolet asymptotic behavior of such theories. We have found that they possess the remarkable feature, perhaps unique among renormalizable theories, of asymptotically approaching free-field theory. Such asymptotically free theories will exhibit, for matrix elements of currents between on-mass-shell states, Bjorken scaling. We therefore suggest that one should look to a non-Abelian gauge theory of the strong interactions to provide the explanation for Bjorken scaling, which has so far eluded field theoretic understanding.

We had a specific theory in mind. Since the deep-inelastic experiments indicated that the charged constituents of the nucleon were quarks, the gluons had to be flavor neutral. Thus the gluons could not couple to flavor. We were very aware of the growing arguments for the color quantum number. Not just the quark model spectroscopy that was the original motivation of Han and Nambu and Greenberg, but the counting factor (of three) that went into the evaluation of the $\pi^0 \to 2\gamma$ decay rate from the axial anomaly,[45] and the factor of 3 that color provided in the total e^+-e^- annihilation cross section.[46] Thus the gluons could couple to color and all would be well. Thus we proposed:[47]

> One particularly appealing model is based on three triplets of fermions, with Gell-Mann's SU(3) × SU(3) as a global symmetry and a SU(3) "color" gauge group to provide the strong interactions. That is, the generators of the strong interaction gauge group commute with ordinary SU(3) × SU(3) currents and mix quarks with the same isospin and hypercharge but different "color." In such a model the vector mesons are (flavor) neutral, and the structure of the operator product expansion of electromagnetic or weak currents is essentially that of the free quark model (up to calculable logarithmic corrections).

The appearance of logarithmic corrections to scaling in asymptotically free theories had already been discussed by Callan and me, in our work on the need for an asymptotically free theory to obtain Bjorken scaling. We also analyzed deep-inelastic scattering in an asymptotically free theory and discovered "that in such asymptotically free theories naive scaling is violated by calculable logarithmic terms."[48] Thus we were well aware what the form of the scaling deviations would be in such a theory. Wilczek and I had immediately started to calculate the logarithmic deviations from scaling. We had already evaluated the asymptotic form of the flavor nonsinglet structure functions, which were the easiest to calculate, at the time our *Physical Review Letter* was written, but did not have room to include the results.

We immediately started to write a longer paper in which the structure of the theory would be spelled out in more detail and the dynamical issues would be addressed, especially the issue of confinement. In our letter we were rather noncommittal on this issue. We had tentatively concluded that Higgs mesons would destroy asymptotic freedom, but had only begun to explore the dynamical consequences of unbroken color symmetry. The only thing we were sure of was that "perturbation theory is not trustworthy with respect to the stability of the symmetric theory nor to its particle content."[49] Politizer's paper appeared with ours.[50] He pointed out the asymptotic freedom of Yang–Mills theory and speculated on its implications for the dynamical symmetry breaking of these theories.

In our second paper, written a few months later, we outlined in much greater detail the structure of asymptotically free gauge theories of the strong interactions and the predictions for the scaling violations in deep-inelastic scattering.[51] Actually the paper was delayed for about two months because we had problems with the singlet structure functions – due to the mixing of physical operators with ghost operators. This problem was similar to the issue of gauge invariance that had plagued us before. Here the problem was more severe. Physical operators, whose matrix elements were measurable in deep-inelastic scattering experiments, mixed under renormalization with ghost operators that could have no physical meaning. Finally we deferred the analysis of the singlet structure functions to a third paper in which we resolved this issue.[52] We showed that, even though this mixing was real and unavoidable, the ghost operators decoupled from physical measurements.

In the second paper we discussed in detail the choice between symmetry breaking and unbroken symmetry and noted that:[53]

Another possibility is that the gauge symmetry is exact. At first sight this would appear to be ridiculous since it would imply the existence of massless, strongly coupled vector mesons. However, in asymptotically free theories these naive expectations might be wrong. There may be little connection between the "free" Lagrangian and the spectrum of states.... The infrared behavior of Green's functions in this case is determined by the strong-coupling limit of the theory. It may be very well that this infrared behavior is such so as to suppress all but color singlet states, and that the colored gauge fields as well as the quarks could be "seen" in the large-Euclidean momentum region but never produced as real asymptotic states.

Weinberg reacted immediately to asymptotic freedom. He wrote a paper in which he pointed out that in an asymptotically free gauge theory of the strong interactions, the nonconservation of parity and strangeness can be calculated ignoring the strong interactions, and thus is of order α, as observed. He also suggested that a theory with unbroken color symmetry could explain why we do not see quarks.[54]

There is a slight difference between our respective conjectures. Weinberg argued that perhaps the *infrared divergences*, caused by the masslessness of the gluons in an unbroken color gauge theory, would make the rate of production of nonsinglet states vanish. We argued that perhaps the growth of the effective coupling at large distances, the *infrared behavior* of the coupling caused by the flip side of asymptotic freedom (later dubbed "infrared slavery" by Georgi and Glashow[55]), would confine the quarks and gluons in color singlet states.

In October 1973 Fritzsch, Gell-Mann, and H. Leutwyler submitted a paper in which they discussed the "advantages of color octet gluon picture."[56] Here they discussed the advantages of "abstracting properties of hadrons and their currents from a Yang–Mills gauge model based on colored quarks and color octet gluons." They discussed various models and pointed out the advantages of each. The first point was already made at the NAL high-energy physics conference in August 1972. There Gell-Mann and Fritzsch had discussed their program of "abstracting results from the quark–gluon model." They outlined various models and asked, "Should we regard the gluons as well as being color non-singlets?" They noted that if one assumed that the gluons were color octets then "an annoying asymmetry between quarks and gluons is removed." In that talk no dynamical theory was proposed and in most of the paper they "shall treat the vector gluon, for convenience, as a color singlet."[57]

In October 1973 Fritzsch, Gell-Mann, and Leutwyler also noted that in the nonrelativistic quark model with a Coulomb potential mediated by vector gluons, the potential is attractive in color singlet channels, which might explain why these are light. This point had been made previously by Harry Lipkin.[58] They also noted the asymptotic freedom of such theories, but did not regard this as an argument for scaling since "we conjecture that there might be a modification at high energies that produces true scaling." Finally they noted that the axial U(1) anomaly in a non-Abelian gauge theory might explain the notorious U(1) problem, although they could not explain how, since the anomaly itself could be written as a total divergence. [It required the discovery of instantons to find the explanation of the U(1) problem.[59]]

The emergence and acceptance of QCD

Although it was clear to me that the strong interactions must be described by non-Abelian gauge theories, there were many problems. The experimental situation was far from clear, and the issue of confinement remained open. However, within a small community of physicists the acceptance of the theory was very rapid. New ideas in physics sometimes take years to percolate into the collective consciousness. However, in rare cases such as this there is a change of perception analogous to a phase transition. Before asymptotic freedom it seemed that we were still far from a dynamical theory of hadrons; afterwards it seemed clear that QCD was such a theory. (The name "quantum chromodynamics," or "QCD," first appeared in a review by Bill Marciano and Heinz Pagels, where it was attributed to Gell-Mann.[60] It was such an appropriate name that no one could complain.) Asymptotic freedom explained scaling at short distances and offered a mechanism for confinement at large distances. Suddenly it was clear that a non-Abelian gauge theory was consistent with everything we knew about the strong interactions. It could encompass all the successful strong interaction phenomenology of the past decade. Since the gluons were flavor neutral, the global flavor symmetries of the strong interactions, SU(3) × SU(3), were immediate consequences of the theory, as long as the masses of the quarks were small enough. (I refer of course to the mass parameters of the quarks in the Lagrangian, not the physical masses that are effectively infinite due to confinement.) Even more alluring was the fact that one could calculate. Since perturbation theory was trustworthy at short distances, many problems could be tackled. Some theorists were immediately convinced, among them Guido Altarelli, Tom Appelquist, Callan, Coleman, Mary K. Gaillard, R. Gatto, Georgi, Glashow, John Kogut, Ben Lee, Luciano Maiani, Migdal, Polyakov, Politzer, Leonard Susskind, S. Weinberg, Zee.

At large distances however, perturbation theory was useless. In fact, even today after 20 years of study we still lack reliable, analytic tools for treating this region of QCD. It remains one of the most important, and woefully neglected, areas of theoretical particle physics. However, at the time the most important thing was to convince oneself that the idea of confinement was not inconsistent. One of the first steps in that direction was provided by lattice gauge theory.

I first heard of Wilson's lattice gauge theory when I gave a lecture at Cornell in the late spring of 1973. He had started to think of this approach soon after asymptotic freedom was discovered. The lattice formulation of gauge theory (independently proposed by Polyakov) had the enormous advantage, as Wilson pointed out in the fall of 1973, that the strong-coupling limit was particularly simple and exhibited confinement.[61] Thus one had at least a crude approximation in which confinement was exact. It is a very crude approximation, since to arrive at the continuum theory from the lattice theory one must take the weak coupling limit. However, one could imagine that the property of confinement was not lost as one went continuously from strong to weak lattice coupling, that is, there was no phase transition. Moreover, one could, as advocated by Wilson, study this possibility numerically using Monte Carlo methods to construct the lattice partition function. However, the first quantitative results of this program did not emerge until the work of Creutz in 1979.[62] The ambitious program of calculating the hadronic mass spectrum has still not attained its goal, and still awaits the next generation of computers.

Personally I derived much solace in the coming year from two examples of soluble two-dimensional field theories. One was the $(\bar{\Psi}\Psi)^2$ theory that Neveu and I analyzed and solved for large N.[63] This provided a soluble example of an asymptotically free theory that underwent dimensional transmutation, solving its infrared problems by generating a dynamical fermion mass through spontaneous symmetry breaking. This provided a model of an asymptotically free theory, with no built-in mass parameters. We could solve this model and check that it was consistent and physical. The other soluble model was two-dimensional QCD, analyzed by 't Hooft in the large N limit.[64] Two-dimensional gauge theories trivially confine color. This was realized quite early and discussed for Abelian gauge theory – the Schwinger model – by Aharon Casher, Kogut, and Susskind, as a model for confinement in the fall of 1973.[65] However, QCD_2 is a much better example. It has a spectrum of confined quarks that in many ways resembles the four-dimensional world. These examples gave many of us total confidence in the consistency of the concept of confinement. It clearly was possible to have a theory whose basic fields do not correspond to asymptotic states, to particles that one can observe directly in the laboratory.

Applications of the theory also began to appear. Two calculations of the β-function to two-loop order were performed, with the result that, in the notation of Eq. 11.2, $b_2 = -[\frac{17}{12}C_A^2 - \frac{1}{2}C_F T_F n - \frac{5}{6}C_A T_F n]$.[66] Appelquist and Georgi and Zee calculated the corrections to the scaling of the e^+-e^- annihilation cross section.[67] Gaillard and Lee, and independently Altarelli and Maiani, calculated the enhancement of the $\Delta I = \frac{1}{2}$ nonleptonic decay matrix elements.[68] The analysis of scaling violations for deep-inelastic scattering continued, and the application of asymptotic freedom, what is now called "perturbative QCD," was extended to many new processes.[69]

The experimental situation developed slowly, and initially looked rather bad. I remember in the spring of 1974 attending a meeting in Trieste. There I met Burt Richter, who was gloating over the fact that $\frac{R=\sigma_{had}}{\sigma_{\mu\mu}}$ in electron–positron scattering was increasing with energy, instead of approaching the expected constant value. This was the most firm of all the scaling predictions. R must approach a constant in any scaling theory. In most theories, however, one cannot predict the value of the constant. But in an asymptotically free theory the constant is predicted to equal the sum of the squares of the charges of the constituents. Therefore, if there were only the three *observed* quarks, one would expect that $R \to 3[(\frac{1}{3})^2 + (\frac{1}{3})^2 + (\frac{2}{3})^2] = 2$. However, Richter reported that R was increasing, passing through 2, with no sign of flattening out. Now many of us knew that charmed quarks had to exist. Not only were they required, indeed invented, for the GIM mechanism to work, but as Claude Bouchiat, John Iliopoulos, and Meyer, and Roman Jackiw and I showed, if the charmed quark were absent the electroweak theory would be anomalous and nonrenormalizable.[70] Gaillard, Lee, and Jonathan Rosner had written an important and insightful paper on the phenomenology of charm.[71] Thus many of us thought that since R was increasing, charm was probably being produced.

In 1974 the J and ψ particles, much narrower than anyone (except for Appelquist and Politzer[72]) imagined were discovered, looking very much like positronium – Coulomb-bound states of quarks. This clinched the matter for many of the remaining skeptics. The rest were probably convinced once experiments at higher energy began to see quark and gluon jets.

The precision tests of the theory, the logarithmic deviations from scaling, took quite a while to observe. I remember very well a remark made to me by a senior colleague, in April 1973 when I was very high, right after the discovery of asymptotic freedom. He remarked that it was unfortunate that our new predictions regarding deep-inelastic scattering were logarithmic effects, since it was unlikely that we would see them verified, even if true, in our lifetime. This was an exaggeration, but the tests did take a long time to appear. Confirmation only started to trickle in in 1975–78, and then at a slow pace. By now the predictions are indeed verified, in some cases to better than a percent.

Nowadays, when you listen to experimentalists talk about their results they point to their lego plots and say, "Here we see a quark, here a gluon." Believing is seeing; seeing is believing. We now believe in the physical reality of quarks and gluons; we now believe in the asymptotic simplicity of their interactions at high energies, so we can *see* quarks and gluons. The way in which we see quarks and gluons, indirectly through the effects they have on our measuring instruments, is not much different from the way we see electrons. Even the objection that quarks and gluons cannot be real particles, since they can never be isolated, has largely dissipated. If we were to heat the world to a temperature of a few hundred MeV, hadrons would melt into a plasma of liberated quarks and gluons.

Other implications of asymptotic freedom
Consistency of quantum field theory

Traditionally, fundamental theories of Nature have had a tendency to break down at short distances. This often signals the appearance of new physics that is discovered once one has experimental instruments of high enough resolution (energy) to explore the higher energy regime. Before asymptotic freedom, it was expected that any quantum field theory would fail at sufficiently high energy, where the flaws of the renormalization procedure would appear. To deal with this, one would have to invoke some kind of *fundamental length*. In an asymptotically free theory this is not necessarily the case – the decrease of the effective coupling for large energy means that no new physics need arise at short distances. There are no infinities at all, the bare coupling is finite – indeed it vanishes. The only divergences that arise are an illusion that appears when one tries to compare, in perturbation theory, the finite effective coupling at finite distances with the vanishing effective coupling at infinitely short distances.

Thus the discovery of asymptotic freedom greatly reassured us of the consistency of four-dimensional quantum field theory. One can trust renormalization theory for an asymptotically free theory, independent of the fact that perturbation theory is only an asymptotic expansion, since it gets better and better in the regime of short distances. We are very close to having a rigorous mathematical proof of the existence of asymptotically free gauge theories in four dimensions – at least when placed into a finite box to tame the infrared dynamics that produces confinement. As far as we know, QCD by itself is a totally consistent theory at all energies. Moreover, aside from the quark masses it has no arbitrary, adjustable parameters. (This is one of the reasons it is so hard to solve.) Indeed, were it not for the electro-weak interactions and gravity, we might be satisfied with QCD as it stands.

Unification

Almost immediately after the discovery of asymptotic freedom and the proposal of the non-Abelian gauge theories of the strong interactions, the first attempts were made to unify all the interactions. This was natural, given that one was using very similar theories to describe all the known interactions. Furthermore, the apparently insurmountable barrier to unification – namely the large difference in the strength of the strong interactions and the electroweak interactions – was seen to be a low-energy phenomenon. Since the strong interactions decrease in strength with increasing energy, these forces could have a common origin at very high energy. Indeed, in the fall of 1974 Georgi and Glashow proposed a unified theory, based on the gauge group SU(5), which remarkably contained the gauge groups of the Standard Model as well as the quark and lepton multiplets in an alluringly simple fashion.[73] (An earlier attempt to unify quarks and leptons was made by Pati and Salam.[74]) Georgi, Helen Quinn, and Weinberg showed that the couplings evolve in such a way as to merge somewhere around 10^{14}–10^{16} GeV.[75]

This theory had the great advantage of being tight enough to make sufficiently precise predictions (proton decay and the Weinberg angle). It was a great stimulus for modern cosmology, since it implied that one could extrapolate the Standard Model to enormously high energies that corresponded to very early times in the history of the Universe. Although the SU(5) theory has been invalidated by experiment, at least in its simplest form, the basic idea that the next fundamental threshold of unification is set by the scale where the strong and electroweak couplings become equal in strength remains at the heart of most attempts at unification.

Notes

1 L. D. Landau, "Fundamental Problems," in M. Fierz and V. F. Weisskopf, eds., *Theoretical Physics in the Twentieth Century; a Memorial Volume to Wolfgang Pauli* (New York: Interscience, 1960), pp. 245–8.

2 R. Feynman, "The Present Status of Quantum Electrodynamics," in *The Quantum Theory of Fields – Proceedings of the 12th Solvay Conference on Physics, University of Brussels, October 1961* (New York: Interscience, 1961), pp. 61–91, on p. 89.

3 L. D. Landau and I. Pomeranchuk, "On Point Interaction, Quantum Electrodynamics," *Dokl. Akad. Nauk SSSR 102* (1955), pp. 489–92.

4 L. D. Landau, "On the Quantum Theory of Fields," in W. Pauli, ed., *Niels Bohr and the Development of Physics* (New York: Pergamon Press, 1955), pp. 52–69.

5 L. D. Landau, "Fundamental Problems."

6 Kenneth A. Johnson, "The Physics of Asymptotic Freedom," in A. H. Guth, K. Huang, and R. L. Jaffe, eds., *Asymptotic Realms of Physics* (Cambridge: MIT Press, 1983), pp. 20–31.

7 V. S. Vanyashin and M. V. Terent'ev, "The Vacuum Polarization of a Charged Vector Field," *Sov. Phys. JETP 21* (1965), pp. 375–80.

8 G. Chew, *S-Matrix Theory of Strong Interactions* (New York: W. A. Benjamin Inc, 1962), pp. 1–2.

9 G. Chew, comment made in discussion session, in *The Quantum Theory of Fields*, p. 229.

10 M. Goldberger, "Theory and Applications of Single-Variable Dispersion Relations," in *The Quantum Theory of Fields*, pp. 179–97, on p. 179.

11 F. Low, "Discussion" (following paper of G. F. Chew), *Supp. Nuovo Cimento 4* (1966), p. 379.

12 Harald Fritzsch and Murray Gell-Mann, "Current Algebra: Quarks and What Else?," in J. D. Jackson and A. Roberts, eds., *Proceedings of the XVIth International Conference on High-Energy Physics, September 6–September 13, 1972* (Batavia: National Accelerator Laboratory, 1972), pp. 135–65.

13 M. Gell-Mann, "The Symmetry Group of Vector and Axial Currents," *Physics 1* (1964), pp. 63–75.

14 M. Gell-Mann, "The Symmetry Group of Vector and Axial Currents."

15 M. Gell-Mann, "A Schematic Model of Baryons and Mesons," *Phys. Lett. 8* (1964), pp. 214–15; G. Zweig, "An SU(3) Model for Strong Interaction Symmetry and its Breaking," CERN Report No. 81821 TH401 (1964), unpublished.

16 Y. Nambu, "Dynamical Symmetries and Fundamental Fields," in B. Kursunoglu, A. Perlmutter, and I. Sakmar, eds., *Proceedings of the Second Coral Gables Conference on Symmetry Principles at High Energy* held 20–22 January 1965 (San Francisco: W. H. Freeman, 1965), pp. 274–85; Y. Nambu, "A Systematics of Hadrons in Subnuclear Physics," in A. De-Shalit, H. Feshbach, and L. Van Hove, eds., *Preludes in Theoretical Physics* (Amsterdam: North-Holland, 1966), pp. 133–42; M. Y. Han and Y. Nambu, "Three-Triplet Model with Double SU(3) Symmetry," *Phys. Rev. 139B* (1965), pp. 1006–10; O. W. Greenberg, "Spin and Unitary-Spin Independence in a Paraquark Model of Baryons and Mesons," *Phys. Rev. Lett. 13* (1964), pp. 598–602.

17 Francis Low, "Dynamics of Strong Interactions," in E. M. McMillan, ed.,

Proceedings of the XIIIth International Conference on High-Energy Physics, August 31–September 7, 1966 (Berkeley and Los Angeles: University of California Press, 1967), pp. 241–9.

18 Stephen L. Adler, "Sum Rules for the Axial Vector Coupling-Constant Renormalization in β Decay," *Phys. Rev. 140 B* (1965), pp. 736–47; William I. Weisberger, "Unsubtracted Dispersion Relations and the Renormalization of the Weak Axial-Vector Coupling Constants," *Phys. Rev. 143 B* (1966), pp. 1302–9.

19 Roger Dashen and M. Gell-Mann, "Representation of Local Current Algebra at Infinite Momentum" *Phys. Rev. Lett. 17* (1966), pp. 340–3.

20 J. D. Bjorken, "Applications of the Chiral U(6) × U(6) Algebra of Current Densities," *Phys. Rev. 148 B* (1966), pp. 1467–78; J. D. Bjorken, "Current Algebra at Small Distances," in *Proceedings of the International School of Physics "Enrico Fermi,"* Varenna, Italy 17–29 July 1967 (New York: Academic Press, 1968), pp. 55–81; J. D. Bjorken, "Inequality for Backward Electron- and Muon-Nucleon Scattering at High Momentum Transfer," *Phys. Rev. 163* (1967), pp. 1767–9.

21 Hirotaka Sugawara, "A Field Theory of Currents," *Phys. Rev. 170* (1968), pp. 1659–62.

22 C. G. Callan and David J. Gross, "Crucial Test of a Theory of Currents," *Phys. Rev. Lett. 21* (1968), pp. 311–13.

23 J. D. Bjorken, "Asymptotic Sum Rules at Infinite Momentum" *Phys. Rev. 179* (1969), pp. 1547–53.

24 E. D. Bloom, et al., "High-Energy Inelastic e–p Scattering at 6° and 10°," *Phys. Rev. Lett. 23* (1969), pp. 930–4.

25 C. G. Callan and David J. Gross, "High-Energy Electroproduction and the Constitution of the Electric Current," *Phys. Rev. Lett. 22* (1968), pp. 156–9.

26 David J. Gross and C. H. Llewelyn-Smith, "High-Energy Neutrino–Nucleon Scattering, Current Algebra and Partons," *Nucl. Phys. B14* (1969), pp. 337–47.

27 Richard P. Feynman, "Very High-Energy Collisions of Hadrons," *Phys. Rev. Lett. 23* (1969), pp. 1415–19.

28 Sidney D. Drell and Tung-Mo Yan, "Partons and Their Applications at High Energies," *Ann. Phys. 66* (1971), pp. 578–623.

29 David J. Gross and J. Wess, "Scale Invariance, Conformal Invariance and the High-Energy Behavior of Scattering Amplitudes," *Phys. Rev. D2* (1970), pp. 753–64.

30 M. Gell-Mann and F. E. Low, "Quantum Electrodynamics at Small Distances," *Phys. Rev. 95* (1954), pp. 1300–12; E. C. G. Stueckelberg and A. Petermann, "La Normalisation des Constantes dans la Théorie des Quanta," *Helv. Phys. Acta 26* (1953), pp. 499–520 (in French with English summary); Curtis G. Callan, "Broken Scale Invariance in Scalar Field Theory," *Phys. Rev D2* (1970), pp. 1541–7; K. Symanzik, "Small-Distance Behavior in Field Theory and Power Counting," *Comm. Math. Phys. 18* (1970), pp. 227–46. N. N. Bogoliubov and D. V. Shirkov, *Introduction to the Theory of Quantized Fields* (New York: Interscience, 1959).

31 G. Veneziano, "Construction of a Crossing-Symmetric, Regge–Behaved Amplitude for Linearly Rising Trajectories," *Nuovo Cimento 57A* (1968), pp. 190–7.

32 G. 't Hooft, "Renormalizable Lagrangians for Massive Yang–Mills Fields," *Nucl. Phys. B35* (1971), pp. 167–88.

33 Kenneth G. Wilson, "Renormalization Group and Strong Interactions," *Phys. Rev. D3* (1971), pp. 1818–46.

34 M. Gell-Mann and F. E. Low, "Quantum Electrodynamics"; *Phys. Rev. 95* (1954), pp. 1300–12; E. C. G. Stueckelberg and A. Petermann, "La Normalisation des Constantes"; N. N. Bogoliubov and D. V. Shirkov, *Introduction to the Theory of Quantized Fields.*

35 Curtis G. Callan, Jr. "Broken Scale Invariance and Asymptotic Behavior," *Phys. Rev D5* (1972), pp. 3202–10; K. Symanzik, "Small-Distance-Behavior Analysis and Wilson Expansions," *Comm. Math. Phys. 23* (1971), pp. 49–86; N. Christ, B. Hasslacher, and A. H. Mueller, "Light-Cone Behavior of Perturbation Theory," *Phys. Rev D6* (1972), pp. 3543–62; Curtis G. Callan and David J. Gross, "Bjorken Scaling in Quantum Field Theory," *Phys. Rev. D8* (1973), pp. 4383–94.

36 Roman Jackiw, Roger Van Royen, and Geoffrey B. West, "Measuring Light-Cone Singularities," *Phys. Rev. D2* (1970), pp. 2473–85; Y. Frishman, "Operator Products at Almost Light Like Distances," *Ann. Phys. 66* (1971), pp. 373–89; H. Leutwyler and J. Stern, "Singularities of Current Commutators on the Light Cone," *Nucl. Phys. B20* (1970), pp. 77–101; David J. Gross and S. B. Treiman, "Light-Cone Structure of Current Commutators in the Gluon–Quark Model," *Phys. Rev. D4* (1971), pp. 1059–72.

37 Kenneth G. Wilson and J. Kogut, "The Renormalization Group and the ϵ Expansion," *Phys. Rep. 12C* (1974), pp. 75–200.

38 G. Parisi, "Deep Inelastic Scattering in a Field Theory with Computable Large-Momenta Behavior," *Lett. Nuovo Cimento 7* (1973), pp. 84–8.

39 Curtis G. Callan and David J. Gross, "Bjorken Scaling."

40 Sidney Coleman and David J. Gross, "Price of Asymptotic Freedom," *Phys. Rev. Lett. 31* (1973), pp. 851–4.

41 A. Zee, "Study of the Renormalization Group for Small Coupling Constants," *Phys. Rev. D7* (1973), pp. 3630–6.

42 Sidney Coleman and David J. Gross, "Price of Asymptotic Freedom."

43 N. K. Nielsen, "Asymptotic Freedom as a Spin Effect," *Am. J. Phys. 49* (1981), pp. 1171–8.

44 David J. Gross and Frank Wilczek, "Ultraviolet Behavior of Non-Abelian Gauge Theories," *Phys. Rev. Lett. 30* (1973), pp. 1343–6.

45 This had been recently emphasized by William Bardeen, Harald Fritzsch, and Murray Gell-Mann. W. A. Bardeen, H. Fritzsch, and M. Gell-Mann, "Light-Cone Current Algebra, π^0 Decay, and e^+e^- Annihilation," in R. Gatto, ed., *Scale and Conformal Symmetry in Hadron Physics* (New York: John Wiley & Sons, 1973), pp. 139–51.

46 M. Y. Han and Y. Nambu, "Three-Triplet Model"; O. W. Greenberg, "Spin and Unitary-Spin Independence."

47 David J. Gross and Frank Wilczek, "Ultraviolet Behavior."

48 Curtis G. Callan and David J. Gross, "Bjorken Scaling."

49 David J. Gross and Frank Wilczek, "Ultraviolet Behavior."

50 H. David Politzer, "Reliable Perturbative Results for Strong Interactions?," *Phys. Rev. Lett. 30* (1973), pp. 1346–9.

51 D. J. Gross and F. Wilczek, "Asymptotically Free Gauge Theories. I," *Phys. Rev. D8* (1973), pp. 3633–52.

52 D. J. Gross and F. Wilczek, "Asymptotically Free Gauge Theories. II," *Phys. Rev. D9* (1974), pp. 980–3.

53 David J. Gross and F. Wilczek, "Asymptotically Free Gauge Theories. I."

54 Steven Weinberg, "Non-Abelian Gauge Theories of the Strong Interactions," *Phys. Rev. Lett. 31* (1973), pp. 494–7.

55 Howard Georgi and S. L. Glashow, "Unity of All Elementary Particle Forces," *Phys. Rev. Lett. 32* (1974), pp. 438–41.

56 H. Fritzsch, M. Gell-Mann, and H. Leutwyler, "Advantages of the Color Octet Gluon Picture," *Phys. Lett. 47B* (1973), pp. 365–8.

57 Harald Fritzsch and Murray Gell-Mann, "Current Algebra."

58 H. J. Lipkin, "Triality, Exotics and the Dynamical Basis of the Quark Model," *Phys. Lett. 45B* (1973), pp. 267–71.

59 A. A. Belavin, A. M. Polyakov, A. S. Schwartz, and Yu. S. Tyupkin, "Pseudoparticle Solutions of the Yang–Mills Equations," *Phys. Lett. 59B* (1975), pp. 85–7; G. 't Hooft, "Computation of the Quantum Effects due to a Four-Dimensional Pseudoparticle," *Phys. Rev. D14* (1976), pp. 3432–50; C. G. Callan, Jr., R. F. Dashen, and D. J. Gross, "The Structure of the Gauge Theory Vacuum," *Phys. Lett. 63B* (1976), pp. 334-40; R. Jackiw and C. Rebbi, "Vacuum Periodicity in a Yang–Mills Quantum Theory," *Phys. Rev. Lett. 37* (1976), pp. 172–5.

60 William Marciano and Heinz Pagels, "Quantum Chromodynamics," *Phys. Rep. C36* (1978), pp. 137–276.

61 Kenneth G. Wilson, "Confinement of Quarks," *Phys. Rev. D10* (1974), pp. 2445–59.

62 Michael Creutz, "Confinement and the Critical Dimensionality of Space-Time," *Phys. Rev. Lett. 43* (1979), pp. 553–6.

63 David J. Gross and André Neveu, "Dynamical Symmetry Breaking in Asymptotically Free Field Theories," *Phys. Rev. D10* (1974), pp. 3235–53.

64 G. 't Hooft, "A Planar Diagram Theory for Strong Interactions," *Nucl. Phys. B72* (1974), pp. 461–73.

65 A. Casher, J. Kogut, and Leonard Susskind, "Vacuum Polarization and the Quark–Parton Puzzle," *Phys. Rev. Lett. 31* (1973), pp. 792–5; "Vacuum Polarization and the Absence of Free Quarks," *Phys. Rev. D 10* (1974), pp. 732–45.

66 W. Caswell, "Asymptotic Behavior of Non-Abelian Gauge Theories to Two-Loop Order," *Phys. Rev. Lett. 33* (1974), pp. 244–6; D. R. T. Jones, "Two-Loop Diagrams in Yang–Mills Theory," *Nucl. Phys. B75* (1974), pp. 531–8.

67 Thomas Appelquist and Howard Georgi, "e^+e^- Annihilation in Gauge Theories of Strong Interactions," *Phys. Rev. D8* (1973), pp. 4000–2; A. Zee, "Electron–Positron Annihilation in Stagnant Field Theories," *Phys. Rev. D8* (1973), pp. 4038–41.

68 Mary K. Gaillard and Benjamin W. Lee, "$\Delta I = \frac{1}{2}$ Rule for Nonleptonic Decays in Asymptotically Free Field Theories," *Phys. Rev. Lett 33* (1974), pp. 108–11; G. Altarelli and L. Maiani, "Octet Enhancement of Non-Leptonic Weak Interactions in Asymptotically Free Gauge Theories," *Phys. Lett. 52B* (1974), pp. 351–4.

69 David J. Gross, "How to Test Scaling in Asymptotically Free Theories," *Phys. Rev. Lett. 32* (1974), pp. 1071–3.

70 C. Bouchiat, J. Iliopoulos, and Ph. Meyer, "An Anomaly-Free Version of

Weinberg's Model," *Phys. Lett. 38B* (1972), pp. 519–23; David J. Gross and R. Jackiw, "Effect of Anomalies on Quasi-Renormalizable Theories," *Phys. Rev. D6* (1972), pp. 477–93.

71 Mary K. Gaillard, Benjamin W. Lee, and Jonathan L. Rosner, "Search for Charm," *Rev. Mod. Phys. 47* (1975), pp. 277–310.

72 Thomas Appelquist and H. David Politzer, "Heavy Quarks and e^+e^- Annihilation," *Phys. Rev. Lett. 34* (1975), pp. 43–5.

73 Howard Georgi and S. L. Glashow, "Unity of All Elementary Particle Forces," *Phys. Rev. Lett. 32* (1974), pp. 438–41.

74 J. C. Pati and A. Salam, "Unified Lepton–Hadron Symmetry and a Gauge Theory of the Basic Interactions," *Phys. Rev. D8* (1973), pp. 1240–51.

75 H. Georgi, H. R. Quinn, and S. Weinberg, "Hierarchy of Interactions in Unified Gauge Theories," *Phys. Rev. Lett. 33* (1974), pp. 451–4.

12

Quark Confinement*

LEONARD SUSSKIND

Born New York City, 1940; Ph.D., 1965 (physics), Cornell University; Professor of Physics at Stanford University; high-energy physics (theory).

In this chapter I present a personal reminiscence of the development of our current ideas about quark confinement. I describe what I remember of my own involvement and that of the people who influenced me. If others remember it differently, I hope they will not be too angry.

By the end of the 1960s our empirical knowledge of hadrons consisted of a vast mountain of data about their spectrum, their low- and high-energy interactions, and their electromagnetic and weak properties. To some extent the story of the eventual interpretation in terms of QCD was like digging a tunnel through the mountain with crews of diggers starting independently at the two ends. At one end was the short-distance behavior of local currents and its interpretation in terms of freely moving quark-parton constituents.[1] At the other end was the low-momentum-transfer Regge structure including a spectrum of highly excited rotational states, shrinking diffraction peaks, and multihadron final states of peripheral collisions, *but no free quarks*.[2] Sometime in 1973 the two tunnel crews discovered that they had met and a complete picture of the strong interactions existed. Of course the two crews were not entirely unaware of each other. The Regge workers were beginning to organize the trajectories by quantum numbers suggested by the quark model. Eventually, the Regge picture culminated in 1968 with a set of scattering amplitudes based on the duality principle of R. Dolen, D. Horn, and C. Schmidt.[3] A beautiful version discovered by Gabriele Veneziano incorporated infinite towers of rotational excitations forming Regge trajectories with angular momentum linear in the square of the

* Supported by NSF grant PHY-8917438.

mass.[4] (Such trajectories had been suggested earlier in the prescient papers of Geoffrey Chew and Steven Frautchi.) It was necessary to provide the amplitudes with isospin and SU(3) (flavor) dependence. A bold hypothesis was made by Jonathan Rosner and Haim Harari to borrow the flavor dependence from the quark model by assuming the Regge trajectories carried $\bar{q}q$ and qqq quantum numbers.[5]

Harry Lipkin described things at that time by dividing hadron physicists into the IBY camp (isospin, baryon number, and hypercharge) and the STU camp (Mandelstam variables). The IBY camp, which had its headquarters at Cal Tech and was led by Murray Gell-Mann, was interested in the symmetries of hadrons and their electroweak currents. The STU camp, located in Berkeley and led by Chew, was concerned with the analytic structure of scattering amplitudes. By 1969 I was not at all in the mainstream of either STU or IBY physics. I had been working on reformulating relativistic quantum mechanics in the infinite-momentum frame (now called light-cone frame), in which I had discovered that there was an underlying nonrelativistic structure and that the Hamiltonian operator was the squared mass.[6] Sometime in early 1969 Hector Rubinstein of the Weizmann Institute came to Yeshiva University very excited about the new Veneziano amplitude. I understood only one thing about his seminar – that there were trajectories of resonances with angular momentum J and mass M satisfying

$$J = \text{const } M^2 .$$

I was greatly intrigued by this because from the "nonrelativistic" light-cone viewpoint, it meant J was proportional to energy; in other words, an harmonic-oscillator spectrum. I immediately set to work on trying to construct a model of hadrons consisting of a $q\bar{q}$ pair interacting by a harmonic force that would scatter according to the Veneziano Model. By the spring of 1969 Yoichiro Nambu and I had independently discovered the string model.[7] Holger Nielsen in Denmark was very close. According to this theory, a meson was a $q\bar{q}$ pair joined by an elastic string with an energy that increased linearly with the length of the string. (This is not in conflict with the fact that in the infinite-momentum frame the energy is quadratic in separation.) The whole picture was exciting because it allowed quarks to exist but never be seen singly.

The fact that the energy of a string was proportional to its length in string theory was to the best of my knowledge first pointed out by Edward Tryon.[8] An equivalent statement is that action for a world sheet (the two-dimensional space swept out by a moving string) is proportional to its area. The point that was to play an essential role in later theories of confinement was suggested independently by three people: Nambu, who was already famous; Tetsuo Goto, who became famous; and Henri Noskowitz, who nobody ever heard of.[9] Noskowitz was a brilliant student of mine and was writing a review of strings with me. One day he showed me that a nice starting point for strings is the area action, which when gauge-fixed gives the usual string equations. I am afraid I told him: "That's nice, but we can avoid all that by just starting with the usual action." Noskowitz left physics and went into raincoats and I feel responsible.

By 1970 I was beginning to become troubled by the sound of the pick axes of the parton crew. The reader will no doubt have noticed the absence in the above history of any concepts from quantum field theory. Indeed the idealized string theory is not consistent with the existence of local currents. Furthermore, more and more evidence was leaking out of SLAC that hadrons were composed of free partons, which carried their momentum and other quantum numbers in discrete finite bits. This was clearly inconsistent with the continuum string hypothesis. Nevertheless, most of the string culture was unmoved by this fact. David Gross told me that string theory could not be wrong because its beautiful mathematics could not be accidental. (Beware, ye superstringers.) Nielsen was one of the earliest to realize that the success of partons would force us to abandon the idealized string.[10] He had been experimenting with the idea that world sheets were really just very high-order planar Feynman diagrams composed of a discrete network of lines. Nielsen and I built a parton model in which the string was built from quarks and scalar partons.[11] But now it was no longer assured that isolated quarks would not appear, especially in deep-inelastic electroproduction in which a single quark impulsively absorbs a very large momentum. The problem was compounded by the fact that the quark-partons were required to be very weakly interacting if the theory was to explain the Bjorken scaling of structure functions.[12]

One way of thinking about the confinement problem was suggested by e^+e^- annihilation into hadrons. Initially, the virtual photon dissociates into a quark q and an antiquark $\bar{q}$ that move with almost the speed of light back-to-back. Feynman had argued that additional $q\bar{q}$ pairs would be produced in the region between them, along the line separating the initially produced $q\bar{q}$. The new pairs and original $q\bar{q}$ would rearrange and become a bunch of outgoing mesons as in Fig. 12.1.

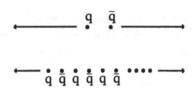

Fig. 12.1. Production of $q\bar{q}$ pairs in electron–positron annihilation.

Feynman's theory of the process was very simple. He would put his hands in front of his face, almost touching. Then he would quickly draw them apart. While doing this he would make a sound something like "Brrrrrrrrrp." I interpreted this to mean that each outgoing quark would radiate a pair, which would almost keep up with their parents. Then one of the newly produced quarks would mate with its parent, producing a meson and leaving an unmated q or $\bar{q}$. The unmated $q\bar{q}$ would still be moving out but with slightly less energy than the original pair. The whole thing would repeat itself until a low-energy pair would emerge at the center and form a final meson. I called this "the outside–inside cascade."

In 1973 John Kogut and I were visiting Tel Aviv University, where I was a half-time faculty member. We had written a paper arguing that the outside–inside cascade took place too slowly at high energy to annihilate the outgoing fractional charges. We soon received a letter from Bjorken who suggested that we think about vector forces producing an inside–outside cascade in which the slowest pair is produced first, symmetrically between the outgoing quarks. Next, as the original quarks separated from the new ones, which were more or less at rest, pairs would be produced between the original ones and the new pair. At each stage the pairs would appear wherever there were large gaps in the one-dimensional invariant phase space. This certainly must have been what Feynman had in mind, but it was a revelation to me at the time. Kogut, Aharon Casher, and I started to study a $1 + 1$-dimensional model in which massless charged fermions interacted with Abelian gauge fields, in the hope that it would do what Bjorken had suggested. The model had previously been studied by Julian Schwinger, who showed that the gauge bosons became massive.[13] Thus the name Schwinger mechanism. Our work showed that another phenomenon took place.[14] The fermions were confined. The physical spectrum contained only massive pseudoscalar bosons, which were composed of fermion pairs. No charged states existed. Furthermore, the annihilation took place exactly as Bjorken suggested and what was even better, deep-inelastic structure functions scaled as if the mesons consisted of free fermions. We speculated that a similar thing could happen in a 3 + 1-dimensional theory of colored quarks with non-Abelian SU(3) gauge forces and that this would account for the absence of free quarks. That aside, we had found the first example of a theory with confined degrees of freedom, and we were happy.

I have been asked whether at the time we were aware of the David Gross, Frank Wilczek, and David Politzer work on asymptotic freedom.[15] I honestly can not remember. However, I am certain that it had no influence on our work, because I didn't understand its implications until later that summer. I was visiting CERN, where I met Gerard 't Hooft, who had independently discovered asymptotic freedom a year earlier but astonishingly had not published it.[16] Perhaps he did not understand its implications at the time, but by the summer of 1973 he certainly did. Most of the emphasis on asymptotic freedom was on the short-distance behavior and how the effective coupling strength diminished with length scale. 't Hooft explained that you could turn that reasoning upside down and say that the same calculation showed that the effective coupling strength was increasing as you go toward the infrared. Thus it supported the view that quarks might become trapped as they tried to escape the environment of a color-neutral system.

I also had the good fortune to run into Ken Wilson at CERN. His work on the renormalization group was revolutionizing our understanding of quantum field theory.[17] Until that time I think most particle theorists had a very confused view of how quantum field theory worked. The path integral was clearly an ill-defined object. Integrals over continuously infinite numbers of variables were not something that mathematicians understood, and the divergence difficulties of perturbation theory made them smell particularly bad. True enough, a formal renormalization procedure allowed the extraction of a perturbation expansion that was order-by-order finite, but the entire series was certainly divergent. Nobody knew exactly what the definition of the theory was, or if there indeed was one. Furthermore, there was a general tendency to view the subject upside down. It was usually assumed that the fine-structure constant that governs the long-range Coulomb force was a truly fundamental constant. Many people hoped to derive its value from some fundamental principle. There was a formula and some weird theory to go with it that expressed α in terms of e and π to incredible accuracy. Steve Adler concocted some equation that was supposed to give α as the root of some function.[18] My point is that a parameter governing the lowest-energy physics was thought of as fundamental while the high-energy short-distance behavior was derived from complicated dynamics. This was like considering elephants as fundamental and deriving the properties of electrons from elephant parameters. But at the time we didn't understand that.

Wilson, on the other hand, was interested in defining field theory as a set of instructions for a computer. Computers require very concrete instructions, and he was forced to give definite meaning to the symbols. His way of making the path integral concrete was to replace space–time by a lattice with a small but finite spacing. At the end you could take the spacing to be very small by comparison with the length scales of interest. From this viewpoint, the fundamental parameters that the computer would directly work with involve the behavior at the cutoff scale. Large-distance physics is thought of as the consequence of complex dynamics and iteration from small to large scales. Today, of course, we are very familiar with this view and only a crackpot would try to derive α from a fundamental formula. Instead, for example, we derive the weak mixing angle $\sin\theta_w = 0.2$ from grand unified theories, in which it has the value $\frac{3}{8}$ at the unification scale. I believe this view of things was almost entirely due to Wilson's influence, although Alexander Polyakov in the Soviet Union had a very similar view.

Asymptotic freedom, from the Wilsonian viewpoint, meant that we should start with a very small coupling at the Planck or cutoff scale, and it would evolve to larger values as the theory became coarse-grained. It was the same lesson that 't Hooft had explained to me but with a very far-reaching generality.

In order to study the implications of a gauge theory of hadrons, Wilson had constructed his now-famous version of non-Abelian gauge theory.[19] Incidentally, Polyakov had also described the same discretization of Yang–Mills theory. Lucky for me, Wilson was in a good mood and patiently explained his model to me at CERN. He claimed the model confined quarks, but the argument as I remember it was not correct. He said that a phase transition separated two types of behavior. In the weak-coupling phase, gauge invariance was spontaneously broken and the fluctuations of the gauge potentials were small. In the strongly coupled phase, the gauge potentials fluctuated over their whole range of variation. This, it was claimed, would force the charged particle (quark) propagator to vanish since it was gauge dependent. I tried to argue that this was wrong, but by that point Wilson was in a less patient mood.

During the fall of 1973 Kogut and I worked on an idea that was suggested by the things 't Hooft and Wilson had explained. It was common knowledge that the effective electric charge varied with scale because the vacuum in QED is a polarizable medium and polarization charge screens the fundamental electron charge. The vacuum has a dielectric constant that in fact depends on scale size. QCD has exactly the opposite behavior. The color charge of a quark is antiscreened. If at large distances the dielectric constant tended to zero, quarks would become confined because the effective interaction strength would diverge. We argued that this would cause an effect similar to the Meissner effect in superconductors, in which magnetic fields are repelled by the superconducting material. In QCD it would be the chromoelectric field that exhibited a Meissner effect. This in turn would squeeze the chromoelectric lines of force between a quark and an antiquark into a narrow tube or fluxoid and, furthermore, the energy of the tube would obviously be proportional to its length. Quarks would be confined in exactly the manner described by string theory. Now, however, the strings were not idealized mathematical strings but objects composed of the gauge field. 't Hooft must have been thinking the same thing, because our papers on the subject crossed in the mail.[20]

Kogut at that time was at Cornell, and I went to visit him in order to complete our paper. Wilson was there also, and by that time he had understood quark confinement in the limit of strong coupling. He showed us that in this limit the amplitude for a quark loop was exponentially small in the area of the loop. I told him that this was just the behavior predicted by the string theory and explained our flux-tube model. We agreed that this was the effect that was happening in the lattice theory.

When I left Cornell, I was completely convinced of three things. The first was that the stringy picture of confinement was correct at large separation. The second was that the strings were really lines of chromoelectric flux and that QCD was the right theory. The last thing was that lattice gauge theory was the tool by which a quantitative theory of hadrons would be built.

In order to better understand the theory, Kogut and I agreed that it would be good to have a Hamiltonian description on a three-dimensional lattice – which we proceeded to build.[21] In this version of lattice QCD, one can discuss the structure of the space of states, the particle spectrum, and particle properties using ordinary quantum mechanics. For example, it became clear that the space of states was described by fermionic quarks that can hop from site to site but are always connected to the end of strings, which occupy the lattice links. Furthermore, the Hamiltonian version proved more useful in making computations of the particle spectrum than the path-integral approach. A computational method was devised in 1975 that gave a spectrum, including many mesons and baryons, that was in good agreement with the empirical spectrum, considering the low-order nature of the calculation.[22] I believe that no Monte Carlo computer computation has done as well. The Hamiltonian calculation was done entirely by hand.

It is interesting to look back and see how much important physics came out of Wilson's attempt to formulate field theory for a computer, and how little came out of actually doing it. The main achievement of the computers in the subject of quantum field theory was to force Wilson to think clearly.

The only other development that I will discuss involves a very interesting view of confinement pioneered by 't Hooft, Polyakov and Stanley Mandelstam. It does not provide a quantitative tool, but it gives a qualitative explanation of confinement that I think has some truth to it. Recall that in certain non-Abelian Higgs theories, magnetic monopoles exist as configurations of the gauge field. These monopoles typically are very massive because the coupling is weak. If you increase the coupling, the monopole mass decreases. At some point the monopoles may condense in the same sense that Cooper pairs condense in the BCS ground state. If this happens, then the vacuum becomes a superconductor, except magnetic and electric fields are interchanged. The electric field would be replaced by the condensate and the electric Meissner effect occurs. In 2 + 1 dimensions, Polyakov was able to show that this indeed happens and confinement follows.

The common theme of a stringlike structure stretched between quarks has become the standard model of confinement, which is not seriously doubted by particle physicists. One may ask how much experimental evidence exists for it. Here I think the only completely convincing evidence comes from the old linear Regge trajectories. The long sequences of rotational excitations with $J \sim M^2$ is a sure smoking gun. Since the original suggestion by Chew and Frautchi, large numbers of excited resonances have been discovered, and they really do form linear trajectories.

Notes

1 J. D. Bjorken, "Asymptotic Sum Rules at Infinite Momentum,"
  Phys. Rev. *179* (1969), pp. 1547–53; J. D. Bjorken and E. A. Paschos,

"Inelastic Electron–Proton and γ-Proton Scattering and the Structure of the Nucleon," *Phys. Rev. 185* (1969), pp. 1975–82; Richard P. Feynman, "Very High-Energy Collisions of Hadrons," *Phys. Rev. Lett. 23* (1969), pp. 1415–17.

2 S. Frautschi, *Regge Poles and S-Matrix Theory* (New York: W. A. Benjamin, Inc., 1963).

3 R. Dolen, D. Horn, and C. Schmidt, "Finite-Energy Sum Rules and Their Application to πN Charge Exchange," *Phys. Rev. 166* (1967), pp. 1768–83.

4 G. Veneziano, "Construction of a Crossing-Symmetric, Regge-Behaved Amplitude for Linearly Rising Trajectories," *Nuovo Cimento 57A* (1968), pp. 190–7.

5 Jonathan L. Rosner, "Graphical Form of Duality," *Phys. Rev. Lett. 22* (1969), pp. 689–92; Haim Harari, "Duality Diagrams," *Phys. Rev. Lett. 22* (1969), pp. 562–65.

6 Leonard Susskind, "Model of Self-Induced Strong Interactions," *Phys. Rev. 165* (1968), pp. 1535–46.

7 L. Susskind, "Harmonic-Oscillator Analogy for the Veneziano Model," *Phys. Rev. Lett. 23* (1969), pp. 545–47; Y. Nambu, "Quark Model and the Factorization of the Veneziano Amplitude," in R. Chaud, ed., *Proceedings of Int'l. Conf. on Symmetries and Quark Models*, Wayne State Univ., June 1969 (New York: Gordon and Breach, 1970), pp. 269–78; Leonard Susskind, "Structure of Hadrons Implied by Duality," *Phys. Rev. D1* (1970), pp. 1182–86.

8 E. P. Tryon, "Dynamical Parton Model for Hadrons," *Phys. Rev. Lett. 28* (1972), pp. 1605–8.

9 Y. Nambu, Lectures at the Copenhagen Symposium 1970; Tetsuo Goto, "Relativistic Quantum Mechanics of One-Dimensional Mechanical Continuum and Subsidiary Condition of Dual Resonance Model," *Prog. Theor. Phys. 46* (1971), pp. 1560–69.

10 H. B. Nielsen, "Connection Between the Regge Trajectory Universal Slope α' and the Transverse Momentum Distribution of Partons in Planar Feynman Diagram Model," *Phys. Lett. 35B* (1971), pp. 515–18.

11 H. B. Nielsen, L. Susskind, and A. B. Kraemmer, "A Parton Theory Based on the Dual Resonance Model," *Nucl. Phys B28* (1971), pp. 34–50.

12 J. D. Bjorken, "Asymptotic Sum Rules"; J. D. Bjorken and E. A. Paschos, "Inelastic Electron–Proton and γ-Proton Scattering"; R. P. Feynman, "Very High-Energy Collisions."

13 Julian Schwinger, "Gauge Invariance and Mass. II," *Phys. Rev. 128* (1962), pp. 2425–29.

14 A. Casher, J. Kogut, and L. Susskind, "Vacuum Polarization and the Quark-Parton Puzzle," *Phys. Rev. Lett. 31* (1973), pp. 792–5.

15 H. David Politzer, "Reliable Perturbative Results for Strong Interactions?" *Phys. Rev. Lett. 30* (1973), pp. 1346–9; David J. Gross and F. Wilczek, "Ultraviolet Behavior of Non-Abelian Gauge Theories," *Phys. Rev. Lett. 30* (1973), pp. 1343–6.

16 The asymptotic freedom of non-Abelian gauge theory was reported by G. 't Hooft at the 1972 Marseilles Conference on gauge theories. See Chapter 10, note 32.

17 See the review article by Kenneth G. Wilson and J. Kogut, "The

Renormalization Group and the ϵ Expansion," *Phys. Rep. 12C* (1974), pp. 75–200.

18 Stephen L. Adler, "Short-Distance Behavior of Quantum Electrodynamics and an Eigenvalue Condition for α," *Phys. Rev. D5* (1972), pp. 3021–47.

19 Kenneth G. Wilson, "Confinement of Quarks," *Phys. Rev. D10* (1974), pp. 2445–59.

20 J. Kogut and Leonard Susskind, "Vacuum Polarization and the Absence of Free Quarks in Four Dimensions," *Phys. Rev. D9* (1974), pp. 3501–12.

21 John Kogut and Leonard Susskind, "Hamiltonian Formulation of Wilson's Lattice Gauge Theories," *Phys. Rev. D11* (1975), pp. 395–408.

22 T. Banks, S. Raby, L. Susskind, J. Kogut, D. R. T. Jones, P. N. Scharbach, and D. K. Sinclair, "Strong-Coupling Calculations of the Hadron Spectrum of Quantum Chromodynamics," *Phys. Rev. D15* (1974), pp. 1111–27.

13

A View from the Island

ALEXANDER POLYAKOV

Born Moscow, 1945; Ph.D., 1969 (physics), Landau Institute for Theoretical Physics; Professor of Physics at Princeton University; high-energy physics (theory).

The "island" in the title of this article means two things – the Soviet Union and my own mind. Partial isolation from the larger physics community had considerable effect on my work, both positive and negative. While the negative aspects of it are obvious, the good thing about isolation is that it gives independence, reduces the danger of being swept up by the intellectual "mass culture."

The beginning of modern field theory in Russia I would associate with the great work by Lev Davidovich Landau, Alexey Alexeevich Abrikosov, and Isak Markovich Khalatnikov.[1] They studied the structure of the logarithmic divergences in QED and introduced the notion of the scale-dependent coupling. This scale dependence comes from the fact that the bare charge is screened by the cloud of virtual particles, and the larger this cloud is, the stronger the screening. They showed that at the scale r, the coupling has the form

$$\alpha(r) \quad \propto \quad 1/\log(r/a),$$

where a is the minimal cutoff scale. Similar, and in some respects stronger, results have been obtained by Gell-Mann and Low, who discovered the "renormalization group" equation:[2]

$$\frac{d\alpha}{d\log(r/a)} = C_1\alpha^2 + C_2\alpha^3 + \cdots$$

The catastrophic consequence of these results was that as $a \to 0$ (no artificial cutoff) one obtains "Moscow zero" – total vanishing of the interaction. Immediately after that the search for different renormalizable theories was started in an attempt to find antiscreening (or as we would

243

say today – asymptotic freedom). The only finding at that time had been the four-fermion interaction in two dimensions.[3] This caused the well-known pessimism toward field theory. For the reasons described below, and also because I was seven years old in 1953, I have never shared this pessimism.

Instead, I was very excited, when entering physics in the early sixties, by this work and also by the marvelous ideas of Yoichiro Nambu and Giovanni Jona-Lasinio and Valentin Vaks and Anatoly Larkin, who traced the analogy between the fermion masses and gaps in superconductors.[4] In the USSR these works were considered to be garbage, but they resonated with my strong conviction, which I still hold today, that the really good ideas should serve in many different parts of physics. Even more than that – the importance of the idea for me is measured by its universality.

As a result, starting from 1963, Sasha Migdal and I were involved in infinite discussions about the meaning of spontaneous symmetry breaking. We had moral support from Sasha's father, a brilliant physicist and great man, who was almost the only one taking these ideas seriously. At about the same time Larkin explained to us the physical origin of massless particles (the "Goldstone theorem") and said that in the case of long-range forces, as in superconductors, they do not occur, although the exact reasons for that were not clear.

Sasha and I started to analyze Yang–Mills theories with dynamical symmetry breaking, and in the spring of 1965 came to the understanding that the massless particles must be eaten by the vector mesons, which become massive after this meal. We had many troubles with the referees and at seminars, but finally our paper was published.[5] We did not know until very much later about the work on the "Higgs mechanism," which was done in the West at about the same time, or slightly earlier.

A little later I became interested in the work on critical phenomena that was done by Valeny Pokrovsky, Alexander Patashinsky, Leo P. Kadanoff, and Vaks and Larkin. It was quite obvious to me that critical phenomena are equivalent to relativistic quantum field theory, continued to imaginary time. I felt that they provided an invaluable opportunity to study elementary particle physics at small distances. The "imaginary time" did not bother me at all; on the contrary, I felt that it is the most natural step, ultimately uniting space and time, and making the ordinary time just a matter of perception.

With the use of the ingenious technique developed by Vladimir Gribov and Migdal in the problem of Reggeons, I found connections between phenomenological theory and the "bootstrap" equations.[6] Sasha Migdal did very similar work independently. There was also something new – I formulated "fusion rules" for correlations, which we now would call the operator-product expansion.[7] I had mixed feelings when I found out later that the same rules at the same time, and in more generality, were found by Kadanoff and Kenneth Wilson.[8]

The paper by Wilson also overlapped with the project in which I was deeply involved at the time. It was an idea to describe elementary particles at small distances using renormalizable field theories. I considered the processes of deep-inelastic scattering and e^+e^- annihilation, and was able to prove that they must go in a cascade way – by forming a few heavy virtual objects, which I called "jets," and then repeating the process with lighter and lighter jets until we end with real particles. The picture was inspired by Kolmogorov's theory of turbulence. I was able to show that these processes are described by what are now called "multifractal" formulas and made predictions for the violations of Bjorken scaling. I considered both a scale-invariant (fixed-point) regime with anomalous dimensions and a logarithmic regime that was easier to deal with.

As a mathematical model I used $\lambda\phi^4$-theory, with the wrong sign of λ. However, looking through my old notes, I see that it was just a toy model for me with no anticipation that asymptotic freedom was a real thing. I thought at that time that anomalous dimensions were just small numbers, as they are in the theory of phase transitions. In any case these papers give a correct picture of the deep-inelastic processes in any renormalizable field theory, predicting the pattern of the Bjorken scaling violation, the jet structure, and the multiplicity distribution (later called KNO-scaling).[9] In the beginning of 1973 I finished the paper on the conformal bootstrap, but postponed its development for ten years because I had heard in May 1973 about the results of David Gross and Frank Wilczek and of H. David Politzer.[10] After a short check it became clear to me that this was *the* theory. All my old statements about deep-inelastic scattering were true in this case, but also could be made much more concrete, since the coupling was small at short distances.

It was not much of a challenge by then to elaborate this side of the subject, and I turned to the large-distance problem. I was impressed by a simple comment by Amati and Testa that if you neglect the $F_{\mu\nu}^2$ term in the gauge Lagrangian, you obtain the constraint that the gauge current is zero, a fact they associated with confinement.[11] In order to make quantitative sense of this argument, I constructed a lattice version of the gauge theory, in which the neglect of $F^2_{\mu\nu}$ is a well-defined approximation. At the beginning of 1974 I gave a few talks on lattice gauge theory, but never published them, since the preprint by Wilson came in at this time. It was clear that he had deeper understanding of the subject of confinement, and I decided to do more work before publishing something.

I kept thinking about the beautiful work by Vadim Berezinskii, in which he showed very clearly how vortices and dislocations in two-dimensional statistical mechanics create phase transitions.[12] It was clear that confinement may be related to the fact that similar "dislocations" disorder the vacuum and create finite correlation length. But what are these "dislocations" in the gauge theory?

At this point I recalled my conversation with Larkin in 1969 about Abrikosov vortex lines. We discussed whether they are normal elementary excitations appearing as poles of Green's functions. As often happens, the discussion led nowhere at that time but was helpful five years later. What also helped was my fascination with the work on solitons in integrable systems, being done by Vladimir Zakharov and Ludwig Faddeev at that time. Actually Faddeev and Takhtajan considered Sine–Gordon solitons as quantum particles. What had been far from clear was the extent to which these results were tied to specific models with complete integrability.

After brief but intense work in the spring of 1974, I arrived at two results simultaneously. First, I found a non-Abelian generalization of the Abrikosov vortex in three dimensions and realized that it must be an elementary particle with nontrivial topology. A question asked by Lev Okun during my talk helped me to realize that the topological charge is in fact magnetic charge. The same work was done simultaneously by Gerard 't Hooft. While the possibility of magnetic poles was envisaged by Dirac in the 1930s, from our work it follows that magnetic charges are inevitable in any reasonable unified theory. I am quite certain that they really exist. How, when, and if they will be found is another matter.

The second result, published only a year later, was that the same monopoles play the role of the "dislocations" mentioned above in the $2+1$-dimensional gauge theories and indeed lead to confinement. It took almost a year to gain confidence in this result and to find the dislocations in four dimensions. In the Abelian case, these dislocations turned out to be just the world lines of magnetic monopoles, and I predicted the phase transition leading to confinement (in the $3+1$ case). In the $(2+1)$ case confinement was the only phase.[13] In the non-Abelian $(3+1)$ case it was necessary to find a novel solution of the Yang–Mills equation in imaginary time and then to investigate its influence on the vacuum disorder. I suggested this problem to my colleagues, Alexander Belavin, Schwartz, and Tyupkin, during a summer school and together we found the required solution. Even before that, when I discussed the problem with Sergy Novikov and asked him about the topology involved in it, he told me about Chern classes.

I had never learned topology before and was somewhat scared by this subject. I thought that my spatial imagination was not adequate for it. At present, I think that in topology just as in physics the more important quality is the "temporal" imagination, also called "intuition," the sense of how things should be related in time.

Anyway, we had a solution (which was later given the name "instanton"), but its effect on confinement turned out to be unclear because of strong fluctuations of large instantons. That is why we do not have a theory of confinement even today. Nevertheless, instantons turned out to be interesting beasts. In the same summer of 1975, Gribov noticed that they can be interpreted as tunneling events between the different vacua. It became clear that the vacua in gauge theory are labeled by the integers, and the instanton is the process of jumping from one vacuum to another.

That was later rediscovered by other people. Inspired by Gribov's remark, I started to analyze the relation of the instantons to the axial anomaly, when I heard about the beautiful results by 't Hooft, who had shown that the tunneling, mentioned above, leads to baryon number nonconservation and to the solution of the η-mass problem. I kept trying to solve the confinement problem, playing with different physical settings. In particular I considered the temperature dependence of the gluon system and found a rather surprising (at that time) deconfining phase transition.[14] Leonard Susskind came to the same conclusion independently.[15]

There are three interesting points about this work. First, it demonstrated that the symmetry group responsible for confinement is the center of the gauge group, which breaks in the process of deconfinement. Second, and more important, is that the natural description of the deconfinement could be given in terms of condensing strings. Third, I realized that temperature can enhance tunneling and increase the baryon number nonconservation via the 't Hooft process.[16] The same idea occurred to Susskind. The details, however, were worked out only in the 1980s by many people. I believe that at present this is the most dramatic manifestation of the instanton structure of the vacuum.

Since strings appeared so naturally in QCD, I turned to string theory. First, I tried to use the equations in the loop space.[17] These loop equations still look interesting to me, although very few results followed from them. In particular, as was noticed by A. Migdal, the equations simplify drastically in the large-N limit. He and Yury Makeenko showed how to reproduce perturbation theory in this approach.[18] Unfortunately, we still don't know how to solve these equations, but expect that the solution must be some kind of string theory.

The fact that 15 years of hard work did not bring the solution should not discourage us. Problems in physics become deeper and more difficult and take more time than before. For comparison, remember how much time it took to solve some of the celebrated Hilbert mathematical problems. This is an inevitable consequence of the maturity of the subject.

Incidentally, the work on instantons, which originated in complete mathematical ignorance, seems to have had an influence on mathematics. In the hands of mathematical grand masters, it helped to solve long-standing problems in the topology of four-dimensional manifolds, and led to the link between quantum field theory and topology. That shows that the notion of "universality" of good ideas should, perhaps, include the realm of mathematics.

We have come (in the proper time of this article) to the end of the 1970s. The 1980s were equally exciting for me, but that is a topic for a different conference. Writing this article brought to my mind the phrase of the old German romanticist, Friedrich Novalis, who said that the greatest magician is "the one who would cast over himself a spell so complete, that his phantasmagorias would become autonomous appearances." I very much hope that there are many beautiful phantasmagorias ahead of us.

Notes

1 L. Landau, A. Abrikosov, and I. Khalatnikov, "On the Removal of Infinites in Quantum Electrodynamics," *Sov. Phys. Dokl. 95* (1954), pp. 497–500.

2 M. Gell-Mann and F. Low, "Quantum Electrodynamics at Small Distances," *Phys. Rev. 95* (1954), pp. 1300–1312.

3 A. A. Ansel'm, "A Model of a Field Theory with Nonvanishing Renormalized Charge," *Sov. Phys. JETP 9* (1959), pp. 602–4.

4 Y. Nambu and G. Jona-Lasinio, "Dynamical Model of Elementary Particles Based on an Analogy with Superconductivity. I." *Phys. Rev. 122* (1961), pp. 345–58; V. G. Vaks and A. I. Larkin, "On the

Application of the Methods of Superconductivity Theory to the Problem of the Masses of Elementary Particles," *Sov. Phys. JETP 13* (1961), pp. 192–93.

5 A. A. Migdal and A. M. Polyakov, "Spontaneous Breakdown of Strong Interaction Symmetry and the Absence of Massless Particles," *Sov. Phys. JETP 24* (1967), pp. 91–98.

6 V. N. Gribov and A. A. Migdal, "Strong Coupling in the Pomeranchuk Pole Problem," *Sov. Phys. JETP 28* (1969), pp. 784–95; A. M. Polyakov, "Microscopic Description of Critical Phenomena," *Sov. Phys. JETP 28* (1969), pp. 533-39.

7 A. M. Polyakov, "Properties of Long and Short Range Correlations in the Critical Region," *Sov. Phys. JETP 30* (1970), pp. 151–7.

8 Leo P. Kadanoff, "Operator Algebra and the Determination of Critical Indices," *Phys. Rev. Lett. 23* (1969), pp. 1430–33; Kenneth G. Wilson, "Non-Lagrangian Models of Current Algebra," *Phys. Rev. 179* (1969), pp. 1499–1512.

9 A. M. Polyakov, "A Similarity Hypothesis in the Strong Interactions. I. Multiple Hadron Production in the e^+e^- Annihilation," *Sov. Phys. JETP 32* (1971), pp. 296–301; A. M. Polyakov, "A Similarity Hypothesis in the Strong Interactions. II. Cascade Production of Hadrons and Their Energy Distribution Associated with e^+e^- Annihilation," *Sov. Phys. JETP 33* (1971), pp. 850–55; A. M. Polyakov, "Description of Lepton–Hadron Reactions in Quantum Field Thoery," *Sov. Phys. JETP 34* (1972), pp. 1177–83.

10 A. M. Polyakov, "Non-Hamiltonion Approach to Conformal Quantum Field Theory," *Sov. Phys. JETP 39* (1974), pp. 10–18; David J. Gross and Frank Wilczek, "Ultraviolet Behavior of Non-Abelian Gauge Theories," *Phys. Rev. Lett. 30* (1973), pp. 1343–46; H. David Politzer, "Reliable Perturbative Results for Strong Interactions?" *Phys. Rev. Lett. 30* (1973), pp. 1346–49.

11 D. Amati and M. Testa, "Quark Imprisonment as the Origin of Strong Interactions," *Phys. Lett. B48* (1974), pp. 227–31.

12 V. L. Berezinskii, "Destruction of Long-Range Order in One-Dimensional and Two-Dimensional Systems with a Continuous Symmetry Group. II. Quantum Systems," *Sov. Phys. JETP 34* (1972), pp. 610–16.

13 A. A. Belavin, A. M. Polyakov, A. S. Schwartz, and Yu. S. Tyupkin, "Pseudoparticle Solutions of the Yang–Mills Equations," *Phys. Lett. 59B* (1975), pp. 85–87.

14 A. M. Polyakov, "Thermal Properties of Gauge Fields and Quark Liberation," *Phys. Lett. 72B* (1978), pp. 477–80.

15 Leonard Susskind, "Lattice Models of Quark Confinement at High Temperature," *Phys. Rev. D20* (1979), pp. 2610–18.

16 A. M. Polyakov, "Models and Mechanisms in Gauge Theories," in T. B. W. Kirk and H. D. I. Abarbanel, eds., *Proceedings of the International Symposium on Lepton and Photon Interactions at High Energies* held at Fermilab, Batavia, IL, 23–29 August 1979 (Batavia, Illinois: Fermi National Accelerator Laboratory, 1979), pp. 520–23.

17 A. M. Polyakov, "Gauge Fields as Rings of Glue," *Nucl. Phys. B164* (1979), pp. 171–88.

18 Yu. M. Makeenko and A. A. Migdal, "Quantum Chromodynamics as Dynamics of Loops," *Nucl. Phys. B188* (1981), pp. 269–316.

14

On the Early Days
of the Renormalization Group*

DMITRIJ V. SHIRKOV

Born Moscow, USSR, 1928; Ph.D. 1953 (theoretical physics), Kurchatov
Institute; D. Sc., 1958 (theoretical and mathematical physics), Steklov
Mathematical Institute; Director, Bogoliubov Theoretical Laboratory,
Joint Institute for Nuclear Research, Dubna, Russia; high-energy physics
(theory and mathematical physics).

In the spring of 1955 in Moscow there was a small conference on QED
and elementary particle theory that took place at the Lebedev Physical
Institute from March 31 through April 7. Among the participants were
a few foreigners, including Ning Hu and Gunnar Källen. I remember it
quite well as it was my first conference on quantum field theory (QFT)
problems with scientists from abroad. My short contribution concerned
finite Dyson transformations for renormalized Green's functions and matrix elements in QED.[1]

The central point of the conference was Lev Davydovich Landau's review talk "Basic Problems of Quantum Field Theory," devoted to the
ultraviolet behavior in local QFT. The point is that a few months earlier,
the problem of short-distance behavior in QED was successfully attacked
by Landau and his brilliant pupils Alesha Abrikosov and Isaak Khalatnikov. They managed to find a closed approximation to the Schwinger–
Dyson equations for two propagators and the three-vertex function that
was compatible with renormalizability and gauge invariance. Besides,
this so-called three-gammas approximation admitted a solution in the
massless limit that, in modern terms, was equivalent to the summation
of leading ultraviolet logarithms.[2]

This solution had a peculiar feature that was controversial from the
physical point of view (the "ghost-pole" in the renormalized photon
propagator amplitude or *Moscow-zero puzzle* in the formal expression
for the "physical electron charge") that attracted attention and excited
one's imagination.[3]

---

* In memory of N. N. Bogoliubov.

At that time I had regular and frequent contacts with my teacher, Nicolaj Nicolaevich Bogoliubov, as we had been working heavily on the final text of our monograph on quantized fields. Its preliminary version had just been finished in the form of two review papers.[4]

Bogoliubov was very intrigued by the Landau results and formulated for me the general problem of finding how to estimate their validity, that is, to construct the second approximation (incorporating, say, next-to-leading logs) to check the stability of obtained ultraviolet asymptotics and the ghost-pole existence.

The birth of the renormalization group

I had rather close contacts with Abrikosov, including our common studentship in the forties. During our discussion shortly after the Lebedev meeting, he told me about the Gell-Mann–Low paper that had just appeared.[5] It was evidently related to the problem, but, he said, was rather difficult to combine with their group analysis. I studied the paper and presented to my teacher a short review of its method and results, which included some general statements about scale properties of the short-distance charge distribution and rather cumbersome functional equations (see Appendix I).

The scene that followed my talk was very impressive. Bogoliubov immediately replied that the Gell-Mann–Low approach is absolutely clear and very important – it represented the realization of *la groupe de normalisation* discovered a couple of years before by Stueckelberg and Petermann. These authors revealed a group structure of the finite arbitrariness in renormalized scattering-matrix elements that arose after subtraction of infinities (see Appendix I).[6]

It became clear that group functional equations analogous to the ones obtained by Gell-Mann and Low should be valid in the general (i.e., not only in the ultraviolet) case, and that combining the Stueckelberg–Petermann ideology with the Gell-Mann–Low equations opens the way for solving the problem of creating a regular algorithm for studying the short-distance QFT behavior. Bogoliubov added that this algorithm should be based upon standard Lie theory – group differential equations. Happily I had some knowledge about group theory.

During the next few days I performed a simple reformulation of finite Dyson transformations that produced renormalization-group functional equations for scalar propagator amplitudes. Each of them turned out to be dependent on a specific object equal to the product of the electron charge squared $\alpha = e^2$, and transverse photon propagator amplitude d. This product, which we named *the invariant charge* of the electron, was an invariant of the renormalization-group transformation. Physically, it was the counterpart of charge distribution mentioned by Gell-Mann and Low and by Landau, and discussed first by Dirac in the early thirties.[7] Written down in the modern notation $\bar{\alpha} = \alpha d$, it satisfies the functional equation:

$$\bar{\alpha}\left(\frac{Q^2}{\mu^2}, \frac{m^2}{\mu^2}; \alpha\right) = \bar{\alpha}\left(\frac{Q^2}{\mu^2 t}, \frac{m^2}{\mu^2 t}; \bar{\alpha}(t, \frac{m^2}{\mu^2}; \alpha)\right). \tag{14.1}$$

[The corresponding massless limit of Eq. (14.1) $\bar{\alpha}(x; \alpha) = \bar{\alpha}(x/t, \bar{\alpha}(t; \alpha))$ precisely coincides with the functional equation one can deduce from the Gell-Mann–Low equation (14.5) given in our Appendix II.] An analogous equation was derived for electron propagator amplitudes.

By differentiating the functional equations, we then obtained differential group equations in the standard Lie (i.e., ordinary nonlinear) form[8] (see Appendix III) that was an explicit realization of the differential equations mentioned in the quotation from the 1953 Stueckelberg–Petermann paper given in Appendix I.[9] These results provided the conceptual relationship between the Stueckelberg–Petermann and the Gell-Mann–Low approaches.

The creation of a renormalization-group method

Our next step consisted in employing this elegant formalism for analysis of the ultraviolet and infrared singularities in QED.[10] The key idea was to use approximate perturbative calculation results for defining group generators β, γ.

Starting, for example, with

$$\bar{\alpha}_{pt}(Q, \ldots) = \alpha + \alpha^2 \beta_1 \ell + \ldots; \quad \ell = \ln Q^2/\mu^2 \tag{14.2}$$

and defining $\beta(\alpha) = \beta_1 \cdot \alpha^2$ via Eq. (14.10) we obtain after solving Eq. (14.8)

$$\bar{\alpha}(x, \alpha_\mu) = \frac{\alpha_\mu}{1 - \alpha_\mu \beta_1 \cdot \ell} \tag{14.3}$$

– the famous controversial result leading in QED to the *ghost-pole* at $Q^2 = \mu^2 \exp(3\pi/\alpha)$.

On the other hand, starting with the next perturbative approximation

$$\bar{\alpha}_{pt,2}(Q, \ldots) = \alpha + \alpha^2 \beta_1 \ell + \alpha^3 \left[\beta_1^2 \ell^2 + \beta_2 \ell\right] + \ldots;$$

we arrived at (see Appendix III for details)

$$\bar{\alpha}(x,\alpha_\mu) = \frac{\alpha_\mu}{1 - \alpha_\mu\beta_1 \cdot \ell + \alpha\beta_2\beta_1^{-1}\ln(1 - \alpha_\mu\beta_1 \cdot \ell)} \, , \qquad (14.4)$$

the famous "log-of-log" two-loop dependence.

From this example (published in 1955)[11] one sees that the *renormalization-group method* is indeed the regular procedure that can produce more and more precise results on the acute problem of short-distance QED behavior.

However, the new method proved to be of greater significance. It gave important results for the infrared problem as well, in particular, the singularity structure of the electron-propagator amplitude around the mass shell – see Eq. (14.11) in Appendix III. All these results were obtained very quickly – during one or two weeks. Both of our papers were submitted to *Doklady Akademii Nauk* on May 16, 1955, or five weeks after the Lebedev Conference.[12]

The ghost-pole story

The last episode I would like to mention is connected with the "ghost-pole" issue. At the above-mentioned Lebedev Institute meeting, Källen presented a paper in which he discussed this problem for the soluble "Lee model" and presented the general argument that the existence of a ghost-pole is the signal of an inconsistency.[13] After his talk he had a furious discussion with Isaak Pomeranchuk about the existence of that pole in QED. Källen argued that any conclusion obtained on a weak-coupling basis cannot be considered rigorous.

This argument was not reflected in the forthcoming publications of the Landau school.[14] However, its validity could be clearly demonstrated on the base of our renormalization group results.

The renormalization analysis of the problem performed on the basis of Eq. (14.6) revealed that any inference based on finite-order perturbation calculations cannot be considered a complete proof. This conclusion precisely corresponds to the impression that can be gained from comparing Eqs. (14.3) and (14.4).[15]

This result, in the mid-1950s, was of major importance, as it restored among the experts the reputation of local QFT. Nethertheless, during the next decade, the applicability of QFT to microparticle physics remained under suspicion in a wide theoretical community. Pomeranchuk was himself so strongly convinced of a deep inner contradiction of local QFT that he closed his seminar on this subject.

(The author is indebted to Profs. Boris Medvedev, Vladimir Fainberg, and Eugenij Feinberg for valuable discussions.)

Appendix I.   The Stueckelberg–Petermann discovery

The renormalization group itself was discovered in 1953 by Stueckelberg and Petermann as a group of transformations exploiting the finite arbitrariness arising in scattering-matrix elements S after removal of ultraviolet divergences and expressed via some finite constants c_i :[16]

... we must expect that a group of infinitesimal operations $P_i = (\partial/\partial c_i)_{c=0}$ exists, satisfying

$$P_i S = h_{i_e}(m, e)\partial S(m, e, ...)/\partial e ,$$

admitting thus a *renormalization* of e.

These authors introduced a "group of normalization" given by infinitesimal operations P_i (i.e., as a Lie group) connected with coupling-constant e renormalization.

Appendix II.   The Gell-Mann–Low asymptotic analysis

In the following year Gell-Mann and Low, by manipulating Dyson renormalization transformations written down in a regularized form, obtained functional equations for QED propagators in the ultraviolet limit.[17] For example, for the renormalized photon propagator transverse amplitude d they got an equation in the form

$$d(k^2\lambda^2, e_2^2) = \frac{d_C(k^2/m^2, e_1^2)}{d_C(\lambda^2/m^2, e_1^2)} , \quad e_2^2 = e_1^2 d_C(\lambda^2/m^2, e_1^2), \quad (14.5)$$

with λ the cutoff momentum and e_2 the physical electron charge. The appendix to their paper contains a general solution (by T. D. Lee) of this functional equation for the photon amplitude $d(x, e^2)$ written in the form

$$\ln x = \int_{e^2}^{e^2 d} \frac{dx}{\psi(x)} \quad \text{with} \quad \psi(e^2) = \frac{\partial(e^2 d)}{\partial \ln x} \quad \text{at} \quad x = 1 . \quad (14.6)$$

The solution (14.6) provided the means for qualitative analysis of the short-distance behavior of electromagnetic interaction in terms of ψ properties. In the 1954 Gell-Mann–Low paper two possibilities were discussed, infinite or finite charge renormalization:[18]

Our conclusion is that the *shape* of the charge distribution surrounding a test charge in the vacuum does not, at small distances, depend on the coupling constant except through the scale factor. The behavior of the propagator functions for large momenta is related to the magnitude of the renormalization constants in the theory. Thus, it is shown that the unrenormalized coupling constant $e_0^2/4\pi\hbar c$, which appears in perturbation theory as a power series in the renormalized coupling constant $e_1^2/4\pi\hbar c$ with divergent coefficients, may behave either in two ways:

It may really be infinite as perturbation theory indicates;

It may be a finite number independent of $e_1^2/4\pi\hbar c$.

The latter possibility corresponds to the case when ψ has a zero at a certain finite point $\psi(\alpha_\infty) = 0$ (α_∞ being the so-called fixed point of the renormalization-group transformation).

It is notable that the 1954 Gell-Mann–Low paper neither paid attention to the group nature of the analysis performed and the results obtained, nor mentioned the 1953 Stueckelberg–Petermann paper.[19] The authors also did not realize that the Dyson transformations they used were valid only for the case of the transverse gauge of the electromagnetic field. Probably due to this, they did not succeed in correlating their results with standard perturbation theory calculations and did not discuss the ghost-trouble possibility.

Appendix III. The Bogoliubov synthesis
Renormalization-group equations

The final step was made in 1955 by Bogoliubov and Shirkov.[20,21] Using the group properties of finite Dyson transformations for coupling constants and field functions, they obtained group functional equations for QED propagators and vertices in the general (i.e., massive) case, for example,

$$d(x, y; e^2) = d(t, y; e^2)d\left(\frac{x}{t}, \frac{y}{t}; \ e^2 d(t, y; e^2)\right)$$

for the transverse photon propagator amplitude (which depends, besides on $x = k^2/\mu^2$, also on the mass argument $y = m^2/\mu^2$, μ being a certain reference momentum value). Here the term "renormalization group" and the notion of "invariant charge" were introduced. (That is widely used now under the name of effective or running coupling.)

In modern notation the last equation [in the massless case $y = 0$, it is equivalent to Gell-Mann–Low Eq. (14.5)] is just the equation for the effective electromagnetic coupling $\bar{\alpha}(x, y; \alpha = e^2) = \alpha d(x, y; \alpha)$:

$$\bar{\alpha}(x, y; \alpha) = \bar{\alpha}\left(\frac{x}{t}, \frac{y}{t}; \bar{\alpha}(t, y; \alpha)\right) . \qquad (14.7)$$

By differentiating the functional equations, these authors first obtained differential group equations in the standard Lie (i.e., ordinary nonlinear) form for $\bar{\alpha}$

$$\frac{\partial \bar{\alpha}(x, y; \alpha)}{\partial \ln x} = \beta\left(\frac{y}{x}, \bar{\alpha}(x, y; \alpha)\right) \qquad (14.8)$$

and for the electron propagator amplitude $s(x, y; \alpha)$

$$\frac{\partial s(x, y; \alpha)}{\partial \ln x} = \gamma\left(\frac{y}{x}, \bar{\alpha}(x, y; \alpha)\right) s(x, y; \alpha) \qquad (14.9)$$

with

$$\beta(y, \alpha) = \frac{\partial \bar{\alpha}(\xi, y; \alpha)}{\partial \xi} , \quad \gamma(y, \alpha) = \frac{\partial s(\xi, y; \alpha)}{\partial \xi} \quad \text{at} \quad \xi = 1 , \qquad (14.10)$$

that was the explicit realization of the differential equations mentioned in the quotation from the 1953 Stueckelberg–Petermann paper cited above in Appendix I.[22] These results provided the conceptual relation between the Stueckelberg–Petermann and the Gell-Mann–Low approaches.

A second important contribution of the 1955 Bogoliubov–Shirkov paper consisted in proposing a simple algorithm for improving an approximate perturbative solution by combining it with the Lie equations: [23] (This quotation from the review paper, that follows the Russian original text, is given here in the usual modern notation.)

> Formulas (14.8)–(14.10) show that to obtain expressions for $\bar{\alpha}$ and s valid for all values of their arguments, one has only to define $\bar{\alpha}(\xi, y, \alpha)$ and $s(\xi, y, \alpha)$ in the vicinity of $\xi = 1$. This can be done by means of the usual perturbation theory.

In the subsequent publication this algorithm was effectively used for the ultraviolet and infrared asymptotic analysis of QED in a transverse gauge.[24] Here the one-loop Eq. (14.3) and two-loop ultraviolet expressions Eq. (14.4) with $\beta_1 = \frac{1}{3\pi}$ and $\beta_2 = \frac{1}{4\pi^2}$ for the photon propagator as well as the infrared asymptotic expression

$$s(x, y; \alpha) \simeq \left(\frac{p^2}{m^2} - 1\right)^{\frac{-3\alpha}{2\pi}} \qquad (14.11)$$

for the QED electron propagator amplitude were obtained.

At that time these expressions at the one-loop level were known from the results of the Landau group. However, the Landau approach gave no simple possibility of constructing the next approximation. The problem was resolved by the newly born renormalization method.

The simplest ultraviolet asymptotic results for QED propagators obtained in the 1955 Bogoliubov–Shirkov paper, such as Eq. (14.3), corresponded precisely to the results of the Landau group. In the renormalization-group approach this equation can be obtained by a few lines of calculation. However, the second renormalization-group approximation, Eq. (14.4), corresponds to the summation of the *next-to-leading* ultraviolet logarithms. This proves that the renormalization-group method is a regular method within which it is rather simple to estimate the region of validity of results obtained. This second-order renormalization-group expression for the effective coupling, first obtained in the 1955 Bogoliubov–Shirkov paper, contains a nontrivial log-of-log dependence that is now widely known as the two-loop approximation for the QCD running coupling.[25]

Notes

1 See pp. 505–16 in N. N. Bogoliubov and D. V. Shirkov, "Probleme der Quantenfeldtheorie," *Fort. Phys. 4* (1956), pp. 438–517, the second German translation of the Russian original (*Usp. Fiz. Nauk 57* (1955), p. 3), as well as section 31.2 in N. Bogoliubov and D. Shirkov, *Introduction to the Theory of Quantized Fields* (New York: Interscience Pub., 1959).

2 L. D. Landau, A. A. Abrikosov, and I. M. Khalatnikov, "On the Removal of Infinities in Quantum Electrodynamics," *Dokl. Akad. Nauk SSSR 95* (1954), pp. 497–500; "An Asymptotic Expression for the Green's Function of an Electron in Quantum Electrodynamics," pp. 773–6; "An Asymptotic Expression for the Green's Function of a Photon in Quantum Electrodynamics," pp. 1117–20; "The Mass of the Electron in Quantum Electrodynamics," *Doklady Akad. Nauk SSSR 96* (1954), pp. 261–4 (in Russian); "On the Quantum Theory of Fields," *Supp. Nuovo Cimento 3*, pp. 80–104.

3 See two papers presented for publication just before the Lebedev meeting: E. S. Fradkin, "The Asymptotic Green's Function in Quantum Electrodynamics," letter in *Zh. Eksp. Theor. Fiz. 28* (1955), pp. 750–2 (in Russian, submitted 27 March 1955); and L. D. Landau and I. Ia. Pomeranchuk, "On Point Interaction in Quantum Electrodynamics," *Dokl. Akad. Nauk SSSR 102* (1955), pp. 489–92 (in Russian, submitted 2 March 1955).

4 N. N. Bogoliubov and D. V. Shirkov, "Voprosy Kvantovoj Teorii Polia," *Usp. Fiz. Nauk 55* (1955), pp. 149–214; "Problems of the Quantum Theory of the Field," *Usp. Phiz. Nauk 57* (1955), pp. 3–92 (in Russian). For German translation see "Probleme der Quantentheorie der Felder," *Fort. Phys. 3* (1955), pp. 439–99 and "Probleme der Quantenfeldtheorie," *Fort. Phys. 4* (1956), pp. 438–517.

5 M. Gell-Mann and F. Low, "Quantum Electrodynamics at Small Distances," *Phys. Rev. 95* (1954), pp. 1300–12.

6 E. C. G. Stueckelberg and A. Petermann, "La Normalisation des Constantes dans la Théorie des Quanta," *Helv. Phys. Acta 26* (1953), pp. 499–520. The pioneer Stueckelberg–Petermann was published in French, a language that was not very popular among particle theorists after the Second World War.

7 P. A. M. Dirac, "Théorie du Positron" in *Septième Conseil de Physique Solvay: Structure et propriété de noyaux atomiques*, Octobre 1933 (Paris: Gauthier-Villars, 1934), pp. 203–30.

8 N. N. Bogoliubov and D. V. Shirkov, "On the Renormalization Group in Quantum Electrodynamics," *Dokl. Akad. Nauk SSSR 103* (1955), pp. 203–6 (in Russian); see also the review paper by N. N. Bogoliubov and D. V. Shirkov, "Charge Renormalization Group in Quantum Field Theory," *Nuovo Cimento 3* (1956), pp. 845–63.

9 E. C. G. Stueckelberg and A. Petermann, "La Normalisation."

10 N. N. Bogoliubov and D. V. Shirkov, "Application of the Renormalization Group to Improvement of Perturbation Theory Formulae," *Doklady Akad. Nauk SSSR 103* (1955), pp. 391–4 (in Russian); see also Bogoliubov and Shirkov, "Charge Renormalization Group."

11 Bogoliubov and Shirkov, "Application of the Renormalization Group."

12 Bogoliubov and Shirkov, "On the Renormalization Group" and "Application of the Renormalization Group."

13 G. Källen and W. Pauli, "On the Mathematical Structure of T. D. Lee's Model of a Renormalizable Field Theory," *Kongelige Danske Videnskabernes Selskab Mathematisk-Fysiske Meddelser 30* (1955), pp. 3–23.

14 I. Pomeranchuk, "Zero Equality of Renormalized Charge in Quantum Electrodyanimcs," *Dokl. Akad. Nauk SSSR 103* (1955), pp. 1005–8 (in Russian, submitted 27 March 1955); and L. D. Landau, "On the Quantum Theory of Fields," in W. Pauli, et al., eds., *Niels Bohr and the Development of Physics* (London: Pergamon Press, 1955), pp. 52–69.

15 N. N. Bogoliubov and D. V. Shirkov, "The Lee Type Model in Quantum Electrodynamics," *Dokl. Akad. Nauk SSSR 105* (1955), pp. 685–8 (in Russian).

16 Stueckelberg and Petermann, "La Normalisation."

17 Gell-Mann and Low, "Quantum Electrodynamics."

18 Gell-Mann and Low, "Quantum Electrodynamics."

19 Gell-Mann and Low, "Quantum Electrodynamics"; Stueckelberg and Petermann, "La Normalisation."

20 For a more detailed historical review see the recent paper, D. V. Shirkov, "Historical Remarks on the Renormalization Group," in L. M. Brown, ed., *Renormalization – from Lorentz to Landau (and beyond)* (New York: Springer-Verlag, 1993), pp. 168–85.

21 Bogoliubov and Shirkov, "On the Renormalization Group"; Bogoliubov and Shirkov, "Charge Renormalization Group."

22 Stueckelberg and Petermann, "La Normalisation."

23 Bogoliubov and Shirkov, "On the Renormalization Group," and "Charge Renormalization Group."

24 Bogoliubov and Shirkov, "Application of the Renormalization Group," and "Charge Renormalization Group."

25 Bogoliubov and Shirkov, "Application of the Renormalization Group," and "Charge Renormalization Group."

Part four

Accelerators, Detectors, and Laboratories

15

The Rise of Colliding Beams

BURTON RICHTER

Born Brooklyn, New York, 1931; Ph.D., 1956 (physics), Massachusetts Institute of Technology; Director, Stanford Linear Accelerator Center; Nobel Prize in Physics, 1976; high-energy physics (experimental) and particle accelerators.

My own career in science has been intimately tied up in the transition from the old fixed-target technique to colliding-beam work. I have led a kind of double life as both a machine builder and as an experimenter, taking part in building and using the first of the colliding-beam machines, the Princeton–Stanford Electron–Electron Collider, and building the most recent advance in the technology, the Stanford Linear Collider. The beginning was in 1958 and, in the more than three decades since, there has been a succession of both electron and proton colliders that have increased the available center-of-mass energy for hard collisions by more than a factor of 1000.

The history of that advance for both electron and proton colliders (constituent center-of-mass energy is plotted versus time of the first physics experiment) is shown in Fig. 15.1. The important number for the experimenter, the constituent center-of-mass energy, has increased by about a factor of 10 every 12 years for both kinds of systems. On the electron line, one can see a kind of complete cycle in accelerator technology from the birth of the colliding-beam storage ring, to its culmination in LEP II, and the beginning of the next technique for high-energy electron collisions, the linear collider. On the proton line, one has gone from the first bold initiative, the ISR at CERN, which used conventional magnets, to the superconducting magnets that are used in all proton colliders built today.

For the historians here, I regret to say that very little of this story can be found in the conventional literature. Standard operating procedure for the accelerator physics community has been publication in conference proceedings, which can be obtained with some difficulty, but

261

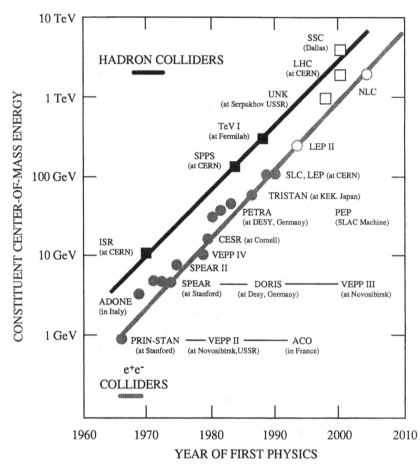

Fig. 15.1. Energy available in the constituent center-of-mass system versus year for the electron and proton colliding-beam machines. The open circles and squares represent machines under construction or in the planning phase.

even more of the critical papers are in internal laboratory reports that were circulated informally and that may not have even been preserved. In this presentation I will review what happened based on my personal experiences and what literature is available. I can speak from considerable experience on the electron colliders, for that is the topic with which I was most intimately involved. On proton colliders my perspective is more that of an observer than a participant, but I have dug into the literature and was close to many of the participants. There are others who can perhaps fill in any gaps that I leave.

The beginnings

The earliest writing that I know of on the construction of machines based on the collision of two particle beams is by Rolf Wideröe, who obtained a Swiss patent on the technique in May 1953.[1] In it he discussed collisions between the same kind of particles (proton–proton), different particles (proton–deuteron), and particles of opposite charge (electron–proton). Wideröe in his memoirs says that the idea came to him in 1943, when he realized that in the nonrelativistic case (colliding automobiles is his example) two particles colliding at equal energy could dissipate four times as much energy as one particle of the same energy would in colliding with a similar particle at rest.[2] In the colliding-beam case, two beams of equal energy have twice as much energy as a single beam colliding with a target at rest, but you achieve four times as much reaction energy.

Nothing came of Wideröe's idea. His patent was not circulated and Wideröe was working in industry at the time with little contact with people in the physics community. His patent was really conceptual in nature and did not address any of the practical questions such as how to bring these beams into collision or inject particles into the machines.

The real beginning of colliding beams comes in a paper by Donald Kerst et al., published as a letter to the editor in the *Physical Review* in early 1956.[3] Kerst was the leader of the Midwestern Universities Research Association (MURA), which was the training ground for so many of the important accelerator physicists of the 1960s and 1970s. The MURA group was working on the design of a new kind of synchrotron, the so-called fixed-field alternating-gradient (FFAG) synchrotron. In this letter Kerst writes,

> The possibility of producing interactions in stationary coordinates by directing beams against each other has often been considered, but the intensities of beams so far available have made the idea impractical. Fixed-field alternating-gradient accelerators offer the possibility of obtaining sufficiently intense beams so that it may now be reasonable to reconsider directing two beams of approximately equal energy at each other. In this circumstance, two 21.6-BeV accelerators are equivalent to one machine of 1000 BeV.

Kerst and his colleagues had recognized in the relativistic case the enormous advantage of colliding beams over the fixed-target technique in attaining very high energy (far greater than the factor of four in Wideröe's nonrelativistic example) and also analyzed the intensity requirements to get sufficient reaction rate to be able to use a colliding-beam machine as a useful physics tool. They also considered the background generated by interactions with the residual gas in the vacuum chamber, circulating beam lifetime, and stacking many cycles to build up the necessary beam intensity.

Kerst and his colleagues' *Physical Review* letter was, of course, the culmination of discussions that had been going on at MURA for some time and that had excited considerable interest in a broader community. Activity built up very quickly, as can be seen in the proceedings of the 1956 CERN Symposium on High Energy Accelerators in Pion Physics.[4] In his conference paper, Kerst expands considerably on the original MURA paper, looking at the complete injection cycle, phase space limitations, space charge effects, and so on.

At this same symposium a new actor came on stage, Professor Gerald K. O'Neill of Princeton University. He too was interested in proton–proton collisions at very high center-of-mass energies and introduced the notion of the accelerator–storage ring complex. Beams would be accelerated to high energy in a synchrotron and then transferred into two storage rings with a common straight section where the beams would interact. Since the beams at high energy need much less space in an accelerator vacuum chamber than is required for beams at injection, the high-energy storage rings would have smaller cross-section magnets and vacuum chambers, adding only little to the cost of the complex, but at the same time enormously increasing the scientific potential. O'Neill noted, "If storage rings could be added to the 25 GeV machines now being built at Brookhaven and Geneva, these machines would have equivalent energy of 1340 GeV or 1.3 TeV."[5] He also observed, "The use of storage rings on electron synchrotrons in the GeV range would allow the measurement of the electron–electron interaction at center-of-mass energies of about 100 times as great as are now available. The natural beam damping in such machines might make beam capture somewhat easier than in the case of protons." That observation was to have a profound effect on O'Neill's career (and mine), as well as on particle physics.

How to realize a colliding-beam machine was the question. The MURA FFAG accelerators discussed by Kerst were enormously complex and none had ever been built at that time (nor has one been built since). There was considerable concern about whether FFAG machines would actually work as well as their proponents claimed. At the same time, the problem of injection into the proton–synchrotron storage ring complex that O'Neill and others discussed was thought to be very difficult. Indeed, O'Neill's original idea of using a scattering foil for injection was

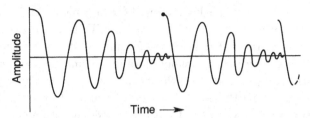

Fig. 15.2. Radiation damping in an electron storage ring with an appropriate magnet configuration leads to a decrease in oscillation amplitude, allowing another injected pulse to damp down on top of the previously injected ones.

soon proved to be impossible. On the other hand, injection and beam stacking in an electron storage ring looked easy because of synchrotron radiation damping, shown schematically in Fig. 15.2. An electron beam could be injected off-axis into a storage ring and would perform betatron oscillations around the equilibrium orbit. These oscillations would decrease exponentially over time in a properly designed magnet lattice because of the emission of synchrotron radiation. When these oscillations had damped sufficiently, another bunch could be injected into the storage ring and it would damp down on top of the first one. Since phase space was not conserved in the presence of radiation, there was, in principle, no stacking problem.

The Princeton–Stanford storage ring

In the mid-1950s the most powerful electron accelerator in the world with an external beam was the then 700 MeV linear accelerator at the Stanford University High Energy Physics Laboratory (HEPL). O'Neill visited HEPL in 1957 to discuss colliding beams with Wolfgang K. H. Panofsky, then the director of that laboratory, and to seek local collaborators. His goal was to develop the new colliding-beam technology as well as to demonstrate it by using the new technology for physics. The energy of the linac was such as to allow an experiment that would go far beyond anything that had ever been done before in testing the theory of quantum electrodynamics; and radiation damping made injection simple, allowing one to get on to confronting the more basic questions of beam stability, beam–beam interaction, and the like that O'Neill felt would be the limitations on large proton colliding-beam systems, which were really dearest to his heart.

O'Neill and Panofsky quickly recruited Carl Barber and myself to join the project. Barber was a senior scientist at HEPL who had built a 40 MeV linear accelerator that was used for nuclear structure studies. He knew the laboratory and, probably more important, he was very good at cost estimating and keeping the project moving. I was a postdoc who had come to HEPL in 1956 because I wanted to use the linac to test quantum electrodynamics. I was actually doing such an experiment at the time, studying large-angle electron–positron pair production that could test QED to 70 MeV/c. While this experiment would be the most sensitive test then done, the opportunity to do it at 1 GeV/c with the new colliding-beam system was too much for me to resist. O'Neill added another physicist, Bernard Gittelman, who was just finishing his Ph.D. at MIT, and we four set out on what we thought would be a great adventure of only a few years' duration. The adventure turned into something more like the voyages of Odysseus, for we were confronting the unknown and uncovered many problems that had to be solved.

The colliding-beam experiment (CBX) would have two 12 m circumference electron storage rings, with one common straight section. It would require the world's largest ultrahigh vacuum system (two cubic meters at 10^{-9} Torr). It needed injection kicker magnets faster than anything that existed at the time (80 ns pulse width, including a reasonable flat top). To do physics, it would require the storage of beam currents in the hundreds of milliampere range. A photo of the partially completed machine is shown in Fig. 15.3.

In the design we thought through many issues. We had a model of the beam–beam interaction, which turned out to be wrong, but which gave us the right limit. We thought the limitation from beam–beam collision effects would come from a shift in the effective focusing strength of the magnet system (betatron tune) to the nearest integer or half-integer resonance. In designing the QED test we arbitrarily derated that number by a factor of 10 and used a tune-shift limit of 0.025, which turns out to be very close to the limit that modern electron colliders achieve. We worried about ion trapping in the circulating electron beams and designed in the electrostatic clearing fields to remove the ions (they were needed). We worried about what are now called chromatic effects (the change in betatron frequency with energy within the stored beam) and designed in correctors to reduce the chromaticity to zero. Interestingly, this turned out to be essential, although we did not know it until subsequent machines without these corrections showed a beam instability.

Fig. 15.3. The partially assembled Princeton–Stanford storage ring in 1962. The lower halves of the magnets are in place and the vacuum chamber is installed. Radio-frequency cavities and beam transport to the rings are yet to be installed.

The Office of Naval Research (ONR), a very imaginative organization that was then the principal supporter of fundamental research in physics, funded the project to the tune of $800,000, thanks to the persuasive powers of Panofsky. At the time this was the largest sum of money ever devoted by ONR to a single experiment. The first beam was stored on March 28, 1962; the first physics results testing QED were presented in 1963; and the facility was finally shut down in 1968. During that time we had to confront many new problems. For example, we found that synchrotron radiation desorbed enormous amounts of gas from the walls of our supposedly clean vacuum chamber, and we had to redesign the system to eliminate all oil diffusion pumps. We found what is called now the long-range wake instability and learned to cure that with octupole magnets. We found a coherent coupled-beam transverse instability, which we fixed by separating the tunes of the two rings. We found that the beam–beam interaction did, in fact, lead to significant beam degradation at tune shifts of 0.025 for head-on collisions and at the same limit applied for crossing-angle operations.

In those early years CBX was a mecca for all who were interested in colliding-beam machines. We had a constant stream of visitors from laboratories in the United States and Europe, three of whom merit special mention because of their important contributions to understanding and solving the new problems that we faced. They were Ernest Courant, David Ritson, and Andrew Sessler. The start of the CBX project encouraged others to think seriously about storage rings. With the storage of the first circulating beams in CBX in 1962, it was clear to all that colliding-beam machines could be built, and plans began to move ahead rapidly for machines at many places.

I close this section on the first of the storage rings with a tribute to O'Neill who, in my opinion, is the real father of colliding beams.

The change to electron–positron colliders

Colliding-beam systems offered the potential for vast increases in the attainable center-of-mass energy that would allow the particle physicists to probe much more deeply into the ultimate structure of matter. While the Princeton–Stanford machine had the double goal of proving out the technology and doing particle physics experiments, the physics potential of the machine was limited. Quantum electrodynamics could indeed be tested to much smaller distances than ever before, but only one or two other specialized experiments (search for $e^- + e^- \rightarrow \mu^- + \mu^-$, for example) could be done.

In the late 1950s and early 1960s, it began to be realized that the electron–positron system offered a much richer vein from which to mine information about the elementary particles. In electron–positron annihilation, a virtual photon is formed that can produce any system that has either charge or magnetic moment. All such final states are accessible. Electron–positron annihilation had the potential not only for studying quantum electrodynamics via the elastic scattering process or the two-photon annihilating process, but also to study hadronic final states as well. In those days one talked about such things as form factors or structure functions of the mesons, for example, and those form factors, as we thought of them then, might have a resonant pole that could be reached by a virtual photon of the appropriate mass giving a large increase in particle production. Studies of different kinds of hadronic final states could reveal the relative structure functions of different kinds of particles.

In the CBX group we had discussed conversion to a machine aimed at the electron–positron system. I came to the realization of the benefits of this system in 1958, and discussed it with the group.[6] We decided on discretion. Electron–positron colliding beams would be more difficult than electron–electron rings, for we would need such things as two beams circulating in one ring, faster kicker magnets, and a positron source. We felt that we had enough problems in developing this technology in the electron–electron system – where we at least had a very high-powered electron beam for injection, and had the flexibility of having the two beams in separate rings.

The first step in the electron–positron direction was taken in Italy, and the key personality was Bruno Touschek.[7] There is a seminal moment in this story that occurred at a seminar by Touschek at Frascati on March 7, 1960, in which Touschek outlined the scientific potential of electron–positron annihilation studies. Giorgio Salvini, then director of the Frascati laboratory, and the high-energy physics community in Italy were immediately convinced by Touschek's arguments and began to work to bring e^+e^- colliders to life.

The first machine, called Anello di Accumulatione or AdA (Fig. 15.4), was brought into operation less than a year after Touschek's seminar. It was a very simple design with a toroidal vacuum chamber and magnet, and could be built rapidly. Injection was made by converting an incoming gamma-ray beam on a target that protruded slightly into the vacuum chamber. The synchrotron radiation process would allow a small fraction of the electrons and positrons pair-produced on the converter to be trapped in the vacuum chamber. Because of this, the machine had a very low injection efficiency, a very low circulating beam current, and a very low luminosity.[8]

In my opinion, AdA was a scientific curiosity that contributed little of any significance to the development of colliding beams (there is one exception; a beam-loss mechanism now called the Touschek effect was discovered). However, the project did keep interest at a high pitch in Italy while a much more important facility called ADONE (for "big AdA") was being designed. While ADONE was to be the first of the high-energy electron–positron colliders capable of getting into the region where many different kinds of hadrons could be produced, the first particle-physics results actually came from two smaller machines that were completed earlier. I will digress briefly before getting back to the important story of ADONE.

Fig. 15.4. AdA, the first electron–positron storage ring.

The two smaller machines were ACO, a 450 × 450 MeV strong-focusing ring built at the Orsay laboratory in France, and the VEPP II, a 500 × 500 MeV weak-focusing machine built at Novosibirsk in the USSR. Both machines were completed in 1966, and the first results of their high-energy physics experiments were submitted for publication around the end of 1967. Both experiments looked at π-pair production and studied the ρ resonance with a precision never before attained. It is hard to know exactly when the Novosibirsk group started on electron–positron work. At the accelerator conference in 1961 there was no mention of any such work in Novosibirsk, whereas in 1963 the VEPP II project was well under way.[9] VEPP II was seriously damaged by a fire in 1968 and the reconstruction of the machine took about two years. By that time, the French group had explored the region around the ρ resonance extensively and Novosibirsk was never again a serious player in particle physics using these colliding beams, but they certainly have contributed enormously to developing the technology.

ADONE

The ADONE project was the real goal of the Italian program that had been stimulated by Touschek's seminar. Serious design work began on this project in 1961 under the direction of Dr. Fernando Amman. The energy was set at 1.5 GeV per beam, high enough for multiple particle production, including meson and baryon resonances. The machine was to be strong focusing with a radius of approximately 16.5 m. Construction was started in 1965 and the project was completed in 1967.

Soon after commissioning of the machine was begun, a new beam instability problem was discovered – the so-called head–tail instability. The instability limited both the positron and electron circulating beams to very low intensity. In 1968 Claudio Pellegrini of Frascati and Matthew Sands of SLAC analyzed the problem and solved it. The instability was driven by what the accelerator physicists called the "chromaticity," that is, the variation of betatron oscillation frequency with momentum. Their analysis also indicated the cure, and the ADONE machine was soon equipped with sextupole magnets with which the chromaticity could be adjusted to the proper sign to cure the problem. It is interesting to note that the CBX collaboration had avoided this instability by building a correction into ends of the bending magnets in that machine. We had no real reason to do it; it just seemed like the right thing to do at the time.

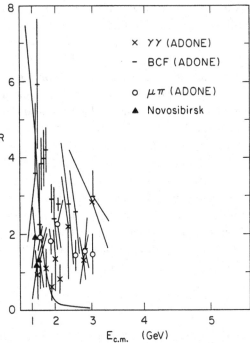

Fig. 15.5. The ratio R of the inclusive cross section for hadron production to the cross section for μ-meson pair production versus center-of-mass energy. The results from the three ADONE experiments differed widely, but all of them were very large compared to the theoretical expectations of that time, shown by the solid line.

Experimental physics began on ADONE in 1968. The early results had a great impact on me and on others in the high-energy physics community, for the cross sections for multiple hadron production were much larger than expected. The early results are shown in Fig. 15.5, where the solid line shows what was expected by most at that time. The cross section should have decreased very rapidly above the ρ resonance and dropped to quite small values by the time one reached the maximum energy that ADONE was capable of, 3 GeV. It clearly did not, but unfortunately the experiments from the four groups working on the machine were inconsistent, and that inconsistency led to a certain skepticism about the validity of the results. I was not skeptical, for the results at high energy disagreed much more with theory than they disagreed with each other.

ADONE's impact on high-energy physics was dulled by the choice of experiments. To quote from Amaldi:[10]

> Between the tendency to assign all, or almost all, the available resources to a single group that thus could have disposed of high-performance equipment and the opposite tendency of dividing the same funds between various groups, each by necessity endowed with an apparatus of limited performance, it was certainly not easy to find the right compromise! The solution finally adopted involved an excessive fragmentation of the financial means, with consequences not completely favorable from the scientific stand-point, and a certain disappointment to Bruno Touschek and Fernando Amman.

After the first results were in, the four groups working on ADONE began discussions with the management of the laboratory on follow-on detectors. These discussions went on for a very long time because of the reasons alluded to by Amaldi. By the time a detector of sufficient capability to do justice to the physics was ready, the science had passed ADONE by.

SPEAR, CEA, and DORIS – the Next Generation

If building CBX was like the voyages of Odysseus, building SPEAR was more like the labor of Sisyphus. We rolled the boulder up the hill seven times (1964–1970) before pushing it over the top.

The project that came to be SPEAR was born in 1961. I mentioned earlier the discussions that the CBX group had had on conversion of the e^-e^- rings to an e^+e^- ring and the decision we made to keep on with our original course. However, I remained convinced of the importance of e^+e^- colliders for the study of hadron physics and in 1961, before the first beam was stored in CBX, I, together with David Ritson (recently come to Stanford from MIT as a member of the Physics Department faculty), began serious discussions on the design of a high-energy e^+e^- collider. Our first problem was to define "high energy," for that would not only define the physics program but set the scale of the project as well. We soon, with the help of the Stanford theoretical physicists, settled on 3 GeV per beam, far enough above threshold (we hoped) to get into the "high-energy" regime where structure could be compared free of threshold effects.

We continued our preliminary work on the machine design and in 1962 Panofsky, by then the director of Project M, the design phase of what would be SLAC, invited me to set up a group to prepare a proposal to be submitted to the Atomic Energy Commission. Panofsky is a man of remarkable vision. He immediately recognized the importance of O'Neill's proposal in 1957, and obtained the necessary funding to build it. He remained the *éminence grise* behind the project, smoothing the fiscal and technical paths when needed. Now, he was betting a great deal on a 31-year-old assistant professor who wanted to look at hadron physics in a new way. I wonder if anyone could take such a risk now? We have now more committees, more detailed reviews, and more conservatism in our field. Even then, it could only be done under the umbrella of a large laboratory, where a small proportion of resources could be devoted to a very high-risk high-payoff gamble.

In 1963 a preliminary proposal was sent to the AEC justifying the project because, "it is in the field of strong interactions that we believe the storage ring can make its main contributions to physics." The proposal already included a full solid-angle coverage (4π) magnetic detector. In 1964 the formal proposal was submitted.

However, physicists at the Cambridge Electron Accelerator (CEA) also submitted a proposal for an e^+e^- colliding-beam project. The AEC now had to deal with two proposals, and they set up a review committee chaired by Jackson Laslett to conduct a comparative review. The committee recommended proceeding with the SLAC proposal, but expressed concern about potential problems from the beam–beam instability that had been observed at CBX. The committee felt that more data from CBX was needed. In 1965 the Laslett committee reviewed new data from CBX and recommended that the AEC proceed with the SLAC project.

Then followed a saga of proposal submission and dashed hopes; redesigns to simplify and lower costs; and modifications to incorporate all of the new ideas generated by colliding-beam studies around the world. Dr. John Rees, an accelerator physicist who had worked on the CEA synchrotron, joined me in 1965 and the two of us kept the group together through the long wait.

In 1965 the remarkable increases in federal funding for the physical sciences, triggered in the Eisenhower years by the Soviet Sputnik spacecraft, came to an end. Our project was not included in the budget. In 1966 the proposal was submitted for the third time and, in spite of the strong recommendation of the advisory committee chaired by George Pake, it was not funded. Similarly in 1967, 1968, 1969, and 1970, in spite of increasingly strong endorsements by the High Energy Physics Advisory Panel, no construction funds were available.

Finally, in 1970, a breakthrough occurred because of an intervention by Mr. John P. Abbadessa, then controller of the AEC. Abbadessa was interested in science as well as in the financial management of the AEC. He became fascinated by the concept of an electron and positron annihilating and turning into other kinds of particles and did what only a great bureaucrat can do – he advised us on how to present the project so that no specific high-level approval was required. A construction project was turned into an equipment project and, with the enthusiastic support of the high-energy physics program people of the Atomic Energy Commission, SLAC proceeded to build the project out of its ongoing budget.

Construction started in October 1970, and the first beam was stored in April 1972 (see Fig. 15.6). Thanks to the early Frascati results, the project still had its 4π magnetic detector, which was so essential to the experimental program that led to the "November Revolution."[11] Those results are described by Gerson Goldhaber in Chapter 4.

During all this time, the CEA group had not dropped out of the colliding-beam business. In the mid-1960s Robinson and Voss of CEA invented the "low-β" interaction region – a vital contribution to the scientific productivity of colliding beams.[12] Low-β allowed much higher luminosity than the previous system within the constraints on beam stability imposed by the beam–beam interaction. Experimenters are always looking for higher yield in any given process, to allow them to study more subtle effects, and low-β allowed an increase of between a factor of 10 and 100 in the yield of a given process. All of the modern colliding-beam machines incorporate this idea.

The CEA group, while not funded for a major colliding-beam project, came up with an idea on how to modify their synchrotron to allow the storage of electrons and positrons and carry out some limited colliding-beam studies. They designed a "bypass" that switched the low-intensity circulating beams that could be accelerated into a section of the synchrotron onto a parallel track to the synchrotron itself that had a low-β interaction region and room for a detector.

John Rees, in a 1986 article on colliding beams, summed up the bypass project very well:[13]

> And even then the luminosity of CEA was not limited by the beam–beam limit; it was limited by the incredible complexity and difficulty of the CEA operating cycle. I think that the saga of CEA is the Book of Job of the accelerator builders. They were afflicted by every handicap that could have been visited upon them, yet they persevered, and in the

Fig. 15.6. SPEAR as it appeared at completion in 1972. The housing is movable shielding blocks, and the buildings are portable. It was the absence of permanent civil construction that allowed the project to be dubbed "an experiment."

end the Lord loved them and they got the right value of R. Of course, nobody believed it. The machine was too hard to operate.

The DESY laboratory, which became such an important player in the colliding-beam business in the 1980s, was not involved in the early developments. In the mid-1960s, the laboratory was discussing the appropriate next step beyond their existing 6 GeV electron synchrotron. There were two camps at DESY; one wanted to increase the energy of the synchrotron, while the other wanted to build an e^+e^- collider, and they were at an impasse.

A critical meeting in the history of DESY took place at SLAC in 1966. Willibald Jentschke, then director of the DESY laboratory, brought the senior staff members who were the strongest advocates of the synchrotron approach to a four-hour meeting with Sid Drell, Panofsky, and me on colliding-beam physics and technology. Jentschke was clearly using us as the sales force to convince his staff to buy into colliding beams.

Table 15.1. *Later electron–positron colliders.*

Project	Beam Energy (GeV)	Location
CESR	6	Cornell
PEP	17	SLAC
PETRA	22	DESY
TRISTAN	35	KEK
LEP	100	CERN

DESY soon decided to proceed with DORIS, a 3 GeV two-ring e^+e^- machine. The double-ring configuration that they chose gave rise to beam instabilities that are understood now, but that seriously limited the performance of the DORIS facility then. With the return of Voss from CEA to DESY, the colliding-beam program at that laboratory began to make great strides. The PETRA e^+e^- machine and the HERA ep machine have made and will make great contributions to physics, but those are stories for the next conference in this series.

The development of colliding-beam storage rings for electron–positron collisions reached a plateau with the completion of SPEAR that has lasted to the present day. The subsequent machines are all scale-ups of SPEAR. There was nothing new until the development of the linear collider, which is generally acknowledged to be the replacement for storage rings for very high-energy electron–positron collisions, and the return of the two-ring machine with the design of the various "factory" machines (*B*-Factory, Tau–Charm-Factory, Phi-Factory). These new developments are not coupled to the rise of the Standard Model and so their stories too can wait until the next conference.

Electron–proton colliders

I want to mention this topic briefly, for although the commissioning of the first electron–proton colliding beam facility only happened in 1992, the story started a long time ago. It began with a meeting in 1971 at SLAC involving Dieter Möhl (CERN), Claudio Pellegrini (Frascati), Andrew Sessler (LBL), and John Rees, Mel Schwartz, and me. Rees presented a paper at the 1971 accelerator conference that aroused great interest.[14] Four proposals soon appeared: from the Rutherford Laboratory (EPIC), from Frascati (Super ADONE), from SLAC–LBL (PEP), and from KEK (TRISTAN). The first two were never built, while the second two turned into e^+e^- colliders when funding limits and lack of experience with the required superconducting magnets forced the elimination of the proton rings.

Electron–proton colliders were proposed again at CERN in the mid-1970s as an upgrade to the SPS, but that project lost out in a competition with the proton–antiproton collider that I will discuss later. Now the HERA project at DESY is operating, and the experimental program has begun – making high-energy electron–proton colliders a reality.

Proton–proton and proton–antiproton colliders

As I mentioned earlier, the first studies on colliding beams were aimed at proton colliders. However, injection, stacking, effects of nonlinear resonances, and so on, were not well understood, and so the actual realization of colliding-beam machines began with the electron colliders. However, the proton machines were not forgotten, and serious studies continued in the early 1960s at both Brookhaven and CERN.

At Brookhaven there were two options: one was to build storage rings to go with the AGS synchrotron, and the other was a major program to upgrade the AGS and greatly increase its intensity as a fixed-target machine. I was not privy to any of the discussions, nor have I had access to any of the minutes of relevant meetings at Brookhaven. The laboratory decided to drop the colliding-beam project and proceed with the AGS upgrade project. It would be interesting to understand why.

CERN took the opposite course and decided to proceed with the construction of what would become the ISR. Serious study of the possibility began in 1960, the CERN council approved the project in 1965, construction began in 1966, and the first collisions were achieved in 1971. Kjell Johnsen tells the ISR story in Chapter 16, and so I will not go into any details here. It was a brilliantly conceived and executed project that should have contributed much more than it actually did to the rise of the Standard Model. The problem came with the choice of experiments that mainly emphasized small transverse momentum phenomena, which turned out not to be very relevant to the Standard Model.

One can say that the discovery of W and Z bosons at CERN was the final step in the confirmation of the Standard Model and so I will go into much more detail on this story. Antiproton–proton ($\bar{p}p$) colliders first came into focus in a talk by G. I. Budker (of the Institute of Nuclear

Physics at Novosibirsk) at the 1966 Saclay conference on storage rings.[15] Budker's talk (presented in the proceedings only in summary form) contains all the key elements of a workable antiproton collider system. He included an outline of the machine design and a brief description of a damping technique that would allow the accumulation of a large number of antiprotons in a small enough phase space to make sufficient luminosity for experimental work. Budker's talk also discussed the physics potential of such machines.

The damping mechanism described by Budker was the so-called electron-cooling technique, in which a beam of electrons with small transverse and longitudinal velocity spreads would co-stream with a proton beam of much larger velocity spread, exchanging momentum with the protons through the coulomb interaction and thus decreasing the velocity spread in the proton beam (cooling) and increasing the velocity spread in the electron beam (heating). A more detailed paper was presented in *Atomic Energy* and a complete description of the project they began to construct in Novosibirsk was given in a paper by Skrinsky in the proceedings of the 1971 International Accelerator Conference.[16] A demonstration of electron cooling was made, but the project was never completed as an antiproton–proton collider both because it went slowly for financial reasons and it proved difficult in practice to get fast enough cooling rates with this co-streaming electron technique. The project was eventually converted to an electron–positron colliding-beam ring (VEPP IV) that has been running for several years.

The next step was the invention in 1968 by Simon van der Meer of CERN of an alternative technique called stochastic cooling. In essence this technique senses density fluctuations in a beam and damps them out by an active feedback system. The first formal report on stochastic cooling was issued in 1972,[17] though the discovery was known throughout the accelerator community soon after it was made, and there probably exist internal reports of the ISR group that describe it. Stochastic cooling had a great potential advantage over electron cooling in that the cooling rate was independent of energy, whereas in the case of electron cooling, the rate decreased as the fifth power of the energy. The optimum energy for antiproton production is much higher than the best energy for electron cooling, and thus to use the electron technique, complex beam manipulations were required to decelerate the antiproton beams to an appropriate energy – typically a few hundred MeV. Stochastic cooling could be applied at the energy where the antiprotons were optimally produced. Experiments were carried out at the ISR in the first half of the 1970s that showed that stochastic cooling worked as van der Meer predicted. Indeed there were informal discussions at CERN about possible antiproton–proton collisions in the ISR, but there was insufficient interest on the part of the experimental community because there was no energy advantage in the antiproton–proton system and the luminosity of the proton–proton collider was much higher.

The next step came from the decision by Robert R. Wilson, then director of Fermilab, to build the energy doubler. Initial discussions centered on the possibility of making a proton–proton collider by making a beam circulating in the FNAL conventional ring collide with a beam circulating in the superconducting ring. The first suggestion of this possibility is, I believe, in a letter from David Cline and myself to Wilson in 1974 or early 1975.

By the time of the Fermilab program committee meeting in June 1976, three very different proposals were in hand. One proposed a 25 GeV high-current proton ring whose beam would collide with the beam in the existing main ring;[18] a second proposed proton–proton collision between a 1 TeV beam in the new doubler ring and a 150 GeV beam in the old main ring;[19] and a third proposed antiproton–proton collisions at energies up to 1 TeV per beam in the new main ring.[20] This last proposal evolved into the CERN $S\bar{P}PS$ collider and the Tevatron collider.

The $\bar{p}p$ concept was detailed in a paper by Cline, Peter McIntyre, and Carlo Rubbia in the proceedings of the 1976 Neutrino Conference at Aachen, describing the possibility of making a very high-energy antiproton–proton colliding-beam facility using one ring of an existing machine.[21] This paper described the full system including the requirements for the antiproton source, the specification for the cooling technique (either electron or stochastic cooling), antiproton yield estimates, accumulation time, and so on. It also described the physics motivation, emphasizing the search for the W and Z.

I have asked at Fermilab about the origins of the $\bar{p}p$ concept, and I have been told by Wilson that the first "bare bones" suggestion came from McIntyre, and that Cline, McIntyre and Rubbia took it from there. The proposal included David Reeder and Lawrence Sulak.

Fermilab was not enthusiastic about proceeding rapidly with any of the proposals. Proposal No. 478 required a new 25 GeV ring and would divert resources from the Tevatron program. Proposal Nos. 491 and 492 required the completion of the Tevatron, and Wilson felt that not enough was known about superconducting magnets to make a firm schedule at that time.

Rubbia was not content with what he regarded as an excessively conservative and slow approach at Fermilab. He returned to CERN and worked with van der Meer and others in the accelerator-physics groups at CERN to produce a detailed design. Leon van Hove, then co-director general, had the vision to recognize the importance of a high-energy antiproton–proton collider to physics and to CERN, and overcame the inertia of the CERN system, gaining formal approval of a two-stage process. The two stages were to include a large-scale test of the cooling schemes and then, if that were successful, the building of a full-scale project.

The role of van Hove is not well known. John Adams, who had brilliantly led the SPS project, was made Director General of the entire lab and van Hove was made co-Director General because of concern that Adams's background was not such as to be able to lead the CERN scientific program. In 1977 CERN had been operating the costly SPS proton synchrotron for only a few years, and discussions were in full swing on the possibility of the LEP project, a 27 km circumference electron–positron collider that dwarfed the SPS. Adams was concerned about the possible reaction of the CERN member states to an expensive SPS, an even more costly LEP, and still another new, although relatively small, $\bar{p}p$ project tucked in between. Van Hove felt very strongly that the scientific potential of the $\bar{p}p$ was such that CERN must move ahead with it if it were feasible.

They had argued about it several times without coming to an agreement. I was present at a meeting with both of them that started as a discussion of LEP and drifted on to the $\bar{p}p$ collider topic. I was the sole audience and the discussion grew quite heated. It reached the point where van Hove reminded Adams that he, van Hove, was the scientific Director General, that in his opinion the case for the $\bar{p}p$ collider was overwhelming and that if Adams did not back the project in the Council van Hove would resign! They then abruptly realized that I was still there, and the meeting ended with embarrassed mumbles.

I have never mentioned this incident except once or twice to van Hove. When the Rubbia–van der Meer Nobel Prize was announced, I not only wrote to congratulate the Laureates, but wrote to van Hove as well telling him that at least one person in the physics community knew that without him there would have been no $\bar{p}p$ collider. In a conference devoted to the rise of the Standard Model, it seems to be appropriate to break my fifteen-year silence.

The rest of the story of the CERN $\bar{p}p$ collider is well known. The cooling experiment worked as predicted by van der Meer. Roy Billinge and van der Meer led the construction of the antiproton source at CERN. Fermilab, now under the leadership of Leon Lederman, decided to stay out of a race with CERN for the W and Z and stick to the Tevatron program and the superconducting magnet technology development that is so important to the proton machines of today. The CERN $\bar{p}p$ collider worked well, culminating with the discovery of the W and Z by UA1 and UA2 experiments and an essential confirmation of the Standard Model. Van der Meer's invention made it possible and Rubbia's drive and determination brought it about.

Conclusion

From the start of the first collider, CBX, to the time of this symposium is thirty-four years, during which colliders have taken over the world of high-energy physics. This paper traces the main threads in the evolution of the technology. It is not a complete history of colliding beams and leaves out important contributions from Orsay, BNL, Cornell, Fermilab, KEK, and Novosibirsk, that advance the art but are not clearly related to the topic of this meeting.

Looking back from now to then, the electron colliders came first because the technology was easier, and relatively small facilities could and did make great contributions to physics. The evolution of the electron machines was very rapid, reaching a plateau with SPEAR wherein essentially all the elements of all the storage rings that have since been built were in place. LEP marks the culmination of the storage-ring technology, for electron machines have a scaling law of costs with energy that is quadratic, and it is too costly to go much further with the storage ring technique. Fortunately, the linear collider, first realized with the SLC at SLAC, has come along to replace the storage ring and an active international R&D program is in progress aimed at the next step in very high-energy electron colliders.

The fact that the early electron machines were low cost and extremely productive created a climate where, for larger and more costly machines, technological and scientific "success" was the expected norm and it was relatively easy, post-SPEAR, to obtain funding for larger projects.

The proton colliders, on the other hand, came more slowly because the technology was more difficult and because, if a collider were to make major advances in physics, the machine had to be large and costly from the beginning. The ISR was a brilliant success as an accelerator project, but the choice of initial experiments virtually precluded the discovery of the new particles and the large transverse momentum phenomena that are the stuff of the Standard Model. Thus there was no "demand pull," as the economists would say, from the physicists until the Standard Model itself began to unfold. The CERN $\bar{p}p$ collider was the first result of this demand pull, and that same demand is driving the programs to realize the SSC and the LHC.

I think we all hope that the next conference in this series will be entitled "Beyond the Standard Model," and if so, it is sure that the high-energy high-luminosity proton machines now being built; the low-energy, high-luminosity electron factories; and the high-energy linear electron colliders will have made essential contributions to whatever unfolds.

Notes

1 Deutsches Patentamt, Patentschrift Nr 876279 Klass 21g Gruppe 36 Ausgegeben am 11 Mai 1953.

2 This story is from a letter from Wideröe to Eduardo Amaldi. It is quoted in Amaldi's biography of Bruno Touschek; see note 6.

3 D. W. Kerst, et al., "Attainment of very high energy by means of intersecting beams of particles," *Phys. Rev. 102* (1956), pp. 590–1.

4 D. W. Kerst, "Properties of an intersecting-beam accelerating system," in E. Regenstreif, ed., *CERN Symposium on High Energy Accelerators and Pion Physics* (Geneva: CERN, 1956), pp. 36–9.

5 G. K. O'Neill, "The storage-ring synchrotron," in E. Regenstreif, ed., *CERN Sympposium*, pp. 64–65, on p. 64.

6 A group of postdocs including me had asked J. D. Bjorken to lead a seminar on how to calculate with Quantum Electrodynamics. He gave us an exercise to calculate the annihilation cross section of an electron and a positron into a pair of spin-zero charged particles. I solved the problem, and suddenly realized that π mesons and K mesons were such particles, that any structure that they had would modify the cross section, and that the colliding-beam technique, if it worked, could be modified to do the experiment.

7 An excellent scientific biography of Touschek is Eduardo Amaldi, *The Bruno Touschek Legacy*, CERN Report No. 81-19 (Geneva: CERN, 1981).

8 The effective intensity of a colliding-beam machine is measured by its luminosity, which is the reaction rate that would be observed for a process with a cross section of one. It is proportional to the product of currents in the colliding beams, divided by the effective overlap area.

9 See, for example, G. I. Budker et al., "Studies of Electron–Electron, Positron–Electron and Proton–Proton Beams at the Institute for Nuclear Physics, the Siberian Branch of the U.S.S.R. Academy of Sciences," *Proceedings of the International Conference on High-Energy Accelerators* (Dubna, USSR, 1963), pp. 334–64.

10 Eduardo Amaldi, *The Bruno Touschek Legacy*, p. 39.

11 Rees and I were under great pressure to reduce the cost of the project. One possibility was to eliminate the magnetic detector in favor of a much less costly detector with no magnetic field. We were traveling home from a meeting at Frascati in 1968 or 1969, where I had had my first opportunity to see the ADONE data and talk in detail with the experimenters. We spent much of the time talking about the results, and I came to the conclusion that the 4π magnetic detector was essential to understanding the physics. No matter what the implications on the SPEAR schedule, we had to preserve it.

12 K. W. Robinson and G. A. Voss, CEA Technical Note, 1965 or 1966. I have been unable to find a copy of the note, and so cannot give the exact citation. It is, however, mentioned, but not referenced, in a CEA report on the bypass project that was issued in later 1966.

13 J. R. Rees, "Colliding beam storage rings – a brief history," in *SLAC Beam Line* (March 1986), p. 8.

14 C. Pellegrini, et al., "A high energy proton-electron positron colliding beam system," in M. H. Blewett, ed., *Proceedings of the 8th International Conference on High-Energy Accelerators* (Geneva: CERN, 1971), pp. 151-8 (presented by J. Rees).

15 G. I. Budker, "Status report of works on storage rings at Novosibirsk," *Proceedings of the International Symposium on Electron and Positron Storage Rings* (Saclay: Presses Universitaires de France, 1966), p. II-1-1.

16 Gersh I. Budker, "An effective method of damping particle oscillations in proton and antiproton storage rings," *Atomic Energy 22* (1967), pp. 346-8; English translation in *Soviet Atomic Energy 22* (1967), pp. 438-40; VAPP–NAP Group, "Proton-antiproton colliding beams," in M. H. Blewett, ed., *Proceedings of the 8th International Conference on High-Energy Accelerators*, pp. 72-8 (presented by A. N. Skrinsky).

17 Simon van der Meer, "Stochastic damping of betatron oscillations in the ISR," CERN Report No. CERN ISR/PO/72-31 (Geneva, August 1972).

18 R. Huson, et al., "Proposal to search for intermediate boson production at 200 GeV in the center-of-mass," Fermilab Proposal No. 478 (1976).

19 C. M. Akeubraut, et al., "Clashing Gigantic Synchrotrons," Fermilab Proposal No. 491 (1976).

20 D. Cline, et al., "Proposal to Construct an Antiproton Source for the Fermilab Accelerators," Fermilab Proposal No. 492 (1976).

21 C. Rubbia, P. McIntyre, and D. Cline, "Producing massive neutral intermediate vector bosons with existing accelerators," in H. Faissner, H. Reithler, and P. Zerwas, eds., *Proceedings of the International Neutrino Conference Aachen 1976* (Braunschweig, West Germany: Vieweg & Sohn Verlag, 1977), pp. 683-7.

16

The CERN Intersecting Storage Rings: The Leap into the Hadron Collider Era

KJELL JOHNSEN

Born Meland, Norway, 1921; Ph.D., 1954 (physics), Norwegian Institute of Technology, Trondheim; retired from CERN; accelerator physics.

The history of colliding-beam devices can be traced back to 1956, when a group at the Midwestern Universities Research Association put forward the idea of particle stacking in circular accelerators. Of course, people who worked with particle accelerators had already speculated about the high center-of-mass energies attainable with colliding beams, but such ideas were unrealistic with the particle densities then available in normal accelerator beams. The invention of particle stacking fundamentally changed this situation. It opened up the possibility of making two intense proton beams collide with a sufficiently high interaction rate to enable experimentation in an energy range otherwise unattainable by known techniques.

A group at CERN started investigating this possibility in 1957, first by studying a special two-way fixed-field alternating gradient accelerator and then in 1960 by turning to the idea of two intersecting storage rings that could be fed from the CERN 28 GeV Proton Synchrotron (PS). This change in concept for these initial studies was stimulated by the promising performance of the PS at the very start of its operation in 1959.

In 1961 the Accelerator Research Division at CERN had gained sufficient confidence to present its first proposal for a 2 × 25 GeV storage ring system. This system was intended essentially for protons, but other particles were mentioned in the proposal. This led to a series of important actions. First, in 1962, France offered a site next to the original CERN site. The European Committee for Future Accelerators was then formed and in 1963 issued a strong recommendation in favor of a pair of 25 GeV proton storage rings, which it named the Intersecting Storage Rings (ISR). The recommendations also included a 300 GeV proton accelerator for fixed-target experiments.

In spite of this strong recommendation in favor of the ISR, there was considerable disagreement in the high-energy community on this project. With some notable exceptions, the theorists and the experimental physicists were against the idea, whereas the accelerator physicists and builders largely supported it. The reception by the members of the CERN Council was in general rather favorable.

In 1964 the Accelerator Research Division prepared a detailed design report for the ISR, which formed the basis for a formal proposal. In his last year as Director General, Victor Weisskopf, who had been one of the ISR's most enthusiastic proponents (strongly supported by Mervyn Hine), saw the CERN Council adopt the ISR as a project by making the decision in principle, in June 1965, to construct this facility. That December the Council accepted the financial plan of the project and voted construction funds to begin on 1 January 1966. The plans quoted a construction cost of 332 million Swiss francs (1965 value) and projected first operation of the facility by mid-1971. Both promises were fulfilled with some margin.

Working principle

Figure 16.1 shows the layout of the ISR and the PS that acted as its injector. The protons were accelerated in the synchrotron to nearly their final energy. A fast ejection system then took all the particles out of the PS in one revolution and guided them into a beam transport system from which a fast injection system put the particles into one or the other of the two storage rings. They were then captured by the ISR radio-frequency system and moved to the stacking region near the outside of the vacuum chamber. Each time the synchrotron accelerated a new pulse to full energy the process was repeated, and the pulses were stacked side by side in the longitudinal phase plane of the storage rings.

Each pulse occupied a very small fraction of the longitudinal phase plane available in the vacuum chamber of the storage rings. It was therefore possible to build up very high beam intensities (in ordinary space) near the middle of the vacuum chambers.

When both rings had been filled in this way, the beam conditions were optimized by, for instance, scraping off the halo and other manipulations, before experimental data collection could start in the experiments around the crossing regions.

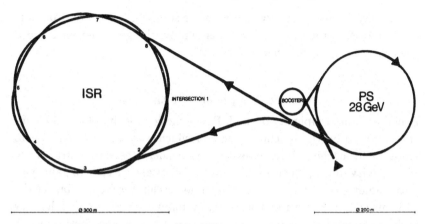

Fig. 16.1. Layout of the ISR and its injector, the PS.

Some features of the ISR construction

Seen in retrospect, the ISR construction proceeded fairly uneventfully, although those involved experienced considerable anxiety at times. No detailed account of the construction will be given, but some features are worth recalling.

It was recognized that the ISR might encounter many unknown phenomena that could limit its future performance. For instance, unlike the case of electron storage rings, there would be no damping to keep beam sizes from growing because of small imperfections in the guiding field or perturbations arising in the beam. In short, it was considered a daring and bold project. Therefore, tight tolerances and flexibility for all components became important guidelines. This paid off handsomely, as tolerances and flexibility had later to be stretched to their limit, and often beyond. Although much care was taken during the construction of the machine, many of the components had subsequently to be developed and upgraded to the extent that the whole facility was a considerably more sophisticated device at the end of its life than when it was started 13 years earlier. Nevertheless, the approach to the design and construction mentioned above gave the ISR builders the reputation of having a conservative attitude.

The management structure during the construction phase was a typical line structure. There was a project head with his deputy with overall responsibility. Their team was divided into groups in charge of the main components of the project, and extended responsibility and authority were delegated to the group leaders. The common problems were then extensively discussed and evaluated in the so-called Parameter Committee, comprising the project head, his deputy, the group leaders, and a few other senior staff. This worked well.

Accelerator physics at the ISR

Because of the complexity of the ISR and the complexity of the phenomena observed on the machine (some of which were unexpected), beam studies and machine development were considered essential parts of the ISR activity during its entire lifetime. It is not always easy to gain sympathy among users for such use of a large accelerator facility, but at the ISR this was never a serious problem, probably because some of the very early beam studies resulted in considerable performance improvements. The fact that over its lifetime about 15% of the "on-time" of the ISR was used for beam and component studies paid off in a rather spectacular improvement in performance over the years, as will be illustrated in what follows. Of course, these results were obtained not only because of the time devoted to such studies, but also because of the competence with which the studies were performed by the staff involved. A few important examples of this kind of activity will be given in the following.

The resistive-wall instability

During the very early operation of the ISR, we were able to stack beam currents only up to 2–3 amperes without difficulties. When we tried to stack to higher beam currents, instabilities arose that resulted in beam losses. Often a partial loss was enough to stabilize the beam, and stacking could continue. The left side of Fig. 16.2 shows a typical example of beam behavior under such conditions. The instability was associated with large coherent transverse oscillations, consistent with the theory of the resistive-wall instability. However, a sextupole field component had been built into the magnet profile to avoid this "brick-wall" instability up to much higher beam currents than 3 A. At first, therefore, we were taken a little by surprise. In fact, what had happened was that space-charge tune shifts had led to the violation of the stability criterion of positive chromaticity *in parts* of the beam although the criterion was satisfied for the global beam. This also explained nicely why only small parts of the beam were often affected. The main cure was a very careful tailoring of the dependence of both tunes on momentum, the so-called working lines. Later such working lines required dynamic compensation

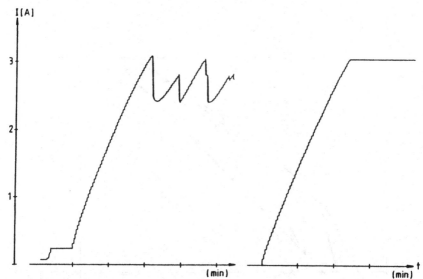

Fig. 16.2. Beam current I versus time plots, showing stacking operation hitting the "brick wall," at left, and an example of good stacking, at right (from CERN Report 83–13).

of the space-charge effect as the beam current increased (Fig. 16.3), and as experience was gained, such compensation was carried out on-line by the control computer. Although this was simple in theory, it required extreme accuracy and flexibility of the components involved.

This remedy was not enough to reach very high beam currents, and so a transverse feedback system was developed and incorporated into the machine for stabilization of the lowest oscillation modes. This kind of development went on over many years and led to a gradual increase of the stability limit to around 60 A.

Pressure bumps

After we started mastering the resistive-wall instability and beam currents of 4–5 A were obtained, another type of beam loss appeared. Again it resulted in a partial loss, as seen in Fig. 16.4. The phenomenon occurred on a much slower time scale than the "brick wall," and there were no associated beam oscillations. However, it was soon observed that dramatic local vacuum deteriorations preceded the beam loss. It was further observed that the pressure effect was not beam-loss dependent, but that it depended sharply on total beam current. Once these

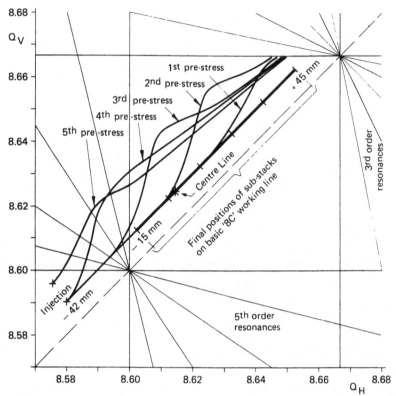

Fig. 16.3. The family of prestressed working lines used at 22 GeV to stack 15 A in five steps of 3 A across the chamber from +45 to −15 mm (α_p average) (from CERN Annual Report, 1973).

facts had been observed, the explanation became simple: The protons in the beam ionize part of the residual gas. These ions are driven into the walls of the vacuum chamber by the beam potential of about 1 kV and penetrate considerably deeper into the wall material than a normal cleaning method does. This ion bombardment leads to desorption from the wall, resulting in more gas to be ionized by the protons. A runaway situation occurs above a certain critical beam current.

In the ISR, the cures were to attack the two parameters, pumping capacity and desorption, by the addition of hundreds of titanium sublimation pumps, by an increase in the bake-out temperature of the vacuum components from 200 to 300°C, and by glow-discharge cleaning of all the critical vacuum components. This was a gradual program. During each shutdown, the weakest point was attacked. The result was a

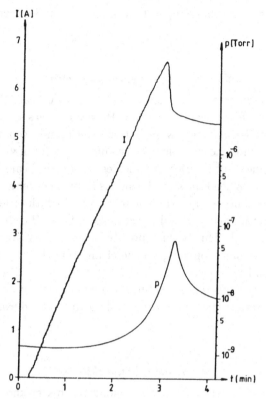

Fig. 16.4. Example of an increase in pressure caused by the stacked beam (from CERN Report 84–13).

steady advance over many years, leading to critical currents above 60 A and average pressures below 10^{-11} Torr.

Development of vacuum chambers for experimental areas

Experimental areas had their special requirements for the vacuum system. For background reasons, it was necessary to produce a pressure well below the average in the rest of the machine, and in the last years of operation, 10^{-13} Torr was achieved. These chambers had to have special shapes, and at the same time the wall thickness had to be kept to a minimum in order not to interrupt the secondary particles. This sometimes conflicted with sound mechanical engineering, and a few collapses of vacuum chamber walls occurred when safety factors were reduced too far. Development over the years nevertheless resulted in very satisfactory designs, giving reliable operation in spite of working very close to the limits.

Backround for physics experiments

Apart from perhaps the first few weeks, the ISR beam lifetime was always adequate for efficient operation. However, it was soon discovered that background conditions were bothersome or unacceptable at loss rates far below those acceptable from the lifetime point of view. This led to the development of a series of remedies to prevent beam losses. The main attack was to keep the stored beam as resonance-free as possible. It was, for instance, necessary to completely avoid the fifth-order resonance (Fig. 16.3), and it was preferable to stay away from all orders up to the eighth order as well. This became possible after the transverse feedback system was developed, opening up the clean area with a betatron tune between 8.9 and 9.

In addition, very good beam collimation was necessary, and it became customary to scrape the halo off the beam at intervals during a physics run.

Longitudinal instabilities

Since longitudinal instabilities were among the better-known phenomena, exceptional care was taken during construction to avoid such difficulties. This effort paid off handsomely, and no real difficulties arose at low frequencies. However, in the microwave range, instabilities were observed when the injected beam was being debunched on the stacking orbit. This led to a reduction in stacking efficiency. A special cavity working on the third harmonic of the main radio frequency was constructed to enhance the nonlinearities of the stacking "bucket." This provided Landau damping, which suppressed the instability.

Phase-displacement acceleration

The magnet system of the ISR had a sufficient safety factor that it could accept particle energies up to 31.4 GeV. However, a high-current stack has an energy spread of up to 3%, which is much more than the rather low-power system of the ISR could rebunch and accelerate from the injection and stacking energy of 26 GeV. During the early operation of the ISR, small currents were rebunched and accelerated to 31 GeV in the normal way, but the results were unsatisfactory for physics owing to the low luminosity that this gave.

A method called phase-displacement acceleration had formed a part of the MURA group's original paper on stacking. The method consists of repeatedly moving "empty buckets" through the beam from above to below. This shift downwards in energy of empty phase space leads to a corresponding shift upwards in energy of the phase space occupied by the beam. No large voltages (buckets) are needed. Small buckets merely require more passages. In 1973 it was decided to try this method at the ISR. It required low-noise operation of the rf system and fine variation of bucket size and magnet power supplies. The working line also required very fine adjustments during the acceleration to avoid instabilities. This was developed over the years into a very sophisticated procedure whereby all the operations are controlled on-line by the computer, with only occasional operator intervention. The result was that 2×31.4 GeV beams were made available for physics with essentially no loss of luminosity or other beam qualities.

Low-β insertions

In the original design of the ISR, some care was taken to make the vertical β in the intersection somewhat smaller than in the rest of the machine. However, a low-β insertion (in the proper sense of the word), was designed and installed only in 1974, consisting largely of borrowed quadrupoles from the PS, DESY, and the Rutherford Laboratory. It increased the luminosity by a factor of 2.3.

At about the same time, however, work began on superconducting quadrupoles for a more powerful low-β insertion. Such an insertion was installed in 1981 and gave a 6.5-fold improvement in luminosity, which brought the operational luminosity to above 10^{32} cm^{-2} s^{-1}, a record that lasted long.

Moreover, this improved insertion yielded other important benefits as it gave both CERN and industry some experience in constructing superconducting magnets of accelerator quality (precision, reliability, etc.). It also provided operational experience under the very stringent conditions imposed by circulating stored beams.

Diagnostics

Many diagnostic methods were used in the ISR, some of which were developed far beyond their planned capability, largely because of the operational demands that arose after the start-up of the machine. Some of these developments were among the most exciting experiences at the ISR, and it is with regret that I can describe only a few typical examples.

Over the years luminosity measurements became more significant than envisaged during the planning and construction period, as these measurements entered so directly into the accuracy of some of the most important experimental results from the ISR. The method of measurement consisted of precision steering for beam separation at the interaction points, giving very accurate values for the effective height of the beams. Together with highly accurate measurements of the circulating currents, this method gave correspondingly accurate values for the luminosity. Considerable development and careful checking were needed to reduce both systematic and random errors to the desired values.

It was foreseen neither in the planning of the ISR nor during its construction to use Schottky noise as an element in beam diagnostics. It came, in fact, as a surprise to observe this noise on the ISR beam in 1972. However, as soon as it was observed, its potential became apparent, both as a tool for diagnostics and for the practical development of stochastic cooling (see below). The fact that a circulating beam consists of a finite number of protons gives rise to statistical fluctuations in the beam current and in the beam's center of gravity. These fluctuations are very small and had not been previously observed in accelerator beams. However, the electronics available in the early seventies had improved, and integration methods could be used with beams that had lifetimes as long as those at the ISR. It thus became possible to observe these fluctuations and to make use of them, both to monitor the distribution of particles in longitudinal momentum and to measure the extremes of the tune values in the stack without any interference with the coasting beam. They also became instrumental in detecting the growth of betatron amplitude at particular orbits in a stack, which helps to detect the presence and strength of various nonlinear resonances. This became an operation tool from about 1974, and has possibly become the most powerful of all beam observation methods, not only for the ISR but also for the other collider projects in operation, under construction, or in the planning stage.

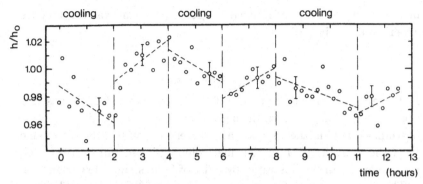

Fig. 16.5. Observation of stochastic cooling in the ISR through measurements of the effective beam height (h/h_o) as a function of time, decreasing when cooling is applied and increasing when not applied (from CERN Annual Report, 1974).

Stochastic cooling

Stochastic cooling was invented in 1968 by Simon van der Meer, but at that time it was considered unrealistic for practical applications. This changed completely in 1972, when Schottky noise was observed and made use of for beam observation. This development made it natural to try an experimental verification of stochastic cooling on an ISR beam. Cooling equipment was built and installed in the ISR, and after some initial difficulties a clear demonstration of the stochastic cooling effect was made, as illustrated in Fig. 16.5.

As is well known, the most spectacular use of this cooling technique so far has been in the antiproton accumulation in the two proton–antiproton colliders at CERN and Fermilab. However, other important applications were also made directly in the ISR. Special cooling systems were built for the circulating antiproton beam in the ISR and for proton beams up to 10 A (the highest intensity ever cooled). This development led first to an initial significant increase in luminosity during proton–antiproton physics runs. Second, it made it possible to keep the antiproton beams circulating for an incredibly long time (the record was 345 hours) without a decrease in luminosity. Third, the antiproton loss rate was unmeasurably small.

Another application of stochastic cooling in the ISR was the hydrogen-jet experiment. A circulating antiproton beam in the range 3.5 to 5.72 GeV was cooled in the transverse and longitudinal planes. In the latter, the relative momentum spread is cooled down to $\pm 3.5 \times 10^{-4}$. Performed in the spring of 1984, this "fixed-target" experiment was the very last experiment on the ISR.

ISR performance

In the above summary of accelerator studies on the ISR, it has been possible to list only some of the highlights. A very much wider spectrum of studies is behind the impressive improvements of the ISR performance from the start in 1971 to the final colliding-beam run in 1983. The performance can be illustrated by a list of figures and a few examples.

The highest center-of-mass energy has been 63 GeV (equivalent to a fixed target accelerator of 2 TeV). This has been an operational energy used for a large fraction of the physics runs and with close to maximum luminosity. The ISR held the center-of-mass energy record from 1971 to 1982, when the proton–antiproton collider of the SPS started up.

The highest stacked current ever seen in a single beam was 57 A. Normal operational currents during high-luminosity runs have been 30 to 40 A. The highest luminosity at the start of a physics run was 1.4×10^{32} cm^{-2} s^{-1}, achieved in December 1982. Loss rates during physics runs were typically kept to one part per million per minute, which rendered very good background conditions. In fact, most of the background that the experiments struggled with came from the unwanted parts of the pp collisions.

Very long uninterrupted physics runs could be provided – of 50 to 60 hours if desired. This was important because all the manipulations, from switch-on of the machine to good beam conditions for physics, typically took 10 hours. Long stable operation was also high – 86% of scheduled physics time during the last year of operation.

The example of phase-displacement acceleration with working line and closed-orbit control also illustrates the level of sophistication reached by the ISR control system.

The ISR has also been used for antiprotons, alpha particles, and deuterons. In 1977 deuterons and protons were stacked in the same ring, and in a rather impressive display, the proton stack was decelerated through the deuteron stack, illustrating how particles with different revolution frequencies can be selectively treated in the same ring.

The quality of the ISR is illustrated by the exceedingly low background, which for instance made it possible to distinguish events in the highly asymmetric case of antiproton beams of the order of milliamperes colliding with proton beams of up to 20 A.

Fig. 16.6. One of the ISR interaction regions before particle detectors were added.

Concluding remarks

The ISR performance improved over the years of operation far beyond the most optimistic hopes of its planners and builders, and the excellent use of this facility made by the physics community was very much appreciated (Fig. 16.6).

In addition, the ISR was the finest instrument one could imagine for research in accelerator physics, and experience in the ISR contributed greatly to collider–detector design. The accelerator studies performed led to technological inventions and developments in such areas as vacuum, diagnostics, stochastic cooling, and controls.

From the general performance of the ISR and the related accelerator development, and from the experience gained by experimental physicists with their detectors on this device, emerged a general confidence in the ability to predict the performance of other hadron colliders. A considerable change in attitude took place. Before the ISR was built, only fixed-target facilities were on people's minds for hadron physics. Nowadays, one talks almost exclusively about colliders. This was a fantastic transformation for those of us who remember the reluctance of many physicists to accept the idea of colliders as a useful physics tool at the time the idea of the ISR was launched.

17

Development of Large Detectors for Colliding-Beam Experiments*

ROY SCHWITTERS

Born Seattle, Washington, 1944; Ph.D., 1971 (physics), Massachusetts Institute of Technology; Professor of Physics, University of Texas at Austin; high-energy physics (experimental).

There is a remarkable similarity among the modern collider detectors operating at many diverse facilities. For example, the experiments running for the past several years at LEP, the SLD detector just beginning to operate at the SLAC Linear Collider, the detectors now coming into operation at HERA, and those planned for the SSC and CERN's Large Hadron Collider all look quite similar to one another even though the colliders on which they function are quite different. I believe that there are simple and understandable reasons for this similarity.

The present situation contrasts markedly with that of the detectors employed in the first collider experiments during late 1960s and early 1970s – essentially the same period we are studying at this Symposium. In the early colliding-beam experiments, many different detector architectures were tried; out of all those ideas came a "standard model" of detectors, the cylindrically symmetric, solenoid-magnet detector that so dominates colliding-beam experiments today. For example, the first detectors at the early storage rings[1] – CBX at Stanford, ACO at Orsay, the VEPP machines at Novosibirsk – were visual detectors, involving spark chambers and other techniques; they were designed to study specific final states over limited ranges of solid angle, with little or no particle identification, limited trigger capability, and no momentum analysis. They were not, as one would say today, general-purpose detectors.

In the very early 1970s, some of these detectors were still functioning and higher-energy experiments began at Frascati's ADONE collider, CERN's Intersecting Storage Rings (ISR), and the Cambridge Electron

---

* This chapter is adapted from a transcription of the talk, which was delivered by remote telephone connection because the speaker could not be present.

Accelerator (CEA) Bypass. While several detectors were commissioned at these facilities, at least the early ones followed the previous trends of limited solid-angle coverage, designed for specific physics processes, and having limited, if any, momentum analysis. Of course, many people recognized the need for large solid-angle acceptance and momentum analysis, but new configurations of detection devices had yet to be invented to respond to the demands of collider physics. Nevertheless, many important results were obtained from the early collider experiments: tests of quantum electrodynamics, the rise in the proton–proton total cross section with energy, and the unexpectedly large production rate for multihadron final states in electron–positron annihilation.

At about the same time, the ideas of Bjorken, Feynman, and others emphasizing the importance of short-distance phenomena in many reactions suggested to many experimenters the need to study precisely the scaling of various single-particle or low-multiplicity spectra over wide ranges of kinematic parameters, but not necessarily over a large fraction of the solid angle. The highly successful deep-inelastic electron scattering program initiated at SLAC certainly brought this approach into vogue. Such measurements generally called for small solid-angle spectrometers often having excellent particle identification and resolution. To state the mentality overly crudely for emphasis, multiparticle physics was largely left to the bubble-chamber types; real-man electronic counter experimenters were busy with their high-resolution, small-acceptance spectrometers!

Then came the construction of the SPEAR electron–positron collider and, with it, the development of the SLAC–LBL magnetic detector (later called the Mark I). This was, in my opinion, truly a watershed event in the development of experimental apparatus as well as in the history of the development of the Standard Model. Why was the Mark I detector (Fig. 17.1) built the way it was and why do so many subsequent detectors look so much like it? To approach these questions, I shall try to relate some of my recollections of our thoughts and concerns at the time – from the point of view of a young postdoc who had the good fortune to have joined the SLAC–LBL collaboration as the new detector was being designed and built. I believe these notions were shared by many, if not all, of my colleagues in that extraordinary collaboration.

First, we were deeply impressed by the then mysterious multihadron events that were being observed at Frascati and CEA. In some quarters, these measurements were not taken as seriously as they should have been, no doubt in part due to the paucity of events and the limita-

Fig. 17.1. The SLAC–LBL detector (later called the Mark I) with author Roy Schwitters standing near its center.

tions of the somewhat disjoint set of detectors providing the information. Furthermore, new experimental possibilities were just opening at that time with the construction of the National Accelerator Laboratory in Batavia and the ISR at CERN, thus relegating, in some people's minds, the electron–positron results to the backwater.

We were also keenly aware of the difficulties and limitations of available detector technology. On the one hand, one could build imaging devices using spark chambers with electronic readout, but for mechanical and other technical reasons, they were invasive in respect to other techniques needed to distinguish identities of particles and there was little successful experience operating them in magnetic fields. On the other hand, we really wanted to measure the momenta and the identities of all particles in annihilation events. It was the strong consensus of the collaboration that we had to solve these problems, that we had to unify, if you will, the visual approach with the electronic approach to give us the image of a complete event as well as good data on the individual particles within the event. This unification was the key technical advance, in my view, that we were able to accomplish with the SLAC–LBL detector. It is also interesting to note that the name Mark I was never attached to that detector, at least in my memory, until after the Mark II had been built!

Let me refer to the conference logo (Fig. 17.2) because I think it is a very good example of the capabilities of the Mark I detector and is also completely representative of what I would call a modern collider-detector.[2] First, the most obvious and significant thing was that it could track essentially all of the charged particles in the event and measure their momenta through their curvature in the magnetic field. The kinds of technical requirements that had to be specified can also be discerned from the logo. In this event picture we observe the decay of the ψ' to the J/ψ with the emission of two low-momentum pions. The J/ψ decays to an electron–positron pair, each member of which has high momentum. The tracking system had to provide sufficient resolution and bending power to measure the high-momentum tracks without curling up and losing the low-momentum tracks. Such requirements led to design choices of the overall size of the detector, the strength of the magnetic field, and the number of layers and the resolution of the tracking system. They determine, to a large extent, the very scope and cost of the detector.

In addition, you see little boxes and other things that provide information on the interactions of the particles in various materials that can give you clues as to the identities of the tracks in the events. At the beginning of the SPEAR program, we did not know for sure that most of the particles in events with more than two tracks would be hadrons. This was the best guess, but we were entering a new physical realm and wanted solid experimental evidence. Thus we invested in a multiplicity of detector types to give overlapping information along each track. We could then combine time-of-flight information, shower-counter pulse

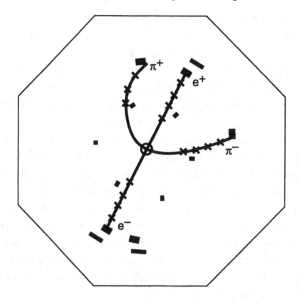

Fig. 17.2. Single-event display of the Mark I detector in which a ψ' particle decays (via the J/ψ) into two pions and an electron–positron pair.

height, and muon chamber information along with the momentum and be able to discern the types of particles giving rise to the observed tracks. All of these detector components put new demands on data acquisition, calibration, and data analysis systems that were solved through innovative application of the revolutionary advances in electronics and computer technologies taking place simultaneously. The Mark I was able to provide, through electronic means, images of collision events with good momentum information and suggested particle identities. Even though many events were scanned by eye, data analysis could proceed without people scanning every event, thus speeding up significantly the analysis time and enhancing the possibilities for exploratory data analysis.

While many of the techniques employed in the Mark I are by now obsolete, such considerations as the necessity for R&D to push particular areas of detector technology, technical compromises that must be taken in order to meet conflicting requirements, computer modeling of performance, and the like, are completely modern.

On more general grounds, I feel that key elements of the SLAC–LBL detector, represented in the logo, will continue to play crucial roles in essentially any collider detector designed to explore unknown territory, irrespective of particular choices of detector technologies. First is the *large solid-angle coverage*. Because any new phenomenon will likely be revealed through a statistically limited sample of events, it is important to have the highest possible detection efficiency. Because one usually observes multiparticle events, the solid-angle coverage strongly affects the detection efficiency. Having good *uniformity of response* is important for several reasons, such as aiding the understanding of systematic errors with small event samples and providing "self-calibrations" of detector components. The azimuthal symmetry inherent in the solenoid magnet design naturally gives excellent uniformity of response; innovative mechanical designs permitted a minimum of inactive regions or excessive material in the paths of particles within the detector volume, thus enhancing both the coverage and the uniformity of response. The third general principle is the importance of *detailed, on-line analysis* of data and detector performance. Collider runs are often long, and the most interesting events are usually extremely rare; the detector must be live and well to not miss potential discoveries.

The SLAC–LBL detector was criticized at the time, because each detector subelement did not represent the best performance or state-of-the-art. However, our emphasis was on integrating the detector as a whole and not trying to do the best possible in any given area. And I think that this strategy worked very well, indeed. The Mark I was probably not superior to others in any given detector element. Some elements were downright lousy. Yet it allowed us to observe all the beautiful things that Nature was kind enough to put in the SPEAR energy range.

One benefit provided by the Mark I's excellent uniformity of response (combined with the astonishing good luck that seemed to surround us) was that we were able to study the azimuthal distribution of hadrons with essentially no experimental bias. In the higher energy data, there was suggestive evidence that hadrons were produced in jets as expected by the quark picture.[3] However, the azimuthal distribution of jet axes appeared highly irregular. Upon examination, we discovered that the electron and positron beams had spontaneously become polarized (an effect anticipated by accelerator physicists at Novosibirsk) and the anomalous-looking azimuthal distribution was actually a dramatic confirmation of the spin-$\frac{1}{2}$ character of the quarks![4]

Since the time of the Mark I detector, there have been highly successful programs using specialized detectors to explore in depth certain questions not accessible to general-purpose detectors. As long as new discoveries are made, there will be need for detectors having special capabilities, but when entering unknown territory, physicists have voted with their feet (and heads) in favor of the Mark I-like "standard model" detector.

The second watershed event in the development of collider detectors that I choose to highlight is the advent of the UA1 and UA2 detectors at CERN's SPS proton–antiproton collider.[5] These devices made extensive use of the relatively new technology of hadron calorimeters in a new colliding-beam environment and proved that one can make new discoveries in this arena. Because the elementary collisions of interest are between proton constituents – quarks and gluons – one effectively loses the longitudinal momentum constraint due to the internal motion of the quarks. Hence, the events of interest have significant and often unknown longitudinal components of momentum in the detector frame. Furthermore, debris from the other quarks and gluons of the colliding protons generate additional particles that are imposed on the event. As opposed to electron–positron collisions, where most of the events are of experimental interest, the hadron colliders must contend with enormous backgrounds of relatively dull events, necessitating the development of sophisticated triggers to select events of interest while ignoring the dull events. These additional complexities and the much higher energies afforded by proton colliders made the task of the UA1 and UA2 teams most challenging. There were also some simplifications arising from the higher collision energies. In many practical cases, hadrons were created in narrow cones, called jets, that could be treated experimentally as single particles. High-energy electrons and muons proved to be relatively easy to identify.

While UA1 and UA2 broke new ground in the development of collider detectors, they also reaffirmed, in my view, the principles that we learned from the Mark I. For example, in designing a significant upgrade of the UA2 detector, the group chose to increase the solid-angle coverage, keeping the same degree of resolution and uniformity of response, rather than to divide up the solid angle into smaller pieces of higher spatial resolution. The dipole magnetic field of UA1 had certain advantages, such as permitting measurement of momenta in the forward direction, but it also had the disadvantage of lacking azimuthal symmetry. Therefore, the momentum resolution varied substantially over the acceptance of the detector, complicating the analysis of certain rare events. This is a minor quibble; the important thing is that UA1 and UA2 paved the way for designing detectors at very high-energy hadron colliders. Their experience combined with the principles learned at SPEAR have permitted new detectors, such as CDF at Fermilab, to carry out rich and productive programs and form the basis of the designs of the huge detectors contemplated for the new generation of supercolliders. Those detectors, while building on the work of previous detector collaborations, will have new challenges, particularly in the areas of very high event rates and particle fluxes.

One kinematic effect that I think is worth noting makes it possible, in my view, to design detectors for a hadron environment that are similar in form to those used to observe electron–positron collisions. The idea relies on the fact that when one organizes the detection elements of a collider detector into roughly uniform bins of rapidity (related to the polar angle of tracks) and azimuth, then jets of hadrons that arise from quarks and gluons will form approximately circular patterns of similar size in this rapidity–azimuth space over the entire detector. Furthermore, background tracks from the spectator quarks and gluons will more or less uniformly occupy rapidity–azimuth space. Thus, despite the inherent complexity of events produced in hadron collisions, organizing the detector elements in a simple, but somewhat subtle, way leads naturally to detected event configurations where information of little interest is spread uniformly over the detector's sensitive volume, while information of significant interest clusters in highly visible ways. This is why, in my view, there is little merit to the "religious" debates that have been known to rage between some proponents of proton–proton or electron–positron approaches based only on appearance of events. There are other concrete bases for such discussions and choices.

Let me close by identifying issues that I think people need to consider for the future of these big detector collaborations. I am referring, of course, to the sociology of large collaborations and the question of how we continue to do physics within the constraints of a very small number of very large detectors. Those of us who were fortunate enough to be part of the Mark I collaboration thought that 35 authors was a pretty big list. I have just received a proposal for a major detector at the SSC with 911 authors! To many scientists, such a thing is preposterous. Yet, at least in my experience, the "doing" of physics inside such a collaboration works very well. There is usually an exciting and dynamic internal literature and debate on physics and technical issues. Insiders know who is doing the work and making the innovations. The exciting physics opportunities and diversity of colleagues can make scientific life in a large collaboration most interesting and rewarding.

There are problems, however – principally on the external side of large collaborations. For example, it is becoming increasingly difficult to promote junior faculty at universities because it is almost impossible for outsiders to understand who is making what contribution. A long publication list of "et al." papers is almost useless. There are other examples, such as a tendency of large groups to become too conservative (or sometimes too bold!) in presenting new results. Taking appropriate risks is an important aspect of scientific discovery; it would be a disaster to "breed" responsible risk-taking out of high-energy physics.

In my view, high-energy physics has a very serious problem here, mainly with our image, but also in substance. Whether real or imagined, if we fail to attract the best young people to our field, then the future of the field is in jeopardy. We are often criticized by colleagues in other areas of science on perceptions exacerbated to some extent by the large size of our collaborations and author lists. Indeed, the author list of a typical experimental paper usually generates ridicule in an audience of scientists from other fields. So I urge the people attending this Symposium, as I am urging our advisory committees and other colleagues, to begin discussing these issues to see if there are ways we can change the way we display ourselves and our work to the outside world and, at least, try to avoid some of the undeserved criticism.

(*Note added in proof:* Although I feel that the loss of the SSC and its consequences are far more threatening to high-energy physics than the matter discussed above, the sociology of large collaborations will remain an issue that should be addressed by the community.)

Notes

1 A collection of review papers with references to the original literature can be found in Robert N. Cahn, ed., e^+e^- *Annihilation: New Quarks and Leptons* (Menlo Park, CA: Benjamin/Cummings, 1985).

2 A "one-event" display from the Mark I detector, showing a view perpendicular to the incident beam direction (and the direction of the detector's magnetic field) of particles produced at the ψ' resonance.

3 Gail Hanson, et al., "Evidence for Jet Structure in Hadron Production by e^+e^- Annihilation," *Phys. Rev. Lett. 35* (1975), pp. 1609–12.

4 R. F. Schwitters, et al., "Azimuthal Asymmetry in Inclusive Hadron Production by e^+e^- Annihilation," *Phys. Rev. Lett. 35* (1975), pp. 1320–22.

5 A general review of hadron collider physics, with references to the original literature, is that of M. D. Shapiro and J. L. Siegrist, "Hadron Collider Physics," *Ann. Rev. Nucl. Part. Sci. 41* (1991), pp. 97–132.

18

Pure and Hybrid Detectors:
Mark I and the Psi

PETER GALISON

Born New York City, 1955; Ph.D., 1983 (physics and history of science), Harvard University; Mallinckrodt Professor of History of Science and of Physics, Harvard University; history of science, high-energy physics (theory).

The broad sweep of theoretical claims and programs commands our attention: even the title of this book, *The Rise of the Standard Model*, points to theory as the capstone of physics. But outside the commitment of theorists to principles of their practice such as causality, determinism, unification, and symmetry breaking, there are commitments built into the hardware of the laboratory. Less dramatic perhaps, less often spoken of without doubt, these traditions of instrumentation shape the practice of experimental physics and embody views about the nature of acceptable empirical evidence. In this chapter, I want to explore the coming together of two great lines of instruments in the twentieth century: on one side, the *image tradition* instantiated in the sequence cloud chambers, nuclear emulsions, and bubble chambers. These devices make pictures, the delicate array of crisscrossed lines that have come to serve as symbols not only of particle physics but of physics more generally. On the other side, there stands a competing *logic tradition*, this one aiming not to make pictures but instead to produce counts – the staccato clicks of a Geiger–Müller counter rather than the glossy print from a cloud chamber. In the line of such counters came a host of other electronic devices that built their persuasive power not through the sharpness of images but through the accumulation of a statistically significant number of clicks. These other devices included the spark chamber – roughly speaking a flattened-out Geiger counter that sparked along the tracks of passing particles – and the wire chambers: a host of instruments that used wires, not plates, to measure the effects of ionic tracks left by the particles under inquiry. This chapter, on one particular detector (the Mark I at SLAC) and one particle (the psi), begins an exploration of the hybrid laboratory I explore in much greater detail elsewhere: the replacement of the "pure" detectors of physics with hybrid teams, hybrid equipment, and hybrid modes of demonstration.[1]

In 1957 Luis Alvarez had been ready to celebrate the demise of the logic tradition. Bubble chambers were producing more data, more discoveries, and more precise information about the subatomic world than their counter counterparts could even imagine. The world of visualization seemed triumphant. But by the early 1970s the bubble era was drawing to a close, and these massive detectors had begun to shut down in laboratory after laboratory. In more than a symbolic act, Alvarez's old 72 inch chamber, expanded to 80 inches when it was sent to SLAC, was ceremoniously shut down at 3:30 p.m. on 16 November 1973.[2] Indeed, the last great hurrah for the boiling liquid was the discovery of weak neutral currents in "Gargamelle," a heavy liquid bubble chamber at CERN. The establishment of neutral currents was a contentious process, one that lasted quite some time but was more or less complete by 1973. But if bubble-chamber physicists of the early 1970s knew they were at the end of the line, the epistemic ideal embodied in these detectors lived on: somehow, within the sphere of logic electronics, rare events would have to be extracted with the indelible tracks that had previously been the purview of the image tradition. The survival of the image amidst the electronics of the logic tradition is my subject, for it is there that one can see at one and the same time the development of a hybrid social structure of the two traditions and a hybrid technical concatenation of techniques, hardware and practices.

Over decades, on both sides of the image/logic divide stood the pure forms of earlier instruments: the cloud chamber, the Geiger–Müller counter, the bubble chamber, or the nuclear emulsion. A single physical process is implicated in each: the cloud chamber predicated on the transformation from a vapor to a liquid; the Geiger–Müller counter from a gas to a plasma; the bubble chamber from a liquid to a gas; and the emulsion from silver halide to its ionic components. And while the instruments in some cases could be used in conjunction with one another – one thinks immediately of the counter-controlled cloud chamber that used counters to trigger a cloud chamber so it "took pictures by itself" – as a broad generalization (with some notable exceptions), one can reasonably characterize experimentation before the late 1960s as grounded on relatively "pure" devices. Bubble chambers were never triggerable, nor was film; the truly complex arrays of Cerenkov counters did not typically exploit film-producing registration mechanisms. Indeed, because of the "background" problem of old tracks superimposed on new ones, and the difficulties associated with fading tracks, the lack of time control on films remained a problem throughout its usage. This instrumental purity was lost as hybrid instruments became the standard fare in the late 1960s and early 1970s. From a purely material standpoint this shift would be a powerful alteration of the material culture of physics in and of itself.

But the shift meant much more than merely combining one instrument from column A (image-making devices) with another from column B (count-producing devices). For accompanying (and indeed in some ways precipitating) the shift there was a concomitant alteration in the social structure of the experimental group and a change in the mode of demonstration. I want here to introduce these concepts with the example of what is arguably the single most important individual instrument high-energy physics ever produced: the detector at the SLAC in the colliding beam facility SPEAR known, by an apallingly un-euphonious name, as the SLAC-LBL Solenoidal Magnetic Detector. For it was with this device (which I will call by its later name, Mark I) that the ψ (psi) particle was co-discovered.[3] Along with its role in ushering in the quark theory of matter through the psi, the Mark I was also used in arguing for the existence of the τ (tau) lepton, and for the production of the D^0 meson that gave credence to the claim that the psi was a bound state of charmed mesons. More than any single device, the Mark I became a prototype for the next several decades of high-energy colliding-beam physics at laboratories around the world, including Fermilab, CERN, and DESY.[4]

The introduction of colliding-beam facilities radically altered instrumentation in several ways. In a fixed-target facility, much of the beam's energy is never converted into new particles; it is simply carried forward along the line of motion set by the projectile. Colliding-beam facilities, by contrast, smash a particle and antiparticle together with equal and opposite momenta. Since the two objects annihilate one another, the energy is fully utilized and the resulting new particles can emerge in any direction – not just along the line of the projectiles' trajectories.[5] Even a moment's thought about the shape of instruments as they passed from fixed targets to colliding beams reveals the enormous shift in thinking about instrumentation that occurred with the Mark I and its contemporary detectors. Fixed-target detectors stood (schematically) like a line of dominoes, with each domino standing to catch particles of particular species. Colliding-beam detectors were, by contrast, shaped like nested cans, with the collision located in the center of the innermost cylinder. Whereas the fixed-target detector was built to nab particles scattered in a narrow cone around the axis of the beam, the colliding-beam detector had to grab particles scattered around any direction from the collision.

The production of a triggerable detector with nearly 4π acceptance was but the latest attempt to join the logic and the image tradition, an attempt to join the *control* over events that characterized the logic tradition with the possibility of a picturelike representation of tracks. Building a 4π detector would be the basis of the imaging. Control, however, would not be ceded as it was in a bubble chamber. As Richter stated:[6]

> Part of my background made me want to know what was happening in the experiment while it was happening. I felt that it was much better when exploring the unknown to know where you are so as to better plan where you can go, rather than after the journey to onlyl know where you have been. [And it's different in the] bubble-chamber business, the physicists took hundreds of thousands of pictures which then had to be developed and scanned, and didn't know anything about what was happening in the experiment until a year or two after the experiment was already over.

Given the views expressed from the late 1950s onward, Richter's remarks are hardly surprising. Wenzel, Charpak, Preiswerk, Macleod, Hine, and others too numerous to cite had emphasized time and again how abhorrent they found the dissociation of data taking from analysis. In every aspect of instrument construction, proponents of the logic tradition continued to mount in hardware their demand for results they could use "while they happened." This insistence on the priority of online results remained central to the Mark I collaboration, reflected most clearly in the extraordinary lengths to which the team went to provide single-event displays in real time.

The hybrid nature of the effort is as visible in the social structure of the collaboration as it was in the material structure of its component parts. In particular, three teams joined to form the Mark I collaboration: one group from Lawrence Berkeley Laboratory brought with it a long tradition of expertise in bubble-chamber track analysis. The other two, from SLAC, carried with them an equally long history of electronic experimentation. Look for a moment at the Berkeley side of the venture. George Trilling was, by the early 1970s, one of LBL's most senior bubble chamber experimentalists; he had come with Donald Glaser, the inventor of the bubble chamber, to Berkeley.[7] Gerson Goldhaber had

begun scientific life in the world of emulsions, culminating in his registration of the first track of an antiproton in 1956. By 1959 he too had
turned to bubble chambers, first with Wilson Powell and subsequently
with Trilling. Together, Trilling and Goldhaber had directed a large and
successful LBL bubble-chamber group, one of the main (internal LBL)
competitors to Alvarez's outfit. Goldhaber also helped recruit William
Chinowsky to Berkeley, first as a visitor and then permanently. Though
Chinowsky's Ph.D. research was done with counters, he too had turned,
by 1954, entirely to track experiments and worked for many years with
bubble chambers.[8]

Like the other LBL members of the team, John Kadyk and Gerry
Abrams were both LBL bubble-chamber veterans; the LBL graduate
students, John Zipse, Robert Hollebeek, and Scott Whittaker filled out
the group. Asked later if the bubblers and electronics people felt they
lived in different communities, Chinowsky laughed, recalling the competition between the two: "The people doing electronics tended to look
down on us in the bubble chambers; on the other hand, it was the bubble
chamber that got all the results. I guess, to some extent, it didn't seem
fair that they should work so much harder and not get a return."[9] In the
search for one particle, the SLAC workers had mounted a vast array of
counters and the Berkeley bubble chamberites had found it trivially.[10]
Bubble-chamber work – certainly once the chamber was up and running – revolved almost entirely around analysis: programming for the
reconstruction of tracks and the identification of particles and processes.
It was precisely these skills that were called for in the elaborate track
structure that would be left in the annihilation of colliding electrons and
positrons. For example, one track program used in the Mark I exploited
the "road" idea from the Hough–Powell device that the Goldhaber–
Trilling group at LBL had used in their bubble-chamber film analysis.
Another program was lifted virtually intact from bubble-chamber analysis to the Mark I.[11] As an indication of just how powerful the highly
honed track-analysis skills were, Abrams, Goldhaber, and Kadyk spent
many hours poring over photographs of computer-reconstructed tracks
that were printed from the magnetic tape.[12] This was a process altogether outside the logic tradition, but in part constitutive of the Berkeley
physicists' credence in (or doubts about) any putative new effect.[13]

The one piece of electronic hardware that Berkeley undertook was the
"shower counter," a sandwiched structure of plastic scintillator (that
measured the energy of electrons and photons by way of the light deposited in it) and lead plates (which converted the photons into electron-

positron pairs). In light of the LBL's prior experience, it is perhaps understandable that the software and tracking programs were a spectacular success and the shower counter was occasionally a source of friction with their cross-Bay, logic-tradition electronic collaborators.[14] By December 1972 it was clear that the scintillators had been badly scratched in fabrication, and a crash program began to try to salvage what could be rescued. Passing a steam jet over the surface failed; the manufacturer's polish did not work; spraying the sheets with acrylic helped not; and heating to 50° C for 24 hours did not aid the cause. What did help (somewhat) was a coat of Johnson's "Glo-Coat" floor wax and standing it up to dry. In the end, foregoing this ad hoc solution, the detector did achieve its design specifications, and even was used in the detector trigger, an unanticipated bonus.[15] Software-embedded image reconstruction and analysis was the Berkeley forte. But aside from the flap over scratched plastic, collaboration between the two institutions worked without much friction.

One link between the Berkeley-image and the Stanford-logic traditions came through Martin Perl, who invited Chinowsky to join the collaboration. Perl himself had been a student of I. I. Rabi, and then (in the early 1950s) had gone to Michigan to work with Donald Glaser on his newly minted bubble chamber. The muon, heavier than the electron but otherwise apparently indistinguishable from it, was a standing rebuke to any motivated understanding of particle physics. Rabi was most adamant on the point, repeating often and insistently his refrain about this heavy new particle: "Who ordered that?" After Glaser departed for Berkeley from Michigan, Perl moved to Stanford and began a six- or seven-year struggle to answer Rabi's question by trying to find some difference other than mass between the electron and the muon. Unable to explain who ordered the peculiar second course of lepton, Perl began ordering more... and more. For years, Perl and his collaborators in SLAC Group E (including J. Dakin, G. Feldman, and F. Martin) had used electronic means to explore possible distinctions between the muon and the electron, and then pushed hard to add muon detectors to the Mark I to allow the collaboration to search for even heavier (and previously unknown) leptons that might be within reach. And so it was that this new detector, while it might have looked like a hadron finder to the LBL group and Richter's Group C at SLAC, was to Perl an instrument for uncovering the "next" in a sequence of heavy leptons. One other experimenter, David Fryberger, came from SLAC's Experimental Facilities Department, formally outside of the structure of groups C and E.

But the majority of the SLAC contingent was affiliated directly with Richter's long-standing research in electron physics. One of Richter's interests had been in photoproduction – using photons produced in electron collisions to make new particles. Out of this effort came one-half of Richter's group, including Adam Boyarski. Boyarski was the computer wizard; it was his job to make sure the data produced by the various detector elements came in with compatible formats, and to ensure an on-line monitoring system for the detector, along with a final data tape that could be accessed by anyone within the large collaboration. Two SLAC researchers, Roy Schwitters (who had been an MIT graduate student with Louis Osborne, working with Richter's group at SLAC's End Station A) and Marty Breidenbach (also an MIT alumnus, who had worked at SLAC and then CERN before returning to SLAC) also moved into the Mark I effort out of this earlier enterprise. On the other side of Richter's (pre-SPEAR) physics pursuits was a series of spark-chamber experiments that included SLAC staff physicists Rudolf (Rudy) Larsen and Harvey Lynch. Larsen became the spokesman for the Mark I collaboration, and he and Lynch carried over their extensive experience with spark chambers, including magnetostrictive readouts and other electronic lore. Some members of the Group C team were entirely devoted to machine building – not to making the detector. John Rees, for example, originally moved to SLAC from the Cambridge Electron Accelerator in 1965–66, bringing expertise in accelerators and storage rings, especially on new ways to handle the RF (radio frequency) facilities and beam-focusing equipment. Gerry Fischer managed the magnet systems for SPEAR.[16]

From the outset, both in the design of Mark I and in the composition of Group C itself, Richter conceived of the experiment as a colliding-beam machine *and* a detector. This was evident on the ground, where, unlike virtually any other experiment in high-energy physics after 1970, Mark I and SPEAR shared a single control room. Even in 1971 this integration was unusual: most laboratories such as Brookhaven and CERN had long since separated the production of particles from their consumption.

In its design, the search for full coverage underlay the design of the Mark I and distinguished it from all previous electronic detectors. By immersing the whole in a magnetic field, the Mark I collaboration aimed to determine the particles' momenta; using time-of-flight or trigger counters (for π/K separation) and the plastic scintillator shower counter (to identify e^+, e^-, and γ), the team could identify all of the resulting particles. Aside from the shower counter, there were three main subdetectors

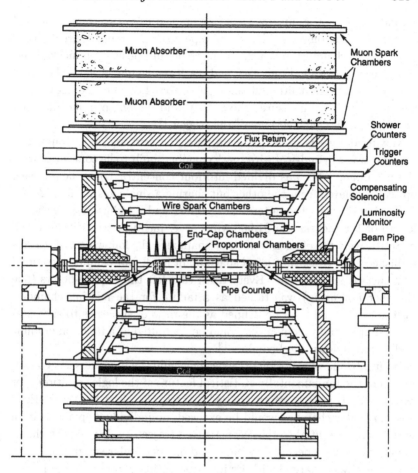

Fig. 18.1. Schematic diagram of the Mark I detector, viewed from the side, as it appeared in the summer of 1974.

within the whole: the cylindrical wire spark chambers that would be the primary track detector, the beampipe electronics that monitored the beam and collision times, and a set of outer spark chambers that identified muons, distinguishing them from strongly interacting particles of similar mass (pions, for example).

The centermost detector, the pipe counter, is there to exclude cosmic rays by triggering the registration of data only when an electron–positron annihilation took place. Effectively, this reduced the number of cosmic ray events from about a thousand per second to about one event per second – a dramatic improvement. Assembled from two semicylindrical sheets of 90 cm long scintillator, the pipe counter wrapped directly over the vacuum chamber in which the electron and positron bunches travelled (see Fig. 18.1). From each end, each semicylinder was linked to a lucite light pipe that brought the light collected outside the detector to where a phototube measured the output, and the phototube could then be linked to the logic circuits of the trigger.[17]

Moving outwards from the beam pipe, the next and most important detector was a cylindrical wire spark chamber, built to lie concentric with the beamline. This was divided into four concentric components, each optically isolated from the other but sharing a single gas volume. As an indication of scale, the outermost chamber had a radius of 53 inches, a length of 106 inches, and 31,900 wires at $\frac{1}{24}$-inch intervals. To establish quantitatively the efficiency of these chambers, the following test was established: any track that left a space-point in all four chambers counted as a success for each one. Every track that left three space-points (one chamber not firing) was tallied as a failure for the missing chamber. "Efficiency" could then be defined as the ratio of successes to successes plus failures, and this quantity could be computed as a function of angle and distance down the beamline.[18]

Built into the very notion of "efficiency" and the programs is a continuation of the fundamentally statistical feature of the logic tradition. For even as the computer reconstructed a track, the absence of a spark in any particular chamber would be excused; no individual spark location was deemed essential to the reconstructed trajectory. Wire spark chambers formed the innermost cylinder and one particular subgroup, led by Schwitters, was charged with its implementation and certification.[19]

Muon detectors formed the outermost layer of the cocoon, insulated with a thick layer of concrete that would stop a large percentage of any strongly interacting particles. Though they were necessarily larger, this fifth set of magnetostrictive wire spark chambers would ferret out the muons. These, like the inner spark chambers, had their own set of tests before they could be trusted. How would they respond with different-strength high-voltage pulses? How frequently would the chambers fire "accidentally" (in the absence of any known signal)? Could "correlated" firings of the chamber (coincident with a bona fide muon event) be avoided by the introduction of alcohol in the chamber?[20]

Each component detector therefore had its own standards, its own tests to pass, its own efficiencies to be measured. But separate functioning was not enough.

Nothing in the epistemological structure of experimental high-energy physics is as important to understand as the coordination between subgroups. For it is this mutual alignment – at once social and epistemic – that undergirds the demonstration of a new effect.[21] Local coordination between diverse approaches to a problem is central to understanding the building of an argument and the cohesion of the larger scientific community. As experimentation grows in scale, this coordinative function is exhibited increasingly within the construction of the instrument itself. This dynamic becomes ever more visible as one moves from the scale of Mark I (20–30 physicists) to the vastly larger collaboration and detector of the Time Projection Chamber at SLAC, those at CERN's LEP (Large Electron Positron) collider, and the planned but never-built detectors of the Superconducting Super Collider. One mortar that holds the components together is the language of computation, for it was ultimately the computer that had to mesh together the output signals from the beampipe detector, the inner tracking chamber, the lead-scintillator sandwich, and the outer muon detector.

I mean quite intentionally to foreground the linguistic character of this synchronization. As one physicist put it after listing various problems, "Assuming that all the foregoing problems could be solved, we are still on the path to Babel unless positive steps are taken to recognize the reality that we are a large group of people who depend upon each other."[22] First priority: computer programs had to give correct answers, and that necessarily involved a check by someone other than the author. This demand for *correctness* was quickly followed by a demand for *reliability* (the programs could not crash) and *intelligibility* (documentation both external and internal to the code had to be clear and well directed). A fourth priority was the constraint of *efficiency*; scarce resources in both core memory and central processing unit time would be swallowed by an inefficient piece of code, and off-line work, while it needed less core, would still tie up crucial CPU time. Fifth (and finally), "The programs must be easy to use.... Since we are a group where people use and depend upon the programs of others, considerable attention should be applied to the 'human engineering' aspect."[23] Technique and social structure were here, as throughout the history of experimentation, inextricably bound.

Even if Babel could be averted, there remained the coordination of pieces in data reduction. On 8 November 1973, for example, Larsen issued a memo that set out the basic problem:[24]

What we all want to do is to extract the physics from our data tapes, right? Right! ... To date, we have not been remarkably successful in this preparation. Many problems have arisen: duplication of effort, conflict over use and status of software routines, lack of definition of problems.... While many of the problems can be attributed to the "lack-of-communication" cliché, the most important void is the absence of a structure that everyone understands and within which we can work.

There were two parts to the process of coordination – event identification and a full characterization of the four fundamental components (cylindrical wire spark chambers, muon spark chambers, trigger counters, and shower counters). As Larsen emphasized, most of the collaborators had been affiliated with the production of one piece of hardware and the attached software for extracting quantitative information. Now that isolation had to end.

Again Larsen: "It is necessary to formalize this existing situation so that all know whom to turn to when they need information; there is clearly a good deal of cross-talk between the various hardware components."[25] As this remark makes clear, the architecture of objects and the social architecture necessarily must move together. Cross-talk between the wire spark chamber (WSC) and the shower counter is dependent on links between the appropriate software, and that meant coordinating the wire spark chamber group with the shower counter group. So the collaboration is subdivided once more, this time into "software" components (see Table 18.1).[26] Significantly, the LBL–SLAC segregation, present in the hardware building, had been crossed; the cylindrical WSC now brought Chinowsky and two Berkeley students (Hollebeek and Zipse) into the collaboration with the builders, Schwitters and Lynch; the shower counter, similarly, now introduced Perl and Feldman, who came from the hardware side of SLAC (muon chambers), into collaboration with the original LBL shower-counter builders, Kadyk, Whitaker, and Friedberg.

But over and above this division there had to be a geographically representative supergroup that would create the integrative "analysis" program. This software would take the output from the component bits of software and weave them into a coherent data set with a clear and consistent set of calibrations. "While everyone is free to maintain his own file, we can't have everyone altering 'the' primary analysis routines at his will."[27]

The next month (December 1973), Richter reported on the Data Analysis Steering Committee's deliberations. There had to be consensus on

Table 18.1. *Subgroups responsible for Mark I software programming*

Cylindrical WSC	Muon WSC	Triggers-Pipe	Showers
Lynch	Bulos	Moorhouse	Kadyk
Hollebeek	Lyon	Feldman	Feldman
Zipse	Dakin	Larsen	Whitaker
Schwitters	Pun		Friedberg
Augustin			Perl
Breidenbach			
Chinowsky			

the first two computer data-crunching programs: PASS 1 and PASS 2. PASS 1 "filtered" out events that were, as one participant wrote, "the most obvious sources of background events."[28] These included demands on the events before they would even be recorded as basic data: a minimum time lapse had to have passed between the annihilation event and the detection of tracks, which guaranteed that the event was minimally plausible as a physical occurrence; another demand insisted that there had to be a prima facie case against the event being of cosmic origin, there had to be enough (at least four) points in the wire spark chambers for a track to be considered viable, and finally there had to be at least one "road," that is a football shaped area formed by two tracks of opposite charge and a minimum amount of energy deposited perpendicular to the beamline. PASS 2 then took the filtered tape produced by PASS 1 and filtered them once more, this time by using the spark information to determine the "hits" in space, then grabbing these space-points to make tracks, and finally sorting the tracks into particle types.[29] Subsequent computer runs fit the points to a helix[30] and filtered the resulting helices on the basis of known physics processes.[31]

Finally, scanners – in the old bubble chamber tradition – pored over the events and by hand vetted the events on the basis of kinks, angles, and timing.[32] Scanners were taught to look for a timing of about 5 or 6 nanoseconds, though they were instructed to be alert for slower particles. There would typically be a low shower-counter pulse, less than 30 units of energy deposited, but the counter can cause a cascade precipitating a much higher pulse. "Obviously, the data is not 'black and white,' " the scanning guide cautioned, "so questions will be the best way to get the feel of what is happening."[33] It is precisely this "feel" that lay behind the demand for hand scanning in the first place, a sense that a human being could pick out events from the thousands recorded that raised specific or systematic questions about the automatic sorting routines. But would it make sense to call the human classification "interpretation" and the computer-based operation merely "data" provision? I think not.

Look at the category "junk." "This category," our authors write, "requires special emphasis and discussion, since there are several different possibilities, and a misjudgment could easily lead to a good event being missed." Here it is explicitly, though everywhere implicitly: judgment. The hand scanning, by both physicists and scanners, was designed to reintroduce the human faculty of assessment as a check on the faculty of algorithmic procedure. One species of junk was, phenomenologically, the activation of a large number of triggers and counters – more, say, than 20 triggers plus shower counters. This would be type A junk. Type B junk is the polar opposite, merely one track visible, counting as a "track" a smooth set of points even where the computer had not actually drawn you (the scanner) a tracklike line. Now type C junk is trickier – events with a time spread of more than 7 nanoseconds. It might be junk, then again it might be hadrons, the very object of the search. Finally, type D junk consisted of background hadron events in which only one beam was involved, since these events occurred outside the fiducial interaction region. Keep them tallied, the scanners were told; the computer would filter them later based on their coordinates and would use the information to sift out later and unwanted events. "Please keep all events when in doubt, and mark those on which you would like to ask the advice of a physicist."[34]

Judgment enters at every stage, whether explicitly – in open debates about the status of a particular event – or implicitly, in the programs, counters, and analysis programs that separated wheat from chaff. Interestingly enough, the statutory convergence of LBL and SLAC subgroups through PASS 2 then stopped. Indeed, after the establishment of the data, the two groups would go separate ways, and then only cross-check their final "physics results," as the Steering Committee members put it.[35] This internal quasi-independence was a nontrivial part of the group establishment of a persuasive argument, as had already been made clear a few weeks before ("It was pointed out by Goldhaber that if these [analysis] programs were not identical but did give the same physics results, it would greatly increase the confidence of the group that our results were correct."[36]) On 5 December that agreement became apparent: "Everyone was extremely gratified to know that the total cross sections as derived at the [SLAC and LBL] laboratories agree to within about 15%, well within the 30% errors that will be assigned to these cross sections in preliminary presentations to the physicists at the two labs." [37]

Even with these cross sections in hand, it was still not entirely clear what they meant to theorists. I put it this way rather than "what they meant theoretically" because theory had already entered, virtually every step of the way. But whether the experimental results could be aligned with *specific models* of the elementary particles remained to be seen. Could, for example, the data be found compatible with the parton model? Or vector dominance? These were theorists' theories, not experimentalists' theories, so to speak.

To find common ground with the SLAC theorists, the collaboration scheduled a meeting for Monday, 12 November 1973, in SLAC's "Green Room." "Subject," the memo read: "What does Inclusive Include? Performers: Various Theorists." [38] This was contact within the laboratory, though outside the group. Typical of internal communication of this sort was a 1972 memo from bj (James D. Bjorken) and Helen Quinn to Richter and Gerry Fischer, where the theorists took various theoretical ideas (such as Weinberg's Z^0 particle, Georgi and Glashow's $J = 0$ particle, or Ne'eman's "fifth force") and calculated the likelihood and signature of neutral resonant states that SPEAR might detect. [39]

A different and sociologically deeper divide had to be crossed with the decision to take results outside the participating laboratories. Indeed, one of the most crucial elements of any collaboration is the boundary crossed when results become "public." Public goes in scare marks because in a world of large collaborations, replete with collaboration meetings, informal talks, e-mail, conference proceedings, faxes, and published "physics" letters, it is problematically definitional just where the private/public divide lies. Here the decision about when to publish was inseparably coupled to the standards of demonstration, standards that had to be set. In early August 1973, Goldhaber had made some rough estimation of the cross sections and wondered if it was worth making a statement at a forthcoming meeting. Larsen wrote Richter on 3 August 1973, horrified at the idea that any disclosure might be forthcoming: [40]

My position is (and I'd like to have it read verbatim): Until we are ready to quote cross sections at the 10% level, we say nothing to anyone, anytime, anyplace.... I think any premature statements are likely to be wrong; they would compromise the ultimate potential of the experiment and, as for the argument that many are awaiting the results, I don't think any theoretician worthy of the title is awaiting any more factor-of-two or is-consistent-with statements.

According to collaboration minutes of 10 August, a "spirited" discussion about disclosure then followed. "A strong case was made that no public statement of any kind be made until we can confidently quote σ_{total} to 10%. Anything less convincing is to be publicly 'denied' to exist. An attempt was made to lower the standard to 20%, but this was quashed."[41] Such discussions were not new. Already in the days of the bubble chamber, "spirited discussions" reigned, as participants struggled over the significance of a bump in the mass plot or teams divided over whether a bump was more like a two-humped bactrian or a simple, dromedarian curve.

The first and most important physics problem for the Mark I collaboration was to march SPEAR through its energy range from 2.4 GeV to 5.0 GeV at 200 MeV intervals and to measure the likelihood that electron–positron interactions would occur at each energy. More specifically, the team wanted to see how the ratio R of two crucial quantities would vary with energy. The numerator of R is the rate of hadron production (strongly interacting particles such as the proton or the pion) in e^+e^- annihilation; the denominator is the rate of production of muons in e^+e^- annihilation. The relatively recent parton model (which represented hadrons as composed of essentially noninteracting pion constituents) militated for an R that would be constant with energy; other, older theories suggested declining values of R as a function of energy.

The first half of 1974 was occupied with understanding the new results. Here and there physicists in the collaboration pointed to oddities in the data; some quickly dissolved, others persisted. John Kadyk at Berkeley had scanned the data in January 1974 and found an anomalous 30% excess of hadron events (defined as events that appeared to have three or more prongs) at 3.2 GeV. Preoccupied as the physicists were with other matters, the excess receded into the background. Then, in June 1974, Martin Breidenbach from SLAC reopened the case and gathered more data at 3.1, 3.2, and 3.3 GeV.[42]

Nothing showed up, and the reason is revealing. As part of an effort to understand the production and energy of gamma-ray photons, a "converter" (essentially a thin steel cylindrical can) was inserted over the interaction region of the beam pipe where electrons hit positrons.[43] When a photon emerged from the collision, it would hit the side of the "can" and convert into an electron–positron pair that could then be detected in the Mark I. To compensate for this increase in electron–positron pair production, the analysis program was modified to take pairs of oppositely charged particles, and classify them as electron–positron pairs (the so-called ECODE 3 events) so they would not be confused with bona fide hadrons (ECODE 5 events).[44] The gamma measurement done, the can was removed, but its compensatory software remained in place. From that moment on, any pair of hadrons produced in the annihilation was reclassified by the computer as an electron–positron pair. Silently, in the heart of the computer, ECODE 5 events became ECODE 3 events. It became *impossible* to observe the production of any hadron pair at 3.1 GeV. Should any evidence for a resonance have arisen, the computer would have instantly killed it.[45] (Shortly afterwards, the conversion computer code was corrected.)

Roy Schwitters, one of the SLAC physicists, took on the task of drafting the "total cross-section paper," and, working with Chinowsky, Feldman, and Lynch, prepared a draft paper that was ready for collaboration critique on 5 July 1974. While numbers were still needed on detector efficiencies and other experimental errors, their conclusion was clear:[46]

> The total cross section is a rather smooth function of C[enter of] M[ass] energy over the range covered in this experiment. There is no strong evidence for resonance peaks or production thresholds. In strong contradiction to the predictions of asymptotic scale invariance, the cross section is essentially constant for C.M. energies between 3 GeV and 5 GeV. As yet, no generally satisfactory theoretical framework encompassing these results has emerged.

No "satisfactory theoretical framework" was putting it rather mildly. When the results were presented at the London Conference later in July, theorist John Ellis declared that "there is no consensus among theoreticians working on electron–positron annihilation, not even about such basic questions as ... whether or not to use parton ideas." The quantity R, Ellis concluded, could be variously deduced to be anywhere from 0.36 to 70,383. Go figure. Richter then presented the SLAC–LBL results which were "in violent disagreement" with the quark model. As Michael Riordan nicely put it, "experimenters thought Theory was pretty confused, and theorists – at least those who trafficked in gauge field theory – felt Experiment was the one befuddled."[47]

Beginning in July 1974 SPEAR was shut down for three months to ready it for higher energies. During that time, Schwitters and Scott Whitaker reanalyzed the data and found that measurements made at 1.6 GeV/beam (3.2 GeV total) indicated a 30% higher cross section and that the measurement at 1.55 (3.1 GeV total) also looked larger than expected. More peculiar yet, in mid-October, Schwitters returned to the logbooks to peruse an odd set of eight runs the team had conducted a few months before. On 29 June 1974, the notebook indicated that the energy scan had been boosted to 3.1 GeV. "Beam dumped, ready for 1.55 [GeV/beam = 3.1 GeV in the center of mass] YOU MAY NO-TICE," the shift inscribed in self-congratulation, "WE ARE EXACTLY ON TIME." "Unfortunate fill, tortuous beam configuration." Run 1381 failed miserably: "Aargh! Some Power Supply has developed a massive leak. Dump, end run." Run 1383 went fairly normally; run 1384: "luminosity is rather disappointing. The boys are studying the situation." So it progressed. Run 1387 "is miserable. Fill is beginning with .7 x 10*30 and we expected [apx] 1.2 x 10*30. Furthermore lifetime only about 1 hour. Furthermore it takes a full hour to fill. This STINKS. Our lead is gone and we sink into the morass."[48] A few hours later: 1389 "CONDI-TIONS STABLE ... KEEP ON TRUCKIN!," and at 14:10 on 30 June, the 1.55 run came to an end: "STOP Run 1389 with 171 μ's logged in at 1.55; on to 1.60. With time precious and running uncertain we defer a background run at each energy. We will try to make a complete scan at 1.6 [GeV/beam]."[49]

Remarkably and inexplicably, two of the eight usable runs (1380 and 1383) at 1.55 GeV yielded an excessive number of hadrons: 1383 yielded a full five times the harvest of hadrons the team expected.[50] On 22 October, Schwitters asked Goldhaber and Abrams to look again at these odd runs.[51] Partly because of the excess hadrons, and in part simply to clean up what had been rather evidently a problematic set of measurements, Goldhaber began on Monday, 4 November 1974, to lobby hard for revisiting the odd energy region around 3.1 GeV. If there was something going on near 1.55 GeV, the varying results from earlier runs suggested they might need all the resolution in energy they could get – Abrams and Goldhaber pressed Richter to know just how precisely the accelerator energies were known. There were, however, counter-currents. Other physicists, including the senior Stanfod experimentalist Bob Hofstadter, wanted to push ahead in the unexplored new regions of energy in order to test the validity of quantum electrodyanimcs. Since Richter was both running the Mark I and in charge of the accelerator he could not easily set aside Hofstadter's program.

Finally, persuaded by a seeming excess of strange particles that Goldhaber and Whitaker had noticed on the microfiches, Richter relented and the machine began taking data near 3.1 GeV.[52]

In the predawn hours of 9 November 1974, new data began to come in. Harvey Lynch, watching the events as they crossed the CRT, put pen to logbook:[53]

> The man was tired, for he had diligently worked the area for weeks. He stooped low over the pan at the creek and saw two small glittering yellow lumps. "Eureka!" he cried, and stood up to examine the pan's content more carefully. Others rushed to see, and in the confusion the pan and its content fell into the creek. Were those lumps gold or pyrite? He began to sift through the silt once again.

Sifting silt meant first of all establishing that the machine was functioning correctly, functioning for known regions of energy (such as 2.4 GeV) as it had previously.

> I. Our first priority is to be sure that the detector is alive, and that the normal analysis program functions properly on the triplex [the three-central-processing-unit computer used to take data]. We should log $\geq$ 100 μ pair equivalents at a beam energy of 2.4 GeV. These results should reproduce our previous result of [the total cross section] $\sigma \approx 18$ nb, with a detection efficiency of 0.63.
>
> II. Having completed the "checkout" phase we can begin the energy scan from 1.5 GeV to 1.6 GeV in 0.01 GeV steps.

This scan would take roughly three hours per energy step, and Lynch expected roughly ten events as a baseline, 40 or so if there was a "good 'bump'." For each run, the team would plot the ratio of ECODE 5 events (hadrons) divided by the number of Bhabas (elastic electron–positron scattering events), indicating the luminosity (to normalize the hadron production to the total number of electron–positron annihilations). Events would pop up on the CRT screens, one by one.[54]

At 8:00 on the morning of 9 November 1974, the log records:

> Filled ring with $E_0 = 1.56$ GeV. Watch data on one-event display. The table below gives the result of this hand scan. A total of 22 hadrons candidates were found along with 37 Bhaba events.... If all this makes sense this means a *trigger* cross section of $\approx$72 nb! Now if the signal just 'disappears' when we run at 1.50 we will be happy.

After a hand tally, Ewan Paterson and Roy Schwitters formalized their observation, "We the undersigned certify that J[ohn] S[cott] W[hitaker]'s above count to be a valid representation of the data."[55] Back to checking, cosmic rays, the shower counter efficiency still "looks o.k." Then, at 15:40 on the 9th, "Frustration! We have had no colliding beams since the 1.56 GeV beams went away. Is there no hope?"[56] More hand scans, back

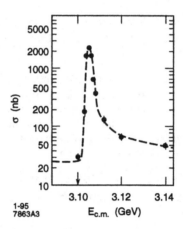

Fig. 18.2. The spike at 3.1 GeV.

to normal levels of hadron production away from the mystery region of 1.56 GeV.

At 1:47 on the morning of 10 November, the crew completed the 1.56 GeV run, and moved on to 1.57 GeV. At 10:05 that morning, as the crew tried to zero in even closer to the peak of this new resonance, they set up run 1460 at 1.555 GeV, a sequence that ended 56 minutes later. Schwitters scribbled:[57]

> This past fill has been incredible. While running 1.55 we saw essentially the baseline value of σ_T [the total cross section]. During the middle of the fill, we bumped the energy to 1.555 and the events starting [sic] pouring in. The visual scan had 61 hadrons in 87 Bhabas. This is a remarkable resonance indeed!

On 13 November, the collaboration submitted its article, "Discovery of a Narrow Resonance in e^+e^- Annihilation," to *Physical Review Letters* (see Fig. 18.2). The word "charm" never appears; perhaps the closest to it is the widely cast remark at the end of the article: "It is difficult to understand how, without involving new quantum numbers or selection rules, a resonance in this state which decays to hadrons could be so narrow."[58]

Charm did, of course, appear elsewhere in the physics community. As Andrew Pickering has nicely shown, there were several alternative explanations that prospered within the theory community, all vying for pride of place in explaining the new peak. For the charm theorists, the psi was a bound state of two quarks, a charm quark and anti-charm quark. What allowed calculations to proceed was the radical contention by David Gross and Frank Wilczek, and by David Politzer and Tom Applequist, that the force tying quarks together got weak at small distances. This doctrine – asymptotic freedom – was a fundamental part of what the gauge theorists meant when they referred to the psi as a bound quark–antiquark pair.

Not so the experimentalists.

Conclusion: no raw data

Physicists have come to speak of the events beginning in the second week of November 1974 as the November Revolution. Half tongue-in-cheek, the allusion to the storming of the Winter Palace nearly six decades earlier evokes a break, a radical discontinuity in physics that accompanied the discovery of the J/ψ. In part, this language of a gap fit the theorists' image of their subculture at a time of uncertainty. It is reflected in talk of an "R-crisis," and the heated ways in which theory was discussed. Theorist John Ellis, for example, referred in the summer of 1974 to the shocking blow inflicted by the experimentalists at SPEAR and elsewhere to theorists, who had been "almost unanimous" in their expectation that R would be constant. The nonconstancy of R, in particular its rise with energy, was nothing short of a "theoretical debacle,"[59] "R crisis," or mini-R crisis.

Such crisis talk, made popular in the 1970s by the oil industry's "energy crisis," permeated the world of high-energy theory. And indeed, there seems to be little reason to doubt that the theorists experienced an inability to account for experimental results. What is problematic is the extension of such a break to the instrument makers and experimentalists. To my knowledge there is not a single reference, among experimentalists, to a crisis that refers to the reliability or efficacy of the methods of their subcultures. By the early days of November 1974, the accelerator at SPEAR had been functioning for a year and a half. The Mark I detector had a year-long track record, including a series of conference papers on the total cross section that had flown in the face of prevailing theoretical work in particle theory. Did judgment and theory enter into the certification of the detector and its functioning? Of course: from the first pretrigger filter to the hand scans on the CRT. Does this mean that the prediction of charm and the associated development of the "new physics" of quantum chromodynamics played a crucial or even a significant role in the spike of Fig. 18.2? Absolutely not.

In his insightful book, *Constructing Quarks*, Andrew Pickering argues along very different lines. It is an intriguing interpretation, one worth pursuing not just for its own sake, but because it nicely illustrates a line of reasoning grounded firmly in the highly influential antipositivist philosophy of science that began in the early 1960s.[60]

> Monitoring the beam energy very precisely, the experimenters obtained curves like that [of the curve in Fig. 18.2]. These they regarded as manifestations of a genuine phenomenon, and accordingly put aside their worries about detector performance. They had not proved that the detector was working perfectly; they assumed this because it produced credible evidence.

What characterized this credible evidence? It became credible just because it fit the preexisting theoretical framework: "theory is the means of conceptualization of natural phenomena, and provides the framework in which empirical facts are stabilised."[61] Adopting Kuhn's argument that "each theory would appear tenable in its own phenomenal domain, but false or irrelevant outside it,"[62] Pickering then turns to his example of the new physics and old physics: "Each phenomenological world was, then, part of a self-contained, self-referential package of theoretical and experimental practice. To attempt to choose between old- and new-physics theories on the basis of a common set of phenomena was impossible: the theories were integral parts of different worlds, and they were incommensurable."[63] It is exactly the point of this paper that the practices of colliding-beam physics at Mark I were *not* uniquely part of the "self-contained, self-referential package" either of the old or the new physics.

Pickering's argument, as I understand it, falls into two parts, each a general thesis and a specific claim about the psi:

1a. Generally, an experimenter's "theoretical construct" precedes and picks out a set of experimental results by means of "tuning" – experimenters adjust their techniques "according to their success in displaying phenomena of interest."[64]

1b. Specifically, the assumption of charm led to a theoretical prediction about the value of R, and this result preceded and determined the experimental result (the narrow resonance at 3.1 GeV depicted in Fig. 18.2): "the discovery of the psi can be seen as an instance of the 'tuning' of experimental techniques ... to credible phenomena."[65]

2a. Generally, experimenters do not establish the proper functioning of their instruments prior to certifying an effect; credibility of the instruments is an outcome of finding the effect that is expected.

2b. Specifically, the SLAC–LBL collaboration did not "prove" that their detector was functioning properly ("working perfectly"); they assumed this because they found the resonance at 3.1 GeV that they expected.[66]

Why are platforms 1a,b and 2a,b important? First, they serve to undermine a view (which Pickering calls the Scientist's Account) that takes theory to be the inductive limit of a series of prior experimental observations. Second, and more importantly, if experiments are tuned to theory and theory suffers discontinuities, then the picture of physics is rent all the way down through experiment itself, and the "world" splits into two incommensurable parts. "The old and new physics constituted, in Kuhn's sense, distinct and disjoint worlds."[67] Philosophically, this amounts to a case for incommensurability; historically it is predicated on a block periodization; and sociologically it is the statement that the cultures of experimentation and theory are sufficiently interwoven to function as a single, nonconfrontational "symbiotically" bound conceptual scheme or paradigm.

The evidence for 1b is that in mid-October 1974, Goldhaber took the excess number of kaons as evidence for an interesting new phenomenon, possibly charm, at 3.1 GeV. Goldhaber's view, however, was by no means universally accepted; I cannot find a single endorsement of the charm hypothesis in any published or unpublished document prior to 11 November 1974. Stranger still, even in the days after 11 November such traces do not appear. Of course, it is possible that the historical rather than interview emphasis given here systematically omits material that would be more evident from oral histories. Here again, however, in a systematic set of interviews with the main actors in the Mark I collaboration, I see little support for the "tuning" of data to a preestablished conviction that charm existed. Schwitters put it this way:[68]

Charm for us never really meant [narrow] resonances. We didn't really think about it – it wasn't in our psyche as I say ("We pronounced it that weekend ... sort of joking with people ... the phi-c's {feces}.") I didn't even really understand [charm] until that Gaillard, [Lee,] and Rossner paper about the strangeness-changing currents. I really wasn't that up on those things. I think that's a fair description of how most of us viewed it. It was much more of an experimental issue of getting in there and understanding things experimentally and then being somewhat intrigued by the possibility that there was something really strange with a constant total cross section. Of course it sounds like such nonsense now – it's embarrassing to mention it! But one does get into that kind of mindset.

Breidenbach recalled that his and his colleagues' main concern in going back to the 3.1 GeV region was to clean up their data: was something wrong with their procedures? "There was never the feeling that this [narrow resonance] was just what the charm theorists ordered. It was nothing like that."[69] Whitaker: "We didn't know this had anything to do with charm until after we'd established the weird behavior and the theorists said o.k. you've discovered charmonium."[70] To Chinowsky, Schwitters' plot of mid-October seemed downright impossible: "Roy showed me this plot, he said 'Oh, look what I did,' ... we looked at each other and I said, 'You know this can't be happening,' and he said 'I know it can't [be] happening.' I said, 'What should we do?' He said, 'We'll run [again at that energy]'."[71]

Downplaying the importance of the charm hypothesis in November 1974, Lynch added this: "Charm was not part of my thinking, not the slightest. We didn't pay a lot of attention to that – we were an iconoclastic bunch. As you know, the total cross section we had found [earlier] was so completely different from what everyone was saying that we didn't pay attention to the theorists." As for the hours following 10:05 on 10 November 1974, Lynch had this to say:[72]

> It was a qualitative difference. You could simply watch them on the CRT. They were very fast and very clean: click, click, click, every few seconds. Pief was in the control room pacing back and forth saying "Oh my God, oh my God"; he was pounding his head in his hands: "I hope we're not making a mistake. I hope we are not making a mistake." Finally, he said, "there's no mistake." You looked and there was simply no question that it was real. We had no idea what it was, but it was real. That's how we all felt.

Even *after* the enormous resonance emerged, Lynch remained dubious about charm. Similarly, Rees, Fryberger, and other physicists on the experiment had essentially no stake in the charm hypothesis before the narrow resonance was established on the days after 9 November 1974.

The underlying difficulty in the "framework" analysis is the binary opposition between an algorithmic move to "prove" that the detector was "working perfectly" and a theory-laden "tuning" of the experiment on the basis of a "theoretical construct." This sort of dichotomy, characteristic of both positivist and antipositivist philosophies of science, has obscured a richer, subtler spectrum of registers in which experimental argumentation proceeds.

Discussions about experiment/theory relations typically contrast "observation" and "theory," and the heart-penetrating intrusion of theory into experiment. One can sympathize with the antipositivist impatience with the putative opposite of this view, the long-discarded doctrine that there are observations that are "independent" of all theory. In the extreme limit lies the notion that observations lead to theory through an inexorable inductive sequence. Here, the more recent sociology of science frequently has drawn on the venerable work of Kuhn, Hanson, Hesse, Feyerabend, and others, who blasted the naiveté of such a picture of cumulative data codified into a whole.[73] In doing so the antipositivists stressed the the psychological and epistemic shaping process that occurred as data was melded by "conceptual frameworks" into self-reinforcing wholes.

I do not see, however, how to even start an analysis of an experiment like the Mark I with such a dimensionless notion of "observation." Like its cousins "raw data" and "theory," the dichotomy between theory and observation obscures the central phenomenon of interest. Such a vocabulary suggests a set of untrammeled data and a more or less coherent theoretical agenda that has the propensity to shape and sustain the data. On the old positivist view, the data were hard and theory ephemeral. On the view espoused by the antipositivists, the data are "tunable" and the theory powerful and controlling. Neither seems adequate.

Data are always already interpreted. But "interpreted" does not mean shaped by a governing high-level theory such as a gauged quantum field theory. The notion that quantum field theories led to asymptotic freedom, and that asymptotic freedom coupled with charm dictated the phenomenon that precipitated a tuning of the Mark I, flies in the face of the continuity of experimental and instrumental practices. Data are interpreted in the PASS 3 visual scan tape; they are already interpreted in the event identification portions of the PASS 2 filter program. So we go back further in search of the raw, but the data are still not in that Edenic state in the early sections of PASS 2, where the program minimizes the least-squared reconstruction of the trajectory helix. So perhaps the raw data were untouched one stage earlier. But here still we find the space-points adjusted, discarded, confirmed in the softly grinding machinery of the PASS 1 filter. Back up again. We find the beam trigger counters excluding cosmic ray events, calculating flux, checking the timing of hits in the outer detector relative to the electron–positron annihilation.

There are no original, pure, and unblemished data. Instead, there are judgments, some embodied in the hard-wired machinery, some delicately encoded into the software. Some judgments enter by way of scanners and physicists peering over CRTs at event displays, others at histograms, and microfiche reproductions. Interpretation and judgments go all the way down.

What follows? One might conclude from the saturation of interpretation that there are no bedrock data, that the physics conclusions drawn from the experiment rest on a bottomless and shifting sea of sand. This seems to me backwards. It is rather the picture of rigid observation and arbitrary interpretation that lies behind a whole class of claims for relativism. What we see here is more like the intercalated picture of partial continuities and discontinuities.[74] Take the spark chamber. It draws its certification, if you will, from a long sequence of device types running back at least as far as the Geiger–Müller counter and up through a long sequence of instruments including Conversi and Gozzini's radar flash tubes, Fukui and Myamoto's discharge chamber, and the massive neutrino detector of Schwartz, Steinberger, and Lederman; in the long-term legitimacy of this tradition, the specific chamber of the Mark I found its general justification. What licenses the PASS 1 filter program to interpolate a "missing" point in a track when one spark chamber fails to fire? It is underwritten by a mesh of processes working simultaneously at a multitude of time scales. At the longest lies the physicists' commitment to the logic tradition's espousal of statistical argumentation: PASS 1 simply does automatically what Bothe and Kolhörster had done self-consciously and dramatically in 1932. In both cases reasoning follows probabilistic lines, from the unlikelihood of independent hits to the conclusion that a single particle had passed through each chamber.[75]

It might be useful to think of the role played by the espousal of the logic tradition as but one in a series of certifications: Validation of the spark chamber rides on the genus of the logic device, on the species of the magnetostrictive chamber, and on the individual instrument (the Mark I). Each level has its own set of certifying processes, and their starting points may be intercalated in complex ways. So to ask the question: Why accept the results of a magnetostrictive wire spark chamber? The answer reaches back in all these directions at once. What sustains commitment to the psi measurement is a multiplication of such commitments across the different subsystems: pipe counters, scintillators, muon chambers. Then there are the myriad cross-talk and coordinative moves that bind the detector into a whole. Does this intercalated history of certification render the whole immune to any possible skeptical challenge? Of course not. As I have stressed, vesting reliability in an instrument does not always come from any single mode of reasoning. There are techniques supported by detailed theoretical justifications, there are techniques adopted without any theoretical understanding. There are pieces of machine bodily lifted from one machine and planted lock, stock, and barrel in another, there are technologies that emerged for this particular machine in this particular application. And importantly, as we have seen briefly here, there is the powerful concatenation of two distinct, often antithetical image and logic traditions, compacted with the computer, into a new and hybrid whole.

Notes

1 An extended version of this paper appears as chapter 6 of Peter Galisons, *Image and Logic: The Material Culture of Microphysics* (Chicago: University of Chicago Press, 1997), with permission of the University of Chicago Press. The contrast between the image and logic traditions is discussed, for example, in P. Galison, "Bubbles, Sparks, and the Postwar Laboratory," in L. Brown, M. Dresden, and L. Hoddeson, eds., *From Pions to Quarks: Particle Physics in the 1950s* (Cambridge: Cambridge University Press, 1989), pp. 213–51. I would like to thank G. Abrams, Adam Boyarski, Martin Breidenbach, William Chinowsky, David Fryberger, Gerson Goldhaber, Robert Hollebeek, J. A Kadyk, Harvey Lynch, Martin Perl, Wolfgang Panofsky, John Rees, Burton Richter, Roy Schwitters, and John Scott Whitaker for helpful discussions. I would also like to acknowledge the National Science Foundation Presidential Young Investigator Award and the American Institute of Physics for partial support of this work. In the following, MBP designates Martin Breidenbach papers (personal) and GGP designates Gerson Goldhaber papers (personal).

2 Invitation to shut down ceremonies, 16 November 1973, GGP, vol. 2.

3 My purpose here is to characterize the emerging nature of hybrid colliding beam detectors, and not to give a comprehensive history of the J/psi discovery which would surely give equal weight to the MIT/Brookhaven experiment led by Samuel Ting. For more on the latter, see M. Riordan, *Hunting of the Quark: A True Story of Modern Physics* (New York: Simon and Schuster, 1987), pp. 262–321; A. Pickering, *Constructing Quarks: A Sociaological History of Particle Physics* (Chicago: University of Chicago Press, 1984), pp. 258–273; and Ting, et al., "Discovery," *Adv. Exp. Phys.* vol. ε (1976), pp. 114–149.

4 I will, in a shocking, ahistorical gesture, call the "SPEAR Magnetic Detector Group Detector" simply the Mark I knowing full well that I sin along with those who brazenly refer to World War I before the second act. One further caveat: my purpose here is to characterize the emerging nature of hybrid colliding-beam detectors, and not to give a comprehensive history of the J/ψ discovery. For more on the

MIT–Brookhaven experiment led by Samuel Ting (as well as further discussion on the competition between the J and psi groups), see, inter alia, M. Riordan, *Hunting of the Quark* (1987), pp. 262–321; A. Pickering, *Constructing Quarks* (1984), pp. 258–73; and Samuel Ting, Gerson Goldhaber, and Burton Richter, "Discovery of Massive Neutral Vector Mesons," *Adv. Exp. Phys.* vol. ε (1976), pp. 114–49.

5 This is strictly true only for electron–positron colliders. In proton–antiproton collisions, for example, the quark–antiquark annihilation may have its center of mass moving with respect to the collider because these constituents are not at rest within the proton and antiproton. See chapters by K. Johnsen and B. Richter, this volume, for more complete discussions of colliding-beam machines and their physics. See also Elizabeth Paris, "The Building of the Stanford Positron–Electron Asymmetric Ring: How Science Happens," unpublished manuscript (1991), on file at the American Institute of Physics. She reviews work at CEA, ADONE, and elsewhere, while concentrating on SPEAR.

6 Burton Richter, interview by author, 18 March 1991, p. 34.

7 See P. Galison, "Bubble Chambers and the Experimental Workplace," in P. Achinstein and O. Hanaway, eds., *Observation, Experiment, and Hypothesis in Modern Physical Science* (Cambridge, Mass.: MIT Press, 1985).

8 Gerson Goldhaber, interview by author, 14 August 1991, pp. 10–11.

9 William Chinowsky, interview by author, 14 August 1991, p. 11.

10 Ibid.

11 For a discussion of the use of roads in the PASS 1 filter program used in primary data analysis, see R. Hollebeek, "Inclusive Momentum Distribution from Electron–Positron Annihilation at $\sqrt{s} = 3.0, 3.8, 4.8$ GeV," Ph.D. diss., University of California, Berkeley, 1975, pp. 21–22; also, Goldhaber, interview by author, 14 August 1991, p. 18. Richter: "This was going to be a 4π tracking detector, and I felt that the experience of the bubble-chamber people in data analysis would be extremely useful in designing the programs that were needed to turn these spark-chamber bits ... into physics." (Richter, interview by author, 18 March 1991.)

12 Goldhaber, interview by author, 14 August 1991, p. 19.

13 My former student, Jonathan Treitel, has made good use of the image/logic distinction in his dissertation, "A Structural Analysis of the History of Science: The Discovery of the Tau Lepton," Ph.D. diss., Stanford University, 1986, which was then published as an article, discussed below, "Confirmation with Technology: The Discovery of the Tau Lepton," *Centaurus* (1987), pp. 140–80.

14 The problem LBL had was that the delicate plastic scintillators had many microscopic scratches in their surfaces due to incorrect wiping to remove excess glue during fabrication. This altered the light-transmitting properties of the plastic, reducing the attenuation length for light transmission. See e.g., Hollebeek, "Inclusive," (Ph. D. diss., University of California, 1975), pp. 11–12. Indeed, as late as August 1973, the team debated replacing the plastic or trying to patch over the difficulty: "new scintillator vs. Johnson's Wax was debated, without decision," in Lynch to SP-1 Physicists, 10 August 1973, GGP, vol. 2. In the end, neither was done, Kadyk to author, 4 December 1995.

15 S. Whitaker to SPEAR Distribution, 5 December 1972, TN-191, "Rejuvenation of Pilot F Scintillator," MBP; also Goldhaber, Abrams, and Kadyk to author, 4 December 1995.

16 John Rees, interview by author, 21 June 1991; Roy Schwitters, interview by author, 17 June 1991, pp. 13–14.

17 Note that a new pipe counter was installed in 1974; see R. Larsen to SP-17 Experimenters, "New Pipe Counter for Magnetic Detector," 23 September 1974, MBP.

18 R. Hollebeek, "Inclusive," pp. 6–10; see also R. Schwitters to SPEAR Detector Distribution, "Inner Spark Chamber Configuration," 16 March 1972, and R. Schwitters to SPEAR Detector Distribution, "Inner Spark Chambers," 2 August 1972, both MBP.

19 R. Schwitters to SPEAR Detector Distribution, "Magnetostrictive Wand Orientation in the Magnetic Field of the SPEAR Magnetic Detector," 31 January 1972, GGP, vol. 2; see also R. Hollebeek, "Inclusive," p. 7.

20 J. Dakin to SPEAR Distribution, "μ Chamber Performance," Detector File, 14 August 1972, MBP; also J. Dakin to SPEAR, "Muon Chambers," 18 September 1973, GGP, vol. 2, describes the muon chambers with cross-sectional view.

21 This is the main thesis of work elsewhere; see P. Galison, *How Experiments End* (Chicago: University of Chicago Press, 1987), chapters 4, 5, and 6.

22 H. Lynch to SPEAR Physicists, "Computer Programming," 21 February 1973, MBP.

23 Ibid.

24 R. Larsen to SP-1, -2 Experimenters, "Data Analysis," 8 November 1973, MBP.

25 Ibid.

26 Ibid. According to Goldhaber, in the end, Perl and Feldman joined the Muon WSC group and Abrams joined the showers group. Goldhaber to author, 4 December 1995.

27 Ibid.

28 R. Hollebeek, "Inclusive," p. 21.

29 Ibid., pp. 22–49.

30 Ibid., p. 35.

31 Ibid., p. 38.

32 Sieh, Kadyk, Abrams, and Goldhaber, "Beginning SPEAR Scanning," TN-194, 27 November 1973, GGP, vol. 2.

33 Ibid.

34 Ibid.

35 B. Richter to Distribution, SP-1, -2, "Meeting of the Data Analysis Steering Committee of 4 December," 5 December 1973, GGP, vol. 2.

36 B. Richter to Distribution, SP-1, -2 Experimenters, "Meeting of the Data Analysis Steering Committee of 21 November," 26 November 1973, GGP, vol 2.

37 B. Richter to Distribution, SP-1, -2, "Meeting of the Data Analysis Steering Committee of 4 December."

38 B. Richter to SP-1, -2 Experimenters, "Special Seminar," 7 November 1973, GGP, vol. 2.

39 J. Bjorken and H. Quinn to B. Richter and G. Fischer, "Search for Neutral Resonant States at SPEAR," 14 June 1972, MBP.

40 [R. Larsen] to [B. Richter], "Miscellaneous Issues," 3 August 1973, in SPEAR FY 74 #1 "Miscellaneous memos," 1 July 1973 through 31 December 1973, B. Richter, Group C leader [91014 box 4], SLAC archives.

41 H. Lynch to SP-1 Physicists, "Minutes of Meeting of 10 August [1973]," GGP, vol.2.

42 G. Goldhaber, in Ting, Goldhaber, and Richter, "Discovery," *Adv. Exp. Phys.*, vol. ε, pp. 114–49, on p. 132.

43 On the converter, see R. Larsen to Distribution, "Inner Package for May–June 74 Cycle," 19 January 1974, GGP, vol. 2.

44 A list of the ECODEs can be found in R. Hollebeek, "Inclusive," Ph.D. diss., 1975, p. 47.

45 G. Feldman, interview by author, 21 January 1994; Schwitters, interview by author, 17 June 1991; and Goldhaber, in Ting, Goldhaber, and Richter, "Discovery," op. cit., p. 135.

46 R. Schwitters to SPEAR Friends, "Draft Paper," 5 July 1974, MBP.

47 M. Riordan, *Hunting* (1987), pp. 259, 261, quoting J. Ellis, "Theoretical Ideas About $e^+e^- \rightarrow$ Hadrons at High Energies," and B. Richter, "Plenary Report on $e^+e^- \rightarrow$ Hadrons," in *Proc. XVII Intl. Conf. on High Energy Physics* (1974), vol. IV, pp. 30, 37, 41, and 54.

48 SPEAR Logbook, 29 June 1974.

49 SPEAR Logbook, 30 June 1974.

50 G. Goldhaber, in Ting, Goldhaber, and Richter, "Discovery," p. 135.

51 Riordan, *Hunting* (1987), p. 272.

52 B. Richter, interview by author, 18 March 1991; and G. Goldhaber, interview by author, 14 August 1991.

53 SPEAR Logbook, 9 November 1974.

54 Ibid.

55 Ibid.

56 Ibid.

57 SPEAR Logbook, 10 November 1974.

58 J. E. Augustin, et al., "Discovery of a Narrow Resonance in e^+e^- Annihilation," *Phys. Rev. Lett. 33* (1974), pp. 1406–8, on p. 1408.

59 Ellis, "Theoretical Ideas," in *Proc. XVII Intl. Conf. on High Energy Physics* (1974), vol. IV, pp. 20–35, on p. 20, cited in Pickering, *Constructing Quarks*, p. 256.

60 Pickering, *Quarks*, p. 274.

61 Ibid., p. 407.

62 Ibid., p. 409.

63 Ibid., p. 411.

64 Ibid., p. 14.

65 Ibid., p. 273.

66 Ibid., p. 274.

67 Ibid., p. 409.

68 R. Schwitters, interview by author, 17 June 1991, p. 20.

69 M. Breidenbach, interview by author, 14 February 1994.

70 J. S. Whitaker, interview by author, 24 February 1994.

71 W. Chinowsky, interview by author, 24 February 1994.

72 H. Lynch, interview by author, 8 March 1994.

73 See P. Galison, "Context and Constraints," in J. Buchwald, ed., *Scientific Practice: Theories and Stories of Doing Physics* (Chicago: University of Chicago Press, 1995).

74 On the periodization question in physics, see P. Galison, "History, Philosophy, and the Central Metaphor," in *Science in Context 2* (1988), pp. 197–212; P. Galison, "Context and Constraints."
75 For a more complete discussion of the problem of raw data, see P. Galison, *Image and Logic*, chapter 6, "Iconoclasm and the New Icons."

19

Building Fermilab: A User's Paradise

ROBERT R. WILSON

Born Frontier, Wyoming, 1914; Ph.D., 1940 (physics), University of California at Berkeley; first Director of Fermi National Accelerator Laboratory and Professor Emeritus, Cornell University; high-energy physics (experimental).

ADRIENNE KOLB

Born New Orleans, Louisiana, 1950; B.A., 1972 (history) University of New Orleans; Fermilab Archivist.

Building Fermilab was a many-faceted endeavor; it had scientific, technical, aesthetic, social, architectural, political, conservationist, and humanistic aspects, all of which were interrelated.[1] Because the emphasis of this Symposium is on the history of science, I intend to highlight the scientific and technical aspects of the design and construction of the experimental facilities, but these other considerations were also important in building the experimental areas (Fig. 19.1).[2] Neither the experiments made at the laboratory, nor improvements such as the Tevatron, made under the aegis of succeeding Directors, will be discussed here.[3]

Before becoming director of Fermilab in 1967, I had been a trustee of URA since its formation in 1965.[4] This experience had sensitized me to the growing number of particle physicists throughout the country who, with no accelerator at their home universities, had become dependent on sharing the use of larger accelerators constructed at national laboratories. It was they who started the revolt against the benevolent rule typified by the University of California's Radiation Laboratory at Berkeley and (on a smaller scale) by my own institution, Cornell University. In 1963 that arch-user Leon Lederman expressed the community's sentiments of wanting the next lab to be accessible by right for all users, that they would have a strong voice in decisions on what was built and how facilities were used, and that it would be a place where they would be "at home and loved."[5] On becoming Director, I was determined that

Fig. 19.1. A view from the north of the experimental areas, with the Neutrino Area in the left foreground, the Meson Area to the center right, and the Proton Area in the center left beneath the grove of trees where our buffalo roam. The Main Ring of the accelerator is at the top of this 1977 photograph.

Fermilab should become just that, a "User's Paradise." Easier said than done.

The above was implicit in the sentiments of the URA. It was also implicitly understood that we at the new laboratory would not set up fiefdoms of research under strong in-house physicists, such as the bubble chamber group under Luis Alvarez at LBL or some of the research groups at Brookhaven National Laboratory (BNL) on Long Island had been. This presented a serious problem, for if users were to make best use of the Laboratory, then they would need the assistance of a core of good Fermilab physicists, not only to set up facilities but also to provide the laboratory's help on the experiments being done. Our first estimate of the optimal fraction of Fermilab physicists participating in this work was that it would be about one-fourth.

In order to attract the best physicists to carry out that fraction of the experimental work, we decided to set up a Physics Department that we hoped would be equivalent in quality to that of a strong university. We would also promise that each research physicist hired could use up to 50% of his or her time doing undirected research.

Promises, promises; my conscience is still troubled. There were many physicists who, in the press of building facilities, never got the chance to do very much undirected research, if any. We did not keep books about who did what, but there was an unspoken agreement that if someone worked one full year on a facility, then the next full year could be spent on undirected research.

We never had to define full-time or part-time. Most of the physicists sorted it out for themselves. Some had a natural talent for administration, some did not. We tried to allow both to flourish. Those with administrative talent were of course much in demand in a laboratory just being built. Many of them managed to accomplish their laboratory tasks and do good research at the same time. How otherwise could they have been good physics administrators? In any case, I am exceedingly proud of that group of superb physicists who were "corrupted" into building Fermilab and making it work – but it was the search for new knowledge that motivated their efforts, just as it was for the visiting physicists.

The Berkeley design

The most immediate concern of our prospective users, apart from the accelerator itself, was the adequacy and the relevance of the experimental facilities that we were to design and then build at Fermilab. These facilities were originally specified in the 1965 Design Study of the 200-BeV accelerator made at the University of California's Lawrence Laboratory at Berkeley under the direction of Edward J. Lofgren.[6] It will be referred to here as the Berkeley Design. The people at Berkeley had done a superb job of bringing together all sorts of potential users to consult with them on the experimental areas, and their plans provided a solid foundation on which we could start the designs of our own facilities.[7]

A reduced-scope plan

It was not helpful that the scope of the Berkeley Design had been reduced in 1966 by fiat (AEC fiat via the Bureau of the Budget) to decrease the estimated construction cost from about $340 million to $250 million. Specifically, the number of experimental target stations were to be reduced from five to three, the designed proton intensity was to be reduced by a factor of 3, and a large bubble chamber (the principal method of event analysis of those days) was to be completely eliminated. The $250 million estimated cost of this "reduced-scope" project was the amount decided upon by Congress as they authorized funds for design studies in 1967. At the same time they challenged physicists to do better – but without extra money. As I was chosen to be the Director, it was "writ in blood" that we, at the very least, come up to that reduced-scope standard. Anything beyond that had to be done within the original $250 million – and not "one penny more."

Dark thoughts in the night

Of course before accepting the directorship of the project, I had thought long and hard about reducing the costs. It would not be just a matter of getting some clever ideas about the accelerator; every aspect would have to be reexamined from the point of view of cost. The buildings, the utilities and especially the experimental facilities seemed to me to be overly expensive in the Berkeley plan, and since more than half the costs would go for such things, I was determined to look particularly hard at them for cost reductions. The trouble was that the success of the lab would depend critically on the quality of the experimental facilities.

Ned Goldwasser

It was a great day for me when Ned Goldwasser agreed to join the project as Deputy Director. I was especially pleased because Ned's experience was complementary to mine. Having worked as a user at BNL and the Argonne National Laboratory (ANL), he had hands-on experience with a bubble chamber and with producing the particle beam that led the protons to it. He had served on the Ramsey Panel, so he knew the problems and he knew the people of the proton physics community to whom we could turn for help. I could count on him to fill in many of the lacunae in my own experience, which had been with modest electron experiments. Ned brought much to the design and use of the experimental areas, but he also participated in every aspect of running the laboratory.

Summer and fall, 1967
Criteria for a new design

The project began on June 15, 1967, when a small but determined group met at an Oak Brook office building to start creating the new laboratory, but this time along austere lines.[8] Our first problem with the design was just that, but at the same time I felt confident that we could build to a proton energy of 400 GeV (maybe even 500 GeV), that we might also exceed the number of experimental areas of the Berkeley Design, and that we might even exceed the originally designed intensity of protons – all within the $250 million limitation. Of course we had no way of knowing how much of this was bravado and how much was real, nor how much time it would take to turn that Illinois cornfield into a sophisticated laboratory equivalent to the Berkeley Laboratory that had taken many tens of years to evolve.

There were many intangibles. Who would join the lab to do the job? How long would it take before we knew how much money we would need? How much money would we really get and at what rate? How many years would we require for the construction? When should we start the experimental areas? The answers to most of these questions were pretty straightforward. We would make very sure to deliver the reduced-scope laboratory within our budget. Money would not be spent on anything nonessential until we knew what we were doing. However, there was one deviation right from the start: we planned to keep our options open to exceed the reduced-scope in every respect. The big question was, by how much?

Even as we were designing the accelerator, we also perforce had to design the experimental areas, if only roughly, because the cost and design of the laboratory would depend crucially on the location and characteristics of those areas. As an example, one of the economies I had expected to make over the Berkeley Design was to have the proton beam extracted from the accelerator at only one position, and then to put tremendous care in the extraction efficiency. The reason for this was that in the Berkeley Design much of the cost had been due to the effects of the radioactivity due to the protons that were not extracted. For example, the radioactivity of the air in the tunnel, the production of nitric acid in the tunnel, the need for cumbersome and expensive equipment for handling radioactive magnets, and the long delays during a shutdown while the radiation level decayed, all would require measures that were far from inexpensive. The solution would be not to lose any protons. My dream was that each proton that left the injector would be made to travel benignly near the center of the vacuum tube. Should a wayward proton strike an object, the resulting radiation would be detected by one of a large array of external detectors and the beam orbit would be adjusted or scraped off so that the offending protons would not strike that object any more. At that time this appeared to be pure fantasy for, quite apart from the beam lost during the acceleration process, the extractors then in use had an efficiency of only about 60%.

Al Maschke

If I was not up to solving the extraction problem, Alfred Maschke was. He had joined the laboratory that summer, coming from BNL.[9] Maschke had a profound influence on all the technical phases of the accelerator and its concomitant experimental facilities. Soon he had invented a new way of cleanly extracting the proton beam from the accelerator for which he claimed an efficiency of 99%, or even greater. I must say that this was pooh-poohed by the experts at the other labs – those at CERN seemed even to be offended by such an extravagant claim. But eventually it turned out just as Al had said, and his invention must have saved millions of dollars from not having tremendous quantities of radioactivity deposited in the Main Ring.

Maschke was in charge of the Beam Transfer Division, which had the responsibility for transferring the proton beam from one accelerator to the next and then to the Switchyard, where the beam would be directed to targets in the various experimental areas. He planned and did this brilliantly. It was tragic for me when in 1971, Al, a feisty guy, left the lab following a serious disagreement with me. I hope he will eventually receive the recognition that is his due.

Jim Sanford

Jim Sanford came to the 1967 Summer Study also from BNL. He brought with him a wealth of experience in experimental areas. I suppose that Sanford was most influential of all in developing the concept of a single external beam that could be switched to a multiplicity of areas where experiments could be done. He worked literally day and night during the summer study with such intensity, and so single-mindedly, that when he returned to BNL, we knew that we should follow his plans. We also knew that we should try our best to recruit Jim to be a permanent member of the Fermilab staff. He returned in a few months as Associate Director in charge of our experimental areas and eventually to coordinate the experiments that were being done at Fermilab.[10] Jim was basically a very conservative person, and was exactly what we needed to balance my own cavalier approach to problems.

It would be unfair to ignore the valiant and valuable efforts of my colleagues who were also developing concepts of the experimental areas, both independently of Sanford or in parallel with him. I am thinking of Art Roberts, Lincoln Read (who had also been an important advisor to me before I became Director), Winslow Baker, James Walker, Timothy Toohig, Edward Blesser, Richard Carrigan, Frank Nezrick, and many others.

Tom Collins

Tom Collins, another Associate Director, had mostly to do with the accelerator, but in addition to being an outstanding expert on all phases of the accelerator, he also played a prominent role in building the experimental facilities. This was because Tom had a deep interest and competence in architecture and the architectural-engineering aspects of the lab. Thus it was one thing to decide on the positions and functions of the experimental areas, but it was quite another to design and construct the tunnels and buildings and bring the necessary utilities to them. Tom was master of all of these aspects of our work, and his participation was crucial to how well the experimental areas functioned and to how much they would cost. Furthermore, he ran a weekly meeting in which those aspects of the laboratory then being built would be reviewed with regard to cost and schedule.

I must emphasize how hazardous these facilities were. A beam of 400 GeV protons at an average intensity of 2×10^{13} per pulse, typical of what we hoped to have, has a daunting one megawatt of power. This can melt a piece of metal almost instantly. Incident on a target, it can make the equivalent of 200,000 grams of radium, whereas one gram would be a serious amount. Clearly we had to handle this fearsome force with great respect and dispose of the radioactivity with great caution. We would have to control access to dangerous parts of the facilities with absolute certainty. Thus we needed a radiation officer and assistants having essentially absolute power. Miguel Awschalom and his assistants, Dennis Theriot, Robert Shafer, Peter Gollen, Larry Coulson, and Andrew Van Ginneken, served among the first radiation protection group. That their plans were good and their vigilance keen is attested to by their excellent record of preventing human exposure to harmful radiation.

The Atomic Energy Commission

How did our safety guys know what to do? For one thing they had experience at other laboratories as well as special training. But our source of confidence in what we were doing was also due to the AEC. I know it is now trendy to bad-mouth the AEC and its ensuing agencies, but we did work closely with them – and did depend upon their expertise and skills about safety. In this respect I must mention K. C. Brooks, Fred Mattmueller, and Andy Mravca, the AEC representatives in residence at the lab.[11] It was no accident that Glenn Seaborg, Chairman of the AEC, and one of my fellow students at Berkeley, kept in close touch with what we were doing. A host of other concerned well-wishers worked equally hard for us at the AEC Headquarters in Washington. Andy asked my permission to attend the meetings at which our construction plans were developed. By being present (not at all usual at other labs), he could make sure that our plans were consistent with the AEC safety standards and requirements, rather than having to go over them seriatim, which was guaranteed to consume months of our time. It was a good deal, and as a result he was able to gain approvals for us from the Washington office within days rather than months. No wonder we held the AEC in such veneration, respect, and friendship – they were very much on our team, or even better, we were on their team!

DUSAF and Parke Rohrer

DUSAF, the architectural-engineering consortium, did the design and construction of all of the conventional structures and utilities.[12] This was differentiated from the accelerator, experimental, and other scientific equipment. Here we lucked out. DUSAF had done the LBL plans which, to my mind, had been far too expensive. Happily, the president of the joint venture, Colonel William Alexander, a man of obvious integrity, promised to provide us with the kind of services we desired and demanded. He named Mr. Parke Rohrer to be the manager in residence, and for that appointment I shall always be indebted to him. Parke was exactly what we needed. He soon demonstrated tremendous expertise in architectural engineering, tremendous experience in administration, and unsurpassed character and compassion. If I am using superlatives, they are absolutely necessary in any appraisal of this remarkable man. He responded to our need to save money and also to our desire to create a workable and beautiful laboratory.

Parke had absorbed the ideals and needs of the laboratory we were to build so well that, instead of setting up a construction division to check all the DUSAF drawings and designs, I simply appointed him to be our Associate Director of Construction. He took this very seriously. I do not know how he managed to serve two masters, but serve us he did to our complete satisfaction – and, I'm quite sure, to the satisfaction of DUSAF as well.

Engineers

From the beginning we had outstanding engineers at Fermilab. For example, Don Young brought with him from Madison, Wisconsin, one group of engineers consisting of Glenn Lee, John O'Meara, Maxwell Palmer, Norma Lau, and Russel Winje. They stayed close to the building of the Linac and were not available to the experimental facilities. To provide for that need and also for making believable cost estimates, I called on an old friend, colleague, and teacher from the U.C. Radiation Laboratory of the thirties whom I knew when I was a student there, Bill Brobeck. Bill is a master engineer; he is Mr. Accelerator personified. He had designed cyclotrons before World War II, calutrons during the war, and the Bevatron after the war. He was renowned as the most conservative estimator of costs in the world, so I knew that if he estimated our technical costs, he would be believed. More importantly, I believed his estimates too!

Heeding my call for help, he came to Illinois and set up a commercial engineering group, using a few key people from his Berkeley firm as a core group and complementing them with a group of local engineers. Bill did important engineering for us. For example, he designed power supplies and made cost estimates of technical components as well. In a sense he did for our technical components what DUSAF did for our conventional facilities. It was because of these two groups that we were able to be "off and running" so rapidly.*

This is how it worked. We would furiously (and I hoped imaginatively) design a particular component or system of the laboratory complex. Brobeck, working quite independently of us, would price it out. It would inevitably cost too much. Since it was taboo to argue with Bill, we would

---

* Of course we were busy recruiting our own corps of engineers. To name a few of those heroes, but not all, were Dick Cassel, Hank Hinterberger, Hans Kautsky, George Mulholland, and Wayne Nestander, in addition to the engineers brought by Don Young from Madison.

go back to the drawing board, hoping to make reductions by making inventions or by cutting more deeply at what fat we could find. This would then go to the cost estimators again; costing less but still costing too much. So this process went on, iteratively, until the cost was within our limit. Every so often we would add up the total expected cost to see how we were doing. Still too much, but always closer to our $250 million goal. Finally, in a few months, we did joyfully hit it – but would we be able to build it?

Theorists

The theorists, too, were important to us in building the experimental facilities, as well as in helping us decide which experiments to do. It was crucial to me to engender a sense of doing physics at the lab at the earliest possible time, a sense that we were doing more than just building an accelerator. It would be easy, out on that Illinois plain, to lose touch with physics, to forget who we were, and why we were there. Theory was something that could be started immediately, and that would give us a sense of doing real physics.

My first scheme was to call on my old friend Bob Serber to come once a week from Columbia University to lecture about the most recent physics or, if he could not come, to arrange for someone else to talk to us. This was in the tradition of Bob at Los Alamos, starting off with a series of lectures in 1943 about neutron physics, or with his "Serber Says" lectures at LBL after the war. Bob has the knack of speaking simply and understandably to experimentalists without patronizing them. It worked again for us at Fermilab.

When Serber tired of his weekly flights from New York, Ned thought we needed a more permanent solution, and he arranged to have Sam Treiman come out from Princeton for a sabbatical year to set up a continuing theory department. Sam enlisted a group of young theorists, all of whom have had outstanding subsequent careers.[13] After his stint Sam asked J. D. "Dave" Jackson to take a turn. Dave then triumphantly recruited Ben Lee (a theorist of exceptional ability) to join the lab as a regular member. We attracted a steady stream of distinguished theorist-visitors. These included Maurice Jacob, William Frazier and Chris Quigg (Quigg stayed on to head the theory department). Eventually Bill Bardeen became the head of this distinguished group and the tradition of excellence continues. The theorists not only brought style and learning to the lab, they also fulfilled their promise to

help with the choice of experiments. We have been proud and fortunate
to have had such a strong group.

The users organization

I have already mentioned a number of institutions, URA, AEC, DUSAF
and the JCAE, with which we interacted closely in building the research
facilities.[14] There were others that were also important. Under the
aegis of Goldwasser and Ramsey, the potential users of the laboratory
organized a Users Organization in 1968. Their first meeting of about
170 members was held at the groundbreaking ceremony on December
1, 1968. The organization reported not to Fermilab but directly to the
URA, consistent with the users' right to have their desired input into the
top management of Fermilab. Before Jim Sanford became a member of
the Laboratory staff, he was the first Chairman of this organization. This
group proved to be of substantial value in facilitating communications
between the users and us at Fermilab, especially with regard to the
research facilities then under consideration. Their Executive Committee
met not only with the URA trustees but directly with me and with other
members of the laboratory. It was an effective method, if occasionally
painful, for learning how short of our aspirations we frequently were.
Indeed the users were positive, if forthright, in their criticism, as well as
praise, for how else would we have known how we were doing? Every
year they organized a general meeting of users. These occasions provided
some of our best opportunities to speak directly to the users about our
mutual hopes and plans.

Physics Advisory Committee

The Fermilab Physics Advisory Committee was a hard-working committee organized by Ned in 1969–70.[15] We came to depend on it heavily
for advice on the research facilities and on the experiments and their
priorities. I appointed the committee members, of which three came
from the East coast, three from the Midwest, and three from the far
West, in order to have a geographical balance. To have an equal representation of the different kinds of physics, each set of three consisted of
one physicist who specialized in experiments that made use of electronic
counters, one who made use of bubble chambers, and one theoretical
physicist. They all had staggered terms of three years. They, as well
as the Users Committee, made recommendations for their replacements,

which I was careful to follow. It was all outrageously bureaucratic, but how else could we strive for a laboratory that would fulfill the Lederman dream of a Users' Paradise? Unfortunately, paradisios do not come cheap. It was about spending money that we necessarily had all too frequent disputes, for our funding was tantalizingly slow in coming.

Meeting the November 1967 deadline

It is hard now to remember the intensity of our labors to meet the November 1967 deadline that would allow us to complete the dread "Schedule 44," a detailed funding plan for the whole project that was required even to have funds for FY1968 authorized by Congress. Not only did we have to redesign the accelerator and the experimental areas, but the whole new design had to pass a URA review. This was held on October 12, 1967. We were also required to prepare and have printed and delivered by the first of January 1968 a complete Design Report for the benefit of the AEC and Congress. Had all this not been achieved, we would have suffered an automatic delay of one year in the project. No wonder we were absolutely ecstatic when we survived these, for this exercise gave us the confidence that we too might know what we were doing. Perhaps just as important was the fact that, in a project such as ours, time was money and a one year delay might have put our cost requirement out of reach.

Research areas, 1968–1969

Once the rough plans and costs of the experimental areas had been fixed and incorporated into our Design Report, our attention was focused on the accelerator. However, we did maintain a low level of design activity in the other areas. Our intention was to have as much input into our needs as possible from the physicists who would use the facilities.

This was rather successfully begun at a 1968 summer study at Aspen, Colorado.[16] About 75 users came to this meeting. Many of them were prepared with proposals or "letters of interest" for their experiments. There was by far more impassioned debate than mountain climbing and, alas, little fishing. The conclusion we reached from listening to the users was that they agreed with, or were neutral about, the single external beam concept. They thought, as we did, that the internal target area was not necessary. What many of them did care about was that we build a large bubble chamber. We returned to Chicago feeling that much

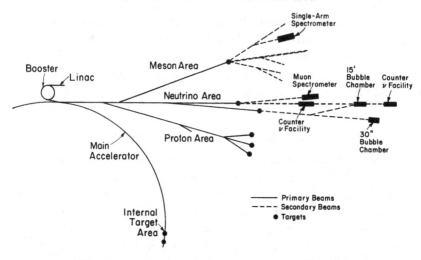

Fig. 19.2. Layout of the major experimental areas and beam lines at Fermilab.

had been learned from the experience, but that we had yet much to learn. The best part, perhaps, was that in Aspen we were removed from the hurly-burly of constant crises at the laboratory, and from endless telephone calls. This allowed us to concentrate on physics. It was not that we did not have many shorter meetings of users at Fermilab, but those were usually directed to some special end and did not have the breadth and depth of our Aspen meetings.

A three-area concept

Out of our deliberations, out of the recommendations of our users, we came to a consensus of what facilities we should build and when we should build them. This was not something we reached lightly. There was no point in building the best accelerator in the world – and we were trying to do just that – if the facilities were inadequate for the experiments our users would conduct; nor, by the same token, was there any point in building the most lavish experimental areas but an inadequate accelerator. No, the facilities had to match the accelerator, and both had to match what our users wanted – and what we both could afford to build and to use with our limited rate of funding.

As illustrated in Fig. 19.2, there would be three areas: a Meson Area, a Neutrino Area, and a Proton Area. These names were chosen to describe the general character, but not the exclusive character, of each of the areas. At first the areas had been given numbers instead of names, but I had a hard time remembering which was called what. The new titles were chosen not only to assist my feeble memory but also so that the names of the particles we were investigating would become familiar to the nontechnical people in the laboratory; I hoped that this might help to engender a sense of participation in the project by everyone working there. Indeed, we tried hard to infuse an understanding of what we were doing to everyone at the lab, and this did much to make an enthusiastic work force.

The Meson Area

The first experimental area to be constructed, the Meson Area (Fig. 19.3), was initially built to accept 200 GeV protons but with the potential of later raising the energy of the protons to 400 GeV. It would be primarily a facility to study secondary particles, such as mesons, that result when the 200 GeV protons from the accelerator strike a target.

The Meson Area lab building was a departure from those previously built at lower energy laboratories in that the whole area was not covered by one huge structure. Because of the higher energy at Fermilab, the range of the secondary particles, as well as the primary protons, would be vastly greater than heretofore, so a building extending from the proton target to the end of some of the envisaged experiments would require a distance of about one kilometer – prohibitively expensive. Instead we had one building, only a few hundred feet in breadth and length, in which targets and experiments would have the luxury of an overhead crane. Experiments extending beyond the building would be contained in corrugated metal tunnels that could be easily moved (but in practice seldom were) to correspond to the physical outlines of an experiment. The building itself, originating from my fevered brain, was a triumph of architecture (well, in my opinion), but it was something of a catastrophe from a practical point of view. I am ashamed to report that the users therein regarded it more as an Inferno than the Paradise I had hoped it to be. The roof, made of corrugated steel culvert plates, leaked seriously and continuously.

Early delays

Construction in the experimental areas was delayed because, just as it was going into the final stages of preparation for experiments, the

Fig. 19.3. The Meson Area, with its laboratory's distinctive roof, contains six particle beam lines.

accelerator went into a series of crises – primarily, many of the Main Ring magnets failed.[17] Two physicists, Rich Orr and Dick Lundy, heroically threw themselves and their comrades, who had been working on the experimental areas, into the breech (or was it the abyss?) to save the day. They were not the only experimental physicists who made this sacrifice and, with everyone on the project pulling together, the accelerator was indeed saved. Within a few months the accelerator belched out its first high-energy beam: 20 GeV in January 1972; 200 GeV in March 1972; and 400 GeV in December 1972. Then they all rushed back to bring

Fig. 19.4. The 15 foot bubble chamber sits at the end of the Neutrino Area. Adjacent is the Bubble Chamber Building, which is covered with a geodesic dome roof.

their respective experimental areas into operation for the patiently (??) waiting users.

But not all of the users were waiting. Consider Bob Walker and Alvin Tollestrup from Caltech experiment E-111 in the Meson Lab. With the Fermilab physicists still off in the accelerator abyss, they put their efforts into the messy work of bringing beams to targets and coping with the floods from my flawed design of the Meson Lab roof. By 1974 they brought their beautiful experiment about meson charge exchange to a successful conclusion.

Neutrino Area

The Neutrino Area, more than a mile in length, is directly in line with the direction of the extracted beam of protons. The bubble chamber was located near the end of the neutrino beam – about one mile from where the proton beam emerges from the accelerator – and protons as well could be brought from one end to the other. Although the Neutrino Area was primarily designed for the study of neutrinos, the muons that are made in the production of neutrinos were also extensively studied there.

A target train upon which proton targets and special magnets were mounted could be pushed into a tunnel in a long, high mound of earth. The pions and kaons that emerged from the proton target were formed into a nearly parallel beam that traveled the length of the 100 meter long evacuated decay pipe. The mesons decayed into neutrinos and muons, and some of these muons were deflected by magnets into the Muon Laboratory. At the end of the decay pipe, the neutrinos and other products of the proton collisions entered a long (800 m) absorption mound where all but the neutrinos were absorbed. Part-way down the mound was located a small "Wonder" building for low-energy neutrino experiments. At the far end of the mound the neutrinos emerged and passed through the 15 foot bubble chamber (Fig. 19.4) and then through a building where the neutrinos were detected by counters for further experiments. Off to the side of the mound a second, smaller (30 inch diameter) bubble chamber was located. High-energy protons could be led either to it or to the 15 foot bubble chamber.

After the intensity and energy of the protons from the accelerator had been increased, it was necessary to increase the absorptive power of the neutrino berm. Since we could not conveniently increase the length of the berm, we increased the average density by burying in it huge pieces of the aircraft carrier *USS Princeton*, recently decommissioned – swords into plowshares and all that. Pieces of the *Princeton*'s deck were used in a sculpture, "Broken Symmetry," located at the main entrance of Fermilab.

The Bubble Chambers

Many of the experimenters who would use Fermilab made it very clear to us at the first summer studies that bubble chambers would be highly desirable for research at Fermilab. Indeed the Berkeley Design Report had included some $60 million that, among other things, would provide for one 2 m^3 bubble chamber and one 100 m^3 bubble chamber, as well as for moving an unspecified, already constructed, large bubble chamber from another laboratory to the new site. Unfortunately, these plans had all been thrown out in arriving at the reduced-scope funds that were to be made available to us.

Of course, eliminating the funds did not eliminate the need. At the first Aspen Summer Study in 1968 there had been general agreement among the users that a 25 foot diameter bubble chamber would be required to do the job of research then anticipated. Until it could be built, a 30 inch chamber from Argonne would be used.[18] We had also hoped to get the large bubble chamber being built by a group at BNL, but naturally they had a strong desire to keep it for their own research, as they did.

A collaboration was soon formed between the Shutt group at BNL (which was just finishing a 7 foot chamber) and our physicists (Nezrick, et al.), to design a 25 foot diameter monster chamber. Despite heroic efforts by Ned Goldwasser and those at Brookhaven, the elegant project was turned down by the AEC, almost surely because it would cost some $15 million, which was just too much for any funds they had then.

I felt that perhaps a lesser sum of money, to be provided from the hard-to-come-by Fermilab construction funds, might be afforded. Goldwasser and Sanford, together with the designers of the 25 foot chamber, eventually arrived at a more affordable design, this time a 15 foot chamber estimated to cost about $7 million. However, in the rush to the new design, I made an obligation to the experimenters that a 15 foot chamber would indeed be built. It turned out that the design of the chamber, for economic reasons, had been reduced to 14 feet. At that time, in 1969, wanting especially to maintain my credibility with the users, I insisted that we stay with the 15 foot size. So as not to have to make a new design, it occurred to me that a small one-foot long conical extension placed at the front end of the chamber would keep the sensitive path length to 15 feet within the chamber. The protuberance was sometimes, and with *lèse majesté*, referred to as "Wilson's nose"!

Bill Fowler, who had led the 15 foot design, came from BNL in early 1970 to head up the construction team. Russ Huson followed about six months later. They recruited a formidable group to do the job, gathering up people like Frank Nezrick, Hans Kautsky, and Wes Smart. John Purcell and his group at the Argonne Lab built the superconducting magnet for the chamber, no small job. Peter Van der Arend and his cryogenic company were responsible for the cryogenics through to operation. Bob Watt and his colleagues at SLAC took on the rapid expansion of the chamber. George Mulholland took over commissioning and operations. Safety was of overriding importance, for after all, the bubble chamber, full of liquid hydrogen, was indeed inherently dangerous. We lucked out in that regard by having Paul Hernandez at LBL serve as my safety officer. Paul had been LBL's chief engineer of Luis Alvarez's 72 inch chamber. He and our own very capable safety group did a magnificent job; there were no accidents.

Andy Mravca has to be celebrated for performing his usual miracles in the AEC. What he did with his consummate mix of science and bureaucratic savvy was just as necessary for the construction of the bubble chamber as it had been for the accelerator.

The 15 foot bubble chamber was started near the beginning of 1970; it was commissioned in September 1973 (remarkably fast for designing, financing, and then building it), and it ran successfully until it was turned off in 1988. That occasion was celebrated by a "15-Foot Fest" at which many of the participants in its construction and operation were able to attend and give voice to poignant memories. Happily these have been gathered into a delightful volume.[19] Therein a cogent review is given by Charles Baltay of some 17 experiments done with various mixtures of hydrogen, deuterium, and neon. Paul Hernandez of LBL also paid a poetic compliment: "after 34 years of Bubble Chamber connections ... I see the 15-foot bubble chamber as the 'Jewel in the Crown.' " It was indeed a good operation.

The Proton Area: life in the pits of the pits

The Proton Area (Fig. 19.5) was the last to be commissioned; it was intended for experiments using protons at the highest energies and the highest intensities. The proton beam from the accelerator could be split there so that it, or any fraction of it, could be guided into any of three underground well-shielded pits of the Proton Area. The pits are named Proton West, Proton Center, and Proton East.

These enclosures are indeed rough-and-ready places. They had the reputation of being, not Paradisios, but rather Purgatorios. Indeed, some of the users were advised by their older colleagues to "abandon all hope, ye who enter here!" I fear that I bear the responsibility for this fiasco.[20] In a frenzy of saving big bucks, I had a fantasy of not putting up (or down) any laboratory building at all. Instead the idea was that, once an experiment had been accepted, an outline of the necessary space would be drawn in an empty field at the end of one of the proton beams, then steel interlocking piles would be driven along the outline down to the necessary depth to protect against radiation. Then the experimental equipment would be lowered to a luxurious graveled floor, and finally a removable steel roof would be covered with the requisite thickness of earth. Once the experiment was finished, the pilings were to be pulled

Fig. 19.5. A 1976 photograph of the Proton Area, which is divided into three subareas providing four particle beam lines.

up, the earth filled in, and then the next experiment would be ready to receive its tailor-made enclosure.

Simple and inexpensive, is it not? I still find it difficult to understand why those users all stopped speaking to me. It is true that there were a few flaws in my logic. The rivers of ground water that flowed through their experiments, the walls of piling rusting away, the impossible access, and all without benefit of toilet facilities. But some of the users had their finest moments down in those proton pits – the discovery of beauty, the bottom quark, where else?! Alas, as far as I know not one piling has been pulled up, not one pit has yet been refilled with earth. How is one to interpret this?

To redress some of the inadequacies of the users' trailers, not to mention their visual blight, the director personally designed a luxurious building, the Proton Pagoda, a double stairway (à la the Vatican), and even toilets were eventually installed – alas, all too little and too late.

Internal target area

The Berkeley Design Report included a rather elaborate internal target area. We did not like it because radiation from the target might contaminate the Main Ring and a separate laboratory building would cost too much. During the summer study, after some debate among potential users, they recommended that we abandon any such area. A few years later we decided that a very thin target would not add too much to the radiation problem, and so we designated the straight section of the accelerator at section C-1 as a possible position for an internal target area, but that any laboratory space there would have to be improvised in the regularly enlarged part of the tunnel at C-1. Actually, the first experiment at the laboratory was an international collaboration of Soviets (V. Nikitin, et al.) working with a group from Fermilab (E. Malamud, et al.) and from Rochester University (S. Olsen, et al.). The Russians had fabricated a gaseous jet of hydrogen that constituted a very thin target when it was fired through the circulating proton beam of the accelerator – the group measured p–p elastic scattering and initiated a continuing and fruitful collaboration with Soviet physicists. This culminated, along with the physics results, in a 1974 performance by the Bolshoi Ballet in our auditorium! It also culminated in a small extension to the tunnel at C-1 to provide a little extra underground space for experiments.

Nooks and crannies

I had a bad conscience for having set up such a formidable bureaucracy to ensure fairness and scientific merit in the acceptance of proposals for experiments, so I tried to improvise a supplemental system of no bureaucracy at all. In this scenario any reputable physicist who could find a vacant nook or cranny for a modest experiment could, without any "by-your-leave," just go ahead and do it. Well, there were obvious flaws in this approach to Nirvana, and soon it was abandoned.

General remarks

I have emphasized the role of the users and of committees of users in the management of the Laboratory, for that had been one of our devices to realize a laboratory where the users "would be in charge" as well as being "at home and loved." But there was an even higher criterion of success to which we, and they, were beholden: the quality and quantity of the physics done. Alas, there is no easy formula, no democratic procedure, that would necessarily ensure our meeting this criterion. There was always a dichotomy between those experimenters who wanted to use the accelerator immediately after attaining 200 GeV and those whose experiments required the highest energy. It was pretty much up to the Director to decide on the basis of his own intuition what energy could be reached within the available funds. Since most of the users had urgent obligations to their students as well as obligations to raise funds for their research, my proclivity to go to the highest energy did not win me many popularity contests.

The accelerator produced its first beam at the design energy of 200 GeV in March 1972, less than five years after we had come to Illinois. Almost immediately our experimental program began. By July an energy of 300 GeV was reached, and then in December it went up to 400 GeV.[21] During that same period the intensity of the beam went from some 10^9 protons per pulse of the synchrotron to about 5×10^{12} protons per pulse – still less than the design intensity by a factor of 10. It took another four years of hard work before the intensity had been pushed up to within a factor of 2 of what we had planned. By then, however, the proton beam was running regularly at 400 GeV, and could, sporadically, run briefly at 500 GeV.

Other factors than just the proton energy and intensity were of equal importance in doing successful experiments; for example, the rate of the pulses and the shape of the pulses in time. Reliability seemed the hardest of all to attain. Far too many times we had to explain to an exasperated group of experimenters who had come from the ends of the Earth that the machine was broken and would take a few days to fix.

As in any adventure where high-spirited people are involved, tempers would occasionally flare and shrill voices would fill the air. Even so, the common goal of producing good physics would soon restore calm. Perhaps an occasional shot of adrenaline helped speed us on our way.

In the beginning we wondered, "Would the users come?" Indeed, they did come – thousands of users, doing hundreds of experiments. Did they feel "at home and loved?" Loved they were by us – in our fashion – but it was not always so evident to the pitiable users. The more relevant question now is whether they were able to use the above experimental areas to do important physics. That is for someone else to say, but I am satisfied that they did. Even one discovery such as the upsilon particle – and there were others, too – made all that effort worthwhile.

Did we construct a foundation upon which those who followed could improve? Apparently the answer to this is also in the affirmative.

It must be emphasized, however, that it was the skill and innovations and dedication and cooperation and good humor and hard work of the Fermilab staff that created the accelerator and its concomitant experimental areas – and then made that infinitely complicated system of tens of thousands of individual subcomponents work together as one system – a *miracol mostrare* – for the use of the experimenters to perform *their* miracles.

Notes

1 From the origin of the laboratory in 1967 until it was dedicated in 1974, it was called the National Accelerator Laboratory (NAL). At the dedication it was renamed the Fermi National Accelerator Laboratory (FNAL). Largely because of my dislike of acronyms, I called it Fermilab, a name that has stuck to it. I shall use that name throughout for simplicity even though it was not called that throughout most of the time about which I am writing.

2 Various aspects of building Fermilab have been described in the 1987 twentieth anniversary issue of the Fermilab Annual Report.

3 Leon Lederman, "Tevatron," *Scientific American 264*, Vol. 3 (March 1991), pp. 48–55.

4 The pronoun "I" is used throughout because this chapter presents the perspective and reminiscences of R. R. Wilson, but its preparation was a joint endeavor between the two of us. Universities Research Association, Inc. (URA) was created in 1965 by the Council of Presidents (of about 46 universities). Norman F. Ramsey was the first President of URA and H. D. Smyth was Chairman of the Board of Trustees, which had 15 members elected by the Council of Presidents from each of 15 groups of neighboring institutions.

5 Leon Lederman of Columbia University, then serving on the Good Committee to comment on the Ramsey Panel's report, presented the paper, "The Truly National Laboratory (TNL)," on June 25, 1963, at

Brookhaven's Super-High-Energy Summer Study, Brookhaven Report No. BNL-AADD-6 (1963).

6 "200-BeV Accelerator Design Study 1965," Lawrence Berkeley Laboratory Tech. Rep. No. UCRL-16000. After the submission of the above report, there were summer studies held in 1965 and 1967 to further explore the ideas of future users. See "200-BeV Accelerator: Studies on Experimental Use, 1966 and 1967," Lawrence Berkeley Laboratory Tech. Rep. No. UCRL-16830.

7 Other people who worked on experimental areas were Dennis Keefe (nominally in charge of this aspect of the Design Study), Robert Ely, William Wenzel, Willim Gilbert, George Trilling, Tim Toohig, Robert Meuser, and of course many others.

8 We had already decided to build as the injector for the synchrotron a near-copy of the 200-MeV Linear Accelerator (Linac) then under construction at BNL. Donald Young, the first NAL employee on May 22, 1967, was in charge of our Linac construction. He brought a group from MURA, the Midwestern Universities Research Association, which was then closing down. Curtis Owen, Cyril D. Curtis, John O'Meara, Glenn Lee, and Maxwell Palmer were the principal members of that group. Soon Philip Livdahl came from ANL to help. Margaret Kasak also came from MURA to be my secretary temporarily. Priscilla Duffield replaced Margaret when she returned to her home in Madison. Priscilla, a no-nonsense super-secretary with invaluable experience in many physics projects, had worked with Lawrence and Oppenheimer. She personally knew many of the people with whom we would be dealing and played an invaluable role in organizing me, and indeed everyone else! Don Getz, Assistant Laboratory Director, and Don Poillon, Purchasing Agent, were there, as was Frank Cole, from the Berkeley project. There were people from the AEC and from DUSAF (an architectural-engineering design team of four companies) . We occupied the tenth floor of the Oak Brook high-rise office building from June 1967 until September 1968, when we relocated to the Village of Weston site. On December 1, 1968, we held the groundbreaking ceremony for the project.

9 Acronyms, acronyms! There I go again, but I shall continue to refer to Brookhaven National Laboratory as BNL, to Lawrence Berkeley Laboratory as LBL, to Argonne National Laboratory as ANL, and to Stanford Linear Accelerator Center as SLAC; otherwise many physicists will have no idea to what I am referring!

10 Jim Sanford has written a good description of the experimental areas in his article "The Fermi National Accelerator Laboratory," *Ann. Rev. Nucl. Sci. 26* (1976), pp. 151–98. He graciously consented to my request to excerpt from his article.

11 K. C. Brooks was a gift from heaven, or rather from Glenn Seaborg, who brought him back from retirement just for us. Casey had a lifetime of experience in the AEC Construction Division. He was a doer and a good friend. His secretary Minerva Sanders and his deputy Fred Mattmueller also kept the fires burning. John Erlewine, Director of Construction for the AEC at Washington, D.C., was also a great source of support.

12 DUSAF, another tiresome acronym, but more justifiable than most since it stands for: Daniel, Mann, Johnson & Mendenhall; Max O. Urbahn; Seelye, Stevenson, Value & Knecht, Inc.; and George A. Fuller Company

– a real mouthful. The latter company had built the Washington Monument! Tom Downs, George Mitchel, Allan Ryder , George Adams, George Doty, William Rowe, etc., were members of Parke Rohrer's team of architects.

13 These have included Henry Abarbanel, Martin Einhorn, Steven Ellis, David Gordon, Emmanuel Paschos, and Anthony Sanda.

14 Well, I should have already mentioned the Joint Committee on Atomic Energy (JCAE) of the Congress. I had to report to them every year. They determined how much money we, or more accurately the AEC, got. Their Executive Director, John T. Conway, was crucially important for us. I got to know him quite well and also liked and respected the Congressional members of the committee: John Pastore, Chet Hollifield, John Anderson, Craig Hosmer, and Melvin Price. I tried to be punctiliously honest and direct with them, and they responded by being most friendly with me.

15 It was originally named "The Program Advisory Committee."

16 After the summer push to get a plausible design for the new laboratory, I was exhausted and went out to Aspen for a few days of fishing. There I ran into David Pines, who showed me around the Aspen Physics Institute. Never at loss for a good idea, David suggested that the Institute would be a good place for our next (1968) summer study that was already planned to be about experimental areas. Having caught a few fish, it was easy to convince myself of the wisdom of his idea – but would it play in Peoria? On my return to Chicago, Ned agreed that Aspen might add a bit of luster to what by many was considered a lackluster site. Although the idea was initially rejected by mid-level AEC officials, Ned felt that we could sell it to the Commissioners themselves on the basis of economy. The pitch was that if we held the Summer Study in the Chicago area, we would have to pay the travel expenses, etc., for whole families, if anyone came at all. If they did not come, it would be difficult to get the kind of user commitment that we wanted and needed. At Aspen we would need only pay the participant's travel because it is a great vacation spot. The attraction to families would be essential to the success of the program. This logic prevailed. The Commissioners overrode previous objections. Excellent people came, and the lab has been saving money there ever since!

17 I have tried throughout this chapter not to be defensive. However, Ned Goldwasser, to whom I passed a draft, has rebelled. He has insisted on the inclusion of the following note: "From the very beginning of the project Bob Wilson realized that to achieve the cost-savings that were required for the authorization of construction, it would be necessary to shave all designs to the bone. In every instance we would have to hew close to the line between a 'go' and 'no-go' design. We realized that in doing this, some systems would turn out to be under-designed and would have to be beefed up at some cost. But we were convinced that this would be far cheaper than designing a comfortable margin of safety for every component. The magnets turned out to be our principal Achilles' heel, but we remain convinced that had we designed conservatively, cost overruns would have seriously compromised the project or lost it altogether."

18 This was the Argonne–University of Michigan bubble chamber. Its

installation was made as a collaboration of Argonne and Fermilab under the general direction of Lou Voyvodic in the summer of 1971.

19 M. Bodnarczuk, ed., "Reflections on the Fifteen-Foot Bubble Chamber at Fermilab," (Batavia, IL: Fermilab, 1988). It is replete with many pictures and the usual enchanting drawings by Angela Gonzales.

20 It could be that it was Maschke who came up with this howler. If so, then he should step forward like a man and accept the blame.

21 Although 500 GeV protons had been produced in 1976 and used in bubble chamber exposures, the nation's energy shortage had reached crisis proportions, so the laboratory was not able to operate above the normal running value of 400 GeV.

20

Panel Session: Science Policy and the Social Structure of Big Laboratories

CATHERINE WESTFALL

Born Loma Linda, California, 1952; Ph.D., 1988 (history of science), Michigan State University; Historian, Thomas Jefferson National Accelerator Facility; history of science and science policy.

The period that witnessed the rise of the Standard Model also saw radical change in the science policy and sociology of large laboratories. In the 15-year span from 1964 to 1979 the science policy climate in Europe and the United States evolved from the post–World War II golden age of strong political support and burgeoning budgets to the current era of political vacillation and uncertain funding. As researchers investigating the fundamental nature of matter used fewer mammoth accelerators and larger, vastly more complicated detectors, requiring larger teams and more specialized workers, the social structure of large laboratories also was transformed.

To help illuminate this pivotal moment, the conference organizers convened a panel on Science Policy and the Sociology of Big Laboratories. I chaired the panel, which included two other historians specializing in big science (Robert Seidel and John Krige), philosopher of science Mark Bodnarczuk, and four physicists who helped administer laboratories during these years (William Wallenmeyer, Wolfgang Panofsky, Maurice Goldhaber, and Norman Ramsey). The panel session, which consisted of 15-minute presentations by each panel member followed by a brief discussion period, was videotaped. Panofsky and Goldhaber also gave me written remarks. At the request of the conference organizers, I reviewed the videotape and written remarks and integrated, expanded, and placed into context common themes from the panel discussion to create this chapter. Panelists are quoted from the videotape of the panel session or from their texts, as indicated. In a few cases, as noted, I quote relevant remarks made by panelists on other occasions. Comments not attributed to other panel members reflect my own interpretations.

While writing this essay I found that Bodnarczuk's comments drew on specialized concepts and language particular to the philosophy and sociology of science, fields that are outside my specialty. Since Bodnarczuk alone addressed the issues currently studied by sociologists of science – a crucial task for a panel covering the sociology of big laboratories – I felt obliged to present his views completely and accurately. Since I was uncertain that I could accomplish this goal on my own, I asked him to write a separate essay based on his panel contribution (see Chapter 21).

The first section of this chapter charts the evolution of the relationship between large laboratories and government, while the second section describes changes in laboratory administration and research. The chapter ends with some reflections on the future of large laboratories in light of the trends evident in the 1964 to 1979 period. My intention is twofold: to present fresh information and provide a point of departure for further scholarly investigation.

The partnership in crisis

Panelists agreed that in the two decades after World War II large laboratories and their government sponsors collaborated in a close "partnership" to accomplish mutually beneficial goals. In the words of former SLAC director Panofsky, this partnership "worked exceptionally well." Panelists disagreed, however, about the terms of the partnership. Panofsky argued that: "The relationship was based on the recognition of a commonality of interests.... During World War II government found that, if adequately funded, physicists are very productive."[1] In Wallenmeyer's words, both partners "expected a payoff" from the federal investment in both pure and applied research dividends, although "there was no way of knowing when or how this payoff would occur." Wallenmeyer added that the government also supported large physics laboratories "in recognition of the wartime contribution made by physicists" and because officials felt that large-scale physics research was so expensive that "the federal government was the most appropriate source of funding."[2]

Krige and Seidel identified other motives for federal support. Seidel insisted that the U.S. government was motivated, primarily by national security objectives, to sponsor the research of large laboratories, in particular the development of accelerators. Government representatives aimed to increase international prestige as well as recruit personnel and develop technology for applied, especially military, projects.

The close connection Seidel finds between the development of accelerators and national security, in his words, "supports the arguments of Daniel Kevles, and others, who note that big science originated in the alliance made between the Army and prominent physicists during the Manhattan Project. Since that time, this argument maintained, the scientific elite has accomplished its goals through ties with the military and other power elites."[3] Krige concluded that European governments did not support large laboratories only for military reasons. They also had scientific and political motives; they wished to bridge the gap between European and U.S. research capabilities and thereby halt and redress the brain drain from the continent.

Although panelists disagreed about the post–World War II "golden age," as Seidel called it, they concurred that the partnership between large laboratories and their governments began to change rapidly in the mid-1960s.[4] By the early 1980s, they agreed, the partners had fewer common interests, less trust, and less contact.

The experience at large laboratories reflected changes encompassing all federally sponsored research. Bruce Smith, Jeffrey Stine, David Dickson, and other science policy analysts report that from the mid-1960s to the mid-1970s a number of factors prompted a "crisis," which transformed the relationship between government and science.[5] By the mid-1960s, public complaint about the highly technological war in Vietnam, the development of civilian nuclear power, and environmental pollution prompted politicians to debate the social value of science. In this critical atmosphere, skepticism rose about the role of scientists in policy making. In September 1963, for example, U.S. political reporter Meg Greenfield remarked: "As presiders over the national purse, are the scientists speaking in the interest of science, ... government, or ... their own institutions? Is their policy advice ... offered in furtherance of national objectives – or agency objectives – or their own objectives?"[6] By late 1963 Congressional investigations were being formed and by mid-decade various aspects of science and technology funding were under close scrutiny.[7] In Europe, scientists were also under fire. As Krige noted, in the midst of "general public disillusionment about the role of scientists ... European policy makers were simply no longer willing to accept the claims of scientists on faith."[8]

Political differences caused further divisions between leaders of the scientific community and top government officials. For example, eminent scientists, including members of the prestigious President's Science Advisory Committee (PSAC), vehemently opposed U.S. President

Richard Nixon's plans to develop antiballistic missiles and supersonic transport. This reaction annoyed Nixon, who was already irritated because many scientists opposed the Vietnam War. Although a 1970 House subcommittee headed by Emilio Daddario and a White House task force recognized the escalating divisiveness between the scientific community and government, and advocated increasing the number and influence of science advisory officers, Nixon abolished PSAC and the Office of Science and Technology and gave the role of science advisor to H. Guyford Stever, who simultaneously directed the National Science Foundation (NSF). The executive branch science advisory system was not reinstated until 1977.[9]

Other forces conspired to widen the gap between large laboratories and government. As Krige explained, in the 1960s and 1970s Europe lost, due to death or retirement, Francis Perrin, Werner Heisenberg, and other top physicists instrumental to the postwar campaign to revitalize European science. Their successors had less political experience and fewer close ties to government leaders and therefore enjoyed less influence in government. One early consequence was the failure of British physicists to forestall Britain's attempt to enforce budget ceilings at the European high-energy physics laboratory, CERN, in the early 1960s. Without an eminent spokesman, "the process of policy formation inside Britain was highly bureaucratized: the mechanisms used by the physicists to transmit their views on CERN to the government were predominantly formal, and so inevitably lacked 'punch' and a sense of urgency."[10]

In the United States institutional changes complicated the administration of large laboratories and further decreased communication between laboratory officials and government leaders. Prompted by the concerns for promoting new nonnuclear energy sources and for separating nuclear development from nuclear safety, President Gerald Ford in 1974 abolished the Atomic Energy Commission, which had supported the nation's largest accelerators since its formation in 1946. The AEC's research and development function was transferred to the newly formed Energy Research and Development Administration, which brought together for the first time major research and development programs for all types of energy.[11] In 1977 ERDA was reorganized into a cabinet-level Department of Energy (DOE). Wallenmeyer, former Director of the Office of High Energy Physics, explained that as the funding agency got larger, accelerator laboratories had to compete for funding with a wider range of programs. Also, with size came "greater bureaucracy and less scientific and technical understanding" at the higher levels of the agency.[12] As a result, laboratory directors had to adhere to more regulations and produce more paperwork to account for their activities. In 1977, in the midst of the transition from the AEC to DOE, the 30-year-old Joint Committee on Atomic Energy (JCAE) was disbanded. Thereafter budget items were considered by established Congressional committees instead of by the JCAE, which had included members well versed in science and technology who were willing to champion deserving accelerator projects through the Congressional funding process.[13]

Ramsey, who helped plan the Fermi National Accelerator Laboratory (Fermilab) as president of the Universities Research Association (URA), explained the challenges faced by laboratory administrators during this period. "In the late 1960s we could go to the top – to powerful Congressmen and even to President Johnson through Glenn Seaborg, who was a scientist and had worked with people like Robert Wilson and Pief Panofsky. But none of us knew the top echelons of DOE," in the 1970s, "and we had less contact with Congress after the JCAE left. With less contact, communication, and understanding, problems were harder to solve and life got more difficult."[14] Problem solving was further complicated, he noted, by the delays induced by greater bureaucracy in Washington and the greater size and complexity of laboratory projects. For example, when Ramsey helped plan Brookhaven National Laboratory, including a proposal for a $25 million research reactor, only 14 months elapsed between planning sessions in late 1945 and the beginning of work at the new laboratory in early 1947.[15] In contrast, when planning began for the $250 million Fermilab accelerator in 1959, 9 years passed before staff members went to work at the new accelerator site. "With greater delay came greater uncertainty – maintaining morale was a real challenge."[16]

As the gap widened between physics and government, national economies worsened both in the United States and in Europe; research budgets soon fell victim to the times. As Seidel noted, although physics had enjoyed an almost exponential increase in funding in the United States from 1945 to 1967, in 1968 funding reached a plateau. This plateau continued with decreases in the mid-1970s and early 1980s (see Fig. 20.1).[17] European research also suffered. The United Kingdom, France, and West Germany reduced research and development expenditures (as a percentage of gross national product) in the late 1960s. Although funding in West Germany increased steadily through the 1970s, the investment slump continued during this period in the United Kingdom and France (see Fig. 20.2).[18]

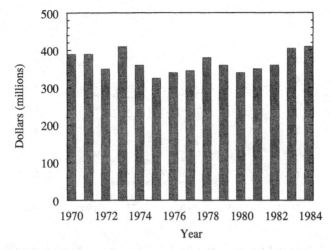

Fig. 20.1. High-energy physics funding in the United States for fiscal years 1970 to 1984. Funding is given in fiscal year 1983 dollars using selected inflation factors. Construction funds are not included.[19]

High-energy physics was burdened with several disadvantages during this period. As budgets shrank, expenses rose for the larger detectors and accelerators needed to advance the field. To make matters worse, proposed high-energy physics projects had to compete with proposals for space science and other large projects of unprecedented expense.[20] Also, as Seidel and Krige noted, at a time when U.S. and European politicians promoted the value of socially useful research, high-energy physics proposals were at a competitive disadvantage because the field promised few immediate practical applications. Large U.S. projects faced further difficulties. The largest funding requests, for example the $250 million proposal for the Fermilab accelerator, were expensive enough to attract considerable public and Congressional attention. Also, as Wallenmeyer noted, "since World War II the AEC and its successors have provided about 90% of the funding for high-energy physics."[21] Thus, even smaller expenses, such accelerator upgrades as PEP at SLAC and the Energy Saver at Fermilab, appeared in a single budget and were therefore more noticeable and vulnerable to budget cuts.

The disadvantageous funding environment left its mark in both Europe and the United States. Krige noted that by the late 1970s, "Britain closed several large facilities, including NINA and Nimrod, France formally announced it would no longer build high-energy physics facilities, and Italy stopped building accelerators. This left [West] Germany as

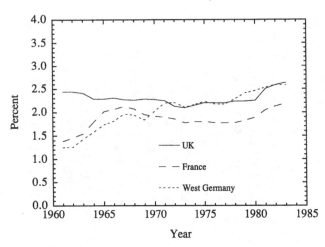

Fig. 20.2. National expenditures for R&D as a percentage of gross domestic product for the United Kingdom, France, and West Germany, 1961–1983.[22]

the sole European country that supported both CERN and a major national laboratory." CERN was also affected. "The laboratory was forced to reduce its budget by 3.5% in real terms per year, and plans to build the 300 GeV Super Proton Synchrotron (SPS) were delayed due to budgetary concerns."[23]

U.S. laboratories also felt the pinch of the shrinking budget. In early 1964 President Lyndon Johnson denied funding for the Fixed Field Alternating Gradient accelerator proposed by the Midwestern University Research Association, a group that had been developing highly innovative accelerator ideas such as colliding beams since 1954. By 1972 the Princeton–Pennsylvania Accelerator was closed and the Cambridge Electron Accelerator (CEA) was no longer used for high-energy physics research.[24]

Panofsky judged that the era that saw the development of the Standard Model was "the most creative" period in high-energy physics.[25] The intellectual achievements of the time demonstrated that high-energy physicists successfully exploited available resources despite the deteriorating relationship with government, tight budgets, and the escalating size and expense of apparatus. In addition, they built the necessary facilities for future research. As Krige noted, for the first time since World War II, European high-energy physicists had the institutions, research expertise, and administrative experience (including procedures for multinational cooperation), necessary to build, maintain, and efficiently use big science facilities. With these resources, Europeans laid the foundation for an experimental program that would challenge U.S. hegemony in high-energy physics. "The Deutsches Electron Synchrotron (DESY) began operation of the Double-Ring Storage collider (DORIS) in 1972 and received funding to build the Positron–Electron Tandem Ring Accelerator (PETRA) in 1975. At CERN, the Intersecting Storage Ring (ISR) reached design luminosity in 1972, the first successful beam cooling experiments were performed at the SPS in 1977, and approval was obtained for constructing the Large Electron Positron facility (LEP) in 1979."[26]

Due to the enduring influence of leaders who successfully adapted to the challenges of the era, U.S. high-energy physicists also constructed the equipment needed to advance research. Fermilab received construction authorization in 1967, thanks to support of the still extant JCAE and AEC and to the vigorous efforts of such leaders as Seaborg, Ramsey, and Frederick Seitz, who consolidated support in Washington and within the physics community for the expensive project. Wilson managed to finish the machine ahead of time and under budget, despite delayed funding allocations, using a U.S. accelerator-building style dating back to the 1930s that deemphasized reliability and solid engineering and celebrated frugal, quickly implemented, clever solutions to technical problems (see Chapter 19). As Ramsey noted, the accelerator, which used small, risky magnets, broke "both the energy and the cost frontier."[27]

Building the Stanford Positron–Electron Asymmetric Ring (SPEAR) required both frugality and creative financing. As Panofsky explained, from 1965 to 1970 plans to build the "pioneering storage ring ... as a formal capital equipment project or construction project" fell prey "to budgetary pressure." SLAC was able to build the collider after Burton Richter drastically cut construction costs, AEC Comptroller John Abbadessa "gave informal acquiescence" to the idea of reallocating "ongoing equipment and operating funds," and Panofsky freed the necessary money by "internal belt-tightening."[28]

Wallenmeyer argued that funding-agency administrators also faced a difficult task when struggling to manage a successful U.S. high-energy physics program. "Administering the program was a juggling act that got harder as budgets tightened and projects got larger. We had to balance the needs of universities versus the laboratories, the needs of each laboratory versus the other laboratories. At the same time, we had to balance the well-being of the current programs, such as the effective operation of existing facilities, future requirements, including R&D for new detectors and accelerators, and R&D on the advanced concepts needed for accelerator development in the very long term future."[29]

Despite budgetary difficulties, SLAC and Fermilab began operation, while two older facilities, the Alternating Gradient Synchrotron (AGS) at Brookhaven and the electron synchrotron at Cornell, were maintained. In addition, other major projects were started, including SPEAR and PEP at SLAC, CESR at Cornell, and the Energy Saver and the Colliding Detector Facility (CDF) at Fermilab. Wallenmeyer noted that the continued vitality of the U.S. high-energy physics program derived in part from the tradition of long-range planning that began in the 1950s with ad hoc advisory panels and culminated with the 1967 formation of the High Energy Physics Advisory Panel (HEPAP), a standing committee of top physicists that judged high-energy physics projects and made funding recommendations.[30] With the help of HEPAP, "which is known as the most powerful [scientific] advisory group in Washington, the funding agencies were able to effectively set priorities and lobby for important projects."[31]

A new social order

Panelists also remarked on the transformation in the social structure of laboratory life that coincided with the new science policy environment. The increase in the scale of research, plus limited budgets, led to radical alterations in laboratory administration and experimentation. In Bodnarczuk's words, "what it meant to do high-energy physics changed forever."[32]

Numerous administrative changes resulted from a shift in the relationship between laboratories and outside users, which began in the 1950s. At this time many small accelerators, mostly cyclotrons, closed as interest shifted to the research capabilities of larger, more expensive machines, such as the AGS at Brookhaven, SLAC at Stanford, and the Proton Synchrotron at CERN.[33] Physicists congregated at a dwindling number of facilities, and complaints arose about the treatment of outside users. As Panofsky pointed out, this caused a major problem for laboratory administrators, who realized that no laboratory can live up to its research potential without a large group of enthusiastic, well-motivated users.[34]

Goldhaber explained that Brookhaven pioneered early efforts to accommodate the entire community of users. After the Cosmotron began operation in the early 1950s, Leland Haworth gathered staff physicists and "occasional outsiders" to form a formal program committee, "a concept which seems," in Goldhaber's words, "to have originated at Brookhaven." To further facilitate the fair treatment of outside users, in 1961 Goldhaber "reconstituted" an existing advisory committee that judged proposals for AGS experiments "to contain comparable numbers of high-energy physicists from Brookhaven and from neighboring universities" in an effort to "balance different interests" in the advisory process, creating a model for the modern program advisory committee.[35] According to Wallenmeyer, Edwin Goldwasser pioneered other attempts to accommodate outside users in the United States. During the construction of the Zero Gradient Synchrotron at Argonne in the late 1950s, Goldwasser organized a users group so that outside users could discuss their common concerns and express these concerns to laboratory administrators. In the next few years, users groups and program advisory committees became standard at U.S. accelerator laboratories. As Krige noted, by virtue of its international character, CERN was forced to respond to concerns of its outside users. The laboratory set up a number of experimental committees based on technique (emulsion, bubble chambers, electronics) "in which visitors were well-represented," and devised procedures for equitable access to experimental resources, such as beams and bubble chamber photographs.[36]

Despite such efforts, outside user discontent intensified in the late 1960s and early 1970s when tightening budgets forced the closure of national laboratories in Europe and major U.S. laboratories, such as PPA. Lew Kowarski, who surveyed users' procedures, identified the problem in a 1967 CERN report. "Practically every accelerator Laboratory has been originally set up in a framework more narrow than the Commonwealth of users it ultimately came to serve." As a result, even those laboratories "which have been set up from the start as co-operative," such as BNL and CERN, had to devise new procedures to ensure that users from institutions outside the framework had equitable access to laboratory resources, including accelerator time.[37]

To forestall outside user discontent, SLAC's contract specified that the laboratory would form a scientific policy committee to assure fair access to the accelerator, which began operation in 1966.[38] "The growing, grass-roots movement for outside user rights," as Leon Lederman later called it, had an even more profound effect on Fermilab.[39] When Lawrence Radiation Laboratory (LRL) physicists received design funding for the new accelerator in 1963, they assumed they would enjoy the traditional prerogative of accelerator builders to manage and build the machine at the site of their choice. Instead, worry that LRL would follow its traditional practice of allowing insiders to monopolize the machine led to the formation of URA, the first accelerator-management consortium with nationwide representation, and an open, AEC-sponsored site contest, which located the machine in Illinois.[40] After his 1967 appointment as director, Wilson chose outside user expert Goldwasser as deputy director, vowed that the laboratory would be "sensitively responsive to the needs of the broad community of scientists," and promised that laboratory physicists would conduct only 25% of the research performed on the new accelerator.[41]

Complaints also surfaced at CERN. As Krige explained, during a series of meetings held by the European Committee for Future Accelerators in the early 1970s, CERN's visitors complained "that the resources and facilities for European high-energy physics were becoming concentrated at CERN" and "this concentration of resources was going along with a concentration of privileges for the in-house staff." In their view CERN staff members had higher pay, more job security, better working conditions, and more decision-making power and obtained funds more readily for experimental equipment than did visitors. To ease such concerns, CERN in the 1970s and early 1980s formed the Advisory Committee for CERN Users, studied decision-making procedures, and surveyed users' attitudes.[42]

When the relationship between U.S. laboratories and users changed in the late-1960s and early 1970s, other aspects of laboratory administration were affected. Wallenmeyer noted that the increased influence of outside users, through users groups and laboratory committees, amplified the voice of universities in laboratory decision making, since most outside users came from universities. "This was very useful because laboratories got the benefit of university leadership and the ties between universities and laboratories got stronger, which was good, since closer collaboration was needed as experiments became longer and more expensive."[43]

Other administrative changes of the era were greeted with less enthusiasm, since the measures that ensured fair decisions in the 1960s and 1970s also made the decision-making process more formal and less flexible. Before the advent of formal program committees, laboratory directors often met promising researchers in the early stages of experimental planning and suggested modifications, perhaps with the help of a few trusted advisors. The obligations of experimenters and their institutions were agreed upon with a handshake. As Fermilab researcher Thomas

Kirk noted in 1970: "The confidential nature of the proceedings avoided unnecessary embarrassment to experimenters.... Very casual proposals were accepted on the reputations of the men responsible."[44] As Goldhaber noted, in later times "funding agencies ... sometimes made a grant to a research group only *after* their experimental proposal had been accepted by a committee." Thus, capable experimenters sometimes faced "the deep psychological impact" of a proposal that failed due to some easily corrected flaw, and other, less capable researchers were allowed to construct costly apparatus, thus obtaining "experiments with tenure."[45]

Outside participation in decision making was not the only factor that increased formality: procedures for processing experimental proposals became increasingly elaborate throughout the 1970s in response to the escalating scale of detectors, which was spurred by the development of the Standard Model, and the decreased technical understanding and trust of funding agencies. For example, by the late 1970s Fermilab had a handbook for users that described the decision-making procedures for proposals, including the roles and responsibilities of decision makers, and "Agreements," which described the obligations and expectations of the institutions involved in experiments.[46]

Increase in scale had other consequences for experimentation. The formation in the late 1970s of the CDF and the LEP detectors, which were comparable in size, complexity, and expense to previous accelerators, ironically reversed some of the trends begun in the mid-1960s in response to increasing scale and tightening budgets. These giant detector projects, which gathered several hundred physicists working in dozens of groups from facilities in the United States, Europe, and Japan, helped to dilute the influence of outside users in experiments (though not necessarily in laboratory decision making) in both Europe and the United States. These collaborations were formed around a core of powerful inside users, who were in a prime position to oversee the efforts of the scattered collaborators and coordinate their work with the activities of the host laboratory. Since a project needed a wide base of enthusiastic support to obtain funding in the late 1970s, due to the increasingly unfavorable science policy climate in Washington, large laboratories faced a new struggle to balance the needs of inside and outside users.[47]

Changing scale had other effects on experimentation in the United States. Greater technological complexity of detectors and other experimental apparatus led to increased reliance on systematic problem solving and engineering skills and decreased emphasis on frugality. In addition, as Ramsey noted, the immense size of collaborations gave rise to a more hierarchical organizational structure and formalized procedures for intracollaboration communication and decision making.[48]

In the era of charm physics, computing brought particularly profound changes to experimentation. As apparatus became more complex, the amount of data grew, and the need to share data among groups increased, high-energy physicists relied more and more heavily on the computer. As Peter Galison has noted, around the mid-1970s the growing importance of computing restructured the organization of research. Whereas previously work was divided into two sequential steps, detector building and data analysis, subsequently provisions were also made for "a third axis of work differentiation around computer programming, spanning the full cycle of data acquisition, maintenance, distribution, and analysis."[49] Computing also increasingly dominated the attention of researchers. As a result, in Kowarski's words, "the idea of an experiment" shifted "from the setting up and running of apparatus to the reduction and analysis of data."[50] In Galison's opinion, this shift "may be the sea change of twentieth-century experimental physics."[51]

The wave of the future

Panelists expressed considerable worry about the future of large laboratories, since troubling trends in the 1964 to 1979 period have accelerated, some previous solutions no longer seem viable, and new challenges have arisen. The chronic funding difficulties experienced by the multibillion-dollar Superconducting Super Collider (SSC), which faced possible cancellation on several occasions in the early 1990s, and was finally canceled in 1993, dramatically illustrate that since 1980 large laboratories have been squeezed more firmly than ever before by tight budgets and the inevitable cost increases that accompany growth in scale. The strategies devised to overcome this problem in the 1970s – creative financing and a quick, frugal, but risky accelerator building style – were of limited utility to those building the SSC, who faced a sometimes hostile reception in Washington, a funding agency that demanded exacting accountability, and very large-scale technology that can only be implemented with careful planning, reliable engineering, and the help of industry.

To accommodate government requirements and industry's new role as a full partner in the construction phase, SSC leaders were forced to invent new approaches to accelerator building, especially for the organization and management of the project, and simultaneously overcome daunting technological hurdles. At the same time they did battle in the new media, on the floor of Congress, with the Department of Energy, and within the physics community to justify the cost and relative value of the facility. The difficulty of these efforts emphasizes the importance of devising better procedures for adjudicating competing funding claims for scientific research.

SSC planning also prompted new concerns about future modes of experimentation. Since the new laboratory, if it were built, would have had 1000-member groups working for over a decade on a single experiment, high-energy physicists were worried about the difficulties of training graduate students, recognizing the contributions of junior collaborators, and encouraging scientific creativity and productivity at this scale of experimentation.[52] Other observers have questioned whether deception, error, and fraud are more likely to occur in such massive collaborations, due to the difficulty of identifying individual responsibility. Another worry is that the informal nature of large teams would undercut efforts to ensure fair treatment for all members, regardless of race, religion, age, and gender.[53]

Perhaps the most troubling aspect of the future of large laboratories is the continuing deterioration of the relationship between government and science. Panofsky complained that every time a mistake is made by one individual within any one large laboratory, all laboratories are burdened with "another layer of oversight, and criticism is leveled at the entire profession of scientists." One result is that laboratories are faced with "ever-increasing pressures for more prior approvals, prior repeated cost analyses and cost reviews," in short, detailed justifications and formal approvals for every step in the research process. Such practices prompt concern about the productivity of large laboratories.[54] As Seidel has warned, "The capabilities of [large] laboratories ... are rich, but they are also easily stifled by the dead weight of a regulatory bureaucracy. A balance must be struck between responsibility and freedom in big science if it is to be a productive enterprise."[55]

Both Seidel and Panofsky felt that the problems of the 1980s and 1990s raised questions about the future of the relationship between government and science. Since he finds a close link between the development of accelerators and national security considerations, Seidel correctly predicted that the government would not be willing to support a project as expensive as the SSC now that the Cold War is over, especially since prominent scientists opposed the development of major military projects, such as the Strategic Defense Initiative. Although Panofsky disagreed with Seidel's interpretation, he agreed about the uncertain future of large laboratories. "We are seeing a shift from the partnership between government and science," he explained, to "'acquisition' of science by government," an approach that is not conducive to creative problem solving and the advancement of scientific knowledge. "Nothing short of restoring a spirit of mutual trust and confidence between government and the scientific community can reverse" the trend and reconcile the partners so that they can continue to accomplish mutually beneficial goals.[56]

Prospects for the future are not entirely gloomy, however. Krige stressed that CERN was in a good position to prosper in upcoming decades, since the laboratory is an important political symbol and provides a unique resource (aside from DESY) for scientific projects to which European physicists have special access. In addition, governments would find it "extremely difficult to withdraw" support, due to the "enormous diplomatic and political consequences."[57]

The very development of the Standard Model also inspires optimism. This achievement testifies to the rich dividends that accrue when physicists and their governments make the sometimes risky investments necessary to continue the search for the fundamental nature of matter. Although large laboratories face a number of formidable problems, these difficulties are, in Ramsey's words, "merely the cost for being able to do one of the most exciting kinds of research known to man."[58]

Notes

1 Wolfgang Panofsky, "Round Table Statement," submitted to the Panel on Science Policy and Sociology of Big Laboratories.

2 William Wallenmeyer, panel session.

3 Robert Seidel, "Summary of Symposium on Science Policy Issues of Large National Laboratories," submitted to the Institute of Government and Public Affairs, University of Illinois, 8 October 1992. See also, Daniel Kevles, "K_1S_2: Korea, Science, and the State," in Peter Galison and Bruce Hevly, eds., *Big Science: The Growth of Large Scale Research* (Stanford: Stanford University Press, 1992).

4 Panel session.

5 Jeffrey K. Stine, *A History of Science Policy in the United States, 1940–1985,* Committee on Science and Technology, House of Representatives, 99th Congress, Second Session (Washington: Government Printing Office, 1986), pp. 57–58; Bruce L. R. Smith, *American Science Policy Since World War II* (Washington: Brookings Institution, 1990), pp. 73–118; David Dickson, *The New Politics of Science* (Chicago: The University of Chicago Press, 1984).

6 As quoted in Daniel J. Kevles, *The Physicists: The History of a Scientific Community in Modern America* (New York: Alfred Knopf, 1978), p. 395.

7 Committees that investigated the appropriate distribution of research

funding included a House subcommittee headed by Emilio Daddario and a House select committee headed by Carl Elliot, both established in 1963, a Senate committee headed by Joseph S. Clark, convened in 1965, and a Senate subcommittee headed by Fred R. Harris, established in 1966. See U.S. Congress, Subcommittee of the Select Committee on Small Business, *The Role and Effect of Technology in the Nation's Economy*, 88th Congress, First Session (Washington: Government Printing Office, 1963); U.S. Congress, *Government and Science: Hearings Before the Subcommittee on Science, Research, and Development of Committee on Science and Astronautics*, 88th Congress, Second Session (Washington: Government Printing Office, 1964); U.S. Congress, Subcommittee on Science, Research and Development of the Committee on Science and Astronautics, *Government and Science: Distribution of Federal Research Funds*, 88th Congress, Second Session (Washington: Government Printing Office, 1964); U. S. Congress, Subcommittee on Employment and Manpower of the Committee on Labor and Public Welfare, *Impact of Federal Research and Development Policies on Scientific and Technical Manpower*, 89th Congress, First Session (Washington: Government Printing Office, 1965); and U.S. Congress, Subcommittee on Government Research of the Committee on Government Operations, Hearings, *Equitable Distribution of R&D Funds by Government Agencies*, 89th Congress, Second Session (Washington: Government Printing Office, 1966). Also see Michael D. Reagan, *Science and the Federal Patron* (New York: Oxford University Press, 1969); Donald R. Fleming, "The Big Money and High Politics of Science," *Atlantic Monthly*, August 1965; and Daniel Kevles, *The Physicists*, pp. 413–414.

8 John Krige, panel session.

9 Jeffrey K. Stine, *A History of Science Policy*; Bruce L. R. Smith, *American Science Policy*, pp. 73–118; Thaddeus Trenn, *America's Golden Bough: The Science Advisory Intertwist* (Cambridge: Oelgeschlager, Gunn, and Hain, 1983), pp. 88–112. For more information on the 1970 Daddario Hearings, see U.S. Congress, Subcommittee on Science and Astronautics, Subcommittee on Science, Research and Development, *Toward a Science Policy for the United States*, 91st Congress, Second Session (Washington: Government Printing Office, 1970). For further discussion on the ABM debate and a review of other key decisions made by the science advisory system, see Gregg Herken, *Cardinal Choices: Presidential Science Advising from the Atomic Bomb to SDI* (New York: Oxford University Press, 1992) and Bruce L. R. Smith, *The Advisor: Scientists in the Policy Process* (Washington: The Brookings Institution, 1992).

10 John Krige, "Finance Policy: The Debates in the Finance Committee and the Council Over the Level of the CERN Budget," in Armin Hermann, John Krige, Ulrike Mersits, and Dominique Pestre, *History of CERN: Building and Running the Laboratory, 1954–1965*, Vol. II (Amsterdam: North-Holland, 1990), pp. 602–603. This source contains details on the dispute and explains the resulting budget policy.

11 In response to criticism that a single agency should not administer and regulate atomic energy programs, Ford assigned regulatory functions to a separate organization, the Nuclear Regulatory Commission. An Energy Resources Council was also established at this time. A. L. Buck, *A*

History of the Energy Research and Development Administration
(Washington: Department of Energy, 1982), p. 2.

12 William Wallenmeyer, panel session.

13 A. L. Buck, *A History of the Energy Research and Development
Administration*, p. 14; "Congressional Science Committees Have a New
Look," *Physics Today 30* (May 1977), p. 109.

14 Catherine Westfall interview with Norman Ramsey, 13 September 1985,
Fermilab History Collection, Batavia, Illinois.

15 Allan Needel, "Nuclear Reactors and the Founding of Brookhaven
National Laboratory," *Hist. Stud. Phys. Sci. 14* (1983), p. 119. For
more information on the founding of Brookhaven, see Norman Ramsey,
"Early History of Associated Universities and Brookhaven National
Laboratory," Brookhaven Report No. BNL 992 (T-421), March 1966.

16 Norman Ramsey, panel session. The first plans for a multihundred-GeV
cascade synchrotron were made by Matthew Sands in 1959. For more
information on this design, as well as earlier designs for other large
synchrotrons, see Catherine Westfall, "The First 'Truly National
Accelerator': The Birth of Fermilab," Ph.D. Dissertation, Michigan State
University, 1988; Lillian Hoddeson, "Establishing KEK in Japan and
Fermilab in the US: Internationalism, Nationalism and High Energy
Accelerator Physics During the 1960s," *Social Studies of Science 13*
(1983), pp. 1–48.

17 Physics Survey Committee, "Organization and Support of Physics," in
Physics Through the 1990s: An Overview (Washington: National
Academy Press, 1986), pp. 119–120.

18 Physics Survey Committee, *Physics Through the 1990s*, p. 77.

19 Data from Physics Survey Committee, *Physics Through the 1990s*, p. 125.

20 For a description of one such U.S. project, see Robert W. Smith, *The
Space Telescope: A Study of NASA, Science Technology, and Politics*
(Cambridge: Cambridge University Press, 1989).

21 William Wallenmeyer, panel session.

22 Data from Physics Survey Committee, *Physics Through the 1990s*, p. 77.

23 John Krige, panel session. For more information on the delay in building
the SPS, see Dominique Pestre, "The Second Generation of Accelerators
for CERN, 1956–1965: The Decisionmaking Process," in Armin
Hermann, John Krige, Ulrike Mersits, and Dominique Pestre, *History of
CERN*, Vol. II.

24 CEA was thereafter used for research into the development of colliding
beams. Physics Survey Committee, *Physics in Perspective*, Vol. II,
(Washington: National Academy of Science, 1972), p. 118.

25 Wolfgang Panofsky, panel session.

26 John Krige, "High Energy Physics Chronology," 13 May 1992
(unpublished).

27 Catherine Westfall interview with Norman Ramsey, 13 September 1985,
Fermilab History Collection, Batavia, Illinois. For more information on
the effect of fiscal stringency on the building of Fermilab, see Catherine
Westfall and Lillian Hoddeson, "Frugality and the Building of Fermilab,
1960–1972," to be published in *Technology and Culture*.

28 Wolfgang Panofsky, "Round Table Statement." For a more detailed
description of attempts to fund SPEAR, see Michael Riordan, *The*

Hunting of the Quark: A True Story of Modern Physics (New York: Simon & Schuster, Inc., 1987), pp. 247–248.

29 William Wallenmeyer, panel session.

30 High Energy Physics Advisory Committee, "Minutes of HEPAP Organizing Meeting," 29 January 1967, FNAL. As Wallenmeyer noted, high-energy physics advisory panels included: a 1954 NSF Panel chaired by Robert Bacher; a 1956 and a 1958 NSF panel, both chaired by Leland Haworth; a 1958 and a 1960 PSAC–General Advisory Committee (GAC), both chaired by Emanuel Piore, and a 1963 PSAC–GAC Panel chaired by Norman Ramsey. For copies of panel reports, see Joint Committee on Atomic Energy, *High Energy Physics Program: Report on National Policy and Background Information* (Washington: Government Printing Office, 1965).

31 Catherine Westfall interview with William Wallenmeyer, 18 October 1992, Continuous Electron Beam Accelerator Facility Archive, Newport News, Virginia.

32 Mark Bodnarczuk, panel session. Bodnarczuk addresses the increase in the scale of experimentation in Chapter 21.

33 For example, from 1958 to 1969 the number of U.S. high-energy physics accelerators was reduced by more than a half, from 15 to 7. See Joint Committee on Atomic Energy, *High Energy Physics Program*, Appendix 3; Physics Survey Committee, *Physics Through the 1990s*, pp. 126–127.

34 Panofsky noted that although Dubna's 10 GeV Synchrophasotron was the most powerful accelerator in the world from 1957 to 1959, it produced few important results. In addition to design problems that hampered machine performance, the accelerator had too few users.

35 Maurice Goldhaber, "The Beginning of Program Committees," 1992 (unpublished).

36 For more information on CERN, see John Krige, "The Relationship Between CERN and its Visitors in the 1970s," in John Krige, ed., *History of CERN, 1965–1980*, Vol. III (Amsterdam: North-Holland, 1994) and Dominique Pestre, "The Organization of the Experimental Work Around the Proton Synchrotron, 1960–1965: the Learning Phase," in Armin Hermann, John Krige, Ulrike Mersits, and Dominique Pestre, *History of CERN*, Vol. II.

37 Lew Kowarski, "An Observer's Account of User Relations in the U.S. Accelerator Laboratories," Report No. CERN 67-4, Geneva, January 1967, p. 3.

38 Richter argues that SLAC felt less pressure from outside users than Fermilab or CERN because fewer people were interested in lepton than proton physics and because initial SLAC experiments clearly required large-scale equipment, which was more easily planned and built by large, in-house groups. Catherine Westfall interview with Burton Richter, 24 June 1992.

39 Catherine Westfall interview with Leon Lederman, 20 July 1984.

40 URA was modeled on the Associated Universities Incorporated, the regional consortium of universities that manages Brookhaven. For more information on this episode, see Catherine Westfall, "The Site Contest for Fermilab," *Physics Today 42* (1989), pp. 44–52.

41 National Accelerator Laboratory, "Design Report" (Batavia: National Accelerator Laboratory, 1968), pp. 2–5, 3–11.

42 For more information on this episode, see John Krige, "The Relationship Between CERN and its Visitors in the 1970s."

43 William Wallenmeyer, panel session.

44 Catherine Westfall, interview with Thomas Kirk, 1970.

45 Maurice Goldhaber, "The Beginning of Program Committees."

46 A measure of the rising complexity of experimental proposal procedures can be taken from the growing number of pages needed to describe them. A 1974 Fermilab handbook had three pages of description; the 1979 handbook had 11 pages of description. See National Accelerator Laboratory, "Procedures for Experimenters," 1974 and Fermilab, "Procedures for Experimenters," 1975–1979.

47 Krige has noted that the struggle between outside and inside users hinges on "conflicts over ownership and control. The form taken by those conflicts will vary depending on the context. The substance will persist" as long as laboratories exist. John Krige, "The Relationship Between CERN and its Visitors in the 1970s." Research at CERN during the transformation of research in the 1970s and 1980s at Fermilab is described in Catherine Westfall, Lillian Hoddeson, Mark Bodnarczuk, and Adrienne Kolb, *Fermilab, 1965–1990: A Case Study in the Emergence of Big Science*, to be published.

48 As Krige noted, the organization of research and the influence of engineers did not change much at CERN, which had traditionally favored tightly organized experiments and solid engineering. See Mark Bodnarczuk, Chapter 21, for more discussion on the implications of the differences between the U.S. and CERN styles. Lillian Hoddeson pointed out that in the late 1970s and early 1980s Fermilab also developed a more formal approach to team organization and a more careful, meticulous approach to building apparatus when faced with the technological challenge of developing superconducting magnets. Lillian Hoddeson, "The First Large-Scale Application of Superconductivity: The Fermilab Energy Doubler, 1972–1983," *Hist. Stud. Phys. Biol. Sci. 18* (1987), pp. 25–54. Also see Peter Galison, "Probe Report on History of the Psi Experiment," *AIP Study of Multi-Institutional Collaborations: Phase I: High-Energy Physics*, Report 4 (New York: American Institute of Physics, 1992), p. 81.

49 Galison, "Probe Report," p. 80; see also Chapter 18, this volume.

50 As quoted in Peter Galison, *How Experiments End* (Chicago: University of Chicago Press, 1987), p. 151.

51 Galison, *How Experiments End*, p. 151. See Mark Bodnarczuk, Chapter 21, for other examples of changes accompanying the increasing scale of computing.

52 High Energy Physics Advisory Committee, "Report of the HEPAP Subpanel on Future Modes of Experimental Research in High Energy Physics" (Washington: U.S. Department of Energy, 1988); High Energy Physics Advisory Committee, "Report of the HEPAP Subpanel on High Energy Physics and the SSC Over the Next Decade"(Washington: U.S. Department of Energy, 1989); American Institute of Physics, *AIP Study of Multi-Institutional Collaborations: Phase I: High-Energy Physics*, pp. 31–32.

53 Jeffrey Stine, "Edited Excerpts from a Smithsonian Seminar Series,"

Knowledge Collaboration in the Arts, the Sciences, and Humanities, (Washington: Smithsonian Institution Press, 1992), pp. 400–406.

54 Wolfgang Panofsky, panel session and "Round Table Statement."

55 Robert Seidel, "Summary of Symposium on Science Policy Issues of Large National Laboratories."

56 Wolfgang Panofsky, "Round Table Statement."

57 John Krige, panel session.

58 Norman Ramsey, panel session.

21

Some Sociological Consequences of High-Energy Physicists' Development of the Standard Model

MARK BODNARCZUK

Born Patterson, New Jersey, 1953; Master of Arts (philosophy), University of Chicago; President, Breckenridge Consulting Group; philosophy of science.

In a scientific discipline that went from experiments with less electronics than a videocassette recorder to 10^5 channels and from collaborations with 5–10 members to 300 during the years 1964–1979, the notion of what high-energy physics *is*, or what constitutes *being* a high-energy physicist, cannot be viewed simply as an immutable category that is "out there" – that remains fixed despite these and other developments. What high energy physics is as a discipline and what it means to be a high-energy physicist are renegotiated by participants relative to the experimental and theoretical practices of the field at any given time. In this chapter I will explore some of the sociological consequences of the decisions made by high-energy physicists as they constructed the edifice that has come to be known as the Standard Model.[1]

Many of these physicists' decisions about the Standard Model have already been carefully documented in Andrew Pickering's sociological history of the development of particle physics, as well as numerous chapters from this volume.[2] I am thinking particularly of factors like the postulation of the notion of quarks and the development of the Eightfold Way and S-matrix bootstrap theory; scaling, hard scattering, and the 1967 MIT–SLAC experiment's evidence for pointlike structure in hadrons; the quark–parton model that was supported by experimental evidence for J/ψ, bare charm, and upsilon particles; the development of gauge theory, the unified theory of electroweak interactions, with the experimental evidence of neutral currents; and finally the development of a theory of strong interactions – quantum chromodynamics. Other than to underscore physicists' decisions to pursue higher and higher energies (as evidenced in the construction of a 200, then 400, GeV proton accelerator at Fermilab), I will not recount these details here. Rather, within the context of such decisions I will attempt to describe how the increases in scale, cost, and complexity mentioned earlier were *consequences of the choices* to go to higher and higher energies in response to the experimental evidence and theoretical constructs that emerged from 1964 to 1979.[3] More particularly, one consequence witnessed at Fermilab was the development of an increasingly complex and bureaucratic organizational infrastructure that I will characterize below as a number of interrelated resource economies, each having its own commodity.[4] Another consequence of larger more complex detectors was the need for larger more complex social structures for the collaborations that designed, fabricated, installed, and operated them, as well as an increased scale and complexity for the computing power needed to collect data samples and bring the results to final publication.

After 1972 Fermilab operated the highest-energy particle accelerator in the world, and consequently competition for use of the wide variety of particle beams it produced was intense. In order to gain access to one of these particle beams, experimentalists had to navigate a number of interrelated resource economies that were embedded within an institutional structure headed by a single scientist, the director, who had ultimate authority in all matters scientific and otherwise.[5] Experimentalists had to learn to trade with and for these commodities in order to participate in the production of knowledge in high-energy physics. Physicists negotiated with these commodities and often fought over them.[6]

One economy at Fermilab was proton economics, based on protons as the commodity. The overall magnitude of the economy was limited by such factors as accelerator flux, efficiencies in primary beam transport, cross sections for secondary beam production, secondary beam transport efficiencies, and expected reaction rates in experimental targets. For example, given the cross section for neutrino scattering (10^{-36} cm^2) and the pion cross section (10^{-26} cm^2), the decision to approve experiments using incident beams of neutrinos was already a major decision that affected proton economics.[7] A neutrino beam was more costly than a pion beam in terms of the number of protons needed to produce it; this was further complicated if the cross section for event production in the proposed experimental target was low and the experiment required a large number of particle events to be competitive with previously accumulated world samples. Given the intense competition for protons, beam management issues became very complex, especially in the kind of user-based environment that typified Robert Wilson's philosophy at Fermilab.[8]

A second economy was based on experimental real estate; here the commodity was possession of an experimental hall at the end of a beam spigot to house the collaboration's apparatus. As detectors became larger and more complex, the lead time needed to assemble and operate apparatus also increased. Consequently, physicists who *were* given a piece of experimental real estate and some beam time tended to move into an experimental hall with the explicit goal of performing that experiment, and the implicit goal of not moving out. Gaining access to an experimental hall, especially when the incumbent collaboration was desperately attempting to hold its place in line, made possession of this commodity one of the most important items to be obtained in an user-based environment.

Another economy, which I call "physicist economics," is based on the commodity of physics expertise. Although the scale and complexity of the experiments during the 1964 to 1979 period continued to increase at an unprecedented rate, the number of high-energy physicists who could commit themselves to perform experiments was constrained by the total number available at that time and the rate at which new Ph.D. graduate students were being produced. Consequently, the enormous increases in the scale and complexity of experiments made physics expertise an increasingly valuable commodity.[9] Larger, more complex and increasingly modularized detectors required larger, more complex and increasingly modularized social structures with the appropriate *number* of physicists and the *distribution* of expertise needed to design, fabricate, install, and operate the apparatus and to develop the computing systems and software programs used to reconstruct and analyze the particle events that were recorded. By the late 1970s, collaborations were characterized by an unprecedented division of labor so that no single member of the collaboration had a detailed knowledge of the entire detector. As pointed out by Galison, this kind of modularization provided each institution with a visible manifestation of its contribution to the experiment.[10] Not only was the modularization of detectors an important aspect of carving out a piece of physics to work on, it was also an important political issue back at the home university. Proposals were increasingly judged on the "physicist design" of the experiment and how well it mapped to the experimental design, with laboratory directors and their advisory committees focusing more and more on whether the collaboration had enough physicist power to make good on its experimental claims.

But the consequences of physicists' choices (increased scale and complexity of detectors, accelerators, and the associated social structures) are most easily seen in a fourth resource economy, computing economics, based on the commodity of on-line and off-line computing power. One example was the attempt to do high-statistics charm production experiments at Fermilab in the late 1970s.[11] On the one hand, the advantages of on-line data reduction using sophisticated trigger processors had to be balanced against the risk of coming up empty handed due to incorrect trigger assumptions and the problems of obtaining the required off-line computing power. On the other hand, the more secure approach to on-line data acquisition (the write-it-all-to-tape approach) had to be balanced against the problem of obtaining immense off-line computing resources, which was difficult given Wilson's belief that the bulk of computing for experiments should be provided by the collaboration's home institutions.[12] There was an abrupt explosion in the number of channels of electronics in detectors after 1980. In terms of the magnitude of computing and number of channels, the period during the development of the Standard Model was the calm before the storm – before the explosion in scale, cost, and complexity of hadron collider detectors (such as CDF) that were conceived after 1977.[13]

A final resource economy was physics economics; the commodity of published physics results was traded back to the laboratory director as a return on investment and was the key to obtaining additional resources to perform follow-up experiments. Within physics economics, the laboratory director's ability to approve or disapprove an experiment was a powerful management tool for leveraging wayward experimenters who failed to make good on their promises, especially when they wanted to move on to the greener pastures of follow-up experiments without first publishing their results.

The study of various Fermilab experiments mentioned earlier also shows that within this socioeconomic–scientific infrastructure of the laboratory, experiments were performed in series of follow-up experiments in which an experiment was performed, then transformed into a second, then a third, or a fourth experiment. I call these series of experiments "experimental strings."[14] Key to describing these transformations is the ability to characterize the continuities between individual experiments in such strings. Evidence that emerged from the previously mentioned study suggests that these experimental strings exhibit well-defined continuities in the physics goals, the detector configuration design, and in the core group of collaborators that participated in 9 or more experiments over a 20 year period spanning three laboratory directors.[15] These continuities transcend a single experiment and provide a method for understanding more complex social structures and research programs that exist for more than 15 years. Each experimental configuration in a string displays a more complex iteration of the original apparatus that leaves the fundamental design of the modularized detector subsystems largely intact. In other words, experimental strings are like mini-institutions within the organizational infrastructure of the laboratory. People outside the laboratory really do not know about them because they do not have formal names.

I believe the experimental string is the preferred and more interesting unit of study for sociological and historical analysis, because the numbers that laboratories such as Fermilab assign to experiments are not at all indicative of what actually constitutes "an experiment." Actually, the experimental numbers assigned by laboratory management are more indicative of such factors as the laboratory's accounting practices, the bureaucratic steps involved in the approval process as defined by a particular director, funding scenarios both inside and outside the laboratory, contrasts between the in-house/facility approach to doing experiments (where strings were largely determined by the laboratory management), and the user-based/non-facility approach (where institutions came together and formed strings more voluntarily). But these numbers do not define what an experiment *is*.

While it has been common practice for philosophers, historians, and sociologists of science to "extract" an experimental "case study" from the organizational infrastructure of the laboratory in which it was performed and attempt to study it as a stand-alone unit, the fact is that experiments like those performed at Fermilab did not exist independent of the organizational infrastructure of the laboratory in which they were embedded. Experiments and collaborations were not closed systems, cohesive entities, or "objects" that had unambiguous boundaries and could be divorced from the dynamics of laboratory life.

Of course, experimentalists did attempt to draw a firm line of demarcation around the "collaboration" or "experiment" and its activities for the sake of defining which names appear on scientific publications, but laboratory personnel often play crucial roles in experiments, and whether or not their names appear on the published paper is a socially negotiated matter that is decided by the people involved.[16] Attempts to "map" the names on various experimental proposals (or the resultant publications) to the collaboration members who actually performed the day-to-day tasks associated with the experiment show that the names on proposals or papers are often not indicative of those who actually performed the work of the experiment. Names of individuals who did not play any substantive role in a particular experiment are included on a proposal or the physics publications because, in some cases, those individuals may have had major responsibility for constructing a portion of the detector in an earlier experiment in the string. In other cases they may have committed a fraction of their overall professional time at the proposal stage but never came through on their commitments because of the heavy load of administrative duties at their home institution, the host laboratory, or their commitments to other experiments that they perceived were producing more important physics results.[17] This is probably related to the problems associated with "physicist economics," and is a fruitful issue for future sociological research.

Physicists' choices to go to higher and higher energies in response to the experimental evidence and theoretical constructs that emerged during the 1964 to 1979 period, and the effect of these choices on increasing scale, cost, and complexity, reveal interesting contrasts between the American and European (CERN) styles of doing physics. During this period many American physicists preferred the more informal, non-bureaucratic, quick-and-dirty, frugal style of doing physics.[18] But the European style was typified by what American physicists considered to be an overly formal, inflexible, bureaucratic, overengineered, "gold-plated" approach to doing physics. Even after the mammoth collider detectors began to be conceived in the late 1970s, both American and European physicists were relatively unreflective about the role that social factors were beginning to play in their work. And consequently the sociological challenges that were intrinsic to collider detector environments with 10^5 or more channels received little or no systematic study by practicing physicists. The sociology of large collaborations just was not viewed as a part of doing high-energy physics and as with the policy of physics journals, the social and human factors were simply *left out*.

But despite this lack of conscious self-reflection on both sides of the Atlantic, the values embodied in European culture more naturally gave rise to a style of physics that was more formal in terms of well-defined roles, responsibilities, and authorities for physicists and engineers, and was more focused on producing robust engineering and physics designs that were less flexible in terms of programmatic changes. As it turned out, these were the very practices, values, and beliefs that became *crucial* to mounting mammoth collider-detector experiments.[19] Conversely, the less formal approach to doing physics put American physicists at a disadvantage in terms of confronting the kinds of organizational and management problems that emerged from this enormous growth in scale, cost, and complexity. While the American style of doing physics may have been an advantage with the scale, cost, and complexity typified by the detectors in most of the 1964 to 1979 period, it became a crucial disadvantage for experiments conceived in the late 1970s, and was absolutely terminal for the proposed SDC and GEM detectors.[20] Also, the European style allowed a more natural transition from the smaller experimental scale that typified the 1964 to 1979 period to the detectors of the present day. In the modern detector environment, not only can social factors no longer be left out of any salient definition of what high-energy physics is, but they become one of the most crucial aspects of doing high-energy physics – they could even become *the* limiting factor of the future of the field.

Notes

1 Currently, there are numerous approaches to the social study of science. For a traditional view of the sociology of science, see Robert Merton, *The Sociology of Science, Theoretical and Empirical Investigations* (Chicago: The University of Chicago Press, 1973). Some of the earliest work in the sociology of knowledge can be found in Karl Mannheim, *Ideology and Utopia; An Introduction to the Sociology of Knowledge* (New York: Harcourt Brace Jovanovich, 1985), and the early development of the "strong programme" of the sociology of scientific knowledge (SSK) is best represented in David Bloor, *Knowledge and Social Imagery*, 2nd ed. (Chicago: The University of Chicago Press, 1991). Some of the more moderate proponents of SSK include Bruno Latour, *Science in Action: How to Follow Scientists and Engineers through Society* (Cambridge: Harvard University Press, 1987) and Trevor J. Pinch, *Confronting Nature: The Sociology of Solar-Neutrino Detection* (Dordrecht: D. Reidel, 1986). Perhaps the most radical SSK position is in Steve Woolgar, *Science, the Very Idea* (New York: Tavistock Publications, 1988). More recently, Andrew Pickering has collected a number of essays that focus on the central role of practice in SSK, in Andrew Pickering, ed., *Science as Practice and Culture* (Chicago: The University of Chicago Press, 1992), and Stephen Cole has provided the first serious critique, by a traditional sociologist, of the SSK position, in Stephen Cole, *Making Science, Between Nature and Society* (Cambridge: Harvard University Press, 1992).
2 Andrew Pickering, *Constructing Quarks; A Sociological History of Particle Physics* (Chicago: The University of Chicago Press, 1984).
3 Pickering claims the relationship between experimental and theoretical research traditions is symbiotic in that each generation of practice within one tradition provides a context within which the succeeding generation

of practice in the other finds its justification and subject matter. Peter Galison claims that the truism that "experiment is inextricable from theory" or that "experiment and theory are symbiotic" is useless because, while vague allusions to Gestalt psychology may have been an effective tactic against dogmatic positivism, experimentalists' real concern is not with global changes of world view. For Galison, the salient issue is where theory exerts its influence in the experimental process and how experimentalists use theory as part of their craft. My point is that once physicists decide to study certain physical phenomena and theoretical constructs at higher and higher energies, such a decision has physical consequences (larger accelerators given the technologies during the 1964 to 1979 era, and larger more heavily instrumented fiducial volumes in apparatus to detect myriad particle interactions) and sociological consequences of the types that constitute the remainder of this chapter. See Pickering, *Constructing Quarks*, pp. 10–11, and Peter Galison, *How Experiments End* (Chicago: The University of Chicago Press, 1987), p. 245, also Chapter 18, this volume.

4 I developed the resource economy model from a case study of several Fermilab experiments. See Mark Bodnarczuk, "The Social Structure of Experimental Strings at Fermilab: A Physics and Detector Driven Model," Report No. Fermilab-PUB-91/63, Batavia, March 1990, pp. 2–7. This model is not unlike Bruno Latour's more generic model of cycles of credit that involve conversions of different types of capital (recognition, grant money, equipment, data, arguments, articles, etc.) into the "credibility" that scientists need to make moves within a scientific field. Bruno Latour and Steve Woolgar, *Laboratory Life: The Construction of Scientific Facts* (Princeton: Princeton University Press, 1986), pp. 187–233.

5 Maurice Goldhaber, in an unpublished article entitled "The Beginning of Program Committees," remarks on the early formation of program committees appointed by laboratory directors for the purpose of obtaining independent assessments of the laboratory's research program.

6 Using numerous case studies, David Hull claims that not only are infighting, mutual exploitation, and even personal vendettas typical behavior for many scientists, but that this sort of behavior actually facilitates scientific development. David Hull, *Science as a Process: An Evolutionary Account of the Social and Conceptual Development of Science* (Chicago: The University of Chicago Press, 1988), p. 26.

7 The pion cross section is roughly constant for energies above 2 GeV at about 40 millibarns. The neutrino cross section is not constant, but is linearly proportional to the energy. For Fermilab, a reasonable neutrino energy to use was 100 GeV, which would give a neutrino cross section of about 0.7 picobarns.

8 In Chapter 20 of this volume, Catherine Westfall notes that Edwin Goldwasser (who later became Wilson's Deputy Director) was one of the first to address user discontent in the United States.

9 For example, a recent study of the research program for the 1990s performed by the High Energy Physics Advisory Panel included a detailed demographic study of "manpower considerations" during the time period under study. See the HEPAP Subpanel, "The U.S. High

Energy Physics Research Program for the 1990s," Report No.
DOE/ER-0453P, Washington, D.C., April, 1990, pp. 68ff.

10 Peter Galison referred to the visibility that modularization provided
    participants in his talk at this Symposium (unpublished).

11 See Bodnarczuk, "The Social Structure of Experimental Strings," for a
    detailed case study of Fermilab experiments E-516, E-691, E-769, and
    E-791 that performed high-statistics photoproduction and
    hadroproduction of charmed particles.

12 Mark Bodnarczuk interview with Robert Wilson, 24 September 1992.

13 The UA1 detector at CERN had about 50,000 channels, the CDF and D0
    detectors at Fermilab, and the SLD detector at SLAC each had over
    100,000 channels, the ALEPH detector at LEP had about 700,000
    channels, and the proposed SDC and GEM detectors at the SSC might
    have had as many as 50,000,000 channels, depending on the available
    technology.

14 See Bodnarczuk, "The Social Structure of Experimental Strings," pp.
    14–20; Joel Genuth, "Historical Analysis of Selected Experiments at US
    Sites," *AIP Study of Multi-Institutional Collaborations: Phase I: High
    Energy Physics*, Report 4 (New York: American Institute of Physics,
    1992); Frederik Nebeker, "Experimental Style in High-Energy Physics:
    The Discovery of the Upsilon Particle" (unpublished), January 1993.

15 The major fixed-target experimental strings at Fermilab were the E-82,
    226, 383, 425, 486, 584, 617, 731, 773 string, the E-531, 653 string, the
    E-8, 440, 495, 555, 620, 619, 756, 800 string, the E-21A, 262, 320, 356,
    616, 770 string, the E-594, 733 string, the E-98, 365, 665 string, the
    E-1A, 310 string, the E-95, 537, 705, 771 string, the E-70, 288, 494, 605,
    608, 772, 789 string, the E-87, 358, 400, 401, 402, 687 string, and the
    E-516, 691, 769, 791 string. By way of comparison with those counter
    experiments, the major continuity between the experiments performed
    with the 15-foot bubble chamber at Fermilab (experiments E-28A, 31A,
    45A, 53A, 155, 172, 180, 202, 234, 341, 380, 388, 390, 545, 564, and 632)
    seems to be the chamber itself. In a less well-defined way, there were
    some continuities in the target substances with which the chamber was
    filled. But the social structures of these collaborations were different from
    the fixed-target counter experiments. Bubble chamber spokespersons
    seemed to draw upon the expertise of the international community of
    bubble chamber physicists each time they formed an experimental group,
    and consequently the collaborations did not exhibit the same type of
    well-defined core-group structure found in large, complex fixed-target
    counter experiments. My preliminary studies show that the relatively
    noncomplex social structure of these collaborations resulted from the
    existence of a Fermilab-based Bubble Chamber Department devoted
    solely to the operation and maintenance of the complex systems of the
    chamber, independent of the collaborations that used it. This type of
    heterogeneous Fermilab/collaboration social structure with a dedicated
    support group is not evidenced in even the largest fixed-target counter
    experiments, but it is interesting to note that a similar phenomenon
    (dedicated support departments) does appear with the advent of the
    mammoth collider detectors such as CDF and D0. For historical details
    see Mark Bodnarczuk, ed., *Reflections on the Fifteen Foot Bubble
    Chamber at Fermilab* (Batavia: Fermilab, 1989).

16 Melvin Schwartz shows how tenuous these socially negotiated walls are for today's large collaborations when he advocates divorcing some of the detector builders from the collaboration, then subdividing the remaining members of these megacollaborations into distinct (smaller) collaborations that would develop their own research programs and compete for time using the detector. In a sense, Schwartz is advocating a return to a social structure that is not unlike that displayed in large bubble chambers as I described in the previous note on the 15-foot bubble chamber at Fermilab. Also, see Faye Flam, "Big Physics Provokes Backlash," *Science 30* (11 September 1992), p. 1470.

17 Some collaborations (such as CDF) required members of the collaboration to run a certain number of data-taking shifts in order to have their name on publications, but many collaborations had no such policies.

18 In Chapter 20, Westfall refers to a similar type of nonbureaucratic, quick-and-dirty, frugal style at SLAC. Also see Catherine Westfall and Lillian Hoddeson, "Frugality and the Building of Fermilab, 1960–1972," to be published in *Technology and Culture*.

19 Kevles attributes the scientist's tendency to leave social factors out of their accounts of science to being accustomed to the literary convention of journal editors and the fact that many scientists consider themselves to be incompetent to write about anything except science itself. Daniel Kevles, *The Physicists: The History of a Scientific Community in Modern America* (New York: Alfred Knopf, 1978), p. x. Pickering claims that references to "judgments" or "agency" on the part of scientists are left out of scientists' accounts so that scientists are portrayed as passive observers of nature, with experiments appearing to be the supreme arbiters of competing theories. Pickering, *Constructing Quarks*, pp. 5–18. Latour claims that there are definable processes that operate to remove all aspects of the social and historical context in order that scientific "facts" do not appear to be socially constructed. See Latour and Woolgar, *Laboratory Life: The Construction of Scientific Facts*, pp. 176–183; and Latour, *Science in Action*, pp. 22–29.

20 It is interesting to note that in an address in honor of the 75th Anniversary of the Max Planck Institute for Physics in Munich, Germany, James D. Bjorken devoted a major portion of his visionary article to the problems associated with the sociology of large collaborations and the possibility that these social factors might have an effect on the physics itself. See James D. Bjorken, "Particle Physics – Where Do We Go from Here?" SLAC *Beam Line*, Vol. 22, Winter 1992, p. 10.

22

Comments on Accelerators, Detectors, and Laboratories

JOHN KRIGE

Born Capetown, South Africa, 1941; Ph.D., 1965, University of Pretoria, South Africa, (physical chemistry) and Ph.D., 1978, University of Sussex, England (philosophy); Historian at European University Institute, Florence, Italy; history and sociology of physics.

The most striking thing about the papers presented in this session is that, aside from Sharon Traweek,[1] the speakers have tended to gloss over or to ignore completely the presence of controversy and conflict in the treatment of their topics.

Of course, it is always dangerous for an historian to draw attention to this dimension of the way scientists present the past. We lay ourselves open to two kinds of charges. First, that we are simply interested in muckraking, in giving physicists a bad press, in seeking to wash dirty linen in public so as to create a sensation and to boost our own visibility. Second, while physicists admit that they do sometimes disagree, they also insist that the community rapidly converges on a shared understanding of events. Historians who stress controversy are simply exaggerating, blowing up out of proportion what are simply normal, unimportant differences of opinion between rational human beings.

For my part let me say at once that yes, we do perhaps have a tendency to concentrate on controversy. Writing history would be pretty boring otherwise! On the other hand, this is done not to titillate, but with far more important aims in mind. Indeed it amounts to a very different way of dealing with the past than that conventionally favored by scientists themselves.

Put crudely, scientists reflecting on their own history tend to start from the present and to cast their eyes back over the past, identifying highlights and allocating credit. History becomes thus a history of successes that follow incrementally one upon the other in a logical succession and lead up to the present state of the art. Historians, or at least historians with my sympathies, turn this approach on its head.

They start from the past and aim to move forward in time with the pro-
tagonists, bracketing what is now known to be "true." They see their
task as that of moving along with the research frontier as it actually
evolves, before the successes were known, before the false starts and
dead ends had emerged. One of their tasks, one might say, is that of
studying decision-making processes from within, of putting themselves
in the heads of a Judd, a Simon, or a Blewett in the early 1960s and of
trying to reconstruct the world as these agents lived it then, in all its
ambiguity and confusion.

There are many dimensions to this "constructionist" project. It has its
"loony left," it has its sober right, and it has its moderate center, where
I roughly situate myself. All these strands have one thing in common: a
rejection of a positivist view that sees scientific knowledge as the product
of cumulative, linear, rational growth on a bedrock of uncontroversial
empirical facts. That view has now been systematically discredited in
a body of literature that has grown steadily in quantity and quality for
the last two decades or more. And while one may disagree with some of
its tenets, there can be little doubt that it is has revitalized the history
and sociology of science, that in these fields there has been a remarkable
flowering of creative and innovative studies of how science is actually
practiced.

That said, let me move on to show, in a practical example, what a
difference this approach can make to how one sees the past. I shall take
Richter's Chapter 15 on the rise of colliding-beam machines to illustrate
the point. As he pointed out, and as Johnsen too mentioned in Chapter
16, one of the first machines of this type was the Fixed Field Alternating
Gradient (FFAG) machine developed in the late 1950s by the MURA
group. Yet if we look at Richter's very helpful and instructive first figure
(Fig. 15.1) we find that the FFAG is nowhere to be seen. The machine
has been written out of the history because it was never built! However,
to understand how it entered the scene and why it disappeared is to
throw important light on the history of colliding-beam accelerators.

An important study relevant to this question has been made by my
colleague Dominique Pestre in Volume II of the history of CERN.[2] I
can only summarize his findings very schematically here. He has explained that when the CERN Accelerator Conference was held in 1956,
the conference to which Richter has also referred, two considerations
dominated the thinking of physicists and machine builders. First, there
was a great interest in building high-intensity rather than high-energy
machines. Physicists were finding that as their experiments became

more sophisticated, they needed intensities far higher than those available on the Cosmotron and Bevatron. This interest was manifested in the decisions taken in the late 1950s to build the ZGS at Argonne (12 GeV, but 10^{12} protons/pulse) and Nimrod (7 GeV, 5×10^{11} protons/sec). Indeed until the summer of 1959 no accelerator with an energy above 35 GeV was seriously considered anywhere in the world. Second, there was a concern not to build a "monster," a machine that simply scaled up existing technologies. The history of accelerators had been the history of constant innovation, of which the most recent example had been the discovery of strong focusing. A new technological principle was sought for the new generation of machines.

It was in this context that the FFAG machine emerged as a strong candidate for support. On the one hand, it involved a new technological idea, the use of a magnetic field that did not change with time. This idea was important, because it opened the possibility of beam accumulation, and so of achieving intensities which, it was thought at the time, might be 1000 times greater than those expected with the CERN PS and the Brookhaven AGS, then under construction. And it had the further advantage that, if one coupled two such machines so that their internal beams intersected with one another, one could reach the much higher center-of-mass energies allowed with colliding beams and still probably have a high enough intensity to do good physics. The MURA group's FFAG machine thus seemed to satisfy the two main criteria required for the next generation of machines as well as opening the way to having both high energy and high intensity in the same device.

The terms of the debate changed dramatically, however, once the PS and the AGS were commissioned at the end of 1959 and in mid-1960. They immediately gave 10^{10} protons/pulse, intensities far higher than anticipated – which were rapidly improved. It became clear then that there was no need to trade high intensity for high energy in a fixed-target machine: one could have both simultaneously.

Once this was clear, a new series of design studies on a high-energy PS got under way. Matthew Sands and his group at Caltech worked on machines in the 100–300 GeV range, while at Brookhaven a group led by John Blewett explored the feasibility of a 300–1000 GeV machine. At the same time the mental block to building such erstwhile "monsters" withered away. Wolfgang Panofsky did the trick for Sands: "It was a little frightening," said Sands, "to think of a one-mile diameter machine but Panofsky has removed most of those psychological disadvantages."

The studies on the East and West Coasts confirmed the feasibility of building a big fixed-target accelerator. Called upon to choose between such a machine and an FFAG-type device, a panel headed by Norman Ramsey effectively buried the MURA option. In 1963 they recommended the construction, by MURA, of a super-current accelerator, but added that this was not to be at the cost of taking steps toward high energies. Given the prevailing financial constraints, this amounted to a kiss of death for MURA.

The point of telling this story is not simply to stress that a history of colliding-beam machines that excludes "drop-outs" like the FFAG machine, as in Richter's diagram, is incomplete. It is also to stress, what is not at all clear from his paper, that the history of colliding-beam machines is intimately interwoven with that of fixed-target accelerators. It is really meaningless to try to treat the one without the other.

There is another interesting "anomaly" in this figure that can be used to bring this point home and extend it. Look at the CERN Intersecting Storage Rings (ISR): the only hadron collider approved in the 1960s and commissioned in the 1970s! What has happened here? Did physicists in Europe think differently from those in the United States? Did they prefer this type of machine to a 300 GeV fixed-target accelerator? Certainly not. As Dominique Pestre again has shown, European physicists insisted in the mid-1960s that if the ISR were built, it was not to be at the expense of a 300 GeV fixed-target machine. They lost the first round of that battle, first because an extremely determined in-house group of CERN engineers were fascinated by the technological challenge of building storage rings, and second because Director General Victor Weisskopf, considering the financial climate of the time, decided on a two-step program in which he would try to sell the ISR to Member States' governments first, and then raise funding for the 300 GeV machine. Member States could hardly believe their luck: here was CERN asking for money for a device, the ISR, that was four times cheaper than the alternative. They jumped at the opportunity – and of course, contrary to what Weisskopf had hoped, took another five or six years to agree to fund a 300 GeV accelerator. The history of the ISR, like that of the FFAG, is intimately tied up with that of fixed-target machines, and by extension, with important choices involving conflicts between in-house groups, rivalries between various national groups, selling tactics to funding agencies and governments, and so on. All of these elements are essential components of the history of colliding-beam machines, and they have simply dropped out of sight in Richter's (and Johnsen's) presentation.

Let me now touch briefly on Robert Wilson's chapter and in particular on his emphasis that Fermilab become a "Users' Paradise." I should like to look at two aspects here: first his remarks on the relationship between in-house staff and outside users, and second, on his comments about the committee structure that he set up to manage the interface between the laboratory and its users – a structure that he apologetically describes as "outrageously bureaucratic." I shall question Wilson's arguments here by drawing on parallel developments at CERN.

I was struck by Wilson's remark that he decided to keep the fraction of Fermilab physicists to outside users to 0.25, adding that this proportion was needed because users would have to have "the assistance of a core of very good Fermilab physicists, not only to set up facilities but also to provide the laboratory's help on the experiments being done." Now what is striking about these claims is that *exactly* the same justifications were used by the CERN management in the 1960s to have a strong in-house staff. And although users accepted them initially, when they were relatively inexperienced and rather overawed by the laboratory, by the early 1970s they were complaining bitterly about being treated as second-class citizens. What is more, it is striking that in the big experiments today at LEP the ratio is more like 0.1, and that in any case this kind of justification for including in-house staff in an experiment no longer has any meaning. In other words, one cannot help feeling that Wilson's justifications were typical of what a laboratory director might have put forward at that time, and while they obviously had some merit, one wonders just how the users reacted to them at the time. Did they accept the arguments for having a relatively high ratio of in-house staff in the experimental teams? Or did they, like the CERN users in the early 1970s, regard the arguments as patronizing and simply a pretext for controlling the experimental program? Again, what conflict there may have been with users – a key question surely for the "would-be users' paradise" (Wilson) – while hinted at, has been sidestepped.

One also wonders why the presence of a committee structure is necessarily to be regarded as "outrageously bureaucratic." It is striking in this connection that at CERN, where there was a similar pressure on the laboratory to serve its users, and where the experiments committee structure was put in place in the early 1960s, users never complained about this aspect of the laboratory except right at the start. Of course, physicists have sometimes found it burdensome to spend so much time away from research at meetings. But the idea of having the committees as such and their modes of functioning have generally been regarded by the community both in and outside of CERN as satisfactory.

In Wilson's spontaneous reaction against "bureaucracy," I think we may have an important difference between American and European physicists. It is a difference which, in American eyes, led to CERN's backwardness vis-à-vis the States, at least until recently. This is a view that I would challenge. There is no doubt that CERN performed poorly compared to similar American laboratories, particularly in the 1960s and 1970s. But this was not, I believe, because the laboratory was too "bureaucratic." It was rather because CERN physicists lacked the experience of their American colleagues in setting up and managing a major experimental program. Remember, they did not have an equivalent to the Cosmotron or the Bevatron on which to learn how to do big physics. Springing from this there was a ratchet effect, in which their self-confidence gradually drained away as they saw one important discovery after another elude their grasp. Finally, and most fundamentally, they had a very different way of doing physics. In particular, there were sharp differences and divisions between physicists, equipment builders, and machine engineers at CERN that one simply does not find in the United States. This fragmentation in skills was mirrored in a fragmentation in organization and labor that had serious effects on CERN's ability to compete. In short, a "heavy" committee structure as such cannot account for CERN's inability to compete; that is rather to be traced back to a number of historical and structural peculiarities of the European physics community. Correlatively, it is no coincidence that it was Carlo Rubbia, an American-style physicist in the sense that he was individualistic, entrepreneurial, ambitious and that he combined these skills in one man, who shared CERN's first Nobel prize.

Notes

1 Sharon Traweek's paper is not published in this volume.
2 Dominique Pestre, "The second generation of accelerators for CERN, 1956–1965: the decision-making process," in Armin Hermann, et al., eds., *History of CERN*, Vol. II (Amsterdam: North-Holland, 1990), pp. 679–780.

Part five

Electroweak Unification

23

The First Gauge Theory of the Weak Interactions

SIDNEY BLUDMAN

Born New York City, 1927; Ph.D., 1951 (physics), Yale University; Professor of Physics and Astronomy at the University of Pennsylvania; theoretical particle and astrophysics and cosmology.

The electroweak sector of the Standard Model[1] contains three logically and historically distinct elements:

1. A chiral gauge theory of weak interactions with an exact $SU(2)_L$ symmetry;[2]
2. The Higgs mechanism[3] for spontaneous symmetry breaking, giving some of the gauge bosons finite masses, while maintaining renormalizability;[4]
3. Electroweak unification through W^0–B^0 mixing by $\sin\theta_W$.[5]

This report is concerned with the early history of the electroweak sector of the Standard Model. I first recall the history of gauge theories in the 1950s and my own motivation for publishing[6] the first chiral gauge theory of weak interactions, predicting weak neutral currents of exact V–A form and approximately the weak strength observed fifteen years later.[7] Then I discuss the evolving appreciation of the fundamental distinctions between global and gauge, partial and exact symmetries, in the weak and strong interactions. Finally, I emphasize that exact gauge symmetry is necessary for the Higgs mechanism for symmetry breaking, but that electroweak unification is not required theoretically: Within the Standard Model, the electroweak mixing angle, $\sin\theta_W$, is not determined, but could have any value, including zero. This leads to an interesting difference between the $\sin^2\theta_W = 0$ limit of the unified electroweak theory and the original $SU(2)_W$ gauge theory of weak interactions alone.

Theoretical consistency requires that a field theory be renormalizable, not necessarily unified. Nevertheless, *historically* the discovery of weak neutral currents with electroweak mixing angle $\sin^2 \theta_W \sim (0.2\text{–}0.3)$ provided evidence for the electroweak sector of the Standard Model and drove the search for massive gauge bosons that were finally observed in 1982–83.

Non-Abelian gauge theory for the weak interactions

Pauli's arguments[8] and the successes of QED in the 1940s had established the importance of electromagnetic gauge invariance. Indeed, in simple enough theories, it led to minimal electromagnetic interactions that were renormalizable. For charged vector mesons, however, minimal electromagnetic interaction was ambiguous[9] and the theory was non-renormalizable. The divergences derive from the longitudinal component of the massive vector meson field and are minimal if the gyromagnetic ratio $g = 2$ and the electric quadrupole moment $Q = -e(\hbar/Mc)^2$. (In the Standard Model, the electroweak scale acts as a regulator for the longitudinal vector meson field, making vector meson electrodynamics renormalizable for just these electromagnetic moments.)

I had always been impressed by the Noether's Second Theorem. While her First Theorem asserted that global symmetries of the Lagrangian implied well-known conservation laws, her Second Theorem was much more powerful: *local* Lagrangian symmetries implied *new* (gauge) fields. This, together with the Yang–Mills theory, led to my first publication, which showed that the then-current pion–nucleon interaction could not be derived directly from a gauge principle.[10] Because I was always motivated only by exact gauge symmetries,[11] I did not think to make the axial current partially conserved, or the pseudoscalar pion a pseudo-Goldstone boson. While approximate flavor SU(3) gauge symmetries led Sakurai[12] to hadronic vector mesons, we now realize that only color is an exact hadronic symmetry and that the approximate flavor symmetries derive from the mass hierarchy of quarks in QCD.

Even before the experimental situation clarified in 1957, Sudarshan and Marshak,[13] Feynman and Gell-Mann,[14] and Sakurai[15] each immediately presented their own derivations of the V–A β-decay interaction. My own derivation followed from what I called Fermi gauge invariance, generated by charge-raising and charge-lowering chiral Fermi charges F^+, F^-.[16] If the algebra of generators is to close, then neutral Fermi charges $2iF^0 = [F^+, F^-]$ are required, that is, SU(2)$_L$ is the minimal symmetry of the Fermi interactions. I went on to impose this symmetry locally and was led to an SU(2)$_L$ triplet of gauge bosons, $W^{\pm,0}$, coupled to a triplet of chiral Fermi currents $F_\mu^{\pm,0}$. This chiral gauge theory predicted weak neutral currents of exact $V-A$ form and the same strength as the weak charged currents. The observed strength of the Fermi interactions, $G_F/\sqrt{2} = g^2/8M_W$, then required, in tree approximation, $M_W = gv/2$, where $v \equiv (\sqrt{2}G_F)^{-1/2} = 246$ GeV. Neither g nor M_W was predicted separately, but, if the field theory was to be perturbative, then $g < 1$, so that $M_W < 123$ GeV was to be expected.

No attempt was made to provide a mechanism for giving the intermediate vector bosons mass, to unify weak with electromagnetic interactions, or to explain the absence of flavor-changing weak neutral currents (WNC). Flavor-changing WNC were known to be absent to $\mathcal{O}(10^{-8})$ and even flavor-conserving WNC were incorrectly reported to be at least 30 times weaker than charged currents.[17] [Ultimately, the absence of flavor-changing WNC at tree level and the reduction of their radiatively induced $\mathcal{O}(G_F\alpha)$ amplitude by a suitably small factor $(m_c^2 - m_u^2)/M_W^2$, where m_c and m_u are quark masses and M_W is the mass of the W boson, was explained by Glashow, Iliopoulos and Maiani.[18]]

My 1958 paper was soon followed by proposals to use accelerator neutrino beams to search for flavor-conserving weak neutral currents.[19] But this search remained very difficult, because of high backgrounds from neutrino-induced charged-current processes in which muons escaped undetected, which were hard to estimate. Thus, neutrino experiments began only in 1968, and were, for several years, preoccupied with deep-inelastic scattering at SLAC and with scaling. These experimental difficulties, together with the need for a consistent theory allowing massive gauge bosons, suggest why chiral weak neutral currents needed to wait from 1958 to 1973 for experimental confirmation.

Spontaneously broken global symmetries

The idea of spontaneous symmetry breaking, better denominated "hidden symmetry," was brought from condensed matter physics to quantum field theory by Heisenberg and by Nambu.[20] It soon led to the Goldstone theorem showing how spontaneous symmetry breaking could produce long-range interactions out of a short-range theory.[21] Klein and I identified the Goldstone bosons expected from different levels of global symmetry breaking and emphasized that Goldstone bosons were not present in theories with long-range interactions. Following Anderson,[22] we suggested that, in an inverse Goldstone theorem, long-range interactions might be converted into short-range. We observed that, although apparently massless, the neutrino could not be a Goldstone particle, because the vacuum could not be macroscopically occupied by fermions. We therefore failed to associate my earlier proposal of a *gauge* theory of weak interactions with the Goldstone theorem.

The 1958 work on chiral invariance was cited by Gell-Mann and by Nambu and ultimately led to current algebras, soft-pion theorems, and PCAC.[23] These successes, however, tended to gloss over the fundamental differences between global and gauge symmetries, and between partial flavor symmetries and exact gauge symmetries, in the strong and weak interactions.

Exact symmetries were useful in classifying fields and particles, and were most satisfying aesthetically. For these reasons I tended to avoid hadron physics and concentrated on weak interactions where, I was convinced, exact symmetries were to be found. At this time, I left the University of California Radiation Laboratory, which was then dominated by dispersion relations and S-matrix theory. I took an academic position at the University of Pennsylvania, and my interests gradually shifted from laboratory to astrophysical particle physics.

Spontaneously broken gauge symmetries

The Higgs mechanism[24] sharply differentiates the role Goldstone bosons play in gauge theories from their role in global symmetry theories. If an exact gauge symmetry is spontaneusly broken by the Higgs mechanism, so that some gauge bosons acquire masses, the symmetry is hidden but the theory remains renormalizable. Indeed, the Standard Model has only exact gauge symmetries, so that massless gauge bosons are usually not manifest: Either the (color) gauge symmetry is unbroken, but the massless gluons are confined, or the gauge symmetry is spontaneously broken, providing masses for the gauge bosons other than the photon.

Weinberg and Salam proposed the Electroweak Standard Model, conjecturing that the theory would remain renormalizable.[25] Nevertheless, the 1967 Weinberg paper was referred to by no one (including Weinberg) in 1967–70 and only once in 1971.[26] Finally, 't Hooft proved that such a theory remained renormalizable.[27] In this way, a complete theory of massive charged vector bosons and of weak interactions was achieved.[28]

In the minimal Standard Model, electroweak couplings enter through the $SU(2)_L \times U(1)_Y$ covariant derivative

$$D_\mu = \partial_\mu - ig\mathbf{T} \cdot \mathbf{W}_\mu - ig'(Y/2)B_\mu,$$

so that: (1) The charged vector bosons couple to the electromagnetic field with magnetic moment $2(e\hbar/2Mc)$ and electric quadrupole moment $-(\hbar/Mc)^2$; (2) the WNC couple to Z^0 with coupling constant $g/\cos\theta_W$, where $\tan\theta_W \equiv g'/g$; (3) charged currents couple to the electromagnetic field with coupling constant $e \equiv g \sin\theta_W \equiv g' \cos\theta_W$. The Higgs mechanism gives the vector bosons (generally) unequal masses $M_W = M_Z \cos\theta_W$. In tree approximation,

$$M_W \sin\theta_W = M_Z \sin\theta_W \cos\theta_W =$$

$$(e/2)\left(\sqrt{2}G_F\right)^{-1/2} = \sqrt{\pi\alpha}v$$

$$\equiv A_0 = 37.3\,\text{GeV}.$$

If the theory were not unified, we could have had either spontaneously broken U(1) symmetry with no weak currents ($g = 0$, $g' = e$, Schwinger's electrodynamics with massive photons) or Bludman's weak SU(2)$_L$ symmetry ($g' = 0 = e$, WNC with coupling constant g).[29] These two examples illustrate the logical possibility of consistent (renormalizable) theories without unification. In a perturbative SU(2)$_W$ theory without unification, only the constraint $g < 1$, $M_W < 123$ GeV obtains.

In a unified theory, however, $e^{-2} = g^{-2} + g'^{-2}$ so that g, $g' \geq e$, and $M_W > 37$ GeV, $M_Z > 74$ GeV. The electromagnetic field exists and B^0 and W^0 mix, but the observed value, $\sin^2\theta_W \approx 0.23$, remains unexplained within the Standard Model. In the electroweak sector of the Standard Model, besides G_F and e, there is only one free parameter

$$\sin^2\theta_W \equiv \left(M_Z^2 - M_W^2\right)/M_Z^2 = (1/2)\left[1 - \sqrt{1 - (2A_0/M_Z)^2}\,\right]$$

in tree approximation, which measures the SU(2)$_L$ symmetry breaking through W^0–B^0 mixing.

This unification condition is nontrivial, even in the $\sin^2\theta_W \to 0$ limit. In a unified theory, holding G_F and e constant as $\sin^2\theta_W \to 0$, makes $g' \to e$ as g, M_W, M_Z all diverge. The $\sin^2\theta_W = 0$ limit of a unified theory would be one with unmixed electromagnetic interactions and weak interactions that are pointlike in tree approximation, but unitary and renormalizable because of huge radiative corrections! But if the theory is to be both perturbative and unified, $e \leq g$, $g' < 1$, $0.0836 < \sin^2\theta_W < 0.916$, and 37 GeV $< M_W < 123$ GeV, 74 GeV $< M_Z < 123$ GeV. In a perturbative theory, unification determines the vector boson masses within a factor of 2 or 4!

For energies $\gg$ 10 GeV, the mass differences between W and Z bosons and among the quarks (other than the top quark) can be neglected and the original $SU(2)_L$ theory will then be a good approximation to leptonic and semileptonic processes, other than top quark decay. Thus, the qualitative effects of unification practically disappear already at energies $\gg$ 10 GeV, much lower than the unification scale at which symmetry-breaking disappears.

Historical Conclusions

The proof that, in an exact gauge theory, renormalizability would persist even as the Higgs mechanism gave masses to some of the gauge bosons, immediately converted many theorists to the Standard Model. Logically, a consistent theory without electroweak mixing was conceivable. Nevertheless, the discovery of weak neutral currents[30] with mixing $\sin^2 \theta_W \sim 0.3$ gave circumstantial evidence for the electroweak Standard Model and predicted $M_W \approx 80$ GeV, $M_Z \approx 90$ GeV. The ultimate discovery of these gauge bosons with unequal masses then directly confirmed the electroweak Standard Model.[31]

Notes

1 S. L. Glashow, "Towards a Unified Theory: Threads in a Tapestry," *Rev. Mod. Phys. 52* (1980), pp. 539–43; A. Salam, "Gauge Unification of Fundamental Forces," *Rev. Mod. Phys. 52* (1980), pp. 525–38; S. Weinberg, "Conceptual Foundations of the Unified Theory of Weak and Electromagnetic Interactions," *Rev. Mod. Phys. 52* (1980), pp. 515–23.
2 S. Bludman, "On the Universal Fermi Interaction," *Nuovo Cimento 9* (1958), pp. 433–44.
3 P. W. Higgs, "Broken Symmetries, Massless Particles and Gauge Fields," *Phys. Lett. 12* (1964), pp. 132–3; "Spontaneous Symmetry Breaking without Massless Bosons," *Phys. Lett. 145* (1966), pp. 1156–63; F. Englert and R. Brout, "Broken Symmetry and the Mass of Gauge Vector Mesons," *Phys. Rev. Lett. 13* (1964), pp. 321–3; T. W. Kibble, "Symmetry Breaking in Non-Abelian Gauge Theories," *Phys. Lett. 155* (1967), pp. 1554–61; G. S. Guralnik, C. R. Hagen, and T. W. B. Kibble, "Global Conservation Laws and Massless Particles," *Phys. Rev. Lett. 13* (1964), pp. 585–7.
4 G. 't Hooft, "Renormalizable Lagrangians for Massive Yang–Mills Fields," *Nucl. Phys. B35* (1971), pp. 167–88; G. 't Hooft and M. Veltman, "Regularization and Renormalization of Gauge Fields," *Nucl. Phys. B44* (1972), pp. 189–213; G. 't Hooft, "Combinations of Gauge Fields," *Nucl. Phys. B50* (1972), pp. 318–53; B. W. Lee and J. Zinn-Justin, "Spontaneously Broken Gauge Symmetries. I. Preliminaries," *Phys. Rev. D5* (1972), pp. 3121–37.

5 S. L. Glashow, "Partial Symmetries of Weak Interactions," *Nucl. Phys.*
   22 (1961), pp. 579–88.

6 S. Bludman, "On the Universal Fermi Interaction."

7 F. J. Hasert, et al., "Search for Elastic Muon–Neutrino Electron
   Scattering," *Phys. Lett. 46B* (1973), pp. 121–4; A. Benvenuti, et al.,
   "Observation of Muonless Neutrino-Induced Inelastic Interactions," *Phys.*
   Rev. Lett. 32 (1974), pp. 800–3.

8 W. Pauli, "Relativistic Field Theories of Elementary Particles," *Rev.*
   Mod. Phys. 13 (1941), pp. 203–32.

9 J. A. Young and S. Bludman, "Electromagnetic Properties of a Charged
   Vector Meson," *Phys. Rev. 131* (1963), pp. 2326-34; G. Feinberg,
   "Decays of the μ meson in the Intermediate-Meson Theory," *Phys. Rev.*
   110 (1958), pp. 1482–3; T. Kuo-Hsien, "Charged Vector Field of Zero
   Proper Mass," *C.R. Acad. Sci. (Paris) 245* (1957), p. 289.

10 C. N. Yang and R. L. Mills, "Conservation of Isotopic Spin and Isotopic
   Gauge Invariance," *Phys. Rev. 96* (1954), pp. 191–5. S. A. Bludman,
   "Extended Isotopic Invariance and Meson–Nucleon Coupling," *Phys.*
   Rev. 100 (1955), pp. 372–5.

11 Many earlier authors: N. Kemmer, "Field Theory of Nuclear
   Interactions," *Phys. Rev. 52* (1937), pp. 906–10; E. Teller, "Scattering of
   Neutrons by Ortho- and Para-Hydrogen," *Phys. Rev. 52* (1937), pp.
   286–95; G. Wentzel, "β-interaction," *Helv. Phys. Acta 10* (1937), pp.
   107–11 had proposed *approximate global* SU(2) symmetries for the *strong*
   interactions.

12 J. J. Sakurai, "Theory of Strong Interactions," *Ann. Phys. (N.Y.) 11*
   (1960), pp. 1–48.

13 E. C. G. Sudarshan and R. E. Marshak, "The Nature of the
   Four-Fermion Interaction," *Proc. Padua–Venice Conference on Mesons*
   and Newly Discovered Particles (Bologna, 1958); "Chirality Invariance
   and the Universal Fermi Interaction," *Phys. Rev. 109* (1958), pp. 1860–2.

14 R. P. Feynman and M. Gell-Mann, "Theory of Fermi Interaction," *Phys.*
   Rev. 109 (1958), pp. 193–8.

15 J. J. Sakurai, "Mass Reversal and Weak Interactions," *Nuovo Cimento 7*
   (1958), pp. 649–60.

16 S. Bludman, "On the Universal Fermi Interaction."

17 M. M. Block, et al., "Neutrino Interactions in the CERN Heavy Liquid
   Bubble Chamber," *Phys. Lett. 12* (1964), pp. 281–5; H. H. Bingam et
   al., "CERN Neutrino Experiment – Preliminary Results," *Proceedings of*
   the Siena International Conference on Elementary Particles (1963), pp.
   555-70; see also Chapter 25, this volume.

18 S. L. Glashow, J. Iliopoulos, and L. Maiani, "Weak Interactions with
   Lepton–Hadron Symmetry," *Phys. Rev. D2* (1970), pp. 1285–92.

19 T. D. Lee and C. N. Yang, "Theoretical Discussions on Possible
   High-Energy Neutrino Experiments," *Phys. Rev. Lett. 4* (1960), pp.
   307–11; B. Pontecorvo, "Small Probability of the $\mu \to e + \gamma$ and
   $\mu \to e + e + e$ Processes and Neutral Currents in Weak Interactions,"
   Phys. Lett. 1 (1962), pp. 287–8; S. S. Gershtein, N. Van Hieu, and
   R. A. Eramzhyan, "On the Possibility of Finding Neutral Currents in
   Neutrino Experiments," *Zh. Eksp. Theor. Fiz. 43* (1962), pp. 1554–6 (in
   Russian).

20 H.-P. Dürr, W. Heisenberg, et al., "Theory of Elementary Particles,"

Zeitschrift für Naturforschung 14a (1959), pp. 441–85; Y. Nambu, "Axial Vector Current Conservations in Weak Interactions," *Phys. Rev. Lett. 4* (1960), pp. 380–2; Y. Nambu and G. Jona-Lasinio, "A Dynamical Model of Elementary Particles Based upon an Analogy with Superconductivity," *Phys. Rev. 122* (1961), pp. 345–58; "A Dynamical Model of Elementary Particles Based upon an Analogy with Superconductivity," *Phys. Rev. 124* (1961), pp. 246–54.

21 J. Goldstone, "Field Theories with 'Superconductor' Solutions," *Nuovo Cimento 19* (1961), pp. 154–64; J. Goldstone, A. Salam, and S. Weinberg, "Broken Symmetries," *Phys. Rev. 127* (1962), pp. 965–70; J. C. Taylor, "ρ-Mesons and the Yang–Mills Field," *Proc. 1962 Intl. Conf. on High-Energy Physics at CERN* (Geneva: CERN, 1962), pp. 670–2; S. A. Bludman and A. Klein, "Broken Symmetries and Massless Particles," *Phys. Rev. 131* (1963), pp. 2364–72.

22 P. W. Anderson, "Plasmons, Gauge Invariance and Mass," *Phys. Rev. 130* (1962), pp. 439–42.

23 M. Gell-Mann, "Conserved and Partially Conserved Currents in the Theory of Weak Interactions," *Proc. 1960 Annual Intl. Conf. on High-Energy Physics at Rochester* (New York: Interscience Publishers, 1960), pp. 508–12; Y. Nambu and F. Lurie, "Chirality Conservation and Soft Pion Production," *Phys. Rev. 125* (1960), pp. 1429–36.

24 P. W. Anderson, "Plasmons"; P. W. Higgs, "Broken Symmetries" and "Spontaneous Symmetry Breaking"; F. Englert and R. Brout, "Broken Symmetry"; T. W. Kibble, "Symmetry Breaking"; G. S. Guralnik, C. R. Hagen, and T. W. B. Kibble, "Global Conservation Laws."

25 S. Weinberg, "A Model of Leptons," *Phys. Rev. Lett. 19* (1967), pp. 1264–6; A. Salam, "Weak and Electromagnetic Interactions," in N. Svartholm, ed., *Elementary Particle Theory, Proc. 8th Nobel Symposium* (Stockholm: Almqvist & Wiksell,, 1968).

26 S. Coleman, "The 1979 Nobel Prize in Physics," Science 206 (1979), pp. 1290–2.

27 G. 't Hooft, "Renormalizable Lagrangians "; G. 't Hooft and M. Veltman, "Regularization and Renormalization"; G. 't Hooft, "Combinations of Gauge Fields"; B. W. Lee and J. Zinn-Justin, "Spontaneously Broken."

28 S. L. Glashow, "Towards a Unified Theory"; A. Salam, "Gauge Unification"; S. Weinberg, "Conceptual Foundations."

29 J. Schwinger, "Gauge Invariance and Mass," *Phys. Lett. 125* (1962), pp. 397–8; S. Bludman, "On the Universal Fermi Interaction."

30 F. J. Hasert, et al., "Search"; A. Benvenuti, et al., "Observation."

31 G. Arnison, et al., "Experimental Observation of Isolated Large Transverse Energy Electrons with Associated Missing Energy at $\sqrt{s} = 540$ GeV," *Phys. Lett. 122B* (1982), pp. 103–16; "Experimental Observation of Lepton Pairs of Invariant Mass around 95 GeV/c^2 at the CERN SPS Collider," *Phys. Lett. 126B* (1983), pp. 398–410; M. Banner, et al., "Observation of Single Isolated Electrons of High Transverse Momentum in Events with Missing Transverse Energy at the CERN $\bar{p}p$ Collider," *Phys. Lett. 122B* (1983), pp. 476–85; P. Bagnaia, et al., "Evidence for $Z^0 \rightarrow e^+e^-$ at the CERN $\bar{p}p$ Collider," *Phys. Lett. 129B* (1983), pp. 130–40.

24

The Early History of
High-Energy Neutrino Physics

MELVIN SCHWARTZ

Born New York City, 1932; Ph.D., 1958 (physics), Columbia University; I. I. Rabi Professor of Physics, and Associate Director, Brookhaven National Laboratory; Nobel Prize in Physics, 1988; high-energy physics (experimental).

The experiment that led to the discovery of the muon neutrino was the largest experiment that had ever been mounted at Brookhaven at its time. The experimental team consisted of only seven people – three professors, three graduate students, and one physicist from the AGS (Alternating Gradient Synchrotron) department. We fashioned the biggest detector that had ever been built at that time, consisting mainly of ten tons of aluminum. It was an experiment in which we ended up having a lot of fun and made some important progress. This chapter will discuss this experiment, the first high-energy neutrino experiment, and mention a few developments that have occurred in neutrino scattering since that time.[1]

What was the state of particle theory in 1959, when planning for this experiment began? In general, theory was in a fairly primitive state: V–A and parity violation were well understood, and there was a general universality among weak interactions involving muons, electrons, nucleons, and neutrinos. And everything was relatively consistent with a simple four-fermion point vertex: the Fermi theory. There had been one prior neutrino experiment, done by Clyde Cowan and Fred Reines – the classic experiment, one of the most beautiful experiments of the 1950s – in which antineutrinos produced in a nuclear reactor gave rise to a reaction in which an antineutrino and a proton yielded a neutron and a positron. The cross section for this reaction was on the order of 10^{-43} square centimeters; it was a real tour de force to be able to find those events and make a measurement of this tiny cross section.

Another significant point in the theory of that time was the absence of the decay $\mu \rightarrow e + \gamma$. Quite a few people had worked on this very

411

intriguing subject. It was an experiment that could be done beautifully at a cyclotron and had been done to a sensitivity of about 10^{-7} or 10^{-8}. Nobody had witnessed such a decay. (Indeed, it's still a very topical subject; people are doing experiments at Los Alamos and at Brookhaven today, searching for flavor-changing weak interactions.) At that time, in the late 1950s, Gary Feinberg had calculated that if a muon decayed into an electron and a photon by means of an intermediate boson,[2] then the branching ratio should be of the order of 10^{-4}. Measurements at that time gave a limit of about 10^{-8}, which was a strong argument against the existence of an intermediate boson – unless (as Feinberg pointed out at the end of his paper) the two neutrinos were somehow *different*: the electron neutrino and the muon neutrino were not the same object.

In November 1959, just at the point I'm talking about here, I happened to be down at Columbia one Thursday afternoon, at the usual time we had our coffee hour. T. D. Lee and a number of others were arguing about how to investigate weak interactions at high energies. Indeed, it wasn't at all obvious. One thing that they suggested was the possibility of using electrons – for example, examining the cross section for electron scattering and trying to polarize the electrons. This experiment was in fact done over a decade later. (See Charles Prescott, Chapter 27.) But none of the ideas that went across the blackboard that day seemed to be at all feasible at the time.

That evening it occurred to me that the simplest way to investigate the weak interactions was to make a beam of neutrinos. In fact, because of the energies of the pion beams that were soon to be available at Brookhaven and CERN, one could actually obtain a natural collimation of the neutrinos coming from pion decay. With a sufficiently high flux, one could observe neutrino scattering and measure the cross section. It required a machine producing 10^{11} or 10^{12} protons per second, at energies of about 10 GeV. At the time there was no machine planned that was expected to achieve this level, except possibly the MURA machine. Remember, the AGS was supposed to have only about 10^9 circulating protons per second. But it eventually reached 10^{11}, which turned out to be a very important factor that allowed us to do our experiment.

Another interesting point, which was very compelling to us, was that in 1960 Lee and Yang observed that any mechanism one could use to salvage unitarity ought to give rise to a $\mu \rightarrow e + \gamma$ decay. Their argument was that the cross section, if it were strictly a four-fermion point interaction, must rise as the square of the center-of-mass energy. On the other hand, the unitarity limit for an S-wave interaction is $\lambda^2/4\pi$ (where

λ is the de Broglie wavelength), which would be reached at about 300 GeV. So somehow there must be – in order to damp that rise and to prevent the limit from being exceeded – some *size* to the interaction region. But once the interaction region had size, they argued, you would then have charges and currents, which would lead naturally to a $\mu \rightarrow e + \gamma$ decay of roughly the same magnitude as from an intermediate boson. Because that decay hadn't been observed, there was a clear implication that there were two different types of neutrinos.

Although I didn't know about it at that time, Bruno Pontecorvo had had many of the same ideas independently. He published one experimental idea (which was not terribly good) along these lines – to stop a beam of π^+ mesons and look at the neutrinos coming from them to see if they made electrons. I mention this mainly to indicate that a lot of thinking was going on at the time about the possibility of there being two different kinds of neutrinos.

In 1960 we began forming an experimental team at Columbia to do this experiment. (It's interesting to review all the records, now that I'm back at Brookhaven. I've accessed a lot of stuff down in the files in the basement that I didn't quite remember.) There were actually *two* independent neutrino groups formed at Columbia. Jack Steinberger and I had a freon bubble chamber that we wanted to use. It had only 250 kg of target material, but first calculations – and, of course, first calculations are always optimistic – indicated we might get an event a day with that bubble chamber. There was also an electronic detector that Leon Lederman and I began considering. We looked at all kinds of possibilities, from sets of Geiger counters with sheets of steel between them to large tanks of liquid scintillator. None of these ideas looked very good; in fact, they didn't look promising at all.

Sometime in 1960, however, I happened to learn while at the Cosmotron that Jim Cronin had a little spark chamber operating down at Princeton. It sounded like exactly the thing we needed, so Leon, Jean-Marc Gaillard, and I went down there to take a look at it. When we saw the chamber operating – and it was a very small chamber, just a set of perhaps a half dozen plates, an absolutely elegant-looking chamber – producing beautiful sparks, it was clear that it could be scaled up into a detector weighing the order of ten tons. From that point on, the electronic detector group was working toward building a spark chamber.

The two groups put together a joint proposal in May 1960 that said we would first try a freon bubble chamber, but if that didn't work we'd move on to a spark chamber. In the fall of that year Steinberger went to

CERN because its proton synchrotron was likely to come on earlier than Brookhaven's, and he felt that he would like to work with the bubble chambers there. So Leon and I continued building the spark chamber. Jack came back in September 1961 after his CERN experiment ran into difficulties because of a miscalculation of the particle fluxes coming out of the 5 foot straight section. It was very clear from that point on that the group would do an experiment using a large spark chamber – even though its exact design was still not clear. The spark chamber eventually built at Brookhaven had aluminum plates and weighed a total of ten tons.

The group consisted of Gordon Danby of Brookhaven, plus Gaillard, Constantin Goulianos, Lederman, Nariman Mistry, Steinberger, and myself – all of Columbia. We set up the chamber at the AGS in late 1961. They turned on the beam a bit, and we were immediately flooded with garbage. The chamber was absolutely white with junk. We had to turn the machine down to roughly half its rated energy, or about 15 GeV. The AGS gave us one pulse every 1.2 seconds with about 10^{11} 15-GeV protons per pulse. We had to make the spill as short as possible because of the cosmic-ray background. In those days, the spill couldn't be any shorter than 30 microseconds, and it consisted of 20-nanosecond bursts spaced 220 nanoseconds apart. For each second of real time, we had only 2×10^{-6} second of beam time. In fact, the entire experiment took only 5 seconds – a very, very cost effective operation. That was important because cosmic-ray background was a serious issue for this experiment.

One anecdote will help me to illustrate how much things have changed in high-energy physics. I was going through all this old paper, and I found this brief agreement between the laboratory and our group (Fig. 24.1). Nowadays that would take many pages of paper, but then it was all on a single page. A curious item labeled "Health Physics and General Safety Requirements" mentions only that "spark chambers require 7 kilovolts to fire." That was the sum total of our concern for health physics and general safety! By the way, we certainly found the right problem. I almost got killed by a 7 kilovolt shock when Leon walked in and turned on the chamber once while I was in the back. So that was the one safety problem that we specified.

Figure 24.2 shows the layout of our experiment. We operated at an angle of 7 degrees out of the G straight section, which was at that time one of the main straight sections at the AGS. These straight sections were 10 feet long. A beryllium target sat at one end of the straight section, and pions came spilling out at the other. The detector sat behind

ALTERNATING GRADIENT SYNCHROTRON

August 22, 1961

Experiment No. 11
Name & Affiliation: Schwartz, Lederman, Steinberger (Columbia)
Beam No. G + 7°

1. Experimental Title: Study of Neutrino Interactions; Search for Intermediate Boson

2. Experimental Operations Group Assignments:
    2.1 Liaison & Supervision - A. Salee, BNL Ext. 2140, ANdrew 5-0814

3. Experimental Planning Group Assignment:
    3.1 Liaison Physicist - G. Danby, BNL Ext.2471

4. Experimental Group:
    4.1 M. Schwartz   - BNL Ext. 718
    4.2 L. Lederman - Columbia
    4.3 J. Steinberger - "
    4.4 J.M. Gaillard -   "
    4.5 N. Mistry -   "
    4.6 C. Goulianos -   "
    4.7 G. Danby - BNL Ext. 2471

5. Description:
    5.1 Some testing inside shielding wall to determine external shielding.
    5.2 Neutrino and Boson search

6. Machine Parameters
    6.1 Standard Target Shape - Any material
    6.2 G-10 location for target
    6.3 20 Bev maximum energy
    6.4 1 millisecond or less spill
    6.5 Fully bunched beam
    6.6 > 10" p/p intensity

7. Services:
    7.1 Electrical power - 5 to 10 kw, 110v, 1 phase

8. Special Requirements:
    8.1 1 Ton "A" frame

9. Shielding:
    9.1 Approximately 5000 tons steel in addition to main machine shield
    as shown on drawings D14-48, 49 and 50.

10. Health Physics & General Safety Requirements:
    10.1 Spark chambers require 7 kv to fire

12. Drawings:
    12.1 D-14-48-4A Second Roof Layer - Neutrino Shielding
    12.2 D-14-49-4A First and Third Roof Layers - Neutrino Shielding
    12.3 D-14-50-4A Neutrino Shielding Floor Layout

AS/hm

Fig. 24.1. Proposal for Brookhaven Experiment No. 11, the "Two-neutrino experiment."

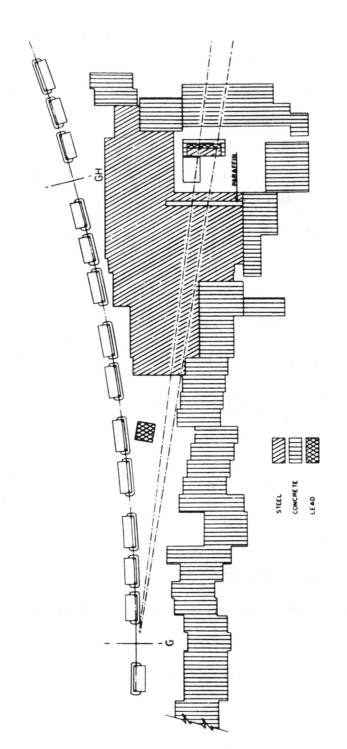

Fig. 24.2. Plan view of the two-neutrino experiment, with the neutrino beam traveling from left to right.

STEEL

CONCRETE

LEAD

PARAFFIN

G

GH

a 42-foot-thick shielding wall consisting mainly of steel, which came not from an old battleship (as has sometimes been reported) but from old cruisers that had been dismantled. With a good dose of serendipity, the Navy was scrapping these ships just at the point that we needed shielding for the experiment.

Figure 24.3 shows what the neutrino spectrum looked like, more or less. The decaying pions gave rise to neutrinos with momenta of around 500–600 MeV/c. The kaon decays took over and became the dominant source of neutrinos at around 1.4 GeV/c. Of course, we had a mixture of neutrinos and antineutrinos, but there were more neutrinos than antineutrinos because there were more π^+ than π^- in the beam. We did not do any sign selection.

Our equipment (see Fig. 24.4) consisted of ten chambers, each chamber containing nine aluminum plates and weighing one ton. These plates were separated by a series of lucite spacers with lenses that allowed us to photograph all of them simultaneously. Between the various chambers there were sandwiches of scintillator – two scintillators and an aluminum plate 3/4-inch thick. There were four such counters between each pair of chambers; these were the trigger counters. There was also an anticoincidence counter covering the front face, which was there to detect any muons that had straggled in or any other charged particles coming from the shielding wall. Anticoincidence counters on the top and the rear of the detector were included to discriminate against cosmic-ray events. The timing on these counters, with a resolution of about 10 nanoseconds, was adequate for the job. We still used a lot of vacuum tubes in those days; there were only a few solid-state circuits available.

Once we began looking through the pictures, it was really quite an exciting time. The first thing we did before we began scanning them was to determine the cosmic-ray background. We studied this background by running the entire apparatus for half a day with the beam off, taking as many event triggers as we could. After the experiment was completed, we realized that, in the course of the 5 seconds of actual running time, we would have triggered on only about five cosmic-ray events that would have looked very much like neutrino events.

There were three different categories of events that we looked for in this experiment. The obvious one is a long track that begins in the detector and makes its way out to the right (see Fig. 24.5); this was clearly something that looked exactly like a muon. We never had any doubt that the bulk of these events were in fact not cosmic rays but were almost certainly neutrino events. They were very convincing in

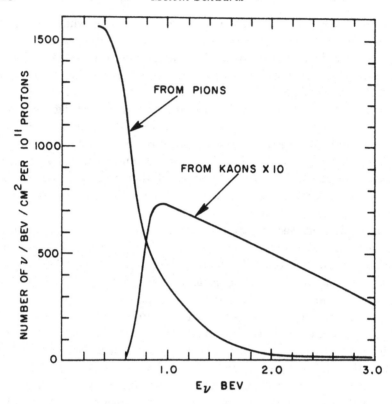

Fig. 24.3. Energy spectrum of neutrinos expected from the AGS running at 15 GeV.

appearance; we observed 34 of these single muon events. Then there were 22 events that had a vertex in the chamber, with other tracks coming out in addition to one long track identified as the muon.

An important test of whether these were really neutrino events or some kind of background was the origin of the events – the direction in which they pointed. In the horizontal plane they essentially pointed right at the target; in the vertical plane there was a slight excess from above, probably due to occasional cosmic ray events. This was a convincing piece of evidence that we really were looking at things that were penetrating through the main line of the shielding wall.

So we had 34 single-muon events with momenta greater than 300 MeV/c (with a cosmic-ray background of 5 ± 1 events) and 22 vertex events. Then there were 6 events, which could have been electron showers, that we had to study more carefully. They had a reasonable number

Fig. 24.4. Photograph of the spark chambers and counters, as viewed from the side.

of single sparks in the chambers, so they could have been neutron events. Very few of these events occurred in the second half of the run; I think there were only 3 out of 30 events after we changed the shielding near the AGS.

Another proof of the neutrino origin of these events came when we removed the first meter of iron shielding. This would have led to a factor of 100 increase in the event rate if they'd been anything except neutrinos, but there was no change. We then erected a shielding wall very near the target, which reduced the pion flight path by a factor of 8. That had the expected effect, reducing the event rate by about a factor of 8. And the long mean free path of those visible tracks was another strong indication. On the average, those candidate muons traveled about 4 pion mean free paths without any interaction. That was fairly convincing evidence about their origin.

To what extent could we be sure that there really were no electron events? That was the real issue: were any electrons being produced by

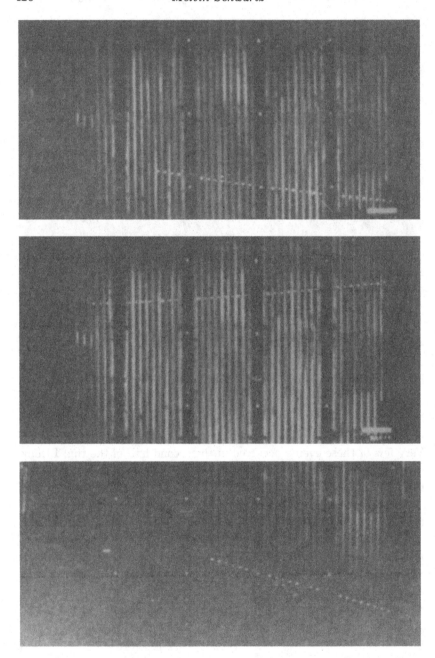

Fig. 24.5. Photographs of three typical single-muon events.

neutrino interactions in our detector? If there had been only one kind of neutrino, there should have been as many electron-type as muon-type events. To see how electron events would appear, we took our chambers over to the Cosmotron and passed a beam of 400 MeV/c electrons through them, which gave events such as shown in Fig. 24.6. They were very showery kinds of events, but they did travel substantial distances. Typically, there were about ten sparks associated with each of these events. This test gave us a clear idea about how such electron events would look if they had occurred during the actual runs.

Finally we made an important plot (see Fig. 24.7) in which we assumed that we had a sample of electron events equal to the sample of muon events and asked what these would look like, in terms of the number of sparks per event. The top histogram told us, roughly speaking, how many sparks per event we should have seen if we had had a sample of events made up of electrons. The six events we actually observed – the so-called shower events – bore no resemblance at all to what we would have seen had there been only one type of neutrino. So at that point we felt certain we really were observing only one type of neutrino in our detector, what is now called the muon neutrino. That was the point at which our first experiment ended.

CERN wanted to do a similar search before Brookhaven, as I mentioned earlier, but it was canceled because the neutrino flux was thought to be too low. I think the real problem at CERN was one of style. In those days, there was very little drive there toward doing adventurous and risky experiments. Most of its program was devoted to "bread-and-butter" physics, such as measuring total cross sections and production rates of various well-known particles. At Brookhaven the neutrino effort always had the highest priority. Besides, as people often said, Columbia "owned" Brookhaven. Or, as Maurice Goldhaber has remarked, "Britannia rules the waves, but Columbia waives the rules!"

In any case, the CERN experiment was set up at a short straight section and was almost ready to run when Guy von Dardel demonstrated that there would be a loss of pion intensity due to defocusing by the fringing field of a magnet. He estimated the loss to be a factor of 5 to 10. The longer straight section would have had no such problem, but that was occupied by a bread-and-butter physics experiment. The Director General of CERN decided to cancel the experiment, an action that I consider to have been a colossal blunder. We all breathed a sigh of relief when we heard about that cancellation.

People have asked why we did not find evidence for neutral currents

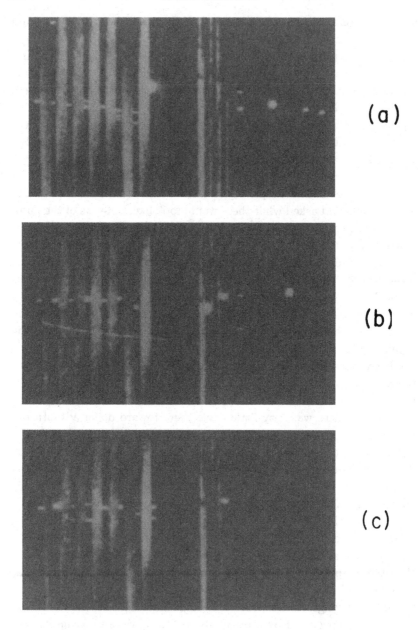

Fig. 24.6. Photographs of typical events produced by 400 Mev/c electrons in a cosmotron calibration run.

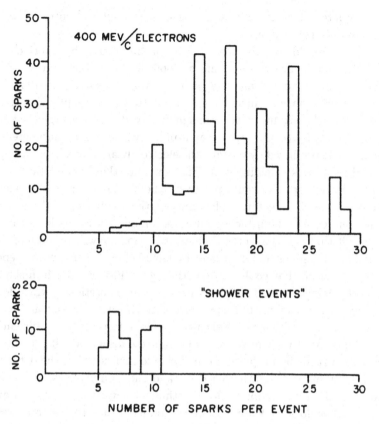

Fig. 24.7. Spark distribution for 400 MeV/c electrons normalized to the expected number of showers assuming $\nu_\mu = \nu_e$ (top). At bottom is shown the observed distribution of "shower" events.

in the early neutrino experiments. In the first experiment that I just described, the trigger was very heavily biased against neutral-current events. Because of the low energy of the neutrinos, it was difficult for them to make pions, and so the typical neutral-current event would have had a short nuclear recoil. The probability of triggering on such an event would have been very small, and the probability of the trigger being accompanied by a recognizable track in the chambers would have been even smaller. In a subsequent experiment that occurred two years later, we triggered the chambers on every pulse of the AGS – and we had more energetic neutrinos, to boot. We should have been able to recognize neutral currents. Years later, after the Gargamelle experiment was completed, I went back and looked at the film. Neutral-current events were indeed there.

Finally, I would like to recall one of the more unhappy periods in my career. It all began sometime in 1969 during a conversation with Al Mann. He reminded me of a paper that I had written in 1962 for a SLAC workshop, in which I pointed out that heavy leptons, if they existed, could be produced electromagnetically at a high-energy electron machine. Because they would decay rapidly (within a few centimeters), they could be produced by plowing the electron beam into a beam dump, followed by a thick shielding wall. The neutrinos that would be coupled to these heavy leptons would penetrate the shield and produce heavy leptons in a large detector. Unfortunately, the sensitivity of the experiment was not very high for lepton masses much above the kaon mass, but we felt that it was a worthwhile investigation anyway because of the possibility of a surprise. Mann and I concluded that a search for penetrating radiation that could make it through a 40-foot-thick iron shield was worth doing – even if it did not have a clear theoretical justification.

In any case, we submitted a proposal with the attitude that an investigation into the unknown was always justified if the cost was sufficiently low. The SLAC Program Advisory Committee disagreed with us, however, and turned the proposal down. I should mention that the detector was already at SLAC gathering dust, and the iron was rusting in the yard. But if I really want to do something, I never take no for an answer. Neither does Al Mann. So we resubmitted the proposal and were turned down again. This time SLAC argued that the experiment was too expensive, noting that steel costs $200 per ton, and that it would take 5000 tons to build the wall. Mann and I pointed out that the steel was rusting in the yard. Pief Panofsky insisted that we had to account for it properly; so the experiment would cost over a million dollars. The proposal was rejected again.

Still, we didn't give up, although I was myself becoming increasingly disenchanted with the field – and indeed went so far as to incorporate Digital Pathways as an alternative to continuing in the academic world. We submitted our proposal a third time and again we were turned down. I finally decided that I would have to invent a theoretical justification for doing the experiment, and I came up with a rather intriguing one I called "strange light."

At the time strangeness was a characteristic of elementary particles similar in some respects to charge. It was conserved in the strong interactions, and very little else was known about it. A natural question was whether there might be some long-range force associated with strangeness just as there is with charge. Would two strange particles a centimeter apart attract or repel one another, depending upon their relative strangeness? If one shakes a charged particle, one produces light. Hence, if there were a long-range force associated with strangeness, one might expect that shaking a strange particle would produce quanta of the field associated with that force. This we would call "strange light."

Now how could we produce strange light? In the process of photoproducing kaons in the SLAC beam dump, we would be doing a lot of shaking. Hence this strange light would be readily produced (assuming it existed). The only way in which this strange light could be detected would be by the pair production of strange particles; if the coupling constant was of the same order as the fine structure constant, the strange light would have no difficulty making it through the wall and being detected in a large detector. Thus the beam dump experiment that we had been promoting for nearly one and a half years would be relevant and acquire what we call "political correctness" today.

Faced with this theoretical "justification" for the beam dump experiment, the committee finally yielded and approved it. But we were constrained against using the steel and had to make do with earth as a shield. What this meant was that we had to locate the detector at 200 feet from the dump rather than at the 50 feet that we had originally proposed. The cost of this move was a factor of 16 in the sensitivity of the experiment.

The experiment began running in September 1970, and we saw a number of events within a few weeks (see Fig. 24.8). Most events were standard charged-current neutrino interactions producing a muon. There were, however, three events that had no visible muon but rather one or more strongly interacting particles.[3] I remember showing these events at the Amsterdam Conference in the summer of 1971 and having them called "mel-ons." In retrospect, they were undoubtedly neutral-current events, but there were just too few of them to conclude anything. In order to demonstrate neutral currents, one must study the spatial distribution of events in the detector to be sure that they did not originate from neutrons produced in the shield. In any case, neutral currents had not yet become an important issue by that summer, and we essentially ran out of steam and support in pushing this experiment. So neutral currents were left for the Gargamelle experiment to discover.

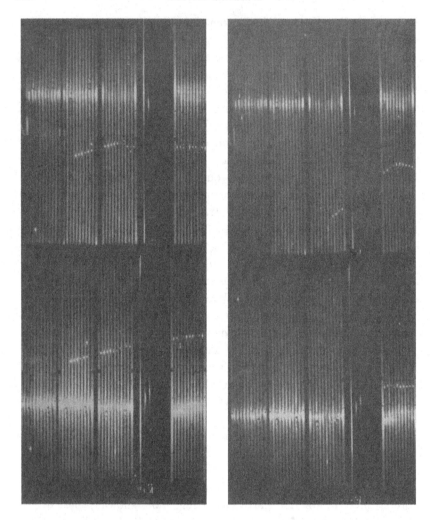

Fig. 24.8. Spark chamber photographs of two "hadronic" events observed in the 1970 SLAC beam dump experiment.

Notes

1 For further information on this experiment, consult M. Schwartz, "The First High-Energy Neutrino Experiment," *Rev. Mod. Phys. 61* (1989), pp. 527–32.

2 G. Feinberg, "Decays of the μ Meson in the Intermediate Meson Theory," *Phys. Rev. 110* (1958), pp. 1482–83.

3 For information on the first SLAC beam dump experiment, see D. Fryberger et al., "A Search for Unknown Sources of Neutral Particles Having No Strong Interaction," SLAC Proposal No. E-56 (January 1970), unpublished; and A. F. Rothenberg, "A Search for Unknown Sources of Neutrino-Like Particles," Stanford University Ph.D. dissertation (1972), unpublished.

25

Gargamelle and the
Discovery of Neutral Currents

DONALD PERKINS

Born Hull, England, 1925; Ph.D., 1948 (physics), University of London; Professor of Physics at the University of Oxford; high-energy physics (experimental).

Several accounts have been written of the discovery of neutral weak currents, mainly by social scientists or theoreticians.[1] Although doubtless well motivated, these authors were themselves not immersed in the experimental situation in neutrino physics in the 1960s and 1970s. There have also been, of course, nonhistorical reviews of the physics of neutral currents.[2] In this chapter I shall present an experimenter's account of the sequence of events in this discovery, based on my own experience and on discussions with other physicists taking part in those experiments. As emphasized earlier in this volume by Leon Lederman (see Chapter 6), discoveries in high-energy physics are frequently stories of false trails, crossed wires, sloppy technique, misconceptions, and misunderstandings, compensated by the occasional incredible strokes of good luck. Certainly this was the case for neutral currents.

It is well known that neutral currents were discovered in 1973 by a collaboration operating with the bubble chamber Gargamelle at the CERN Proton Synchrotron. Gargamelle (shown in Fig. 25.1) was a large (4.8 m long, 1.9 m diameter) heavy-liquid chamber filled with freon (CF_3Br) with 20 tons total mass. For the neutral-current investigation, a relatively small fiducial volume of 3 m^3 (4.5 metric tons) was employed. The chamber was conceived and constructed by André Lagarrigue with the help of engineers from Saclay, and funded largely by the French atomic energy commission. Other physicists participating included André Rousset and Paul Musset, who were prominent in the subsequent physics program at CERN. The assembly of Gargamelle took place at CERN during 1970, and physics runs started in early 1971. The original plans for Gargamelle had been laid following the Siena Conference in 1963, at

428

Fig. 25.1. Photograph of the Gargamelle heavy-liquid bubble chamber.

which results were reported from the first neutrino experiments in the CERN 1.2-m heavy liquid chamber. There was a proposal for another large chamber by MURA at Wisconsin, but that was never built (much to the relief of Victor Weisskopf, then Director-General at CERN, who considered that two giant heavy-liquid chambers in the world would be one too many).

The results from the small CERN chamber included a limit on the ratio of elastic neutral-current (NC) to charged-current (CC) cross sections:

$$\sigma\left(\nu_\mu + p \rightarrow \nu_\mu + p\right) / \sigma\left(\nu_\mu + n \rightarrow \mu^- + p\right) \leq 0.03 \qquad (25.1)$$

a value that is some 4 times smaller than the presently accepted ratio.[3] (The actual limit at 90% confidence level was ≤ 0.075.) The explanation is simple: the result (25.1) was wrong because of a stupid bookkeeping error. This error was actually discovered by a research student from University of Strasbourg, Michel Paty, in the course of writing his thesis! The intention upon discovering this mistake was to publish an erratum, together with a new limit from a forthcoming propane run. Propane contains free protons so that one could exploit the kinematics of a νp collision to reduce neutron background. The propane run was, however, delayed by about two years, and the corrected limit for the ratio (25.1), 0.12 ± 0.06, was not published until 1970.[4]

This was hardly an auspicious beginning to the neutral-current story, and Sakurai very rightly castigated us in the CERN group for delaying publication of the correction.[5] I believe, however, that he overstated the case when he claimed that not only had some theorists been put off study of electroweak models but that journals had refused publication of theoretical papers incompatible with the limit (25.1). If this were the case, one wonders how Weinberg's papers ever got published! In fact, until 1970 at least, the main impediment to the theorists was the observed total absence of strangeness-changing neutral currents: for example, the branching ratio for $K^+ \to \pi^+ \nu \bar{\nu}$ was less than 10^{-5}.

The confused situation in weak interaction physics through the 1960s can be illustrated by showing the priority list for the forthcoming neutrino runs in Gargamelle, and reached at a meeting in Milan in November 1968, and repeated in 1970. As shown in Table 25.1, ten topics appeared in the list. Numbers 1 and 2 were the search for the $W^{\pm}$ and the investigation of the partonlike behavior of the inclusive neutrino cross sections. The search for neutral currents was number 8 (!) on our list, whereas the study of "diagonal interactions" was number 4.

Here we have to remember that in the late 1960s there were, in addition to the electroweak gauge models described in the papers by Glashow, Salam and Ward, Weinberg, and Salam, several other proposed cures for the divergence problems of the Fermi theory.[6] One of these, by Gell-Mann, Goldberger, Kroll, and Low, proposed that cross sections for "non-diagonal" processes, involving unlike currents (for example in muon decay or beta decay) would be finite and given by the first-order Fermi term, while the divergences appeared in the "diagonal" or like currents, for example in $\nu_e + e^- \to \nu_e + e^-$, for which the cross sections could be arbitrarily large.[7] There was even support for this model from a measurement by Reines and Gurr, who found that the cross section for this scattering process at a reactor was over 100 times the V–A value – yet another wrong experimental number, although we did not know it then.[8] The paper of Cundy et al. yielding the revised limit for (25.1) also set a limit for the diagonal coupling constant for $\nu_e + e^- \to \nu_e + e^-$ of $g_{\text{diagonal}} \leq 18 G_{\text{F}}$, and in fact gave this result pride of place before the neutral-current limit.

The fact is that, up until 1973, there was no firm evidence in favor of neutral currents and plenty of evidence against them (from the absence of strangeness-changing processes mentioned above). And until 1971 there

Table 25.1. *Priority list for Gargamelle neutrino experiment, Milan collaboration meeting, November 1968*

1. $W^{\pm}$ search.
2. Deep inelastic scattering, scaling.
3. Current algebra sum rules, CVC, PCAC.
4. Diagonal Model.
5. $\Delta S = 1$ processes, inverse hyperon decay $\bar{\nu}_\mu + p \to \Lambda + \mu^+$
6. Inverse muon decay, $\nu_\mu + e^- \to \mu^- + \nu_e$.
7. Electron-muon universality, $\sigma(\nu_e)/\sigma(\nu_\mu)$
8. Neutral current search.
9. Form factors in exclusive reactions.
10. Search for heavy leptons.

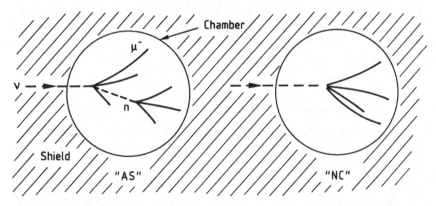

Fig. 25.2. Associated stars (AS) and neutral-current candidates (NC), as viewed in the CERN bubble chambers.

seemed also to be little conviction about the electroweak models in the bulk of the physics community. I can assure you nevertheless that some experimenters at CERN were keenly aware – and had been since 1964 – of the fact that "neutral events," containing hadrons only, were present in numbers in the bubble chamber and that there were difficulties in trying to account for all of them as due to neutrons generated by neutrino interactions in the shielding material surrounding the chamber.

Single, unassociated "hadrons-only" events in the chamber were called NC events, as neutral-current candidates. Another category was called AS events (for "associated stars"), which were interactions containing hadrons only, but produced by a neutron from an upstream charged-current neutrino event in the same picture (see Fig. 25.2). The ratio,

NC/AS, was calculable in principle if the NC events were all due to neutrons. The point is that unlike neutrinos, neutrons would be rapidly absorbed in the bubble chamber and surrounding material. The neutron flux would therefore be fed by, and in equilibrium with, the parent neutrino beam, and the death rate of neutrons in the chamber, given by the rate of NC events, would be equal to the birthrate, which could be calculated from the AS event rate, taking into account the chamber geometry and estimates of the neutron attenuation length in the chamber surroundings. Always, however, there seemed to be about 3 times as many NC events as expected, but it was never possible to *prove* the existence of a signal. Our usual conclusion in those days was that we did not understand the neutron-cascade problem, or that there must be an extra source of neutrons, perhaps from the proton beam cascading through the shield. A research student, Enoch Young, was set to work on the latter problem but could not find any evidence for such an extra source.[9]

Gargamelle and leptonic neutral currents

As we now know, the correct explanation of the absence of strangeness-changing neutral currents was given by Glashow, Iliopoulos, and Maiani in 1970, by invoking a fourth (charmed) quark, to be discovered four years later at SLAC and BNL.[10] But as I recall, this paper did not have too much impact on experimenters at that time. The reason is that they looked on the Glashow–Salam–Weinberg theory as a field theory of *leptons* and their electromagnetic/weak interactions. It was not clear how hadrons might be incorporated; for all we knew, the strong interactions might well mess things up. Of course, we knew about the quark-parton model, and Robert Palmer, visiting from BNL, had even worked out the details of neutral-current cross sections in this model. The problem was to know how well it could be applied for neutrino energies of the order of 1 GeV.

The situation changed dramatically in 1971, following the proof by Gerard 't Hooft that the electroweak theory was indeed renormalizable.[11] Then everyone had to take the electroweak theories seriously. But the 't Hooft paper was again in terms of the interactions between leptons. In the Gargamelle collaboration – consisting of groups from Aachen, Brussels, CERN, Ecole Polytechnique, Milan, Orsay, and University College, London – it is not surprising that some members felt that the effort should concentrate on a search for the leptonic neutral-current processes

$$\bar{\nu}_\mu + e^- \to \bar{\nu}_\mu + e^- \tag{25.2}$$

$$\nu_\mu + e^- \to \nu_\mu + e^-, \tag{25.3}$$

characterized by a single electron projected at a small angle $\theta \simeq \sqrt{2m/E_\nu} \sim 2°$ to the beam at GeV energies. The first process (25.2) was preferred, because the principal background is from the reaction

$$\nu_e + n \to e^- + p \tag{25.4}$$

at very low momentum transfer, so that the recoil proton is invisible and the electron is at a small angle. Even allowing for Pauli suppression factors, the cross section is much larger than for the corresponding process with an electron target. However, there is a factor going the other way from the fact that the focusing system would enhance negative pion fluxes (to give $\bar{\nu}_\mu$) and defocus positive kaons, which would be the source of ν_e (via the decay mode $K^+ \to \pi^0 e^+ \nu_e$), with the result that the $\nu_e/\bar{\nu}_\mu$ flux ratio is at the 10^{-3} level. So the background to reaction (25.2) was expected to be only 1%.

Charles Baltay and I calculated that, with a 1.4 million picture run in the antineutrino beam, we should record between 5 and 30 events of reaction (25.2), depending on the value of the mixing angle, $\sin^2\theta_W$. Of course, it was to turn out that the actual value of $\sin^2\theta_W \simeq 0.23$ is such as almost to minimize the cross section. After cuts on the electron recoil energy (≥ 0.3 GeV), just three events of type (25.2) were found.[12] Anyone can imagine what a nightmare it was, from the point of view of the scanning of the film, to dig out these very rare processes. But they really were gold-plated and unmistakable events, of which the first is shown in Fig. 25.3. This was found at the University of Aachen at the end of 1972.

I can remember the occasion very vividly. On December 30, 1972, Helmut Faissner, Jurgen von Krogh, and Donald Cundy came to Oxford to write a beam-dump proposal for the BEBC chamber at the forthcoming CERN SPS, and I remember meeting them at London Airport. They showed me the picture, and I had to ask only one question: neutrino or antineutrino film? On being told the latter, I suggested a dash to the bar to celebrate, while Helmut Faissner proposed putting it in to Christie's for auction. We had then scanned only about 100,000 pictures, so I knew the background must be very much less than 0.01 events, and from that time onward I simply believed in neutral currents. (Fortunately, I did

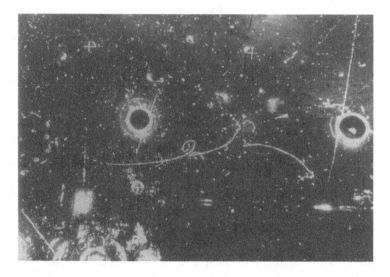

Fig. 25.3. First bubble-chamber photograph of the neutral-current process $\bar{\nu}_\mu + e^- \to \bar{\nu}_\mu + e^-$.

not know then that, in scanning 14 times as much more film, we were to find only 2 more such events.) After this memorable episode, everything that subsequently happened in the neutral-current story was for me something of an anticlimax.

Gargamelle and the hadronic neutral current

The establishment of a neutral-current signal for events containing only hadron secondaries was a different and more complicated story, which of course involved the scintillator/spark-chamber detector at Fermilab as well as Gargamelle at CERN. Examples of candidates for charged-current and NC events are shown in Fig. 25.4.

The first major step for the Gargamelle collaboration was to resolve to study *inclusive* events, that is of the type $\nu + N \to \nu +$ hadrons, rather than the much rarer and more difficult-to-identify exclusive processes such as $\nu + p \to \nu + \pi^+ + n$. That meant that – depending on how well one could represent the hadronic current – the interpretation of any NC signal in terms of $\sin^2 \theta_W$ might be suspect, but the higher cross section offered the possibility of establishing the existence of neutral currents on the basis of *hundreds* of events, rather than two or three as in the lepton case. From our previous experience with the old 1.2 m chamber,

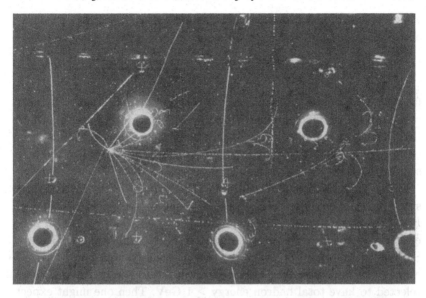

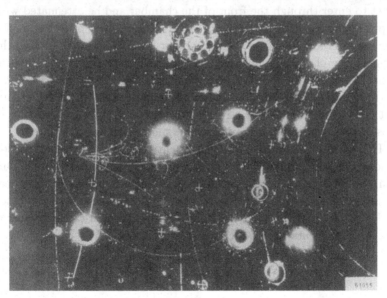

Fig. 25.4. *Top*: Hadronic charged-current event with muon leaving chamber at right. *Bottom*: Neutral-current event with all secondaries identified as hadrons. In both pictures, the neutrino beam enters from the left.

we knew that the maximum possible inclusive neutral-current signal was appreciable: at the 1972 Fermilab Conference this limit was given as,

$$R = \frac{\sigma\left(\nu_\mu + N \to \nu_\mu + \text{hadrons}\right)}{\sigma\left(\nu_\mu + N \to \mu^- + \text{hadrons}\right)} \leq 0.17, \qquad (25.5)$$

from which I quote: "the inclusive processes apparently offer the best possibility of proving or disproving the Weinberg theory as applied to hadronic weak currents.... [However,] one can criticize the results from existing experiments on the grounds that the events are not in the true scaling region."[13]

In the Gargamelle collaboration in late 1972 and early 1973, several subgroups worked intensively on the study of hadronic neutral currents. First, there were qualitative but nevertheless sound and powerful general arguments based, for example, on the spatial distributions of events. Suppose many or all of the NC events are due to neutrons. The NC events, like the CC and AS events with which they were compared, were selected to have total hadron energy ≥ 1 GeV. Then one might expect most to enter through the front of the chamber and be attenuated with depth in the chamber along the beam axis (the total depth was about 5 nuclear interaction lengths). On the other hand, if neutrons entered the sides of the chamber, one would expect a characteristic radial dependence. As shown in Fig. 25.5, the NC events showed, on the contrary, a fairly uniform spatial distribution, very similar to that of the CC events, but of course the statistical errors were quite large. The provisional conclusion was that most of the NC events must be neutrino-induced.

To obtain an actual value for the neutron background contribution among the NC events, a more quantitative analysis was required. One can give first an argument based on a one-dimensional model, with neutrons traveling along the neutrino beam and chamber axis, as indicated in Fig. 25.6. Then a simple calculation gives for the ratio of neutron background events B to associated-star events AS:

$$\frac{B}{AS} = \frac{\lambda_{att}\left(1 - e^{-L/\lambda}\right)}{\left(L - \lambda\left(1 - e^{-L/\lambda}\right)\right)} \simeq \frac{\lambda_{att}}{(L - \lambda)}, \qquad (25.6)$$

where $L \simeq 600$ g/cm^2 is the length of the fiducial volume of the chamber, $\lambda \simeq 135$ g/cm^2 is the neutron interaction length in the liquid, so that $L \gg \lambda$. Here $\lambda_{att} \simeq 2.5\lambda \simeq 300$ g/cm^2 is the attenuation length of neutrons in the material (iron) surrounding the chamber, which can be calculated from the known inelasticity distribution in neutron–nucleus interactions and the slope of the (approximately power-law) hadron spec-

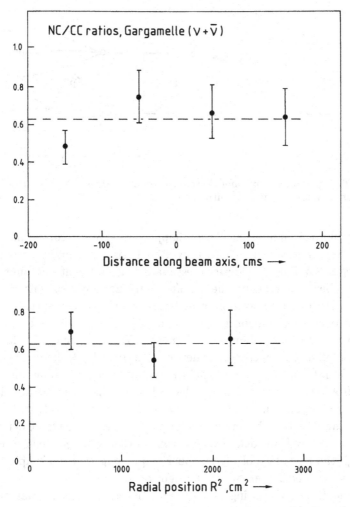

Fig. 25.5. Ratio of candidate NC to CC events in Gargamelle. Distributions along the beam axis (top), and along the radial direction (bottom).

trum. The meaning of equation (25.6) is fairly self-evident. Neutron background from upstream of the chamber comes from a layer of material of thickness λ_{att}, while inside the chamber, associated stars can be generated anywhere over a distance L, less an amount λ for neutron detection. Using (25.6) one obtains $B/AS \simeq 0.7$, compared with an observed ratio for $NC/AS \simeq 6.0$. Thus only some 10% of the NC candidates could be ascribed to neutrons.

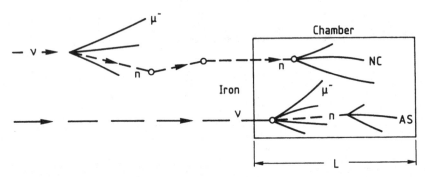

Fig. 25.6. One-dimensional model used in estimating the neutron-induced contribution to hadronic neutral-current candidates.

Although this simple calculation happens to give the correct answer, it was hardly good enough to provide a basis for a major discovery in physics. A fully quantitative estimate must take into account the neutrino beam divergence, the angular distribution of neutrons at production, the exact disposition of material around the chamber, nuclear cascade effects on neutron propagation, and the dependence of the inelasticity and other parameters on energy. This required a full Monte Carlo simulation, which was undertaken in 1973 by William Fry and Dieter Haidt, and formed the main basis for the quantitative numbers on hadronic neutral currents.[14] Independent calculations were also done by other subgroups in the Gargamelle collaboration, notably in Milan, Orsay, and CERN. All came to similar conclusions. These more detailed estimates showed, incidentally, that some 75% of the neutrons entered through the front face of the chamber, explaining why the simple one-dimensional model gives about the right answer.

At the time (the spring and summer of 1973), I recall that opinions on procedure inside the Gargamelle collaboration were somewhat divided. Confidence in the existence of neutral currents was by then fairly solid, but attitudes toward publication of the results varied from wild enthusiasm to extreme caution. A democratic vote, taken in May 1973, narrowly decided to continue further analysis in order to refine the numbers before announcing the discovery.

The matter of publication was eventually decided by a letter from Carlo Rubbia to André Lagarrigue in July 1973, stating that the HPWF collaboration had some hundred neutral-current events, and that they would like to refer to the Gargamelle experiment if we would reciprocate! So it was that the publication of the first Gargamelle results on hadronic neutral currents was perhaps made with undue haste: a few weeks more polishing of the data analysis and of the presentation of the arguments would have greatly improved the paper.[15] The full paper was published in the following year.[16]

The results were presented by Gerald Myatt at the Bonn Conference in August 1973, giving the ratios of NC to CC events, after background corrections, of[17]

$$R = \left(\frac{NC}{CC}\right)_\nu = 0.21 \pm 0.03, \qquad \bar{R} = \left(\frac{NC}{CC}\right)_{\bar{\nu}} = 0.45 \pm 0.09. \quad (25.7)$$

The Fermilab Experiments

Obviously, the most important neutrino experiment at Fermilab during the period 1973–1974 was the HPWF (Harvard, Pennsylvania, Wisconsin, Fermilab) experiment, using a wide-band neutrino beam of the same type as at the CERN PS, but with the advantage of much higher proton energy (300 GeV versus 26 GeV) and the disadvantage that it was unfocussed (i.e., a mixture of neutrinos and antineutrinos). However, it is important to note that another experiment at Fermilab by a Caltech group was coming along; this used a focused, narrow-band beam of higher average neutrino energy but much lower intensity.[18] It was interesting because it was to provide not only supporting evidence for neutral currents but also to exclude an alternative model proposed by Georgi and Glashow, in which the divergent terms of the Fermi theory were canceled by introducing a new heavy lepton instead of a neutral current.[19]

Fig. 25.7(a) shows the initial disposition of the HPWF experiment. It consisted of a liquid-scintillator detector acting as the neutrino target and hadron calorimeter, instrumented with four spark-chamber (SC) layers as shown, followed by a muon spectrometer consisting of magnetized iron toroids also instrumented with spark chambers to measure the muon trajectory. A charged-current event would be characterized by a hadron shower recorded in the liquid scintillator and spark chambers, and a muon penetrating through the magnetized iron of the spectrometer. On the other hand, a neutral-current event would consist of a hadron shower recorded in the liquid scintillator only. From the observed angular distribution of the recorded muons in CC events, it was possible to compute the true muon angular distribution at production, and hence to estimate, in terms of the position of the event vertex, how many true

CC events would appear as NC events because the muon was at wide angle and missed the spectrometer. After subtracting this correction, the resulting excess of genuine NC events was found to give an NC/CC ratio of

$$R = \left(\frac{NC}{CC}\right)_{\nu,\bar{\nu}} = 0.29 \pm 0.09 \qquad (25.8)$$

in fair agreement with the Gargamelle result (25.7), taking account of the antineutrino/neutrino flux ratio, of order $1:2$ in the HPWF experiment. The above result was quoted by Carlo Rubbia on behalf of the HPWF collaboration at the Bonn Conference in 1973, but was not actually published for several months.

Following the Bonn Conference the HPWF collaboration undertook a major modification to the experiment, in an effort to enhance the NC signal by reducing the number of "lost" muons from CC events. The modified apparatus is shown in Fig. 25.7(b). A large spark chamber, SC4, which had originally belonged to the hadron calorimeter, was now made part of the muon detection system by inserting in front of it a 30 cm thick steel block. Previously the first spark chamber of the spectrometer had been placed behind 1.25 m of steel. The area of the muon spectrometer spark chambers was also increased, and both changes substantially increased the proportion of muons from CC events intercepted by the spectrometer. Unfortunately, however, the reduced thickness of steel in front of SC4 led to a dramatic increase in punch-through of hadrons: genuine NC events gave high-energy hadrons that penetrated the steel, simulating muons. These events would therefore be wrongly classified as CC events.

The seriousness of the punch-through problem was not fully appreciated at the time. Consequently, the apparent NC signal now fell essentially to zero (numbers such as $R = 0.05 \pm 0.05$ were being quoted). Even this result was uncertain, however, because – in the absence of any real knowledge of the hadron energy spectrum – it was difficult to make reliable calculations. In fact, the punch-through probability used toward the end of 1973 turned out to be a factor of 2 too small.

These problems were eventually resolved during the spring of 1974, and the final result (25.8) of the first HPWF experiment published.[20] The results from the modified detector, using also separate (horn-focused) neutrino and antineutrino beams, were published shortly

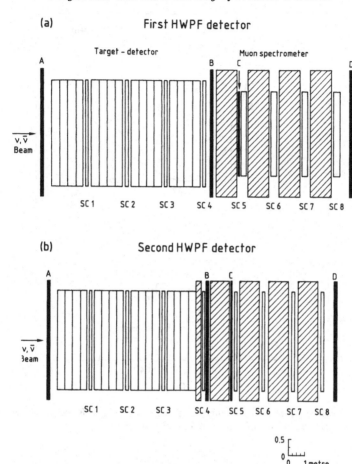

Fig. 25.7. The HWPF detector in its two configurations: (a) before the 1973 Bonn Conference, and (b) after this conference.

thereafter.[21] The corresponding R values were

$$R = \left(\frac{NC}{CC}\right)_\nu = 0.11 \pm 0.05, \qquad \bar{R} = \left(\frac{NC}{CC}\right)_{\bar{\nu}} = 0.32 \pm 0.09. \quad (25.9)$$

Figure 25.8 shows a plot of $\bar{R}$ versus R, for both the Gargamelle and HPWF experiments. The curve shows the prediction of the Salam–Weinberg theory, based on the quark model, but neglecting the small antiquark content of the nucleon as well as any effects due to the experimental hadron energy cuts.

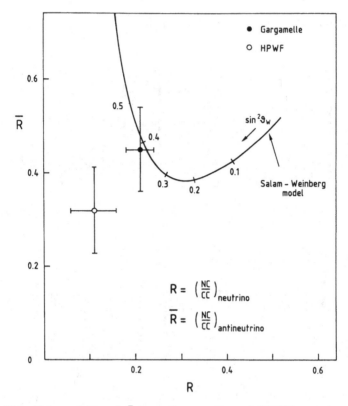

Fig. 25.8. Values of R and $\bar{R}$ from Gargamelle and HPWF experiments, as compared with prediction of Salam–Weinberg model.

It is interesting to note that the HPWF result is actually inconsistent with the Salam–Weinberg theory, while the Gargamelle result shows a value of R that is only about two-thirds of the present-day value, as found with higher energy beams. The value deduced for $\sin^2\theta_W = 0.38 \pm 0.09$ has to be compared with the present value of 0.23. This discrepancy may be connected with the fact that for Gargamelle, the neutrino energies were only a few GeV, so the 1 GeV hadron energy cut was quite important, and systematic uncertainties could have arisen from assessing these effects using the quark–parton model, which is of somewhat doubtful applicability at these energies. However, both experiments now saw a clear neutral-current signal, regardless of whether it agreed with any particular model.

Following the publication of the Gargamelle and HPWF results, some physicists were still unconvinced about the existence of neutral currents.

A number of confirmatory experiments then started to come in during 1974 and 1975, notably from Caltech at Fermilab, from BNL, and from ANL, after which neutral currents could be regarded as established.[22] There then followed a period of consolidation, with more detailed and precise measurements of the vector and axial-vector couplings of the leptons and the various quarks. Many people stressed the importance of making model-independent analyses of the data. In time, the results zeroed in on the simplest, Glashow–Salam–Weinberg model of electroweak interactions, involving a single neutral Higgs. That simplest model has, even today, defied all experimental efforts to discover deviations from it. The principal prediction of the model – once neutral currents were established – was of course the existence of the massive bosons $W^{\pm}, Z^0$, to be found at the CERN $\bar{p}p$ collider almost a decade later.

The Gargamelle experiment and CERN

Nowadays, everyone is impressed by the precision, predictive power and seeming inevitability of the Standard Model. But it was not always like that. We can now look back at the discovery of neutral currents as a crucial step in support of the Standard Model, but 20 years ago people could hardly have been expected to see the matter so clearly.

Today, CERN prides itself on being the world's leading high-energy physics laboratory. Whether or not this is so, it is clear that 20 years ago, things were rather different. At that time, although recognized for the very high quality and reliability of its accelerator engineering, CERN unfortunately did not have a similar reputation in its physics, and it was still recovering from disasters such as the "split A2" affair. CERN always seemed to be second best behind the leading U.S. laboratories, with their vastly more experienced physicists. And during the 1960s it had been repeatedly beaten into the ground, for example, over the discoveries of the Ω^- hyperon, the two types of neutrinos, and CP violation in K^0 decay. All these things could and should have been found first at CERN, with its far greater technical resources, but the Americans had vastly more experience and know-how. Even today, the scoreline in Nobel laureates in high-energy physics (counting from the end of World War II) tells the story: United States 26, Europe 6.

It is important to understand this legacy of inferiority in considering the attitudes at that time of people in CERN over the Gargamelle experiment. When the unpublished (but widely publicized) negative results from the HPWF experiment started to appear in late 1973, the

Gargamelle group came under intense pressure and criticism from the great majority of CERN physicists. Part of this was presumably just prejudice against the technique: people could not believe that such a fundamental discovery could come from such a crude instrument as a heavy liquid bubble chamber. After all, the past discoveries with bubble chambers had been in hadron resonance physics, achieved with the great measurement precision attainable with liquid hydrogen filling. By contrast, the hadron energy resolution in Gargamelle was typically only 20%.

But equally important, many people believed that, once again, the American experiments must be right. One senior CERN physicist bet heavily against Gargamelle, staking (and eventually losing) most of the contents of his wine cellar! The CERN management was obviously very worried, and there were intense discussions between members of the CERN Directorate and the leading Gargamelle physicists. Despite these pressures, the Gargamelle physicists stuck to their claims; they had spent the best part of a year in exhaustive and detailed analyses by several independent subgroups, to convince themselves that there was a clear signal, and they were certainly not going to back down on the basis of rumors from Fermilab. It is indeed a dramatic testimony to the rapidly changing fortunes in the world of high-energy physics that what was undoubtedly *the* principal discovery during the first 25 years of the CERN laboratory was to be greeted initially with total disbelief by the vast majority of CERN physicists. This has to be contrasted with the later observation of the $W^\pm$ and Z^0 bosons in the UA1 and UA2 experiments at the CERN $\bar{p}p$ collider in 1982–83. Although a magnificent achievement, this was hardly a bolt from the blue. Once neutral currents were established and the value of $\sin^2\theta_W$ had been measured, the mass, production rate, and decay modes of these weak bosons were accurately predicted; when they were found, the discovery was greeted with relief rather than disbelief.

Toward the end of 1973, the CERN physicists in the collaboration carried out runs with the 25 GeV proton beam from the PS, transported directly to the neighborhood of Gargamelle, in order to check experimentally the predictions of the cascade calculations that had been so critical in determining the neutron background in the neutral current experiment. These results, appearing in early 1974, fully exonerated the Monte Carlo calculations, and at last the critics were satisfied.

Finally, it is appropriate to record that some years after the neutral-current episode, while operating at the CERN SPS, Gargamelle suffered a major mechanical failure and experiments with it came to an abrupt end. During its construction, the welds on the chamber body had caused problems, and these were the cause of its demise. For years the carcass of Gargamelle rotted away in obscurity, but very recently, I am happy to say, the chamber body has been refurbished and is displayed, alongside components of the hydrogen chamber BEBC, as an exhibit near the CERN reception area.[23]

Notes

1 See, for example, P. Galison, "How the First Neutral-Current Experiments Ended," *Rev. Mod. Phys. 55* (1983), pp. 477–509.

2 For example, P. Musset and J. P. Vialle, "Neutrino Physics with Gargamelle," *Phys. Rept. 39C* (1978), pp. 2–130; F. Sciulli, *Prog. Part. Nucl. Phys. 2* (1979), p. 41; D. Haidt and H. Pietschmann, *Electroweak Interactions* (Berlin: Springer–Verlag, 1988).

3 H. H. Bingham et al., "CERN Neutrino Experiment – Preliminary Bubble Chamber Results," *Proc. of the Siena International Conf. of Elementary Particles* (Siena, Italy, 1963), Vol. 1, pp. 555–70; M. M. Block, et al., "Neutrino Interactions in the CERN Heavy Liquid Bubble Chamber," *Phys. Lett. 12* (1964), pp. 281–5.

4 D. C. Cundy, et al., "Upper Limits for Diagonal and Neutral Current Couplings in the CERN Neutrino Experiments,"' *Phys. Lett. 31B* (1970), pp. 478–80.

5 J. J. Sakurai, "Neutral Currents and Gauge Theories – Past and Present," in J. E. Lannutti and P. K. Williams, eds., *Current Trends in the Theory of Fields* (New York: American Institute of Physics, 1978), pp. 38–80.

6 Sheldon L. Glashow, "Partial-Symmetries of Weak Interactions," *Nucl. Phys. 22* (1961), pp. 579–88; A. Salam and J. C. Ward, "Electromagnetic and Weak Interactions," *Phys. Lett. 13* (1964), pp. 168–71; Steven Weinberg, "A Model of Leptons," *Phys. Rev. Lett. 19* (1967), pp. 1264–6; Abdus Salam, "Weak and Electromagnetic Interactions," in N. Svartholm, ed., *Elementary Particle Theory* (New York: Wiley, 1968), pp. 367–77.

7 Murray Gell-Mann, Marvin L. Goldberger, Norman M. Kroll, and Francis E. Low, "Amelioration of Divergence Difficulties in the Theory of Weak Interactions," *Phys. Rev. 179* (1969), pp. 1518–27.

8 F. Reines and H. S. Gurr, University of California, Irvine Report No. UCI-10P19-28 (1969); see also F. Reines and H. S. Gurr, "Upper Limit for Elastic Scattering of Electron Antineutrinos by Electrons," *Phys. Rev. Lett. 24* (1970), pp. 1448–52.

9 Enoch C. M. Young, "High-Energy Neutrino Interactions," CERN Yellow Report No. 67-12 (Geneva: CERN, 1967) .

10 S. L. Glashow, J. Iliopoulos, and L. Maiani, "Weak Interactions with Lepton–Hadron Symmetry," *Phys. Rev. D2* (1970), pp. 1285–92.

11 G. 't Hooft, "Renormalization of Massless Yang–Mills Fields," *Nucl. Phys. B33* (1971), pp. 173–99.

12 F. J. Hasert, et al., "Search for Elastic Muon–Neutrino Electron

Scattering," *Phys. Lett. 46B* (1973), pp. 121–4; J. Blietschau, et al., "Evidence for the Leptonic Neutral Current Reaction $\bar{\nu}_\mu + e^- \to \bar{\nu}_\mu + e^-$," *Nucl. Phys. B114* (1976), pp. 189–98.

13 D. H. Perkins, "Neutrino Interactions," in J. D. Jackson and A. Roberts, eds., *Proceedings of the XVI International Conference on High Energy Physics* (Batavia, Illinois: National Accelerator Laboratory, 1972) Vol. 4, pp. 189–247.

14 W. F. Fry and D. Haidt, "Calculation of the Neutron-Induced Background in the Gargamelle Neutral Current Search," CERN Report No. 75-1 (Geneva: CERN, 1975).

15 F. J. Hasert, et al., "Observation of Neutrino-Like Interactions Without Muon or Electron in the Gargamelle Nutrino Experiment," *Phys. Lett. 46B* (1973), pp. 138–40.

16 F. J. Hasert, et al., "Observation of Neutrino-Like Interactions Without Muon or Electron in the Gargamelle Neutrino Experiment," *Nucl. Phys. B73* (1974), pp. 1–22.

17 G. Myatt, "Neutral Currents," in H. Rollnik and W. Pfeil, eds., *6th International Symposium on Electron and Photon Interactions at High Energies* (Amsterdam: North-Holland, 1974), pp. 389–406.

18 B. C. Barish, et al., "Neutral Currents in High-Energy Neutrino Collisions: An Experimental Search," *Phys. Rev. Lett. 34* (1975), pp. 538–41.

19 Howard Georgi and Sheldon L. Glashow, "Unified Weak and Electromagnetic Interactions Without Neutral Currents," *Phys. Rev. Lett. 28* (1972), pp. 1494–7.

20 A. Benvenuti, et al., "Observation of Muonless Neutrino-Induced Inelastic Interactions," *Phys. Rev. Lett. 32* (1974), pp. 800–3.

21 B. Aubert, et al., "Further Observation of Muonless Neutrino-Induced Inelastic Interactions," *Phys. Rev. Lett. 32* (1974), pp. 1454–7; B. Aubert, et al., "Measurement of Rates for Muonless Deep Inelastic Neutrino and Antineutrino Interactions," *Phys. Rev. Lett. 32* (1974), pp. 1457–60.

22 B. C. Barish, "Results from the Cal. Tech.–FNAL Experiment," pp. IV-111–IV-113; W. Lee, "Observation of Muonless Neutrino Reactions," pp. IV-127–IV-128; P. Schreiner "Results from the Argonne 12-Foot Bubble Chamber Experiment," pp. IV-123–IV-126, all in J. R. Smith, ed., *Proceedings of the XVII International Conference on High Energy Physics* (Chilton, England: Rutherford Laboratory, 1974).

23 Dieter Haidt of DESY and Don Cundy of CERN have memories of distant events that are much better than mine, and it is a pleasure to thank them for a number of informative discussions and for providing me with original material.

26

What a Fourth Quark Can Do

JOHN ILIOPOULOS

Born Calamata, Greece, 1940; Doctorat d'Etat, 1968 (theoretical physics), University of Paris (Orsay); Director of Research, National Center for Scientific Research, France.

I do not claim any deep understanding of *Finnegans Wake*, but I believe that, had Murray Gell-Mann known the existence of more than three elementary constituents of hadronic matter, he would have chosen a different name. This paper is my recollection of the events that led to the conjecture about a lepton–hadron symmetric world. I want to warn the reader that, as I discovered experimentally, my memory is partial and selective. I would have been particularly worried by this discovery, had I not discovered at the same time that I share this human defect with practically all my colleagues. The difference is that most people are not aware of it, as I was not a couple of years ago, and, furthermore, different people forget or distort different things.[1]

As far as I am concerned, the story begins around 1967 or 1968. I was on a postdoctoral fellowship at CERN coming from the University of Paris at Orsay, where I had done my thesis work under the direction of Philippe Meyer and Claude Bouchiat. I came to CERN in September 1966 and started working on current algebra, one of the most fashionable subjects at that time. Together with other postdocs and visitors, we formed a band of joyful youngsters, enjoying tremendously both physics and skiing, mountaineering, eating, drinking, and so on. We were not doing much in terms of physics, but as David Sutherland, a member of the band, put it, we were doing it in great style. We had organized an informal study group in which we discussed each other's work as well as the literature. I think I remember the day we discussed Steven Weinberg's 1967 paper. I believe Bruno Renner was reporting on it, and we unanimously decided it was totally uninteresting. I promptly forgot everything about it. I also remember a seminar by Robert Brout on the

Brout and Englert mechanism. It all sounded Chinese to me (I wished it were Greek!).

At that time weak interactions were thought to be described by the Fermi current-current theory. The existence of an intermediate vector boson was accepted as a possibility although, of course, there was no evidence for it, and it was assumed to be only charged. This theory is nonrenormalizable, which means that higher orders are not computable, but at least it has a very elegant structure. Few people worried about its lack of mathematical consistency and, in fact, I remember many famous physicists claiming that it would be useless to try to solve the problem of weak interactions before solving that of strong interactions. For the latter, we had neither consistency nor elegance. It was through the successes of current algebra that a few people began realizing that strong interactions could not provide a cutoff for the weak. I do not know who was the first to make this important observation, but I learned it from a paper by B. L. Ioffe and E. P. Shabalin.

I shall present their argument in a slightly more general form that was explained to me a bit later by T. D. Lee, who was spending a year at CERN and was also interested in the divergences of perturbation theory. Let me introduce a cutoff for the higher order weak interactions called Λ, which has the dimensions of a mass and determines the scale up to which the theory can be trusted. The Λ dependence of the nth order diagrams can be written, up to logarithmic corrections, as $C_0 \left(G\Lambda^2\right)^n + C_1 G \left(G\Lambda^2\right)^{n-1} + C_2 G^2 \left(G\Lambda^2\right)^{n-2} + \cdots + C_n G^n$ where G is the Fermi coupling constant and the Cs are in principle calculable coefficients. What Ioffe and Shabalin showed is that, at least to the lowest order, the coefficient of the leading divergence is nonzero, provided the strong interactions satisfy current algebra. This was quite disturbing. Most people were assuming that Λ was quite large, of the order of a few hundred GeV. But then $G\Lambda^2 \sim 1$ and the leading divergent terms become, effectively, of the order of the strong interactions, the next-to-leading terms become of first order in the weak interactions. However, weak interactions are known to violate strangeness as well as parity. This raises the spectre of strangeness and parity violations in strong interactions, as well as of $\Delta S = 2$ transitions in first-order weak processes. Why did most people not worry about it? I suspect that the main reason was a widespread mistrust of field theory in general and higher-order diagrams in particular. Since we had no theory, why bother about its higher-order effects? However, if we follow Ioffe and

Shabalin, we see that the usual selection rules of both strong and weak interactions yield a remarkably small cutoff of the order of 5 GeV.

I did not know what to do with this problem until one day Bouchiat, who was also visiting CERN, walked into my office and told me that he had a way to solve the problem of the leading divergences. I got very excited. Jacques Prentki was also interested in the same problem, and so all three of us set down to work. Bouchiat's idea was very simple. He had noticed that to lowest nontrivial order, the term proportional to Λ^2 contained an operator whose properties depended crucially on the symmetries of the strong interaction Hamiltonian. In particular we were able to prove that, under the assumption that the chiral SU(3)⊗SU(3) symmetry-breaking term in the Hamiltonian belongs to a $(3, \bar{3}) \oplus (\bar{3}, 3)$ representation, this operator is diagonal; that is, it does not connect states with different quantum numbers, strangeness, and/or parity. Therefore, all its effects could be absorbed in a redefinition of the parameters of the strong interactions and no strangeness or parity violation would be induced. On the other hand, the particular form of the breaking we had assumed was the simplest possible. In the language of the quark model, it corresponds to an explicit quark mass term, and it was the favorite one to most theorists, so this was considered a welcome result.

I remember that this paper was very well received, both at CERN and abroad. I was very excited because, for the first time, I had the impression of participating in something important, and so I assigned myself the task to generalize this result to all orders of perturbation. It took me some time of hard work, especially because, as I realized later, my method was not the most efficient one. I remember in particular a whole night of lengthy combinatorial calculations, only to realize by dawn that I had discovered the world's most complicated method to compute the coefficients of the binomial expansion. At least I was sure I had made no mistakes.

While I was struggling with my higher-order terms, the subject of weak interaction divergences started attracting considerable attention. T. D. Lee and Gian-Carlo Wick invented a mechanism to cure all these diseases, but in a very unorthodox way. For years physicists had used what they called "regulator fields." These were massive fields quantized with negative metric whose role was to provide a cutoff in intermediate steps of the calculations. They were not supposed to represent physical particles because, it was argued, their negative metric would yield violations of the conservation of probabilities. They were purely mathematical tools. Lee and Wick took the opposite point of view. They remarked that such massive fields could represent physical particles because they would be necessarily unstable. The higher the mass, the shorter the lifetime, so all effects of violation of unitarity and/or causality could be confined at very short times and be unobservable. This approach required a modification of the usual Feynman rules, which turned out to be rather complicated and, although (to my knowledge) nobody ever proved that it was inconsistent, it was eventually abandoned.

From Italy came two groups of papers, the first by R. Gatto, G. Sartori, and M. Tonin and the second by Nicola Cabibbo and Luciano Maiani. They all imposed the absence of divergences and claimed to compute the value of the Cabibbo angle. Gatto, Sartori, and Tonin in their first paper obtained a set of consistency conditions that involved the Cabibbo angle, but they had to assume that the angles for the vector and the axial current were different. Cabibbo and Maiani, as well as Gatto et al. in their subsequent papers, assumed one angle, but they imposed a cancellation between electromagnetic and weak divergences. Although none of these approaches sounded very convincing to me, I was fascinated by the equations. They were expressing the Cabibbo angle in terms of ratios of quark masses. It was a simple and elegant relation in very good agreement with experiment. Furthermore, Cabibbo and Maiani found an extremely interesting and intriguing result, namely that isospin violation in the quark masses could be very large. I often discussed these problems with Sheldon Glashow, who had come to CERN and had immediately joined our gang. We soon found many independent ways to rederive these equations, but none was satisfactory. We discovered instead that we had many common tastes in physics as well as in eating and drinking.

At the end of 1968 my stay at CERN came to an end, and I had to go back to Greece to join the Navy for my military service. Before leaving, I applied to several universities in the United States for a second postdoc starting September 1969. My first choice was Harvard because I wanted to continue my collaboration with Glashow, so I was very happy when I received the offer. In the summer of 1969, just as I was getting out from military service, I met Cabibbo, who was visiting Greece. I was glad to hear that he liked our mechanism for solving the problem of the leading divergences and considered the $(3, \bar{3})$ symmetry breaking as an established fact. Nicola showed me a new method he had to organize the divergences of weak interactions in terms of a parameter he called ξ. This ξ was in principle calculable by summing higher-order diagrams, and I decided to try it. I spent a few weeks computing more and more complicated diagrams, trying to extract the coefficients of the various divergent terms, but eventually I gave up. The problem looked hopeless.

I again met Glashow in Cambridge that September. Our eating and drinking habits had not changed, and we became regular customers of the various restaurants and liquor stores of the area, but the next-to-leading divergences of the weak interactions remained untamed. One of the first things I remember upon arrival is an absolutely remarkable seminar at MIT by Francis Low, who was interested in the same problem. A few months before he had written a paper together with M. Gell-Mann, M. Goldberger, and N. M. Kroll – something like the dream team in particle physics. They had invented an extremely complicated but ingenious scheme, involving a large number of vector and scalar intermediaries with degenerate couplings, that reproduced all known weak processes but developed dangerous divergences only in diagonal matrix elements. It was based on the following observation: divergences arise in perturbation theory because of the asymptotic behavior of Feynman propagators in momentum space. Unitarity prevents you from arbitrarily modifying the behavior of a single propagator. However, for the case of a matrix field, you may improve some of the propagator matrix elements without violating unitarity. If you are sufficiently clever, you can construct a model in which all bad divergences appear only in harmless places. The result looked like a generalization of our solution of the leading divergence problem with Bouchiat and Prentki, although the actual mechanism was completely different. As is often the case, the idea was very simple, but the implementation turned out to be very complicated. To quote one of Shelly's remarks: "Few would concede so much sacrifice of elegance to expediency."

In his seminar Low did not present any model in detail, but gave instead a general review of the subject. I have rarely been so much impressed by a lecture as that afternoon at MIT. I had been working in this field for over a year and I thought that I had a very thorough understanding of all the problems involved. However, by listening to Low's extremely clear and beautiful review, I could see my problem in a new light. For the first time I started thinking that the solution may not lead to a determination of the Cabibbo angle. During that year, Low had a high position at MIT with many responsibilities that did not leave him much time for research and prevented him from working any further in this problem.

A few weeks later we were joined at Harvard by Luciano Maiani. I had never met him before, but we became very good friends almost immediately. Shelly and I had no trouble convincing him on two points: that the problem of the next-to-leading divergences was a very serious one, and that he should join us in solving it.

Our collaboration soon developed a standard pattern: each day one of us would have a new idea and invariably the other two would join to prove to him it was stupid. Then we would change roles. Often the discussions would continue over dinner in one of our two favorite restaurants: the Peking on the Mystic (Chinese) or a fish place by the strange name Legal Sea-Food. We made no discernible progress for several weeks. One day, for some reason, Luciano arrived late. I was with Shelly and trying to defend my new idea, which involved the introduction of new species of leptons (I was mainly worried about the $K_L^0 \to \mu^+ \mu^-$ decay). Under Shelly's attacks I kept on changing my scheme and drawing various diagrams on the blackboard, but every time Shelly would find a new flaw. I remember that at a certain moment I drew a new diagram and asked "What about this one," to which Shelly replied, "It is lovely except that it does not exist." When I asked why, he replied, "You idiot, this does not take a new lepton, it takes a new quark!" By mistake I had drawn a diagram with a lepton in the place of a quark. At this moment Luciano entered the room and asked: "What is this new theory with four quarks?" We both stared at him. The magic word had been pronounced.

Shelly promptly remembered that a few years earlier he and James D. Bjorken had considered a model with four quarks. Such studies had been made by many people, for no apparent reason. He wrote down the couplings. It did not take us long to check everything and to prove the cancellations to all orders. By early afternoon we had convinced ourselves that we had the solution. It turned out to be so simple that we are still ashamed it took us so long to find. Even today, when I am teaching the subject, I have a hard time convincing students that this was anything but a trivial exercise. The answer was already implicit in the Cabibbo construction of the weak current. Although we discovered it by looking at diagrams, it is in fact a symmetry property of the weak interactions.

At the limit of exact flavor symmetry, quark quantum numbers such as strangeness are not well defined. Any basis in quark space is as good as any other. By breaking this symmetry, medium strong interactions choose a particular basis, which becomes a privileged one. Weak interactions, however, define a different direction, which forms an angle θ_C with respect to the first one. Having only three quarks to play with, one can form only one doublet of weak SU(2). It will contain the Cabibbo rotated combination $d_c = d\cos\theta_C + s\sin\theta_C$. The orthogonal combination $s_c = -d\sin\theta_C + s\cos\theta_C$ will be necessarily a singlet. The neutral component of the current contains $\bar{d}_c d_c$ and, therefore, has flavor-changing pieces. The only way out is to add the $\bar{s}_c s_c$ term in order to form a flavor invariant. This implies that s_c should also belong to a doublet, that is, one needs a second up-type quark. In this case $\bar{d}_c d_c + \bar{s}_c s_c = \bar{d}d + \bar{s}s$ for all values of θ_C, and the Cabibbo angle remains undetermined.

This argument is exact in the limit of flavor symmetry. In the real world one expects corrections proportional to the quark mass differences. Therefore, Ioffe and Shabalin's estimation can be translated into a limit for the new quark mass and yields an upper bound of a few GeV for the masses of the new hadrons. This fact is very important. A prediction for the existence of new particles is interesting only if they cannot be arbitrarily heavy.

We chose to present our scheme as a lepton–hadron symmetry. Indeed, as far as the weak interactions are concerned, quarks and leptons must behave the same. Notice that this argument alone does not impose the complete family structure. For example, when the τ lepton was later discovered, one could not infer the existence of b- and t-quarks based on the absence of flavor-changing neutral currents. The complete argument requires also the cancellation of triangle anomalies, which was found two years later.

We were obviously very excited about this discovery. We jumped immediately into my car and drove to MIT. An informal seminar was improvised in Low's office, which was attended by most of our colleagues. I remember Low, of course, but also Steven Weinberg, Sergio Fubini, Ken Johnson, Roman Jackiw, Gabriele Veneziano, and probably others. We presented our solution to the problem of next-to-leading divergences and went through all diagrammatic proofs in detail. We proved that weak interactions could be formulated as a Yang–Mills theory and even raised the problem of the origin of the gauge boson masses. And then the most unbelievable thing happened: nobody seemed to be aware of Weinberg's 1967 article, including Weinberg himself, who showed great interest in our work.[2] Nobody mentioned that this question had been answered less than three years ago. It seems that Steve, as he explained to me later, had a psychological barrier against his paper.

We passed the MIT test with flying colors. Low praised our solution, and there were no objections we could not answer. We felt very confident. The same evening we all had dinner at the Legal Sea-Food, and Pucci Maiani (Luciano had been married a few months before) remarked that we looked very happy. Shelly told her that we expected our work to be part of future physics textbooks. In the following days we wrote up the paper in Shelly's best English and sent it to the *Physical Review*. The answer came a little later in the form of a referee's report. I still do not know who he was, but he knew what he was doing. He had read our paper very carefully and said it was interesting and worth publishing, but he raised one objection. The power counting we were using was based on the intermediate-boson model without self-couplings. In this case the nth-order diagrams give a leading divergence of $(G\Lambda^2)^n$. In our paper we were implicitly assuming that the same remains true even in the presence of the Yang–Mills interaction. However, as the referee pointed out, a naive power counting gives instead $(G\Lambda^6)^n$. He then went on to remark: "This behavior can undoubtedly be improved, but the assertion that it can go down to $(G\Lambda^2)^n$ must be either proven or deleted." He did not say it was wrong, which proves that he knew the problem very well. We immediately realized that proving the Λ^{2n} behavior was not easy, and so we decided to change one or two sentences and resubmit. This time the paper was accepted.[3]

Soon after we had finished, Luciano and Pucci decided to return to Italy. Through some friends they succeeded in finding a luxurious first-class cabin in one of the ocean liners that were still crossing the Atlantic for regular passenger service. The day of their departure we organized a farewell champagne party on board. Several friends were there. I remember that, at one point, we were talking to Sam Ting, an experimental physicist from MIT. We tried to convince him to look for charmed particles, but he wouldn't listen.

After Luciano's departure Shelly and I still felt too excited to start a completely new project. We worked essentially in two directions – to advertise our work by giving seminars in various places, and to answer the question raised by the referee.

The first task turned out to be difficult and frustrating. Spoiled by the warm welcome with which our work had been met by Francis Low and the theorists at MIT, we were unprepared to face the skepticism we found almost everywhere else. I now see that people were bothered by almost every point in our theory; they did not even appreciate the importance of the issue itself. As I said earlier, the question of the divergences of weak interactions was considered to be a hopeless one since the theory was nonrenormalizable. The very concept of ordering them, and the fact that we worried about a subclass of them – questions that were obvious to us – were thought to be mathematically unsound and physically meaningless. And more importantly, our proposed solution contained too many high-risk predictions. We were proposing the existence of a whole new hadronic world, the charmed particles, as well as new and as-yet unobserved weak processes, the neutral-current ones. And all that, just to fix an obscure and ill-defined higher-order effect. We also discovered that most people, including many experimenters, had completely wrong ideas concerning the then-current experimental limits on these questions. We were often told, for example, that the existence of charmed particles with masses of a few GeV and normal hadronic production cross sections was already excluded experimentally. Or that weak neutral currents with normal strength would produce intolerably large parity violation in nuclear physics. Going through the literature and checking all these random assertions kept us busy for some time.

Studying the divergence structure of massive Yang–Mills theory turned out to have all the characteristics of a good game of chess: intellectually very challenging and totally useless. We sat down to prove that the naive power counting that gave divergences of the form Λ^{6n} could be reduced to at least Λ^{2n}. It took only a few hours to go from Λ^{6n} to Λ^{4n}. The next step, however, looked much harder. We tried to go through the literature but we found so many contradictory claims that we gave up. The most complete work in the subject seemed to be a paper by Martinus Veltman. I knew that he had been the first to study the divergences of Yang–Mills theories, but Tini's papers have never been easy to read, and furthermore in the introduction he stated explicitly that he was studying only the case of a massive Yang–Mills theory coupled to conserved fermionic currents. He considered this case as simpler to analyze before looking at the real one. Since we were interested in the actual nonconserved weak currents, we decided to prove it ourselves.

This was probably a mistake. Although we succeeded, this very success put us definitely on the wrong track, namely the study of a pure massive Yang–Mills theory. Tini had also started from the same point, but at that time he was very close to finding the truth, that is, the introduction of additional scalar fields. And during all that time the solution was lying in Weinberg's paper without anybody noticing it. We found the main step of the proof, which was essentially combinatoric, during the summer when we were both visiting the National Polytechnic

Institute in Mexico City. We fixed all the details a little later and we wrote the paper while we were visiting Brookhaven in September.[4] I still believe that it is by far the most intelligent paper either Shelly or I have ever written and, until some years ago, I was convinced no one had read it. In fact, I was assured by Veltman that he did read it, in which case he must be the exception to the rule.

I spent most of my second year in the United States traveling around – Harvard, of course, but also Marseille, Rockefeller, and SLAC. I tried desperately to go one step further to the divergence cancellation that would have proven the theory renormalizable, but in the end I convinced myself it was not true. Incidentally, even today one finds in the literature all sorts of wrong and/or misleading statements, although there now exist rigorous proofs that settle the issue: massive Yang–Mills theory, without any appropriate physical scalar fields, is hopelessly nonrenormalizable. But at that time, nobody, Weinberg included, spoke of physical scalar intermediate bosons. I remember that I gave a set of lectures on Yang–Mills theories during my visit to Rockefeller. Abraham Pais asked me whether one could find a generalization of the Stückelberg formalism to non-Abelian theories. I wish I had paid more attention to this question. In any case I had no time. In a few months, we all learned of a young student of Tini's by the name of Gerard 't Hooft. Then all hell broke loose!

I came back to Europe in October 1971 and joined the department in Orsay where I had studied. These were times of great expectations. We had been through so many lean years in particle physics that most of us had lost hope to ever experience the excitement of great discoveries. I am not even sure that anybody was aware of the revolution we were witnessing.

Back in Orsay I started again my collaboration with Bouchiat and Meyer. There were so many new things to learn, so many questions to answer. Gauge theories had revolutionized our way of thinking. While we were trying to understand the proof of renormalizability in various gauges, we realized the vital importance of the Ward identities. A change of gauge produces a completely new theory. All these theories that look, and in many respects are, so different, are linked together through the Ward identities. But then a new problem appeared. Weak interactions involve both vector and axial currents. In many cases one cannot enforce the conservation of both. This is due to the famous triangle anomalies, which had been discovered a few years previously by Stephen Adler, John Bell, and Roman Jackiw.[5] We checked rather easily that Weinberg's model for leptons did indeed suffer from such an anomaly. The Ward identities were broken and the renormalizability and unitarity proofs did not apply. I felt like a child from whom someone takes away his most wonderful toy. I was discussing this problem with Bouchiat and Meyer. It was obvious that adding electrons and muons wouldn't help.

At this moment, the thought appeared simultaneously to all of us. We stopped in the middle of a sentence and stared at each other. It was obvious: the answer was leptons versus hadrons. The essential point is that the anomaly is universal. Its coefficient depends on the charges of the fermions but not on their masses. Light or heavy contribute the same amount. We first checked a model with protons and neutrons. It worked. Then we knew that all realistic quark models would also work. Soon we found a general formula that contains all acceptable models.[6] It includes the standard one with three colored fractionally charged quarks as well as many others, like Han–Nambu quarks with integer charges. I was particularly happy to see that the introduction of the fourth quark was, once more, essential. In fact, it is through the requirement of anomaly cancellation that the family structure is imposed. Families must be complete. The last obstacle for a consistent electroweak theory was removed.

It is often said that progress in physics occurs when an unexpected experimental result contradicts the theoretical beliefs. It forces physicists to change their ideas, and eventually it gives rise to a new theory. But the revolution that brought geometry into physics had a theoretical, in fact an aesthetic, motivation.

Obviously, these notes are not meant to tell the entire story. I consider myself extremely fortunate to have lived through this most wonderful adventure. Here I recall only the little corners of the puzzle that I helped put together. But pleasure and excitement came mostly from following closely the whole enterprise – from sharing the deceptions and taking part in the expectations.

I have not mentioned a memorable meeting we organized at Orsay in January 1972. It was probably 't Hooft's first public appearance, and I still remember the marathon lectures he gave that lasted several hours every day. Then there were the long hours we spent talking with André Lagarrigue and the Gargamelle team. I witnessed all their struggles against neutron backgrounds, and I will not forget the glorious days of their success, when for the first time we had a clear proof that we were on the right track. I have occasionally had the good fortune to taste some great wines, and I can truly appreciate the ones Jack Steinberger served us at his home after having lost a famous bet on neutral currents. Finally, I have not mentioned the years of expectation that preceded the discovery of charmed particles or my bet at the London Conference, which nobody has yet paid; the noon telephone call in which Lagarrigue woke me up to announce the discovery of the J/ψ; or the evening, two years later, during which Haim Harari was searching for me in Rehovoth to tell me the good news from SLAC. All these are certainly part of my story, but their place in the history of the Standard Model is questionable. Actors make poor historians, so I deliberately limited my recollections to those which, to a large extent, may be supported by published documents. To the best of my knowledge they are correct, but I would take no bets.

Notes

1 Except as otherwise cited, references to publications mentioned in this paper can be found in S. L. Glashow, J. Iliopoulos, and L. Maiani, "Weak Interactions with Lepton–Hadron Symmetry," *Phys. Rev. D2* (1970), pp. 1285–92.
2 S. Weinberg, "A Model of Leptons," *Phys. Rev. Lett. 19* (1967), pp. 1264–6.
3 S. L. Glashow, J. Iliopoulos, and L. Maiani, "Weak Interactions."
4 S. L. Glashow and J. Iliopoulos, "Divergences of Massless Yang–Mills Theories," *Phys. Rev. D3* (1971), pp. 1043–5.
5 S. L. Adler, "Axial Vector Vertex in Spin or Electrodynamics," *Phys. Rev. 177* (1969), pp. 2426–36; J. S. Bell and R. Jackiw, "A PCAC Puzzle: $\pi^0 \to \gamma\gamma$ in the τ-model," *Nuovo Cimento A60* (1969), pp. 47–61.
6 C. Bouchiat, J. Iliopoulos, and P. Meyer, "An Anomaly-Free Version of Weinberg's Model," *Physics Letters 38B* (1972), pp. 519–23.

27

Weak-Electromagnetic Interference in Polarized Electron–Deuteron Scattering*

CHARLES PRESCOTT

Born Ponca City, Oklahoma, 1938; Ph.D., 1966 (physics), California Institute of Technology; Professor of Physics at the Stanford Linear Accelerator Center; high-energy physics (experimental).

In 1978 a team of twenty physicists performed an experiment at SLAC that demonstrated convincingly that the weak and electromagnetic forces were acting together in a fundamental process, the inelastic scattering of polarized electrons. This result showed that the electron was a normal partner in the model of electroweak interactions as first spelled out by Steven Weinberg in 1967.

The work I describe here was done mostly by other persons as part of a team effort. In this paper I have tried to give credit to the many excellent contributions from this group. I had hoped to point out all of the important individual efforts that were so critical to the overall success, but I feel that this summary falls short of that goal. This chapter should be taken as a personal recollection of the work that occurred over a period of eight years at SLAC, Yale University, and elsewhere.

As a part of this chapter, the organizers asked that I summarize the work in atomic physics to seek out parity-violating effects in atomic levels. I reluctantly agreed to attempt this, even though I had no involvement in those experiments. What I present here is only a brief history of the search for optical rotation by bismuth vapor, as reported in the literature. I have not attempted to extend this summary to cover the work on the other atoms – thallium, lead, and cesium – which came somewhat later. A proper talk on the history of parity violation in atomic physics would include those contributions as well. The work with bismuth began in the mid-1970s, and so events were occurring during the time work was

* Work supported by Department of Energy contract DE–AC03–76SF00515.

under way at SLAC. Some of those events had significant impact on our work.

Physicists love symmetries. Among the important symmetries, parity (the symmetry of mirror reflection) was assumed to be valid for Nature's forces and fundamental processes until the mid-1950s, when the weak force was shown to violate maximally the parity symmetry in β-decay processes.[1] This unexpected result came as a shock and a surprise. The experimental observations were made in charged-current processes (mediated by the $W^{\pm}$, as we now know). In those days it was conjectured that weak neutral-current processes should exist, but no experiments had access to such processes, and advances in the state of knowledge were slow. In the 1960s, however, progress on the theoretical side was beginning.

Central issues through the 1960s and early 1970s that related to the weak neutral currents were: (1) Do they exist? (2) If so, what are their characteristics? and (3) If so, are they maximally parity violating, like the charged currents? Underlying the theoretical speculation was the desire for a common theory that would unify the weak and electromagnetic forces. It was the growing interest and debate in the theoretical community over the connections between the weak and electromagnetic forces that stimulated a number of ventures in the experimental community to look for neutral-current effects in electromagnetic processes.

The early experiments at SLAC

The interest in searching for parity violation at SLAC began in 1970. I was at that time working at the University of California, Santa Cruz, and often visited SLAC. Richard Taylor's Group A at SLAC was heavily involved in the inelastic scattering program. I knew that the Taylor group had recently performed a time-reversal measurement in electron scattering, and so I discussed with Taylor and members of his group my interest in looking at the recoiling protons for $\vec{\sigma_p} \cdot \vec{p_p}$ terms in electron scattering (specifically elastic scattering) as an experimental approach to parity violation.[2] Although Taylor's group showed considerable interest, the experimental underpinnings of the ideas being considered were too weak to permit a sensitive measurement, and so the interest died.

Taylor's group was performing a series of experiments in deep-inelastic scattering, and so I joined that effort in 1970 as a collaborator from Santa Cruz and a year later moved to SLAC. Among the experiments to come in the near future was experiment E61, which studied 4° scattering from hydrogen and deuterium, providing the basic information later used for the parity-violation work. Cross sections, counting rates, and backgrounds were measured, and E61 was the beginning of a learning curve for me: the facility, the equipment, the beams and monitoring, and the people who inhabited the lab.

In late 1970 Vernon Hughes from Yale visited SLAC and presented a proposal to build a polarized electron source for the linac and to accelerate these polarized electrons to high energies. The proposed source was based on a Yale prototype, which stripped electrons from a polarized atomic beam of ^{6}Li using an ultraviolet flash lamp.[3] The physics motivation for this proposal was to study the spin of constituents inside polarized protons. That proposal was soon presented to SLAC and was formally accepted and designated E80. The E80 proposal was the beginning of a long and successful SLAC program on spin structure that continues to be active today.

I attended Hughes's seminar in 1970. I was still interested in searching for parity violation, and perhaps this source could be used to look for $\vec{\sigma_e} \cdot \vec{p_e}$ terms using the polarized electron beam. I took this idea seriously and began to study the feasibility of a parity-violation measurement in the End Station A facility. I remember taking my idea to Taylor and later Sid Drell, looking for support and encouragement, which I got from them. Early in 1971 I arranged a visit to Yale to talk to Hughes and his group. I wished to form an experimental collaboration and felt I needed the involvement of the Yale group. We discussed the physics possibilities and various strategies, and identified three possible approaches: (1) to utilize the planned E80 experiment and to study parity violation by averaging over the E80 target polarization; (2) to extend the E80 running to provide dedicated time for a parity-violation search; (3) to propose an independent dedicated search for parity violation. We agreed to collaborate and to pursue (1) and (3).

As the E80 experiment was already planned, the first item required no action on my part. I focused my attention on a new experimental proposal, E95, whose objective was solely the search for parity violation. Unlike that in E80, the E95 target was chosen to be *unpolarized* hydrogen, eliminating a potentially serious systematic error from polarized protons. The E95 target was optimized for the parity-violation test. If it existed, parity violation would show up as a nonzero asymmetry A_{pv}, defined as

$$A_{pv} = \frac{1}{P_e} \frac{N_\ell - N_r}{N_\ell + N_r},$$

where N_ℓ and N_r are the numbers of scattered electrons for left- and right-handed helicity polarized beams, and P_e is the magnitude of the polarization.

Motivation for E95 was not easy. It was (quoting from the proposal) "not sensitive to weak neutral currents." We knew that the statistical error on A_{pv} would be too large. Weak-electromagnetic interference required an error $\Delta A_{pv} < 10^{-4}Q^2/M_p^2$, whereas our estimated statistical error was an order of magnitude larger. Weak neutral currents were simply not reachable by the techniques we had at hand. (The Yale–SLAC source, called PEGGY, was too low in intensity, 2×10^9 electrons/pulse at 80% polarization.) Furthermore, helicity reversals required reversal of a magnetic field to flip the electron spin. The action of spin reversal affected the beam parameters and introduced worrisome systematic errors as well. In spite of these limitations and concerns, we proceeded with the E95 proposal.

The formalism for inelastic scattering of polarized electrons was not available in the literature. We knew that polarized inelastic electron scattering and inelastic neutrino scattering were kinematically very similar. With the aid of a paper by Stephen Adler we showed that relaxing parity invariance introduced a third structure function $W_3(\nu, Q^2)$ in addition to the usual W_1 and W_2.[4] Furthermore, requiring that current conservation be valid led to $W_3(\nu, Q^2) \to 0$ as $Q^2 \to 0$. We argued in the E95 proposal that such parity-violating terms *may* exist and could have escaped detection in former experiments at low energies, for example, in nuclear physics studies that were the most sensitive. We could find no experimental work that ruled out such terms at the level of sensitivity achievable in E95.

Speculation on the existence of parity-violating effects in electromagnetic processes could be found in the literature. In a 1957 paper entitled "Electromagnetic Interaction with Parity Violation," Zel'dovich speculated on such terms with a particular model.[5] One year later he wrote a remarkable paper anticipating future experiments with high-energy longitudinally polarized electrons and with optical rotation of linearly polarized light in atoms.[6] Weinberg's paper "A Model of Leptons" appeared in 1967, and 't Hooft's demonstration of renormalizability appeared in 1971.[7] It was in the context of these ideas that E95 was proposed and defended before the SLAC Experimental Program Advisory Committee. It was approved in June 1972.

In the years 1972 to 1975, work on E80 and E95 proceeded along with many other activities at SLAC. Deep-inelastic scattering in the

End Station A facility continued actively. The SPEAR program came into full swing during this period, leading to the discoveries of the J/ψ and ψ', the τ-lepton, the charmed mesons, and charmonium states. The PEP program was starting up. During this very busy time, E80 ran (in 1975) and shortly thereafter so did E95 (in 1976). Neither experiment saw any parity-violating signals. E80 established a limit $A_{pv} \leq 5 \times 10^{-3}$ at $Q^2 \approx 1.4$ and 2.7 $(\text{GeV}/c)^2$ (1976).[8] A limit $A_{pv} \leq 0.8 \times 10^{-3} Q^2$ at $Q^2 \approx 4$ $(\text{GeV}/c)^2$ was established by E95 (1978).[9] These null results were not surprising. Neutral currents had been observed in 1973–74, including those involving the electron (in 1976).[10]

The lessons of E80 and E95 were many. They taught us a lot about techniques for doing this kind of experiment. Equally important, they taught us a lot about how *not* to do this kind of experiment.

In 1974, long before E80 and E95 took data, plans were begun to develop a new kind of source, one that would enable us to obtain the statistical samples needed to observe the weak effects. Charles Sinclair and I wrote a letter of intent to Wolfgang Panofsky concerning a future experiment to look for weak neutral currents at the level of the Weinberg–Salam model in inelastic electron scattering, and we sought his support. We proposed to replace the PEGGY source with a new polarized-electron source that would be laser driven. We had discussed more than one type of device and were considering photoionization of cesium as one possibility. We were also interested in using semiconductor materials for a cathode and had discussed our needs with Ed Garwin of SLAC. During 1974 Garwin visited ETH Zürich and while there proposed with H. C. Siegmann and Dan Pierce the use of negative-electron-affinity gallium arsenide as a suitable cathode material for a high-intensity polarized-electron source.[11] It was their proposal that turned out to be a crucial step for success. The combination of a laser (at moderately high power) and a solid-state cathode material (having high electron density) promised to provide the large electron currents needed to observe the elusive weak-electromagnetic interference effects. Polarization of the photoemitted electrons resulted from circularly polarized laser light exciting valence-band electrons to the conduction band. Electrons near the surface could escape. Polarization values near 50% were expected as a consequence of the angular-momentum selection rules. Polarization reversal could be accomplished optically, by reversing the circular polarization.

Thus in 1974 a new experiment E122 emerged, designed to test the Weinberg–Salam model via parity violation with a sensitivity $\Delta A_{pv} \leq$

1×10^{-5} near $Q^2 \approx 1 \; (\text{GeV}/c)^2$. The E122 proposal was developed from the experience with E95. Improvements over the E95 rates would be large: (1) the polarized beam current was up by a factor of 250; (2) a new spectrometer using magnets from the 8 and 20 GeV spectrometers was designed to have a large acceptance, an improvement by a factor of 5; and (3) a 30 cm long deuterium target was planned, for an additional factor of 3. The overall gain over E95 was the product of these factors (approximately 3750), which would allow us to reach 1σ sensitivity to Weinberg–Salam neutral currents in as little as 15 minutes of beam time.

The proposed E122 experiment was approved in June 1975. During the next two and a half years, work on the PEGGY-II source was under way. In December 1977 the new source was ready, and it was tested in a brief run on the SLAC linac. Before describing the E122 experiment, however, I now want to turn to developments in the field of atomic physics that were progressing rapidly at the time.

Parity violation in atomic physics

Zel'dovich was perhaps the first to suggest looking for optical rotation of the plane of linear polarization of light passing through a gas vapor.[12] He concluded in 1960 that optical rotation by hydrogen would be too small to detect. The subject of optical rotation was revitalized by the 1974 work of the Bouchiats in Paris and by Khriplovich at Novosibirsk.[13] They pointed out that in high-Z atoms, the optical rotation is enhanced by an approximate Z^3 factor and that the sought-after parity-violation effects could become measurable in atomic systems using reasonable laboratory techniques. With this stimulation, several groups at widely separated institutions proposed experiments in 1974. Bismuth ($Z = 83$) was identified as a particularly promising atomic system. Four groups – at Oxford, Seattle (University of Washington), Novosibirsk, and Moscow – proposed generally similar measurements based on optical rotation of the plane of polarization in bismuth. The specific details of the four proposed experiments differed considerably. At Berkeley a thallium ($Z = 81$) experiment was proposed to measure circular dichroism of a light beam. (Circular dichroism, the unequal absorption of left and right circularly polarized light, is closely related to optical rotation of linear polarization.) A Paris group proposed studying circular dichroism in cesium ($Z = 55$).

The bismuth experiments start with crossed linear polarizers. In a hypothetically ideal experiment with perfect optical elements, a light beam is not transmitted. Introducing a cell of bismuth vapor between the crossed polarizers should lead to a rotation of the polarization plane and thus a net transmission of light. In the real world, however, one has to deal with imperfect optics, so the experiments become somewhat more elaborate.

A bismuth atom has three *p*-wave electrons outside fully closed shells. Two suitable optical-absorption lines can be excited from the ground state by magnetic-dipole excitation, one at $\lambda = 648$ nm and one at $\lambda = 876$ nm. Through the weak interactions between the electrons and the nucleus, parity admixtures of these states are expected to exist, leading to a small electric-dipole amplitude in these transitions. This leads to an optical rotation proportional to the imaginary part of the ratio of these two amplitudes $R = \text{Im}(E1/M1)$. The optical rotation ϕ is expected to be $\approx 10^{-7}$ radians. This extremely small rotation can be seen by scanning the light frequency across the line. The absorption by a line is symmetric about the line center. Faraday rotation (which can be induced by a longitudinal magnetic field) is symmetric about the line center and could be used to calibrate the equipment. In contrast, the parity-violating signal is antisymmetric about the line center, and the experiments were designed to look for an antisymmetric piece in the absorption. The bismuth experiments approached the problem with different techniques. Oxford, Moscow, and Novosibirsk chose to work on the 648 nm line, while Seattle studied the 876 nm line. Oxford, Seattle, and Moscow chose to modulate ϕ by using Faraday rotators in the light beam, while Novosibirsk modulated λ. Seattle initially did not resolve the hyperfine splittings, while the others did. Each experiment had to deal with a set of systematic effects, which were somewhat different from those of the other groups. The experiments also had to deal with statistical errors. Fluctuations in photon-counting statistics required averaging over long runs.

Tests of systematic errors required careful studies and null tests. Problems common to these early measurements included: (1) difficulties in obtaining suitable lasers; (2) molecular species in the bismuth vapor that masked the desired spectral lines; (3) Faraday rotations induced by stray residual fields; (4) extra undesired materials in the optical path, such as cell windows; (5) thermal drifts; (6) scattering and reflections leading to laser-beam interferences; and (7) undesirable influence on the laser beam due to the scanning or modulation techniques used.

On the theoretical side, considerable work was under way to understand the proper approach needed to deal with the complicated electronic structure of bismuth. The uncertainties were exacerbated by the lack of a well-established value for $\sin^2\theta_W$.

In 1977 Seattle and Oxford completed their first measurements and published adjacent articles in *Physical Review Letters*.[14] Both experiments reported null results for optical rotation, with experimental precision substantially better than needed for testing the Weinberg–Salam model predictions. These groups announced that its prediction for the electron's neutral current interaction was wrong. In hindsight we know that these experiments were wrong, not the theory. However, the simultaneous publication of two separate groups at that time created considerable turmoil and controversy in the physics community.

During this period when the atomic physics experiments were active and being discussed at conferences and meetings, the work at SLAC had been proceeding steadily. It was at this time, shortly following the publication of null results by the Seattle and Oxford groups, that the SLAC experiment was finally ready. A polarized source suitable for testing the Weinberg–Salam effects in deep-inelastic electron scattering was completed. The source was tested on the SLAC linac in December 1977, just before the scheduled February 1978 start of the E122 experiment.

In March 1978 the Novosibirsk group reported seeing evidence for parity violation on the 648 nm line in bismuth.[15] The initial reports were accompanied by somewhat large systematic errors, which were subsequently reduced in 1979 without affecting the reported value. In 1980 the Moscow group reported a null result on the same 648 nm line in bismuth in their experiment, in agreement with the earlier Oxford and Seattle results. By 1981 Seattle had improved and repeated their experiment and reported new results. The Seattle group now reported seeing evidence for parity violation, but somewhat smaller in magnitude than the Novosibirsk result. In 1984 Moscow and Oxford reported results of their improved experiments, which agreed with the 1981 results of Seattle. The Novosibirsk group apparently did not report any new measurements in the years after 1979. This history is summarized in Fig. 27.1, where the bismuth (but not the thallium, lead, or cesium) results are shown.

The theory of parity violation in atomic bismuth was sufficiently uncertain in the early years that calculations provided only general guidance. The authors of the papers reporting parity violation all reported agreement with the theory. As the experimental results were improved, the results settled down to approximately one-half the value reported by

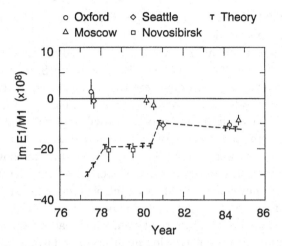

Fig. 27.1. Test of parity violation in bismuth by measurement of the optical rotation of the plane of linear polarization. The parameter $R = \text{Im}(\text{E1}/\text{M1})$ is plotted versus the date of publication. Where possible, the value of the theoretical prediction *as quoted by the authors at that time* is also shown. The Seattle group studied the 876 nm bismuth absorption line, while the other groups studied the 648 nm line.

the Novosibirsk group. The theory was refined and remained in agreement with the experiments, as shown in Fig. 27.1.

In 1987 a group of authors published a global analysis of weak neutral-current experiments.[16] In that report, regarding the early atomic parity-violating experiments, they say:

> We have omitted the early null experiments, the Novosibirsk bismuth (648 nm) experiment which is clearly inconsistent with the Moscow and Oxford results, and the original Berkeley thallium result.

Clearly, unknown systematic uncertainties dominated the atomic parity-violation results in the years before 1981. Today the cesium experiments continue to be refined and now offer the best prospects of precision measurements of parity violation in atoms.

SLAC experiment E122

Preparation for a new experiment sensitive to the weak neutral currents predicted in the Weinberg–Salam model began in earnest in 1975 following approval of E122. Development of a new source and design of a new spectrometer quietly occupied the efforts of a number of people during the next several years, during which time the null results of E80 and E95 were obtained.

The frame of mind in the group preparing E122 was certainly colored by the null results obtained in mid-1977 by the Seattle and Oxford experiments. The expectations were that our experiment could likely provide a null result. The consequences of that concern was to force a redoubling of the experimental effort to provide "proof" that even if the experiment were to see no parity-violation signature, the experiment *could* see one if it were there. The experiment had to show it would be sensitive to such effects even if they were not seen. Sensitive beam monitors were developed. Feedback controls on the beam position on target, on the beam angle on target, and on the beam energy were developed and installed to stabilize the beam, which had a natural tendency to drift around. Beam-polarization monitors were installed and backup monitors were added to provide a redundancy. A Mott polarimeter was installed at the source and a Møller polarimeter before the target.

The spectrometer was instrumented with two detectors, which operated independently to measure asymmetries. Two independent computer codes were developed to check the analysis (ultimately the data were processed in two computers). This rather elaborate preparation before the experiment reflected our internal concerns that the experiment would be a very difficult one to prove, first to ourselves, but then ultimately to the rest of the physics community.

By February 1978 the E122 experiment was scheduled to run, and checkout of the beam and the spectrometers began using unpolarized electrons from the thermionic gun. The checkout procedures were rather lengthy, involving looking at each component and carefully testing its performance. These tests typically utilized low beam repetition rates while beams to other experiments were in use. By late March, the tests were mostly complete. Taylor had arranged, through earlier negotiations within the laboratory, to run E122 without any beams being sent to other experiments. This dedicated mode – with the sole use of the linac for our experiment – was exceptional, but it proved to be very important to the experiment by contributing to the stability of the beams. It also contributed to an improved confidence in the crew of experimenters, and to the undivided attention of the accelerator operators to E122. In April 1978 the experiment began dedicated-beam operation with polarized beams.

The polarized-electron source delivered longitudinally polarized electrons to the linac at the rate of 120 pulses per second. The source was

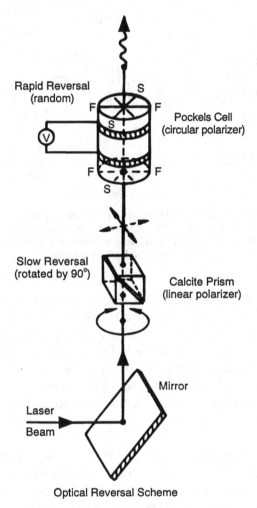

Polarized Laser Beam to the Ga As Cathode

Rapid Reversal (random)

Pockels Cell (circular polarizer)

Slow Reversal (rotated by 90°)

Calcite Prism (linear polarizer)

Mirror

Laser Beam

Optical Reversal Scheme

Fig. 27.2. Rapid reversals of electron spin were achieved by a voltage-driven Pockels cell, which generated 100% circularly polarized laser light. The pattern of voltages was randomized to avoid synchronization with potential harmonic components in the beam parameters. A calcite prism (linear polarizer) was rotated periodically by 90° and 45° to study systematic effects associated with the rapid reversals. Raw asymmetries were constructed using the sign of the Pockels-cell voltages (see Figs. 27.3 and 27.4).

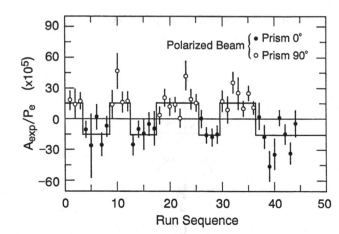

Fig. 27.3. The raw experimental left–right asymmetry divided by the measured beam polarization, A_{exp}/P_e, shown for a sequence of 44 runs, each lasting from 1 to 3 hours. The solid line represents the expected result (on average), taking into account the actual prism orientation.

driven by circularly polarized light from a dye laser at 710 nm wavelength. Circular polarization was achieved by use of a calcite prism linear polarizer followed by a Pockels cell, as seen in Fig. 27.2. Reversal of the biasing voltage on the cell would reverse the direction of circular polarization. The linac accelerated the beams with little depolarization. The experiment could be operated with left (e_L) or right (e_R) circularly polarized electron beams, at the choice of the experimenters. Throughout most of the running, e_L and e_R pulses were mixed with a randomized pattern.

The spectrometer measured forward scattering at 4° from the 30-cm-long deuterium target. Scattered electrons that entered the spectrometer aperture and fell within the momentum acceptance were detected in the two independent counters, a gas Čerenkov counter and a lead-glass calorimeter. Up to 1000 electrons per linac pulse were detected in the spectrometer. To analyze such a high counting rate, signals were integrated and digitized for the Čerenkov counter and the lead-glass counter. Signals from each of the beam monitors were also digitized for each beam pulse. The data-acquisition computer stored the normalized signals (the digitized counters divided by the digitized beam charge) for each pulse, sorted by the beam polarization, one for e_L and one for e_R. After a period of running (typically 1 to 3 hours), the run was ended and summarized.

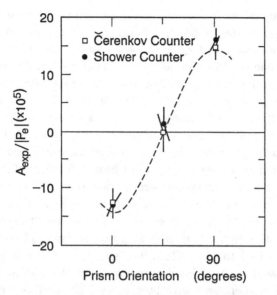

Fig. 27.4. The raw asymmetry divided by the beam polarization, A_{exp}/P_e, for the three prism settings. Measurements made with two independent detectors in the spectrometer, a gas Čerenkov counter and a lead-glass shower counter, are shown and seen to agree. The null values at 45° indicate there are no helicity-dependent systematic errors at the level of the statistical errors shown.

The linear polarization of the laser light reading the Pockels cell was periodically rotated by mechanically rotating the axis of the calcite prism by 90°. This rotation had the effect of reversing the circular polarization and hence interchanging e_L and e_R. However, the data-acquisition computer was not informed of these prism reversals, but continued summarizing the data referenced to the sign of the voltage on the Pockels cell. Thus the on-line asymmetries were expected to reverse sign. Figure 27.3 shows the series of 44 runs taken during the April running. The combined data show a clear pattern of asymmetries that follow the prism rotation. Figure 27.4 shows the combined data for the prism oriented at 0°, 45°, and 90°. For each angle the asymmetries measured independently by the Čerenkov and the lead-glass counter are shown, in excellent agreement.

Systematic errors were studied at considerable length. The most serious of systematic errors arose from beam parameters that changed with helicity reversals. Such effects could induce false asymmetries indistinguishable from real ones. The experiment was set up to monitor six beam parameters, horizontal and vertical positions x and y at the target, θ_x and θ_y at the target, Q (charge per pulse), and E_{beam}. Each parameter was read and logged for each beam pulse. Asymmetries, or differences, generated by the helicity reversals were an important part of the monitoring and analysis. The most important contribution to the systematic errors arose from ΔE_{beam} due to the helicity reversals. This effect arose due to minor changes in Q generated at the source by the Pockels-cell modulation. Beam loading in the linac coupled changes in Q to changes in E_{beam}. Since the deep-inelastic scattering cross section is strongly dependent on the energy of the beam, the effect of ΔE_{beam} was the most serious. The combined errors amounted to 4% of the observed asymmetries and were treated as part of the overall systematic errors.

By the end of April, the experiment was observing clear evidence for a parity-violating signal. The next step invoked was to lower the beam energy from 19.42 to 16.18 GeV. The spin motion through the beam transport system (a 24.5° bend angle) was such that the spin precessed ahead of the momentum vector by π every 3.237 GeV of energy. That is, at 19.42 GeV, the spin precessed 6π before reaching the target, and at 16.18 GeV, only 5π. The on-line asymmetries would be expected to reflect the change in spin orientation.

Data were also taken at 17.80 and 22.20 GeV. Figure 27.5 shows the measured asymmetries for the two counters, the Čerenkov and the lead-glass devices. The asymmetries clearly followed a $g - 2$ curve, which was expected if the asymmetries were not dominated by false effects. The null points, one at 45° in the prism-rotation curve and one at 17.80 GeV in the $g - 2$ scan of beam energy, were important evidence that the false effects must have been small. A short run on hydrogen was also taken that showed evidence for parity violation in agreement with the deuterium results.

Evidence for parity violation in deep-inelastic scattering of polarized electrons was announced at a colloquium at SLAC in June 1978 and a week later in Europe at Trieste.[17] The results agreed with the Weinberg–Salam model for a value of $\sin^2 \theta_W = 0.20 \pm 0.03$.

In the summer and fall of 1978, plans for further measurements were made. During this period, many talks were given at many places. I would like to tell one story that occurred at Caltech. Richard Feynman was in the audience and listened to the talk I gave on the careful tests and checks that were done. At the end he asked a typically astute question: "How do you know that the detectors respond equally to e_L and e_R beams?" He sought an experimental test we had done to exclude that

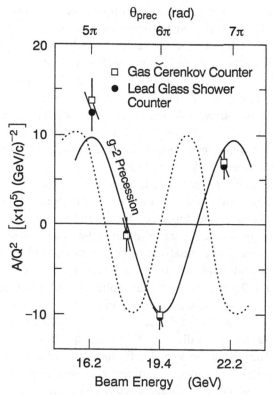

Fig. 27.5. By varying the beam energy, the spin orientation at the target varied due to the $g-2$ electron spin precession in the 24.5° of bending by the beam transport system. The quantity $A/Q^2 = A_{exp}/P_e Q^2$ is plotted for two independent detectors in the spectrometer. The solid curve is the best fit to the expected variation. The dotted curve represents the spin orientation of the electrons as they passed through the detectors.

possibility. I explained the usual arguments, that soft-electromagnetic processes were responsible for light produced in the Čerenkov and lead-glass counters, and these processes were known to be helicity independent. We had not performed experimental checks because we did not have the facility to do so. He was not satisfied. He preferred to see checks with experimental tests. Upon returning to SLAC, I looked into the question of the spin in the detectors. The spectrometers deflected the scattered electrons an additonal 14° bending at a lower energy E'. The spin at the detectors precessed even faster than the $g-2$ curve (see Fig. 27.5, the dotted curve). I argued in a letter to Feynman that the dotted curve showed that there was no evidence for his conjectured systematic effect. In a subsequent conversation he told me he believed our results, even without that argument, but felt we should have made tests to rule out *experimentally* that possible systematic error.

By the fall of 1978 we resumed our running of E122 to extend our data sample. The goals of the spring 1978 running had mostly been met. Existence of parity violation in deep-inelastic electron scattering had been demonstrated. However, questions regarding the Weinberg–Salam model remained open, and we used the extended running period to pursue the answers. Let me explain.

In a parity-violating process such as deep-inelastic scattering, where the neutral-current interaction is mediated by a vector boson (the Z^0), there are two electron couplings, one for e_L and one for e_R. These couplings, g_L and g_R, are necessarily different for parity nonconservation to exist. The vector and axial-vector couplings g_V and g_A are defined to be the sum and difference of g_L and g_R; $g_V = (g_R + g_L)/2$, and $g_A = (g_R - g_L)/2$. It turns out that while deep-inelastic scattering is sensitive to both coupling terms, the atomic-parity violation in bismuth is sensitive only to g_A. Could it be that both the SLAC and the Oxford/Seattle results were valid? (At that time the Oxford/Seattle results had not been proved to be wrong.) Perhaps $g_A \approx 0$ and $g_V \neq 0$, thus agreeing with the experiments (but not the Weinberg–Salam model). The purpose of the fall 1978 running was to measure both g_V and g_A in deep-inelastic electron scattering in order to investigate that question.

The fall running extended the kinematic range of the data. Separation of the g_V and g_A terms required measurement at different values of $y = (E_{beam} - E')/E_{beam}$, which in a simple quark model is related to an angular distribution. Figure 27.6 shows the results of the fall running. The "model-independent" fit corresponds to a $g_A \neq 0$, ruling out any possible agreement with Oxford/Seattle null measurements. (The "hybrid" curve in Fig. 27.6, which was one such model satisfying a null value in atomic-bismuth parity violation, was excluded by the fall 1978 data.) Our data were also in excellent agreement with the Weinberg–Salam model.

With the ending of the fall 1978 running, the E122 experiment was concluded. It had been a great success. From the combined running in 1978, the SLAC data were consistent with the Weinberg–Salam model for a value

$$\sin^2\theta_W = 0.224 \pm 0.020 \,(\text{stat. and syst. errors combined}),$$

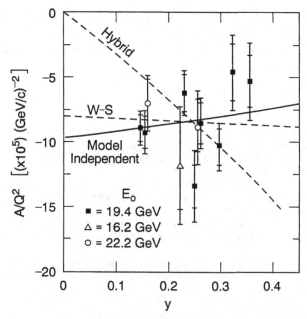

Fig. 27.6. The quantities $A/Q^2 = A_{exp}/P_e Q^2$ plotted versus the kinematic variable $y = (E_{beam} - E')/E_{beam}$. The model-independent fit is in good agreement with the Weinberg–Salam model, but not with the hybrid model (see text). These data established values for the neutral-current couplings g_V^e and g_A^e, and showed that the SLAC results were incompatible with the earlier reports of no parity violation in bismuth.

which agreed with then-existing measurements from neutrino scattering and significantly improved on the errors at that time. This observation that the weak neutral-current process interfered with the photon-exchange process demonstrated that the neutral currents were mediated by a spin-1 boson. Within the context of the Weinberg–Salam model, the electron behaves as a normal partner; from the measurements of the couplings, e_R is placed in a weak-isospin singlet assignment, whereas e_L and ν_e are given a doublet assignment.

In subsequent months we considered further extension of this experiment. Several factors argued against further running. We had achieved essentially all the goals for E122; the experiment had occupied nearly six months of SLAC's beam time and made a heavy hit on other experiments trying to run; and significant improvement over E122 would be difficult, requiring new developments in the source and experimental apparatus. Finally, SLAC was beginning to embark on its linear collider project, and it seemed best to participate in that project to understand better the physics of the Z^0.

Thus ended the eight-year search at SLAC for parity violation in the electromagnetic processes.

Notes

1 T. D. Lee and C. N. Yang, "Question of Parity Conservation in Weak Interactions," *Phys. Rev. 104* (1956), pp. 254–8; C. S. Wu, et al., "Experimental Test of Parity Conservation in Beta Decay," *Phys. Rev. 105* (1957), pp. 1413–15.

2 S. Rock, et al., "Search for T-Invariance Violation in the Inelastic Scattering of Electrons from a Polarized Proton Target," *Phys. Rev. Lett. 24* (1970), pp. 748–53.

3 M. J. Alguard, et al., "A Source of Highly Polarized Electrons at the Stanford Linear Accelerator Center," *Nucl. Inst. Meth. 163* (1979), pp. 29–59.

4 S. L. Adler, "Sum Rules Giving Tests of Local Current Commutation Relations in High-Energy Neutrino Reactions," *Phys. Rev. 143* (1966), pp. 1144–55.

5 Ya. B. Zel'dovich, "Electromagnetic Interaction with Parity Violation," *Zh. Eksp. Theor. Fiz. 33* (1957), pp. 1531–3, in Russian; English translation in *Sov. Phys. JETP 6* (1958), pp. 1184–8.

6 Ya. B. Zel'dovich, "Parity Nonconservation in the First Order in the Weak-Interaction Coupling Constant in Electron Scattering and in Other Effects," *Za. Eksp. Theor. Fiz. 36* (1959), pp. 964–6.

7 S. Weinberg, "A Model of Leptons," *Phys. Rev. Lett. 19* (1967), pp. 1264–6; G. 't Hooft, "Renormalizable Lagrangians for Massive Yang–Mills Fields," *Nucl. Phys. B35* (1971), pp. 167–88.

8 M. J. Alguard, et al., "Deep Inelastic Scattering of Polarized Electrons by Polarized Protons," *Phys. Rev. Lett. 37* (1976), pp. 1261–5.

9 W. B. Atwood, et al., "Search for Parity Violation in Deep-Inelastic Scattering of Polarized Electrons by Unpolarized Deuterons," *Phys. Rev. D18* (1978), pp. 2223–6.

10 F. J. Hasert, et al., "Observation of Neutrino-Like Interactions Without Muon or Electron in the Gargamelle Neutrino Experiment," *Phys. Lett. 46B* (1973), pp. 138–40; J. Blietschau, et al., "Evidence for the Leptonic Neutral Current Reaction $\bar{\nu}_\mu + e^- \to \bar{\nu}_\mu + e^-$," *Nucl. Phys. B114* (1976), pp. 189–98.

11 E. L. Garwin, "Polarized Photoelectrons from Optically Magnetized Semiconductors," *Helv. Phys. Acta 47* (1974), p. 393.

12 Ya. B. Zel'dovich, "Parity Nonconservation."

13 M. A. Bouchiat and C. C. Bouchiat, "Weak Neutral Currents in Atomic Physics," *Phys. Lett. 48B* (1974), pp. 111–14; I. B. Khriplovich, "Feasibility of Observing Parity Nonconservation in Atomic Transitions," *Pis. Zh. Eksp. Theor. Fiz. 20* (1974), pp. 686–9, in Russian; English translation in *JETP Lett. 20* (1974), pp. 315–17.

14 L. L. Lewis, J. H. Hollister, D. C. Soreide, E. G. Lindahl, and E. N. Fortson, "Upper Limit on Parity-Nonconserving Optical Rotation in Atomic Bismuth," *Phys. Rev. Lett. 39* (1977), pp. 795–8; P. E. G.

Baird, et al., "Search for Parity-Nonconserving Optical Rotation in Atomic Bismuth," *Phys. Rev. Lett. 39* (1977), pp. 798–801.

15 L. M. Barkov and M. S. Zolotorev, "Observation of Parity Nonconservation in Atomic Transitions," *Pis. Zh. Eksp. Teor. Fiz. 27* (1978), pp. 379–83, in Russian; English translation in *JETP Lett. 27* (1978), pp. 357–61.

16 U. Amaldi, et al., "Comprehensive Analysis of Data Pertaining to the Weak Neutral Current and the Intermediate-Vector-Boson Masses," *Phys. Rev. D36* (1987), pp. 1385–1407.

17 C. Y. Prescott, "Inelastic Scattering of Polarized Electrons at SLAC," in *Proceedings of the Sixth Trieste Conference on Particle Physics*, Trieste, Italy, 26–30 June 1978, pp. 163–92; see also SLAC Report No. SLAC-PUB-2218 (1978); C. Y. Prescott, et al., "Parity Non-Conservation in Inelastic Electron Scattering," *Phys. Lett. 77B* (1978), pp. 347–52.

28

Panel Session: Spontaneous Breaking of Symmetry

LAURIE M. BROWN

Born Brooklyn, New York, 1923; Ph.D., 1951 (physics), Cornell University; Professor Emeritus of Physics and Astronomy at Northwestern University; high-energy physics (theory) and history of physics.

ROBERT BROUT

Born New York City, 1928; Ph.D., 1953 (physics), Columbia University; Professor of Physics at the Université Libre de Bruxelles; statistical mechanics and high-energy physics (theory).

TIAN YU CAO

Born Shanghai, China, 1941; Ph.D., 1987 (history and philosophy of science), University of Cambridge; Assistant Professor of Philosophy, Boston University; history and philosophy of science.

PETER HIGGS

Born Newcastle-upon-Tyne, United Kingdom, 1929; Ph.D., 1954 (physics), King's College, London; Professor of Theoretical Physics at the University of Edinburgh; high-energy physics (theory).

YOICHIRO NAMBU

Born Tokyo, Japan, 1921; Sc.D., 1952 (theoretical physics), University of Tokyo; Professor Emeritus of Physics, Enrico Fermi Institute, University of Chicago; high-energy physics (theory).

This panel was intended to function as a discussion, but instead it emerged as a series of short presentations by the participants Robert Brout, Tian Yu Cao, and Peter Higgs, with an introductory discussion by the chair. The present chapter consists of a revised and edited version of those reports and also includes a later submission by Yoichiro Nambu, who was scheduled to be on the panel originally but was unable to attend.

Introduction

The two sectors of the current Standard Model of particle physics, the strong color and the electroweak sectors, are distinct and are tied together only by ontology. Together, they describe the interactions, other than gravitation, of the three generations of quarks and leptons. The dream of representing the strong and weak "nuclear" interactions (as they were known before the acceptance of the quarks) as quantum field theories (QFT) goes back to the 1930s. The first such QFT, other than quantum electrodynamics, was Enrico Fermi's weak-interaction theory of 1934. This theory was almost immediately extended by Werner Heisenberg in 1935 to include the strong interactions (thus making it the first unified QFT) whose exchanged "quanta" were those of the electron-neutrino "Fermi-field." In 1935, Hideki Yukawa invented "U-quanta," now called pions, to represent the field of strong interactions, adjusting their mass to fit the range of nuclear forces. This was again a unified QFT, as the U-quanta were also intended to serve as intermediate bosons of the weak interaction. The discovery by Carl Anderson and Seth Neddermeyer (who did not know about Yukawa's theory) in 1937 of cosmic-ray particles of nearly the right mass was taken to be a triumph of that theory, which had been almost totally disregarded for two years. In spite of the general unwillingness to accept a larger number of fundamental particles (even the neutrino had been strongly resisted), the history of particle physics since Yukawa's theory has functioned, at least partly, in what Nambu has called the Yukawa mode, the introduction of new particles to represent genuinely new effects, and partly in what he has called the Dirac mode, the search for beautiful equations of physics.

The Standard Model, with its 48 quarks and leptons and its 12 vector bosons, certainly illustrates the power of the Yukawa mode. What about the Dirac mode? And what constitutes a beautiful QFT? Clearly, such a theory should be self-consistent, thus implying that it should be finite or renormalizable.[1] It should be capable of fitting the range, strength, and other characteristics of the interactions. But most beautiful of all, it should possess a principle that restricts the form of the interaction, so that the theory contains no arbitrarily specified functions and the smallest possible number of parameters, preferably only one. (The last condition is sometimes referred to as "universality.")[2] In other words, it should resemble as closely as possible the prototype QFT, which does possess such beauty, namely QED. The characteristic property of QED responsible for its powerful economy is gauge invariance, the local invariance under the combined gauge transformation of classical electrodynamics ($A \rightarrow A + \partial_\mu \varphi$), where φ is any scalar function, and the quantum mechanical phase transformation belonging to the Abelian gauge group SU(1). The two sectors of the Standard Model, based upon non-Abelian gauge groups, are beautiful theories, in the sense of the Dirac mode.[3]

The idea of using a non-Abelian gauge theory to fix the form of the nuclear interactions goes back at least to Oskar Klein in 1938, but the modern usage began with the seminal work of Robert Mills and Chen Ning Yang.[4] However, one major problem prevented its direct application: The nuclear forces, strong and weak, have finite range, and therefore, as shown by Yukawa, their field quanta must have nonzero mass. However, it is a property of field theories possessing gauge invariance (and QED is a good example) that they have quanta of zero mass. If the theory's Lagrangian, and thus the field equations derived from them, contain terms that explicitly represent the mass of the quanta, these terms always violate gauge invariance. In the case of the SU(3) color gauge group sector of the Standard Model, the solution to this "mass problem" was suggested by the discovery that non-Abelian gauge theories (in contrast to QED) have a property called asymptotic freedom. That means that the effective coupling strength decreases with increasing squared invariant momentum transfer q^2, vanishing at $q^2 = \infty$. This property suggests that the quanta of the gauge field, called gluons, together with their sources, the quarks, are very strongly coupled at small q^2, and indeed permanently confined within the fundamental particles they inhabit.[5] Thus the effective nuclear forces between hadrons are actually carried by quasi-elementary particles of nonzero mass, the mesons. In the electroweak sector, however, the solution to the zero-mass problem depends upon another process – the subject of this panel – the spontaneous breaking of symmetry (SBS).[6]

Since the color gauge group, being unbroken, is a much simpler structure than the spontaneously broken electroweak theory, one might have expected it to have been used first, especially since the Yang–Mills theory was originally thought of as a candidate theory of the strong interactions. In 1960 Jon J. Sakurai proposed a "vector-dominance" model in which the strong interactions were carried by a set of vector "gauge" bosons (but with gauge-violating explicit mass terms), and in which the (virtual) photon was mixed with the neutral vector mesons.[7] This vector-meson dominance theory gave rise to a phenomenology that was able to claim a number of successes. Nambu proposed an unbroken color gauge theory (which was really QCD, but with quarks of integral charge, and without using the name "color") as early as 1965, but it received little attention until after 1973. The electroweak theory, utilizing SBS, was already in place by 1967, and while it, too, was ignored for several years, it was taken up again seriously after 1971, when 't Hooft proved the renormalizability of non-Abelian gauge theories.[8]

In his contribution to this symposium, Robert Brout has described some of the original applications of SBS in condensed matter physics, and we will mention here only the familiar example of the 1928 Heisenberg theory of the infinite ferromagnet, in order to recall the main idea.[9] If S_i and S_j are spin operators on neighboring lattice sites of a ferromagnetic, the Heisenberg theory assumes for their magnetic interaction energy

$$H_{ij} = S_{ij}S_{\mathbf{i}} \cdot S_{\mathbf{j}}.$$

Although this interaction, as well as the rest of the Hamiltonian function, is clearly spherically symmetric, the ground state of the ferromagnet (below the critical Curie temperature) has a magnetization that points in a particular direction. This is not surprising, because the symmetry of the Hamiltonian does not require a similar symmetry of the ground state unless it is unique. In the case of ferromagnetism (and other similar examples) there exists an infinitely degenerate set of ground, or so-called vacuum states. A complete set of quantum states can be built upon each ground state, and each describes the same physics. The magnetization is an example of a parameter that measures the extent of symmetry breaking, a concept that Lev Landau generalized into that of "order parameter," which he introduced in his 1937 theory of phase transitions, and which he developed, especially in his postwar work on superconductivity with V. L. Ginsburg.[10]

In 1958 Heisenberg tried to carry over this idea into particle physics, proposing a nonlinear spinor theory of elementary particles, in which the known breaking of isospin, for example, was to be of the spontaneously, rather than explicitly broken type. Heisenberg suggested: "When it appears impossible to construct a fully symmetric 'vacuum' state," it should "be considered not really a 'vacuum,' but rather a 'world state,' forming the substrate for the existence of elementary particles."[11] Because Heisenberg's nonlinear spinor theory came to be held in low regard, the importance of including SBS in particle theory was not appreciated until it was reintroduced, via a detour into the field of superconductivity taken by Nambu.

The relevance to particle physics of the kind of symmetry breaking that lies at the heart of the Bardeen–Cooper–Schrieffer (BCS) microscopic theory of supeconductivity (as opposed to the macroscopic Landau–Ginsburg theory) arose in connection with its apparent breaking of electromagnetic gauge invariance. As Brout emphasizes later in this chapter, the gauge invariance of the theory is not really broken, but rather "hidden," and it is only a matter of the ground state of the system not possessing the full symmetry of the Hamiltonian. In the course of his treatment clarifying the logical relations in BCS theory among gauge invariance, the energy gap, and the collective excitations, Nambu made use of a formulation of the problem by Nicolai N. Bogoliubov.[12]

In Bogoliubov's theory, the elementary excitations in a superconductor are coherent superpositions of electrons and holes (thus not charge eigenstates), and obey the equations of motion:

$$
\begin{aligned}
E\psi_{p+} &= \epsilon_p \psi_{p+} &&+ \phi \psi_{p-}^* \\
E\psi_{p-}^* &= -\epsilon_p \psi_{-p-}^* &&+ \phi \psi_{p+}
\end{aligned}
$$

with $E = (\epsilon_p^2 + \phi^2)^{\frac{1}{2}}$. Here $\psi_{p\pm}$ is the wavefunction of an electron with momentum p and spin + or −, so that ψ_{-p-}^+ effectively represents a *hole* of momentum p and spin +; ϵ_p is the kinetic energy measured from the Fermi surface; ϕ is the energy gap arising from the phonon-mediated attractive force between the electrons.

Although the theory so formulated violates gauge invariance, the quasiparticles not being eigenstates of charge, it must be recalled that the superconducting phase is only a part of the complete system. For the latter, of course, gauge invariance of the interactions must hold, and that it does was shown by Nambu and others, who showed that the energy gap itself is gauge dependent and that gauge invariance is restored when certain collective excitations are taken into account.

Nambu used the insight he gained from his study of superconductivity to observe that the same sort of hidden symmetry might be present in the apparently gauge-violating ground state of a model, similar to Heisenberg's nonlinear model or to a composite model of hadrons proposed by Fermi and Yang in 1949, and generalized by Shoichi Sakata in 1956 to include strangeness. In the Fermi–Yang model, the pion is composed of a bound nucleon–antinucleon pair. A theory based upon massless spin-half particles possesses a symmetry called "chiral" (since it is symmetric between left- and right-handedness). The chirality operator is the product of the four Dirac gamma matrices and is denoted by γ_5. The equations of γ_5-invariant Dirac theory are remarkably similar to Bogoliubov's quasi-particle equations:

$$E\psi_1 = \sigma \cdot p\psi_1 + m\psi_2$$
$$E\psi_2 = -\sigma \cdot p\psi_2 + m\psi_1$$

where $E = \pm \left(p^2 + m^2\right)^{\frac{1}{2}}$ and ψ_1 and ψ_2 are the two eigenstates of the chirality operator.

Nambu and his collaborator, Giovanni Jona-Lasinio, noticed that the analogy went much further than this similarity of the equations, as Nambu summarized in 1960 with the following table:[13]

superconductivity	elementary particles
free electron	bare fermion (zero or small mass)
phonon interaction	some unknown interaction
energy gap	observed nucleon mass
collective excitation	meson (bound nucleon pair)
charge	chirality
gauge invariance	γ_5 invariance (rigorous or approximate)

The Nambu–Jona-Lasinio model constructs the pion as a composite of fermion and antifermion of small (nonzero) mass, in order to avoid the problem of having a conserved axial vector current.[14] However, most of the pion's mass is obtained through the spontaneous breaking of chiral symmetry. Thus it is an example of a spontaneously *and* explicitly broken symmetry, as discussed by Higgs.[15] Nambu chose to use a Heisenberg model to illustrate that SBS might be useful in QFT "because the mathematical aspect of symmetry breaking could be mostly demonstrated there," and not because he took the model seriously.[16] However, Nambu and Jona-Lasinio were influenced by Heisenberg's idea of a "world state," saying: "Our γ_5-invariance theory can be approximated by a phenomenological description in terms of pseudoscalar mesons.... The reason for this situation is the degeneracy of the vacuum and the world built upon it."[17]

The next major step was taken by Jeffrey Goldstone, who generalized Nambu's work, using as his example a renormalizable theory of a complex spin-zero quantum field, with a quartic self-interaction $V(\varphi)$ that has a line of minima in the complex plane. These nonzero minima are possible vacuum expectation values for the field in its ground state, another example of Heisenberg's degenerate vacua.[18] The theory possesses invariance under the transformation $\varphi \rightarrow e^{i\alpha}\varphi$, and a unique vacuum is selected by choosing a value of α. Alternatively, the two degrees of freedom(magnitude and phase of φ) can be represented by two real fields, one of which turns out to be massless. Goldstone conjectured, and later proved with Abdus Salam and Steven Weinberg, that to realize a continuous symmetry that is spontaneously broken in a relativistic QFT, the theory must contain a massless particle of zero spin, the famous Goldstone or Nambu–Goldstone boson.[19] Such massless spinless particles are conspicuously absent in nature.

For this reason, attempts to explain the broken hadronic (flavor) symmetry of SU(3) by the introduction of self-interacting scalar fields to induce SBS were bound to fail.[20] However, several authors were able to show that it was possible to "evade" the Goldstone theorem in a most important class of relativistic QFTs, namely gauge theories, which are important, if for no other reason, because QED is one of them. The physical reason that this is so is analogous to the situation in superconductivity where, as Philip Anderson pointed out, the Coulomb repulsion turns the massless phononlike excitation into an effectively massive plasmon. Anderson also speculated that a similar effect might occur in relativistic gauge theories, citing the plasmon as a "physical example demonstrating Julian Schwinger's contention that under some circumstances the Yang–Mills type of vector boson need not have zero mass."[21] The reason in the gauge QFT is that in passing from the Lagrangian of the theory to the specification of quantum mechanical field operators, it is necessary to choose a gauge, just as one specifies the Coulomb gauge in treating the superconductive state. And when the manifest gauge invariance is lost in this manner, the axiomatic basis for proving the Goldstone theorem is no longer present.

While that is not the same thing as saying that physical models in QFT can actually exhibit SBS, such models were soon provided by a number of authors. The most important example (given by François Englert and Brout and by Higgs) is the so-called Higgs mechanism, in which a complex scalar field φ is self-coupled by a quartic potential and is also coupled to a massless vector gauge field.[22] If, say, the real part of the scalar field φ_1 (where $\varphi = \varphi_1 + i\varphi_2$) is chosen to have a nonvanishing vacuum expectation value, this creates a spontaneous symmetry breaking. By a redefinition of the fields, the vector field acquires a new (longitudinal) degree of freedom as well as a mass, so that one has now two massive fields, one vector and one real scalar, replacing the original massless (and hence purely transverse) vector field and two real scalar fields. This Higgs "miracle" makes possible the electroweak sector of the current Standard Model by imparting mass to the gauge bosons $W^{\pm}$ and Z.[23] The evolution of the electroweak model is discussed in Chapter 27, and by Tian Yu Cao later in this chapter.

Before ending, I cannot resist quoting a part of Anderson's concluding remarks in note 21, with respect to the Goldstone theorem:

> This theorem was conjectured, one presumes, because of the solid state analogues, via the work of Nambu and of Anderson. The theorem states, essentially, that if the Lagrangian possesses a continuous symmetry group under which the ground or vacuum state is not invariant, that state is, therefore, degenerate with other ground states. This implies a zero mass boson. Thus, the solid crystal violates translational and rotational invariance, and possesses phonons; liquid helium violates (in a certain sense only, of course) gauge invariance, and possesses a longitudinal phonon; ferro-magnetism violates spin rotation symmetry, and possesses spin waves; superconductivity violates gauge invariance, and would have a zero-mass collective mode in the absence of long-range Coulomb forces. It is noteworthy that in most of these cases, upon closer examination, the Goldstone bosons do indeed become tangled up with Yang–Mills gauge bosons and, thus, do not in any true sense really have zero mass.

Notes on spontaneously broken symmetry
Comments by Robert Brout

These notes are conceived to help delineate some of the issues that were the subject of the panel discussion on spontaneously broken symmetry. In addition to addressing certain conceptual questions, I shall make a few historical comments and include some personal reminiscences that may be of historical interest as well.

The first of these concerns Mark Kac. It was he who brought me to Cornell in 1956 after we had met in Brussels. During my five-year stay at Cornell, Mark had a strong influence on my thinking about SBS. This influence will be reflected in the following and I shall point this out at the right time. We developed a strong personal affection for each other during this period, and whenever Mark passed through Brussels we renewed both our friendship and our scientific dialogue. His last visit was shortly before his death, when he received a Doctor Honoris Causa from the University of Brussels. After having spent an afternoon together, he said it was a strange thing to say good-bye to old friends when one knows that it is not very likely that they shall ever meet again.

(He had already undergone a major cancer operation at that time.) With humility, respect, and affection I dedicate these few notes to the memory of Mark Kac, a profound scientist, mathematician, and humanist.

SBS goes back to the end of the nineteenth century through the understanding that Weiss brought to ferromagnetism and his use of the self-consistent molecular field. But I think that the first use of SBS to get a physical result, other than the fact itself, probably goes back to Debye at the beginning of this century, when he corrected Einstein's observation on the quantum origin of the deviation of the specific heat of crystals from the law of Dulong–Petit. Rather than an exponentially small specific heat at low temperature, Debye pointed out that translational symmetry of the lattice gives rise to excitations that are lattice waves, whose frequency is proportional to wave number; hence the gap in Einstein's exponential is smeared out into a band. At low temperatures the long wavelength excitations dominate and give a power law specific heat ($C_v \sim T^3$). Debye did not think of this phenomenon in terms of SBS, nor do most of us these days. But, in fact, this is an example of what is now covered by the sobriquet "Goldstone's theorem." (The symmetry broken is translation and rotation – a three-dimensional Poincaré group.)

Further development came in the twenties with Heisenberg's fitting out Weiss's phenomenology with the exchange potential, and then with his suggestion to Ising to look at a discrete version of ferromagnetism in which spins were either up or down. At the same time x-ray experiments turned up the superlattice structure in alloys (like Cu–Au). These were explained by Bragg–Williams in Weissian terms, and it very soon became apparent some time in the 1930s that the Ising model problem and the superlattice were one and the same. Unlike Debye's construction, this class of SBS, built on discrete symmetries, does not give rise to a continuum of low-energy, long-wavelength excitations, but rather a gap, as in Einstein's original lattice model.

In the 1930s there was further rich development due to investigations of Bloch on Heisenberg's ferromagnet, which did possess the continuous rotational symmetry. To my knowledge this was the first fully conscious use of SBS to get a continuous spectrum, that is, an explicit derivation of "Goldstone's theorem" in the context of a particular model. In addition, London's investigations of superfluids and superconductors in the thirties were beginning to creep into the domain of spontaneously broken gauge symmetry, one of the all-important leitmotifs of this conference.

So about fifty years ago, there was implicitly a fairly well-developed understanding that SBS is a fact of nature; at that time it came in two versions, according to whether the broken symmetry was discrete or continuous. I now want to discourse a bit on the character of each of these before going into the later development concerning broken gauge symmetries (in the sense of local gauge).[24]

Broken discrete symmetries

The prototype is the Ising model for which there is a Hamiltonian

$$\mathcal{H} = -\frac{1}{2} \sum_{i,j} v_{ij} \mu_i \mu_j - \sum \mu_i H \tag{28.1}$$

in which μ_i is a two-valued variable ($= \pm 1$). For the spin case its meaning is obvious. For the superlattice (say 50% A and 50% B), the state of lowest energy on a cubic lattice is: As surround Bs and Bs surround As. Thus if one fixes the position of one atom, all the others sites have their A-ness or B-ness determined. Call this $\mu_i = +1$. Then $\mu_i = -1$ have A-ness and B-ness interchanged on site i. An amusing version is the lattice gas. Here a site is filled or empty. [Define $\epsilon_i = (1 + \mu_i)/2 = 0, 1$ according to $\mu_i = \pm 1$.] The magnetic field H plays the role of a chemical potential. The resulting theory is that of a liquid–gas system wherein the liquid (gas) phase has more (less) than half the sites occupied. (This analogy is due to Yang and Lee in the fifties.)

The potential v_{ij} is short-ranged and negative, and so it favors all spins of the same sign. H is an external field ($H > 0$ favors upness). The molecular field on i is

$$H_i = H + \sum v_{ij} \mu_j \tag{28.2}$$

and the Weiss (Bragg–Williams) approximation is

$$H_i = \langle H_i \rangle = H + \left(\sum_j v_{ij} \right) \langle \mu_j \rangle = H + \sum_j v_{ij} \langle M \rangle$$

where $\langle M \rangle$ is the mean magnetization (independent of i by translational symmetry) and $v(0) = \sum_j v_{ij}$. Then Weiss's equation is

$$\langle M \rangle = B \left(\frac{H_{mol}}{kT} \right) = B \left(\frac{H + v(0)\langle M \rangle}{kT} \right) \tag{28.3}$$

where B is the Brillouin function ($= \tanh$ for our case). For $H \neq 0$,

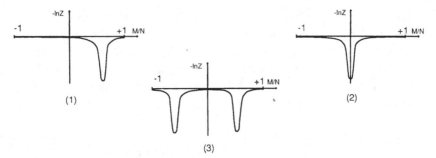

Fig. 28.1. Effective potentials for (1) $H > 0$; and (2) $H = 0$ and $kT > v(0)$; and (3) $H = 0$ and $kT < v(0)$.

$\langle M \rangle \neq 0$ and bears the sign of H. For $H = 0$, $\langle M \rangle = 0$ for $kT > v(0)$, and for $kT < v(0)$, one finds $\langle M \rangle \neq 0$ (with both signs allowed) or $\langle M \rangle = 0$.

To understand the latter circumstance, one can develop the concept of an effective potential: the free energy as a function M at fixed T and H (or alternatively $- \ln Z$ where Z is the partition function). One finds for $H > 0$, the picture in Fig. 28.1(1). $H = 0$ and $kT > v(0)$ is shown in Fig. 28.1(2) and $H = 0, kT < v(0)$ is shown in Fig. 28.1(3). The width of each peak is $O(1/\sqrt{N})$, infinitesimal. So for $kT < v(0), H = 0$ one gets $\langle M \rangle = \pm M_0(T)$. But $\langle M \rangle = 0$ is excluded. SBS is the choice $+$ or $-$.

Kac called Fig. 28.1(3) the potential with donkey's ears; he defined SBS mathematically as $\lim_{H \to 0} \lim_{N \to \infty} \langle M \rangle / N$ so as to choose the left or right ear. This is a convenient definition and is quite physical. One cools in a magnetic field and then turns off the field. Because N is large, it is nigh to impossible to get to the other ear. Often the magnetization is locked in by extraneous small forces. Equally, one can also project into the one-ear sector by fixing the value of one spin – say, with the finger of an angel.

Though all of this discussion is based on Weissian thinking, the exact solution gives the same type of situation. What changes is the value of the critical temperature $[kT_c|_{true} < v(0)]$ and the analytic characterization that describes the approach to the critical point $(T = T_c, H = 0)$. This is the theory of critical phenomena, essentially solved by Wilson and Fisher in the seventies, based on the phenomenology of Kadanoff and Widom, which was elaborated in the 1960s.[25]

As beautiful as the modern development of critical phenomena is, this conference is not the place for a detailed or even qualitative account. The only important point to bring out here is the important role of dimensionality. For $d > 2$, SBS occurs in usual spin models. For $d > 4$, the molecular field theory handles critical fluctuations correctly, whereas for $2 < d \leq 4$, the Wilsonian considerations based on the existence of scaling and the infrared fixed point give an elegant account of critical phenomena. For $d = 2$, things are touch and go. The Ising model and similar discrete models undergo an ordered phase transition, described in modern times by the extremely elegant conformal theoretical methods of Alexander Polyakov and collaborators. For the case of continuous symmetries such as $O(n)$ spin models ($n \geq 2$), there is no ordering possible, essentially because the number of spin waves diverges (logarithmically) so that order is unstable against small fluctuations. But for $n = 2$, the Abelian case, a Kosterlitz–Thouless topological ordering resulting in a very weak transition does occur. These facts find analogous counterparts in gauge theory where, by and large, d is replaced by $2d$.

Continuous SBS (global gauge)

For $T > T_c$, the theories of continuous SBS and discrete SBS are qualitatively the same, though they differ in dynamical details when the refinements of critical phenomena are included (how to weight intersections of chains). One important point is that even if the variables are quantum operators – e.g., replace μ_i of the Ising model by the matrix representation of a spin operator $T_a(i)$; a labels a group direction [$a = 1, 2, 3$ for SU(2)] and the sum over configurations is a quantum trace – the critical fluctuations are still classical,[26] that is, the infrared behavior is governed by the classical sector. Quantum effects contain a "gap" of $O(kT_c)$.

But for $T < T_c$ the systems differ radically. This is because the order parameter $\langle M \rangle$ then becomes a multidimensional vector $\langle \vec{M} \rangle$, for example, the donkey's ears of Fig. 28.1(3) get rotated about the vertical axis for $\langle \vec{M} \rangle$ having two dimensions, so as to give a surface of revolution (and is even more complicated for higher-order groups). SBS then picks out one direction instead of simply one sign. Clearly if one applies a transverse field, as small as it may be, the vector $\langle \vec{M} \rangle$ just obligingly moves over into that direction. [For groups of higher dimensionality than SU(2), these remarks require qualification.] This means that the susceptibility in directions orthogonal to $\langle \vec{M} \rangle$ is infinite, a signal for the existence of zero-mass excitations in this group direction. This latter can be understood as follows:

Fix $\langle \vec{M} \rangle$ at a minimum of $-\ln Z$ in some direction. Then if the system rotates en masse to another $\langle \vec{M} \rangle$, $\ln Z$ does not change. So if half the system rotates, the only energy that it costs is localized to the region where the two halves rub against each other. Similarly if three regions rotate against each other, there are two regions where energy is localized, and so on. We conclude that there are modes of excitation whose energy grows with wave number. This process of course stops once the length of a region (wavelength of the excitation) is comparable with the range of the force, for then all degrees of freedom in one region rub against all in the next, and so it costs no extra energy to make more nodes. Thus we have a spectrum that grows in energy with some power of wave number q until $q = O(R^{-1})$, where R is the range of force. If the range is infinite, $q = 0$ is an isolated point. One is beginning to sniff out interesting exceptional circumstances in the presence of long-range forces.

These results can be formalized by going into a field-theoretic formulation, valid for many body systems on a scale large compared to microscopic distances and intrinsic to relativistic field theory. If the field is a scalar, one considers Lagrangians of the type

$$L = (\partial_\mu \vec{\varphi})^2 + V(\vec{\varphi}) \tag{28.4}$$

where $V(\vec{\varphi})$ is a group invariant, φ_i is the basis of a representation, and $\delta \varphi_i \to T_{aij}\varphi_j$ is an infinitesimal group transformation. T_a are the generators in the representation. If there is a minimum of $V(\varphi)$ at certain values of $\langle \varphi_i \rangle$, then we are in SBS. The zero-mass bosons are then the linear combinations

$$\langle \varphi_i \rangle T_{aij} \varphi_j \tag{28.5}$$

once correctly normalized. This is Goldstone's theorem.[27] Some elementary examples are:

1. SU(2): regular representation: $\vec{\varphi} = \varphi_1, \varphi_2, \varphi_3$ (real)
   $\langle \varphi_3 \rangle \neq 0$, then φ_1, φ_2 have zero mass
2. SU(2): spinor representation $\vec{\varphi} = \varphi_1, \varphi_2$ (complex)
   $\langle \operatorname{Re} \varphi_1 \rangle \neq 0$ then $\operatorname{Im} \varphi_1$ and φ_2 have zero mass
3. SU(3): regular representation $\varphi_1, \ldots \varphi_8$ (real)
   $\langle \varphi_8 \rangle \neq 0$, then the presence of trilinear coupling $(d_{ijk}\varphi_i\varphi_j\varphi_k)$ implies that only

$$\varphi_4, \varphi_5, \varphi_6, \varphi_7 \tag{28.6}$$

have zero mass.

The elementary examples are most easily proven by direct computation, using simple forms for $V(\varphi)$. The general theorem is nicely explained in the papers of Goldstone and Goldstone, Salam, and Weinberg.[28]

In 1962 I called Salam in London concerning his paper with Goldstone and Weinberg. I was bothered because their proof seemed so general. Yet I knew a counterexample: infinite range forces. I was not yet into gauge theory, but it was being hatched in the thoughts of Englert and myself at that time. Salam said he would look into it. Apparently he didn't, since he forgot my call.

These considerations of Goldstone were preceded (and presumably prompted) by the work of Nambu and Gell-Mann and Lévy on spontaneously broken chiral symmetry (SBCS) introduced in 1960. I recall vividly a remark of Victor Weisskopf in a seminar at Cornell (1960): "Particle physicists are so desperate these days that they have to borrow from the new things coming up in many-body physics – like BCS. Perhaps something will come of it."

SBCS in one sense is just a special case of SBS in which $\langle \bar{\psi}\psi \rangle$ and $\langle \bar{\psi}\gamma_5\psi \rangle$ form a representation of $U(1)|_{chiral}$ under which $\psi \to e^{i\gamma_5\theta/2}\psi$. It is a symmetry of massless Dirac theory that allows for vector and axial vector interactions. Under the action of the symmetry, $\langle \bar{\psi}\psi \rangle$ and $\langle \bar{\psi}\gamma_5\psi \rangle$ rotate one into the other, so that if there is a solution $\langle \bar{\psi}\psi \rangle \neq 0$ (i.e., mass $\neq 0$), then we are in a case of SBS. Nambu used this fact to explain the smallness of the mass of the pion, a pseudoscalar, since this is the direction orthogonal to the scalar (the relevant parameter is $\alpha' m_\pi^2 \simeq O(0.02)$, where α' is the universal Regge slope), and from thence to the success of the Goldberger–Treiman relation and soft pion physics – one of the dominant and successful theoretical movements of the 1960s, especially in the hands of Weinberg. It is appropriate here to sketch the main ideas.

First of all, Goldstone's theorem in the presence of small external breaking gets modified into the existence of "would-be" Goldstone modes. They have small mass. For example, for a Heisenberg ferromagnet where

$$\mathcal{H} = -\frac{1}{2}\sum v_{ij}(\vec{\sigma}_i\vec{\sigma}_j) + H\sigma_i^Z, \tag{28.7}$$

the elementary spin wave excitation with all spins up has energy[29]

$$\omega(q) = v(0) - v(q) + H \simeq \left[q^2 + (H/v(0)) \right] v(0) \qquad (28.8)$$

If $|H/v(0)| \ll 1$, the mass is "small" since $v(0)$ is the natural mass scale of the problem.

In hadron physics the mass^2 scale is $(\alpha')^{-1} \simeq 1$ GeV^2. If there is a small perturbation $m_0 \bar{\psi}\psi$ in an otherwise chirally symmetric Lagrangian, then

$$\omega(q) = \sqrt{q^2 + m_\pi^2} \qquad (28.9)$$

where $m_\pi \sim m_0$ (i.e., $m_\pi^2 = O[m_0(\alpha')^{-1/2}]$) as in an antiferromagnet.[30] This means $m_0 = O$ (10 MeV), suggesting that its origins are electroweak rather than hadronic.

Consider now the Goldberger–Treiman relation. To show its relation to SBS in general, I will announce a theorem related to Goldstone's theorem of (28.5). The matrix elements of a conserved current $j_{a\mu}$ between scalar mesons i, j of momenta $p_{\mu,i}$ and $p_{\mu,j}$ (with $p_i^2 = \mu_i^2$) is of the form

$$\langle i \mid j_{a\mu} \mid j \rangle = [F_1(q^2)]_{ij} [p_i + p_j]_\mu + [F_2(q^2)]_{ij} q_\mu$$

where $q_\mu = (p_j)_\mu - (p_i)_\mu$. Current conservation then implies

$$\lim_{q^2 \to 0} [F_2(q^2)]_{ij} = \frac{[F_1(0)]_{ij} [m_j^2 - m_i^2]}{q^2} \qquad (28.10)$$

The pole at q^2 is due to the Goldstone pole, and factorization of its residue gives a coupling constant relation

$$f_a g_{aij} = (m_j^2 - m_i^2) [F_1(0)]_{ij} \qquad (28.11)$$

where $q_\mu f_a = \langle 0 \mid j_{a\mu} \mid \varphi_a \rangle$ and φ_a is the ath Goldstone boson and g_{aij} is its coupling to mesons i, j.

In SBCS the difference $m_j^2 - m_i^2$ is the mass itself, that is, $j_{a\mu}$ is replaced by $j_{\mu 5}$ and the states are spinors. The form factor $F_1(q^2)$ is multiplied by $\gamma_\mu \gamma_5$ and F_2 by γ_5. Multiplication by q_μ then converts (28.10) into

$$F_2(q^2) = \frac{2M F_1}{q^2} \qquad (28.12)$$

whence the conversion of (28.11) to

$$f_\pi g_{\pi NN} = 2M F_A, \qquad (28.13)$$

where F_A is the axial vector form factor and f_π is measured from the weak pion decay of charged pions.

Suppose now that $m_0 \neq 0$ such that $m_\pi^2 \neq 0$ but "small." Then the pole at $q^2 = 0$ is displaced to $q^2 = m_\pi^2$, but all other form factors hardly move between $q^2 = 0$ and $q^2 = m_\pi^2$ since they are expected to vary on the scale of $[\alpha']^{-1}$. Since the Goldberger–Treiman relation is between the nonsingular parts of the form factors, we expect it to hold to $O(\alpha' m_\pi^2)$. And it does! This recounts PCAC and the birth of soft-pion physics. The K meson may be considered approximately soft as well, since $m_K^2 < \Lambda_{QCD}^2$. There remains the U(1) problem, which I touch upon in the gauge section. It is relevant here to recall the fantastic wealth of physical results that resulted from PCAC and current algebra under the impulse of Gell-Mann, Weinberg, and Fubini.[31]

Gauge theories

Superfluidity of liquid He^4 and superconductivity were both explained in terms of spontaneously broken gauge theory in the 1950s. A rigorous characterization of the former was given by Penrose and Onsager who defined (generalized) Bose–Einstein condensation by $\langle r|\rho^{(1)}|r'\rangle = \langle \varphi^*(r)\rangle\langle\varphi(r')\rangle$.[32] Here $\rho^{(1)}$ is the single particle density matrix and φ takes on the meaning of the "wave function of the condensate," for example, $\vec{\nabla}\langle\varphi\rangle \simeq \rho\vec{\nabla}\theta$ is the velocity of the superfluid, where ρ is the absolute value of the order parameter and θ its phase. The theory can be investigated in terms of the second quantized field φ, wherein the action is a function of $\varphi^*\varphi$ only. Thus $\langle\varphi\rangle \neq 0$ is SBS of the gauge group $\varphi \to \varphi e^{i\theta}$. As previously explained, the critical phenomena can be reduced to the classical theory (in the sense of non-quantum). It is fascinating to see how spin problems and the many-boson problem takes the same form. The role of Landau's mass parameter $\mu^2(\sim |T - T_c|)$ is played by the chemical potential; the all-important special role of intersections among chains of correlating paths is played by the two-particle interaction – more precisely a scattering matrix characteristic of net repulsive interaction. With these identifications everything goes through as in the previous section. In particular the implementation of Goldstone's theorem in this case has an interesting and important history.

In the forties Bogoliubov discovered how free-particle excitations got converted into collective modes in the presence of condensation ($\langle\varphi\rangle \neq 0$).[33] It occurs due to a mixing of annihilation and creation operators of particles of finite momentum, which is induced by the presence of the condensate. In this formulation then, particle number is not conserved, that is, there is SBS of the gauge group.[34] The long-wavelength modes are phonons, that is, $\omega(q) = cq$ where $c = $ velocity of sound. (This macroscopic identification is in fact rigorous, as shown by Nozieres and Pines and also by Feynman.) This mixing of positive- and negative-frequency operators is called the Bogoliubov transformation; it was independently discovered in the later 1950s by Bruckner and by Lee and Yang. What is so important in relativistic field theory is the connection between the Bogoliubov transformation, the Klein paradox, and vacuum instability. For the case of liquid helium the instability is not so dramatic; it simply concerns the instability of the trivial ground state in which all particles are in the state of zero momentum. But in other contexts this effect is dramatic. Such is the case in Hawking's blackbody radiation, wherein vacuum fluctuations in the interior of a star whose geometry is flat prior to infall get converted into physical on-shell particles as the star falls toward its Schwarzschild horizon.[35] Particle production in a static electric field (Heisenberg–Euler, Schwinger) can be understood in similar terms.[36]

Superconductivity is an even richer domain. Here the interaction of the condensate (which is charged) with the electromagnetic field gives rise to a wealth of new phenomena. The fact of condensation itself is independent of this part of the theory. It is a result of bound state formation into Cooper pairs induced by the attractive interaction due to phonon exchange among electrons. The whole macroscopic system is coherently correlated (the BCS wave function). Since the condensate can be described in terms of electron pairs, phenomena concerned with the condensate are fruitfully described in terms of an effective (complex) bosonic field. It has charge $2e$. The theory then reduces to that of a charged bose field with SBS $\langle \varphi(r) \rangle = \langle \rho e^{i\theta} \rangle \neq 0$ in interaction with the electromagnetic field.

The first result is an explanation of the Meissner effect. The existence of a length parameter ρ^{-1} of atomic dimensions supplies the penetration length, beyond which a magnetic field cannot penetrate a superconductor. It is screened out by supercurrents (London's equation $\vec{j} = \Lambda^2 \vec{A}$ where Λ^2 is proportional to $|\langle \varphi \rangle|^2$).

At first sight it would appear that the BCS theory gives rise to a Goldstone boson, but Anderson and Nambu independently showed that the Coulomb force among the charged particles kills the Goldstone boson.[37] It becomes the familiar plasmon, that is, a longitudinal photon. Now the existence of this Coulomb force is not a simple extraneous afterthought. On a fundamental level, gauge invariance implies that if there are magnetic effects (transverse), then there are electric ones (longitudinal) as well. In fact the mass parameter, the inverse of London's penetration depth (due to $\langle\rho\rangle \neq 0$) is a pure dimensionless factor times the plasma frequency ($\hbar = c = 1$). The fact that they are not identically the same is due to the fact that the vacuum (ground state superconductor) is not charge conjugation symmetric, a remark that I think is due to Nambu. Be that as it may, it is herein that lay the origins of what is now called the Higgs mechanism. Before going into this, I mention in passing two phenomena that emerge from the theory and that have been so beautifully confirmed in the laboratory: the Josephson effect and the existence of quantized flux in superconductivity of the second kind. The latter will come up again in subsequent discussion.

During the fall of 1963, Englert and I were actively studying Nambu's work and PCAC. There was a meeting in Rochester on SBS about that time, and some remarks of Ken Johnson were very stimulating as well. What with Goldstone's theorem, Nambu's work, and our knowledge of superconductivity, our discovery of the existence of the generation of a gauge vector mass was in the offing. We held up publication of the idea until July 1964 because we wanted to make sure that we were not making fools of ourselves by violating gauge invariance. Once we were happy that it was the eating-up of the Goldstone boson into the photon's self-energy dressing that guaranteed its transversality, and that this could be proven from the Ward identity without any predjudice to this or that mechanism (such as the necessary existence of elementary scalar fields), we published.[38] Apparently Higgs followed a similar route, and our publications were essentially simultaneous.[39]

Apart from the existence of mass (now isotropic since the vacuum is C invariant) two interesting points came up:

1. The dressing was transverse in our approximation, and through our use of Ward identities this seemed to us quite general. But it was not general enough for the non-Abelian case where the then unknown, more extended Ward–Slavnov identities were required. So we could only strongly conjecture renormalizability and show in some nontrivial graphs how it worked. This was written up in a paper of Englert, Thiry, and myself.[40] In the 1966 Solvay Conference Weinberg professed some skepticism as to the renormalizability of this type of theory.[41] We were rather surprised, and Englert gave our arguments for renormalizability.[42] I went up to my office and got our paper on the subject and gave it to Weinberg. Apparently, it did not make the

impression we wanted since renormalizability came only with 't Hooft some five years later.

2. I had firmly imprinted in my mind some remarks made by T. D. Lee in a seminar at Cornell in 1960, which damned the intermediate meson model of weak interactions because of the nonrenormalizability of the theory (due to the $q_\mu q_\nu / m_W^2$ term in the numerator of the propagator). Here we had a theory with $q_\mu q_\nu / m_W^2$ replaced by $q_\mu q_\nu / q^2$. And from the outset we had the non-Abelian generalization. Through use of Goldstone's theorem, one finds in channel a the mass formula

$$\mu_a^2 = \sum \langle \varphi_j \rangle T_{ajk} T_{ak\ell} \langle \varphi_\ell \rangle.$$

So it was natural to think that we had a good candidate for $W^\pm$ mesons. It was also natural to identify the neutral piece of a triplet with the photons. We worked for six months in 1964 trying to get a model and failed. Why? For one thing I was obsessed by the existence of three quarks and four leptons. So I tried to get ν_μ and ν_e into a four-component object and work out everything in terms of the mystical number 3. Needless to say, I failed. The imagination wasn't there to be able to invent a neutral current and a GIM mechanism, nor did we take sufficiently seriously the existence of two independent neutrinos.[43] Furthermore, we were unaware of Glashow's very early paper on unification.[44] Then again, one must remember the strong pull of dispersion relations and Chew's persuasiveness in bootstrap physics. It took a great deal of fortitude in those days to construct a field theory. This was one of the great merits of Weinberg and Salam in those days.[45]

I should like here to recall a thought expressed in Weinberg's *First Three Minutes*, which I paraphrase: "It is not so much that we theorists take our theories too seriously. We don't take them seriously enough." Need more be said?

Pure gauge theories

An equally fascinating chapter is the role of SBS or lack thereof in pure gauge theory. The most interesting problem here is, of course, quark confinement. Wilson pointed out that the essential part of this problem is pure gauge field without quarks, the signal for confinement being the area law of the Wilson loop.[46] Indeed the area law is equivalent to the existence of a correlation function with finite range in the analogous spin problem, hence no long range order. Hand in hand with confinement goes asymptotic freedom, and I would like to explore the two together in the sense of SBS (i.e., the absence of SBS). I like to call it the maintenance of disorder.

Asymptotic freedom is a property of gauge theories in four dimensions and spin theories in two dimensions, provided the former is non-Abelian (Gross–Wilczek, Politzer, 't Hooft) and the latter is made of spins representing $O(n)$ for $n \geq 3$ (Polyakov), hence in a sense non-Abelian as well.[47] The physics of the two is, however, rather different.

In the spin theory what happens is this. Take two spins at a distance one from the other, sufficiently close to make them almost parallel in group space. At low temperatures $[kT \ll v(0)]$ they can still be quite far one from the other. Intermediate spins tend to line up in the plane between them. For $O_n(x)$ with $n \geq 3$, these latter can wobble outside of the plane so that the effective interaction between distant spins is less effective than in the case $n = 2$. The result is that the effective spin–spin interaction between distant spins is described by a larger temperature than the bare temperature $T(\Lambda) \simeq T(\Lambda_0)[1 - \frac{n-2}{2\pi} \ln \frac{\Lambda}{\Lambda_0}]$ with Λ/Λ_0 being the ratio of the inverse of the two scales.

The result is that a catastrophe happens on the length scale $\Lambda_0^{-1} \times \exp[(\frac{n-2}{2\pi})J/T(\Lambda_0)]$, where J is the strength of the interaction between neighbors. At this scale the perturbative calculation sketched above breaks down and non-perturbative effects are called into play. It is still not known for the case n finite but > 3 whether or not the system orders in some sense. For $n = 2$ there is no asymptotic freedom and the system does order topologically, as was shown in the remarkable work of Kosterlitz and Thouless.[48]

A case that is exactly solvable is $n = \infty$, where the system reduces to the Kac–Berlin spherical model (as Stanley showed in the sixties).[49] One can then show that $g(r) \sim e^{-\mu(T)r}$ for all finite T where $\mu(T) \sim \exp(-[(n/2\pi)J/T])$. It is interesting that this response to the infrared catastrophe implied by asymptotic freedom is mathematically the same as that used by Onsager in the 1930s to explain why liquid solutions of HCl do not order into a ferroelectric.[50] In fact, the spherical model method is the same as Onsager's reaction-field correction to the molecular field.[51]

For finite n, there is a tendency to strong local correlation, as witnessed by a peak in the specific heat at a temperature near what would be the critical temperature, had one occurred. It is thought that those local correlations come in the form of quasi-stationary local configurations (skyrmions). The upshot of all of this is that for $n \geq 3$, disorder seems to be the order of the day at all finite temperatures, in keeping with one's expectations from asymptotic freedom, but we are not sure.

For the gauge theory in four dimensions, asymptotic freedom is realized differently. To describe the physics it is most convenient to work in the Euclidean version of the action formalism – and also to absorb the charge into the vector potential $A_{a\mu}$. Then the action is

$$S = \frac{1}{4e_0^2} \int d^4x \sum_a F_{a\mu\nu} F_a^{\mu\nu} \tag{28.14}$$

where

$$F_{a\mu\nu} = \partial_\mu A_{a\nu} - \partial_\nu A_{a\mu} + f_{abc} A_{b\mu} A_{c\nu}.$$

Thus e_0^2 plays the role of a bare temperature and the functional integral over $A_{a\mu}$ is a partition function: $Z = \int DA_\mu e^{-S}$. (We forgo here the discussion of the restriction over the measure to distinct physical configurations given by the Fadeev–Popov procedure.) In this formalism asymptotic freedom is exhibited by dividing A_μ into slow and fast parts and integrating over the latter to get an effective action for the slow part.[52] Thus, write $A_\mu = \mathcal{A}_\mu + a_\mu$ and integrate over the a_μs. They are coupled to the $\mathcal{A}_\mu$s as follows:

$$e_0^2 S(a) = \sum_a \left(D_\mu a_{a\nu} - D_\nu a_{a\mu}\right)^2 + m_{a\mu\nu} \mathcal{F}_{a\mu\nu} + O\left(a^3, a^4\right) \tag{28.15}$$

where

$$D_\mu a_{a\nu} = \partial_\mu a_{a\nu} + f_{abc} \mathcal{A}_{b\mu} a_{c\nu}$$

and

$$m_{a\mu\nu} = f_{abc} a_{b\mu} a_{c\nu}.$$

These mimic diamagnetic and paramagnetic couplings to the background, respectively. The latter is antiferromagnetic in character, favoring antialignment of m and $\mathcal{F}$, where m is the effective magnetic moment carried by the a fields. When integrating over a_μ the result for small e_0^2 is proportional to $\sum_a \mathcal{F}_{a\mu\nu} \mathcal{F}_a^{\mu\nu}$. The contribution to this term from the paramagnetic piece is 12 times the diamagnetic contribution and is negative in sign. This all-important sign comes from the antipolarization of m with respect to $\mathcal{F}$. It results in an increase in e^2 in the effective action describing the slow modes. This is the mechanism of asymptotic freedom in physical terms. It is more of a classical effect rather than due to fluctuations as in the spin case. This is seen when one writes the expression for the value of $\langle m \rangle$ induced by the background F. Dropping irrelevant constants and indices

$$
\begin{aligned}
\langle m \rangle &= \int_\Lambda^{\Lambda_0} d^4q \left[\tfrac{1}{q^2+e^2F} - \tfrac{1}{q^2-e^2F} \right] \\
&= -e^2 F \int_\Lambda^{\Lambda_0} \tfrac{d^4q}{q^4-e^4F^2} .
\end{aligned}
\tag{28.16}
$$

This approximation is obtained using only the quadratic terms of S and to this order the term in e^4F^2 in the denominator of (28.16) should be dropped. Then the free energy due to the fast modes $\simeq \int_0^F \langle m \rangle dF$ gives the correction to the bare term in F^2, which is the paramagnetic piece of the asymptotic freedom correction to e^2.

Note, however, that (28.16) as it stands is to be taken quite seriously. In particular, since $|F| = O(\Lambda^2)$ a catastrophe develops when upon successive iterations e^2 gets built up to be $O(1)$. Then the nonquadratic terms in (28.15) are essential for stabilization. But the problem becomes impossible. I have made some progress by building up a mean field approximation that essentially converts F to B $(B = F + \langle m \rangle)$. This does stabilize the theory. At the same time it makes the long-wavelength fluctuations wildly chaotic [long scale means values of Λ smaller than Λ_{QCD} where $e^2(\Lambda_{QCD}) = O(1)$] – so chaotic, in fact, that these fluctuations are insensitive to source terms whose scale is less than Λ_{QCD}^{-1}. This means that there is little or no correlation of the degrees of freedom of the field beyond $O(\Lambda_{QCD}^{-1})$. This condition is sufficient for confinement – the Wilson loop area law. One sees in this way the strong parallel between how $d = 2$ spin systems and $d = 4$ gauge systems avoid SBS. Beyond numerical computation, which confirms such notions, little progress has been made to make the theory quantitative. In my opinion the strong-coupling approximation of Wilson's lattice gauge theory to get the area law, interesting as it may be in itself, is insufficient to get a line on the true dynamics that must be built up from the small- to large-length scales.[53] Nevertheless, lattice gauge theory may be a useful guide on the long-length scale once the fact of a correlation length is established.

A further interesting phenomenon is that of Coleman and E. Weinberg.[54] As SBS is approached from above the critical temperature, the coupling of the order parameter with the gauge field in four dimensions provokes a discontinuity, thereby converting the transition from second to first order (the donkey's ears split apart discontinuously). Similar phenomena also occur in ferromagnets due to magnetostriction-phonon, spin-wave interactions. The Coleman-Weinberg effect may have very important cosmological consequences since the universe does cool in the course of its expansion.

Finally, it is fitting to conclude these notes with one of the most elegant developments of all: the solution of the U(1) problem. From the fact that there are three light quarks (on the scale of Λ_{QCD}), one should get nine soft pseudoscalars. There are eight; the ninth, η', is "hard." Why?

First, there is an important symmetry breaking that is not SBS. This is the chiral anomaly. The flavor-singlet chiral current is not conserved due to quantum effects. The mechanism in QED was originally displayed as an ultraviolet effect (Adler, Jackiw–Bell), but it is one of those things that can also be understood in the infrared. That one influences the other can be understood by appeal to Levinson's theorem on how phase shifts are influenced by bound states. And, indeed, it is the bound-state aspect that has been so fruitful in QCD. It turns out that there are normalized modes in which quarks are bound to instanton configurations of the $A_{a\mu}$ fields. These latter were discovered by Polyakov and collaborators and their application to the U(1) problem is due to 't Hooft.[55] The effects of these bound-state configurations is such as to make the ninth current unconserved. Behind all of this lies the remarkable mathematical elegance of the Atiyah–Singer theorem, which indexes the instanton configurations in terms of topological mapping of the group on to space in terms of the number of zero-mass states of Dirac particles bound to them.

't Hooft has pointed out that in the context of the Standard Model this theoretical structure gives rise to a (very small) violation of baryon number. Current research has promoted the idea that certain metastable structures (sphalerons) – sort of half instantons – can serve as catalysts of this effect. There is some hope that appeal to this phenomenom, along with CP violation, can explain the matter predominance over antimatter in the cosmos.

Spontaneous symmetry breaking

Comments by Tian Yu Cao

I wish to make two brief remarks and then raise a question. The first remark is on the relationship between spontaneous symmetry breaking and partial symmetry. In recent years there has been a claim by some physicists that in their work on PCAC they had discovered the Nambu–

Goldstone boson independently. Let me just give you one quotation: "In the limit of exact conservation, the pion would become massless, and thus we found the Nambu–Goldstone mechanism independently."[56] I would like to suggest that this claim is not justifiable and that the conflation of the limiting case of PCAC with the Nambu–Goldstone mechanism is somewhat misleading. It is misleading because the physical ideas underlying the two are incompatible. The Nambu–Goldstone mechanism is based on the idea of degenerate vacuum states, which are stable asymmetrical solutions to a nonlinear dynamical system. So the symmetry is broken at the level of solutions rather than dynamical law. In the case of PCAC, neither nonlinearity nor the degeneracy of the vacuum is a characteristic feature, and the symmetry is broken at the level of the dynamical equation. More illuminating is the fact that in the framework of PCAC, when the symmetry is broken there is no massless spinless boson; once you obtain the massless boson there is no symmetry breaking at all. In sharp contrast with this situation, the massless scalar bosons in the Nambu–Goldstone mechanism occur as the result of symmetry breaking. These massless bosons are coexistent with the asymmetrical solutions so that the symmetry of the whole system can be restored.

Closely related with this is another claim by Gell-Mann, which asserts that the Higgs mechanism is a solution to the soft mass problem. The idea of soft mass was indeed a response to the major preoccupations of the time with renormalizability and symmetry breaking, including obtaining approximate global symmetries. Yet the original formulation of the soft mass mechanism was simply to add a gauge-boson mass term to a gauge-invariant Lagrangian, a term that destroyed both the gauge invariance and renormalizability. Only at its second stage (if we wish to call the later developments a second stage) was a mechanism found by which gauge bosons acquired masses without breaking the symmetry of the Lagrangian. The second stage began with an insightful remark by Julian Schwinger in 1962, to the effect that gauge bosons may acquire nonzero masses through their strong coupling with a conserved current. Schwinger's idea was elaborated by Philip Anderson, who introduced the idea of gauge bosons interacting with a matter system that contains the Nambu–Goldstone modes. In this manner he obtained a gauge-invariant system in which the gauge bosons were massive and the Nambu–Goldstone bosons were absent. Anderson's work opened a new direction, along which intensive investigations were made by Bludman, Klein, Lee, Englert and Brout, Higgs, Guralnik, Hagen and Kibble, and others. This pursuit eventually led to the very sophisticated Englert–Brout–Higgs–Kibble mechanism. Nevertheless, the masses that gauge bosons acquired through this mechanism are by no means soft. On the contrary, they are very, very hard.

My second remark concerns the contributions of symmetry breaking to the rise of the Standard Model. Three things were crucial for this development: a conceptual framework, model building, and renormalizability. The general conceptual framework was provided by the quark model combined with the idea of generalized gauge couplings, which was suggested by Yang and Mills, Utiyama, Schwinger, and others. Quarks and leptons have been taken as the basic ingredients of the microstructure of Nature, thus providing us with an ontological basis for theorizing about the physical world. Then Nature's dynamical structure has been supposed to be specified by gauge couplings. All these have nothing to do with symmetry breaking. The second component, phenomenological model building, was attempted at an early stage by Bludman, Schwinger, Glashow, Salam and Ward, Weinberg, and others. The models were largely dictated by observations, which could thereby be explained and new ones predicted. Yet spontaneous symmetry breaking also played a role here, because it provided a mechanism for obtaining diverse phenomena from a unified dynamical system. The third requirement was a proof of renormalizability. An empirically adequate yet nonrenormalizable model had only limited use. As far as the electroweak part of the Standard Model is concerned, the introduction of the Higgs mechanism was an important step in the right direction, although not a decisive one. It was important because with the Higgs mechanism, the massiveness of gauge bosons, which was crucial for saving the phenomena, can be made compatible with an exact rather than approximate or partial dynamic symmetry. And this made the proof of renormalizability much easier. However, the recognition of this advantage, combined with a widespread conviction that a Yang–Mills theory with an unbroken symmetry was renormalizable, has produced a naive belief that the original Weinberg model was already renormalizable. From this belief has come another widely held opinion that all the later theoretical developments in the area of the electroweak interactions amount to nothing more than the invention of the renormalizable gauge by 't Hooft, which merely facilitated the proof. It seems to me that this view trivializes the intellectual history of the genesis of the Standard Model.

The crucial point here is that the conviction that a Yang–Mills theory with an unbroken symmetry is renormalizable was based solely on the naive power-counting argument. Therefore this belief was not justified in 1967; in fact, it was unjustifiable. Here we have to remember that the relationship between gauge invariance and renormalizability is more subtle than was thought to be the case in the 1950s and 1960s by Yang and Mills, by Glashow, by Komar and Salam, by Weinberg, and by others. Feynman, on the contrary, had already fully realized the subtlety in 1963, when he published his seminal work on gravity and Yang–Mills theory. In that article Feynman said that he knew the symmetry argument, "but I don't believe in it, I have to check it in a problem." For a theory to be renormalizable, the gauge invariance is neither sufficient (for example, gravity) nor necessary (for example, the neutral vector meson theory), although it certainly places severe constraints on model building. Thus a proof of the renormalizability of a Yang–Mills theory was a great challenge, and far from a trivial task. It required serious and difficult investigations, which were carried out by a group of theoreticians, from Lee and Yang, and Feynman, through DeWitt, Englert and Brout, Faddeev and Popov, Mandelstam, Boulware, to Veltman and 't Hooft. In addition to the invention of the renormalizable gauge, these investigations (1) derived the Feynman rules and Ward identities; (2) proved unitarity, which involves the introduction of a complex system of nonphysical degrees of freedom required by accepted physical principles; and (3) invented a proper regularization scheme. Without these investigations, no proof of renormalizability of any Yang–Mills theory would be possible, and all the convictions and conjectures based on symmetry worship of an a priori kind would be groundless and empty. Ignoring the contributions of these investigations and placing too much weight on the introduction of the Higgs mechanism distorts the theoretical structure of the Standard Model and its conceptual history.

Also relevant in this regard is the anomalous symmetry breaking. The original Weinberg model of leptons contained the chiral anomaly, thus was not renormalizable. This fatal defect was remedied with ease by Bouchiat, Iliopoulos, and Meyer, and also by Gross and Jackiw, with the introduction of a quark sector so that anomalies from the quark sector and the lepton sector canceled each other.

The anomalous symmetry breaking also contributed to the rise of QCD. The scale anomaly, suggested by Wilson and elaborated by Jackiw, Coleman, Callan, and Symanzik, provided a conceptual basis for the revival and reformulation of renormalization-group equations, without which no idea of asymptotic freedom would have occurred to anyone, nor would the idea of QCD have been sustainable.

A question

The question I have is related to Goldstone's work. As far as the basic ideas are concerned, there is no difference between Nambu and Goldstone. Yet Goldstone takes the scalar bosons as elementary particles and explores the conditions and results of spontaneous symmetry breaking in this boson system, while in Nambu's framework the scalar bosons are derivative because these are composite modes that appear only as a result of symmetry breaking in a fermion system. An advantage of Goldstone's model is its renormalizability, which makes it much easier to find conditions for the existence of asymmetrical solutions to a nonlinear system. More profound than this, however, are some new features brought about by the introduction of an independent boson system in the study of symmetry breaking.

First, an indication of symmetry breaking in Goldstone's boson system is the occurrence of an incomplete multiplet of massive scalar particles. In Nambu's framework no massive spinless boson is possible without explicit symmetry breaking. Thus Goldstone's approach, as compared with Nambu's, has brought out a surplus theoretical structure, the massive scalar boson. It is surplus in the sense that it is not required by the fermion system for symmetry restoring. In the Standard Model, this surplus structure has various implications. For example, it may be responsible for the cosmological constant and cosmic phase transitions.

Second, in the Standard Model the spontaneous symmetry breaking of the fermion system is not specified by its own nonlinear dynamical structure but is induced, through Yukawa coupling and gauge coupling, by the symmetry breaking in a primary system of scalar bosons. This double structure of symmetry breaking has brought a peculiar feature to the Standard Model. Apart from a theoretically fixed dynamical sector that explains and predicts observations, that is, there is an arbitrarily tunable sector that makes a phenomenological choice of actual physical states. Thus we find that Goldstone's introduction of a scalar system has opened new possibilities for our understanding of the physical world, from the spectrum and interactions of elementary particles to the structure and evolution of the Universe.

Now my question is this: What is the ontological status of this scalar system that is solely responsible for the spontaneous symmetry breaking of the Standard Model? Notice that there is a big difference between this system and a superconducting system. In a superconducting system the asymmetrical phase, the symmetrical phase, and the phase transition between the two phases are all real. In the scalar system, however, the Goldstone boson is nonphysical, the Higgs boson escapes our observation, and the symmetrical solution attracts little or no attention from physicists. The nonphysical Goldstone scalars, together with other nonphysical scalars, covariant ghosts and Faddeev–Popov ghosts, are deeply entangled in the theoretical structure of the Standard Model, namely in the description of gauge bosons. What is their relation to Nature? Are they representative? Or just auxiliary constructions for coding some information without direct physical relevance? The instrumentalist takes them as ad hoc devices for obtaining the required observations, W-particles, neutral currents, and so on, and does not take all of their implications, including the Higgs boson, seriously. Then further questions face the instrumentalist: What is the status of the information coded in these constructions? Is it possible for us to have direct access to the information without resort to fictitious devices? Can we construct a self-consistent theory, that is, with the synthesizing power and the powers of explanation and prediction equal to those of the Standard Model, without all the nonphysical degrees of freedom that are deeply entrenched in it?

If we take a realist position, then the searches for the Higgs particle, for the symmetrical solution of the scalar system, and for the agent that drives the system from its symmetrical phase to its asymmetrical phase will be serious physical problems. Then a further question faces the realist: What is the status of the scalar particles? Are they elementary or composite? Some physicists feel that only in a phenomenological model can the scalar particles be taken as elementary, and in a "fundamental" theory they should be derived from fermions. For them, Goldstone's approach is a retreat from Nambu's more ambitious program, and the idea of dynamical symmetry breaking, including the idea of technicolor, seems to be more attractive than Goldstone's approach. Yet the primary role that the scalar system has played in the Standard Model seems to support an alternative view, which was extensively explored in the late 1950s by Nishijima, Zimmermann, and Haag, and in the early and mid-1960s by Weinberg and others. That is, as far as the theory of scattering is concerned, there is no difference between elementary and composite particles.

I am extremely curious to know the positions on the various aspects of this issue – concerning the status of the scalar system – taken by physicists and by philosophers.

Spontaneous breaking of symmetry
and gauge theories
Comments by Peter Higgs

My interest in spontaneous symmetry breaking was stimulated in 1961 by reading Nambu's papers.[57] His idea was to generate hadronic masses and mass splitting within multiplets by spontaneous breaking of the relevant symmetries. His field-theoretic models were inspired by the Bardeen–Cooper–Schrieffer (BCS) theory of superconductivity, so the scalar field vacuum expectation values were generated by fermion pairing – hadron–antihadron pairing in fact, since this was before the invention of the quark model.

I found this program very appealing; I had always been puzzled by broken symmetries in particle physics, and it seemed to me that they would be rather less puzzling if they were unbroken at the level of Lagrangian field theory. So I began to study models of this type.

However, it soon became clear that there was an obstacle to the realization of Nambu's program. This was the Goldstone theorem, which says that if a manifestly Lorentz-invariant local field theory exhibits spontaneous symmetry breaking, its spectrum will include massless spin-zero bosons.[58] Since no such massless hadrons had been detected, it seemed as if spontaneous symmetry breaking was not enough: there would have to be explicit breaking as well to generate mass for would-be Goldstone bosons, such as the pion in Nambu's original model. This was rather disappointing.

Evading the Goldstone theorem

During the years 1962 to 1964 a debate developed about whether the Goldstone theorem could be evaded. Anderson pointed out that in a superconductor the Goldstone mode becomes a massive plasmon mode due to its electromagnetic interaction, and that this mode is just the longitudinal partner of transversely polarized electromagnetic modes, which also are massive (the Meissner effect!).[59] This was the first description of what has become known as the Higgs mechanism.

Anderson remarked that "the Goldstone zero-mass difficulty is not a serious one, because we can probably cancel it off against an equal Yang–Mills zero-mass problem." However, since he had neither found an error in the proof of the Goldstone theorem nor discussed explicitly any relativistic model, Anderson's remark was disbelieved at the time by those particle theorists who read it, myself included! In March 1964 a letter by Abraham Klein and Benjamin Lee provided the first clue to how the theorem might be evaded.[60] They studied the structure of the commutator function that had played the central role in the theorem for a relativistic theory, but in the more general context of a condensed matter system, such as a superconductor, made to look formally relativistic by using a timelike unit four-vector n to specify the rest-frame of the system. They found that n-dependence of the function allowed the theorem to be evaded, and speculated that truly relativistic models might exist where this would occur. I was encouraged by this, but could not see how to construct such a model.

Three months later a response to Klein and Lee appeared.[61] Walter Gilbert (who was at that time in transition between theoretical physics and molecular biology) contended that a relativistic field theory with a Lorentz-invariant vacuum could not depend on a special four-vector n, and so the Goldstone theorem could not be evaded this way. In 1964 Edinburgh University's copy of *Physical Review Letters* came by surface mail. The part containing Gilbert's letter reached our library on July 18. My own letter, which identified the loophole in Gilbert's argument, was received by the Geneva editor of *Physics Letters* on July 27, and so my reaction must have been quite prompt![62] What struck me within a day or so of reading Gilbert's letter was that I did indeed know an example of a fully relativistic field theory with a quite harmless dependence on a special timelike vector – quantum electrodynamics in a Coulomb gauge. Here was the way to evade the Goldstone theorem. The speed of my reaction owed a lot to the influence of Schwinger, whose papers on gauge invariance and mass I had been following with interest.[63] At this point, the most relevant thing that I had learned from him was how various vacuum expectation values in a relativistic theory could depend on the frame of reference in which a Coulomb gauge is imposed.

The relativistic Anderson mechanism

By July 24 I had also written down the simplest relativistic gauge theory with spontaneous symmetry breaking by scalar fields (now known as the Higgs model) and had verified, by linearizing the classical field equations, that the Anderson mechanism did indeed occur. Before writing up this piece of work, I spent a few days searching the literature to check whether it had been done before. In particular, I thought that it might have been done by Schwinger, who had shown that gauge invariance alone does not prevent the photon from being massive and had already invented a model in 1+1 dimensions (where there are no transverse modes) in which this does occur.[64] When I had satisfied myself that he had not noticed that spontaneous symmetry breaking could generate a photon mass, I wrote a second letter, "Broken Symmetries and the Masses of Gauge Bosons," and sent it off to *Physics Letters* by the end of July.

It was rejected. The Geneva editor (Jacques Prentki) wrote that it was not considered suitable for rapid publication in a letters journal but that a more extended version might well prove acceptable to *Il Nuovo Cimento*.

I was indignant. I believed that what I had shown could have important consequences in particle physics. Later, my colleague Squires, who spent the month of August 1964 at CERN, told me that the theorists there did not see the point of what I had done. In retrospect, this is not surprising: in 1964 the European particle theory scene was dominated by S-matrix theorists. Quantum field theory was out of fashion, and I had rashly formulated my description of the mass-generating mechanism in terms of linearized classical field theory, quantized by invoking the de Broglie relations. What relevance could this possibly have to the brave new particle world of S-matrices, bootstraps, and Regge poles?

Realizing that my paper had been short on salestalk, I rewrote it with the addition of two extra paragraphs, one of which discussed spontaneous breaking of the currently fashionable SU(3) flavor symmetry, and sent it to *Physical Review Letters* (which I knew still published letters on field theory). This time it was accepted, but the referee (who, twenty years later, revealed himself as Nambu) invited me to add a comment on the relation of my letter to that of Englert and Brout, which had been submitted in June and published on the day that my revised letter was received.[65] Their paper, which eventually reached me late in September, contained for the first time the general spin-one mass formula in the tree approximation for spontaneously broken non-Abelian gauge theories.

Higgs bosons

The final paragraph of my revised letter drew attention to the scalar and vector modes, that are not mixed by spontaneous symmetry breaking. "It is worth noting that an essential feature of the type of theory which has been described in this note is the prediction of incomplete multiplets of scalar and vector bosons. It is to be expected that this feature will appear also in theories in which the symmetry-breaking scalar fields are not elementary dynamic variables but bilinear combinations of Fermi fields."

The existence of the characteristic massive spin-zero modes had not been noticed by Anderson or by Englert and Brout. Indeed, the theory of what particle physicists would call the Higgs mode in a superconductor was not published until 1981, after it had been detected in the Raman spectrum of superconducting $NbSe_2$![66]

Some early reactions

I have already described the response of theorists at CERN to my first brief account of a simple Abelian model. I shall now describe some reactions from groups on the other side of the Atlantic.

During October I had discussions with Guralnik, Hagen, and Kibble, who had discovered how the mass of noninteracting vector bosons can be generated by the Anderson mechanism, and with Streater, who was involved in the more rigorous proofs of the Goldstone theorem.[67] But it was not until September 1965, when I arrived in Chapel Hill on sabbatical leave at the invitation of Bryce DeWitt, that I settled down to work out the details of my Abelian model. The result of this work was my *Physical Review* paper, which appeared as a preprint in December 1965.[68] In the New Year I received an invitation from one of the recipients of that preprint, Freeman Dyson, to give a colloquium in March at the Institute for Advanced Study at Princeton. The previous summer, at the General Relativity Conference in London, Stanley Deser had invited me to give a talk at the joint seminar at Harvard sometime during my year in the United States, and so I took the opportunity to arrange this for the day following my Princeton talk.

At tea before my Princeton talk the axiomatic field theorist Klaus Hepp told me that there must be an error in my work, since Kastler, Robinson, and Swieca had just proved the Goldstone theorem by C*-algebraic methods – the ultimate in rigor! Nevertheless I survived the questions of the Princeton axiomatists. Encouraged by this experience, I was ready for a rather different style of discussion the next day at Harvard. Years later, when I met Sidney Coleman, he told me that he and his colleagues "had been looking forward to some fun tearing to pieces this idiot who thought he could get round the Goldstone theorem." Well, they did have some fun, but I had fun too!

Failures of Communication

By then I had already spent some time fruitlessly trying to construct a realistic model. The trouble was that, like so many people at that time, I was too preoccupied with the breaking of hadronic (flavor) symmetries: I was aware that leptonic symmetries had been proposed by various people, but I had not appreciated their significance. Shelly Glashow, in his Nobel lecture, said of Goldstone, Kibble, and myself:[69] "These workers never thought to apply their work on formal field theory to a phenomenologically relevant model. I had many conversations with Goldstone and Higgs in 1960. Did I neglect to tell them about my SU(2) × U(1) model, or did they simply forget?"

I first met Glashow in 1960 at the first Scottish Universities Summer School in Physics, where he was a participant and I was a member of the executive committee with the duties of steward. I do not recall hearing about the SU(2) × U(1) model there: my duties kept me from taking part in the discussions, which continued far into the night (lubricated by wine – which it was my job as steward to conserve – from a cache in a grandfather clock) among Glashow, Cabibbo, Veltman, and others.

Later in his Nobel lecture Glashow said, about the failure of Bjorken and himself to solve the problem of strangeness-changing neutral currents in 1964, "I had apparently quite forgotten my earlier ideas of electroweak unification." His amnesia unfortunately persisted through 1966, for he was at my Harvard seminar but did not spot the relevance of the mass-generating mechanism to his model.

And so it was left to Weinberg and Salam, the following year, to shake off the preoccupation with hadronic symmetries which had been preventing progress and graft spontaneous symmetry breaking onto Glashow's electroweak model.

Spontaneous symmetry breaking
Comments by Yoichiro Nambu

Spontaneous breaking of symmetry is already a very old subject; it now belongs to the daily vocabulary of particle physicists, and it might sound rather corny to bring it up anew. In recent years, I have in fact had numerous occasions to talk about the subject from my own perspective, to the point that it has become repetitive and boring. Nevertheless, each talk or article has been addressed to a different audience and different purposes, and is therefore slightly different in emphasis and coverage. By their very nature, these papers deal mainly with my own contributions to the subject, rather than being extensive and balanced reviews. The present one is yet another addition to the series, and I hope this does not go against the kind intentions of the organizers of the session in which I could not participate in person. So my emphasis will be not on the impressive triumph of the Standard Model, where SBS plays a vital part in the weak sector as represented by the Salam–Weinberg (SW) theory. No doubt this has thoroughly been covered by this symposium. Rather I will mainly focus on general developments in other fields. Inevitably I will have to refer to these earlier reports of mine and some historical reviews by others for a more complete picture.[70]

The symmetry principle

First some brief words about symmetry. The symmetry principle, as it is used and formulated by physicists today, is about a century old, dating back to Pierre Curie almost a century ago.[71] He applied group-theoretical considerations to the symmetries of crystals and of electric and magnetic fields that may be present, and derived general rules under which the various physical effects can occur.

In the present century, with the emergence of relativity and quantum mechanics, the symmetry principle has come to be regarded as one of the cornerstones of physics, not just a useful tool for physicists. Here I have in mind the space–time and internal symmetries to which the names Maxwell, Poincaré, Einstein, and Weyl are closely associated.

In physics, symmetry is a mathematical statement. According to our general belief, the body of physical laws is ultimately reduced to the action principle, although this point does not seem to be often emphasized. It is the invariance of the action under symmetries that makes the concept of symmetry important, as it leads to the various conservation laws through the Noether theorem (and its analogs applicable to discrete symmetries).

I also emphasize here the distinction between local and global symmetries. The gauge principle is the embodiment of local symmetries, and the success of the Standard Model seems to reinforce our conviction that the gauge principle is the dynamical basis of all physical laws. On the other hand, the global symmetries, appearing not to be associated with any gauge fields, are still somewhat mysterious. Is it because the gauge fields are not seen at the present energies? Are they just accidental, or do they follow something other than the gauge principle? An important example that confronts the Standard Model is the symmetry (or the lack thereof) among the fermion generations. This assumes, of course, that the actual differences among the various fermions are due to SBS, a topic to which I now turn.

Spontaneous breaking of symmetry

The phenomenon of SBS is even older than the symmetry principle, if one includes such examples of macroscopic bodies as were discussed by Euler and Jacobi. Even in condensed matter phenomena, it is almost as old as the work of Curie. The theory of spontaneous magnetization in a ferromagnet goes back to Weiss, although the recognition of it as a typical instance of a general principle is of a more recent origin.[72] Heisenberg, in his nonlinear fermionic unified theory, invoked the concept of degenerate vacua in order to generate more quantum numbers than are manifest in the original fields, referring to ferromagnetism for analogy.[73] After all, it was he who had formulated the quantum theory of ferromagnetism.[74]

I would like to draw here a distinction between an SBS in infinite media in the sense of the thermodynamic limit, which I consider to be the proper one, and a similar phenomenon in finite systems. Jahn and Teller had shown that the ground configurations of polyatomic molecules may not in general have the highest possible symmetry.[75] These two cases are different in some important repects. In infinite media, there exist degenerate vacua, which are orthogonal to each other and separated by a superselection rule; in finite systems the degeneracy is in general lifted due to tunneling between asymmetric states by a finite-system analog of the Nambu–Goldstone mode. (For example, imagine the NH_3 molecule, an asymmetric nucleus, or a skyrmion.) This distinction is my way of resolving some controversies around the definition of SBS.

As is well known, the BCS theory of superconductivity has turned out to be the prototype of the modern theory of SBS.[76] The name SBS is due to Baker and Glashow.[77] In 1957 Robert Schrieffer gave a seminar on the BCS theory at Chicago before its publication. Despite the brilliant successes of the theory, its lack of gauge invariance in the treatment of the Meissner effect was most upsetting. Maintaining charge-current conservation and the associated Ward identity required the existence of a gapless mode, and it took me some time to understand its dynamical basis to my own satisfaction, independently of the arguments of Bardeen, Anderson, and Rickayzen.[78] For this, the work of Bogoliubov

and Valatin on the fermionic eigenmodes was of great value.[79] It led me
to two general propositions:

1. Analogy between the Bogoliubov–Valatin equation and the Dirac
   equation, meaning that the mass of a Dirac particle can be of a dynamical origin.
2. The general existence of a massless bosonic mode accompanying an
   SBS, as well as its quenching by a gauge field, turning the latter into
   a massive plasmon mode. The organized plasma modes in ordinary
   ionized matter go back to the pioneering work of Irving Langmuir,
   while its more modern theoretical treatment was later developed by
   David Bohm and collaborators.[80]

Application of the above two propositions to the baryon led to a natural explanation for the Goldberger–Treiman relation, with the identification of the pion as the "massless" mode.[81] The latter is of course
an additional step beyond the SBS of a rigorous chiral symmetry. The
finite mass μ of the pion could not be due to quenching because it would
destroy the relation, so it had to be intrinsic, that is, due to a bare
mass m_0 of the nucleon, which explicitly broke chiral symmetry. It is
small compared to the actual nucleon mass $M : m_0 \sim \mu^2/4M$ (a kind
of see-saw relation) ~ 5 MeV, a fact which was comforting to me, as
it suggested the nucleon and electron (bare) masses to be of a common
unknown origin. In the Standard Model, the bare masses of fermions
(current masses) come from the SBS of the Higgs sector, which breaks
the $SU(2)_L \times U(1)$ symmetry, whereas the large "constituent" masses of
quarks are believed to arise from the SBS in the $SU(3)$ dynamics.

A few historical remarks are in order.

1. The above set of ideas was gradually developed after 1959, and reported in various places in increasing detail.[82] It is to be noted that
   at the Rochester Conference of 1960 where Heisenberg and I spoke,
   Vaks and Larkin submitted a brief paper that contained essentially
   the same model as the Nambu–Jona-Lasinio model (NJL), pointing
   out the presence of a massless mode.[83] A preliminary version of the
   NJL model, reported at the Midwestern Theoretical Physics Conference of 1960, and quoted by the celebrated paper of Goldstone, also
   contains discussions on a QED-like model (see below).[84]
2. There were two points that required caution in presenting the theory. The first was the concept of degenerate vacua, and the second

was the choice of a model. Under the influence of the Yukawa–Sakata school, I had always tended to seek physical substance under formal mathematical statements. Thus Dirac's assumption of the filling of the negative energy states in the vacuum, combined with his remark about a Lorentz-invariant ether, was to me not a mathematical trick but a reality.[85] Still I thought it necessary that the arguments for the existence of a multiplicity of vacua be made as convincing as possible. Initially a QED-like model was considered for mass generation. This would have been most natural and relevant in view of the long-standing questions, dating back to Lorentz, about the dynamical origin of mass. But the Dyson–Schwinger equation for the self-energy could be handled only in the ladder approximation; I did not know how to control higher order corrections, especially the vacuum polarization effects. Under these circumstances, the demonstration of the existence of nonperturbative vacua, which was my primary goal, would be obscured by nonessential technical issues. So I switched, rather reluctantly, to a nonlinear four-fermion interaction model similar to the models of BCS and Heisenberg, because it still seemed sufficiently realistic. Yet the mathematics was clean-cut, once the quadratic nonrenormalizable divergences were dealt with by a straightforward cutoff (not by a trick like indefinite metric, as had been done by Heisenberg, which would distract one's attention from the central theme). It is interesting to note that the four-fermion interaction model is now regarded as an effective theory for the QCD of quarks.

3. From my analysis of the problem of gauge invariance in superconductivity, it was more or less obvious to me that the existence of massless modes, as well as their quenching in the presence of a gauge field, is a general phenomenon associated with an SBS. Spin waves (magnons) in ferromagnets and phonons in crystals were familiar examples that came to my mind easily. The theory of spin waves goes back to Bloch, and the version relevant in the present context is the work of Holstein and Primakoff, with which I was familiar.[86] I do not know if there exists a literature of similar nature on phonons. At any rate, I thought it would be interesting to collect other examples in condensed matter phenomena before writing a general paper on this subject. Regrettably, however, I was at the time not sufficiently familiar with the theory of Ginzburg and Landau (GL), and had not thought of its relevance as a model for SBS.[87] As a matter of fact, the project still remains unfinished. The problem of plasmons in the

relativistic context, on the other hand, was clarified by the work of Schwinger, Anderson, Higgs, and Brout and Englert.[88]

4. There were also theories of more phenomenological nature: a series of papers by Gell-Mann and collaborators on the sigma model, starting with one by Gell-Mann and Lévy (GML), as well as the ideas of Nishijima on the massless phase field and of Gürsey on the nonlinear sigma model.[89] Again, as it has turned out, these theories may be considered as an effective description, similar to the GL theory, of the SBS in an underlying dynamical theory. But these papers had not recognized an underlying dynamical principle like SBS.

5. The experimental tests of the idea of SBS were foremost on my mind for the following years. It is for this reason that the soft-pion theorems for pion production in strong and weak processes were developed, starting with the collaborative work with Lurié and Shrauner.[90] I will skip here, however, a discussion on chiral perturbation theory and current algebra, which have since evolved with great sophistication. One important outcome of them is the discovery of chiral anomaly, which has taken on a life of its own.

6. Among the later developments, I would like to pick a few topics here for emphasis. The NJL model predicts the existence of a massive scalar boson besides the massless pseudoscalar "pion." As was shown in the NJL paper, its mass is twice the fermion mass, and this relation is rather insensitive to the details of the dynamics as long as the interaction is short range, and thus it is generic to the BCS-type mechanisms. It was only in 1980 that this bosonic mode was discovered in superconductors, and subsequently given the above theoretical interpretation.[91] I regret that the course of events need not have been so late and in such an order. In terms of the GL–GML–Higgs–SW theories, the two modes are nothing but the Goldstone and Higgs (or π and σ) bosons.

A new element that emerges in this line of reasoning is the possible role of the composite bosons associated with SBS in causing (a) a tumbling chain of SBSs, as was originally proposed by Raby, Dimopoulos, and Susskind, or (b) bootstrapping an SBS, in which the σ boson, formed as a bound fermion pair in the s-channel, generates the attractive interaction in the t-channel that is responsible for the SBS in question.[92] Following the latter line of reasoning, I have recently postulated that the Higgs boson in the Standard Model is a composite of top and antitop quarks, and should be approximately twice as heavy as the top quark (subject to renormalization corrections).[93] This gives a natural explanation for the large mass of the top quark, but on a quantitative level the mass comes out a bit on the high side of the current expectation.

A similar bootstrap-type mechanism may be at work in other phenomena as well. I have in mind in particular the high-T_c superconductivity. As for the tumbling, examples of this already exist in the chains: (a) crystal formation $\longrightarrow$ phonon $\longrightarrow$ superconductivity, and (b) QCD chiral transition $\longrightarrow$ sigma meson $\longrightarrow$ formation of nuclei, nuclear pairing, and nuclear collective modes.[94]

Finally just a word about a supersymmetry-like structure built into the nonrelativistic BCS mechanism.[95] It is not real supersymmetry, but a broken symmetry between fermions and composite bosons leading to simple mass relations among them. This observation allows one to explore relativistic extensions in various ways. There might exist new connections between SBS and supersymmetry.

Notes

1 However, since beauty is in the eye of the beholder, some physicists find it sufficient for the theory to hold effectively only in a restricted energy range, without being renormalizable. This does not violate any observations, since behavior at infinite energy is not in principle susceptible of experimental test.

2 The term is sometimes much more loosely applied. For example, the "Universal Fermi Interaction," which is sometimes said to have originated in the 1940s (the Puppi triangle, etc.), could claim little more than a rough order-of-magnitude similarity beween β-decay, μ-capture, and μ-decay, since the types of interaction (scalar, vector, etc.) were not securely known, even for β-decay.

3 But perhaps they would not have been so regarded by Dirac, who never accepted the idea of infinite renormalization.

4 C. N. Yang and R. L. Mills, "Isotopic spin conservation and a generalized gauge invariance," *Phys. Rev. 95* (1954), p. 631; "Conservation of isotopic spin and isotopic gauge invariance," *Phys. Rev. 96* (1954), pp. 191–5; R. Shaw, "The problem of particle types and other contributions to the theory of elementary particles" (Ph.D. thesis, Cambridge University, 1955).

5 The color gauge group as a basis for strong interactions (though not the name color) was proposed in January 1965 by Yoichiro Nambu at a conference and further elaborated in "A systematics of hadrons in subnuclear physics," in A. De Shalit, H. Feshbach, and L. van Hove, eds., *Preludes in Theoretical Physics* (Amsterdam: Academic Press, 1966), pp. 133–42. See also Chapters 12 and 13, this volume.

6 For a history of SBS with special reference to particle physics, see L. M. Brown and T. Y. Cao, "Spontaneous breakdown of symmetry: Its

rediscovery and integration into quantum field theory," *Hist. Stud. Phys. Biol. Sci. 21* (1991), Part 2, pp. 211–35.

7　J. J. Sakurai, "Theory of strong interactions," *Ann. Phys. 11* (1960), pp. 1–48.

8　As Bludman points out in Chapter 23, Weinberg's 1967 paper, "A model of leptons" (see note 23), was cited only once in the period 1967–71.

9　W. Heisenberg, "Zur Theorie des Ferromagnetismus," *Z. Phys. 49* (1928), pp. 619–36.

10　V. L. Ginsburg and L. D. Landau, "On the theory of superconductivity," *Zh. Eksp. Teor. Fiz. 20* (1950), p. 1064, in Russian; English translation in D. ter Haar, ed., *Collected Papers of L. D. Landau*, (London: Pergamon Press, 1965), pp. 546–68.

11　H. P. Dürr, W. Heisenberg, H. Mitter, S. Schlieder, and K. Yamazaki, "Zur Theorie der Elementarteilchen," *Zeitschrift für Naturforschung 14A* (1959), pp. 441–85.

12　N. N. Bogoliubov, "A new method in the theory of superconductivity. I, II and III," *Sov. Phys. JETP 34* (1958) Vol. 7, pp. 41–55; Y. J. G. Valatin, "Comments on the theory of superconductivity," *Nuovo Cimento 7* (1958), pp. 843–57; Y. Nambu, "Quasi-particles and gauge invariance in the theory of superconductivity," *Phys. Rev. 94* (1960), pp. 648–63. I have emphasized the role of Bogoliubov and Nambu because of their relevance to particle physics. Others who dealt with the puzzle of the breaking of gauge invariance in superconductivity between 1958 and 1960 included J. Bardeen, P. W. Anderson, D. Pines, J. R. Schrieffer, G. Wentzel, J. G. Rickaysen, J. M. Blatt, T. Matsubara, and R. M. May. See note 6 for more details.

13　Y. Nambu, "A 'superconductor' model of elementary particles and its consequences," in *Midwest Conference on Theoretical Physics, Proceedings* (Purdue University, 1960); Y. Nambu and G. Jona-Lasinio, "Dynamical model of elementary particles based on an analogy with superconductivity," *Phys. Rev. 122* (1961), pp. 345–58.

14　For the theory of the partially conserved axial vector current see, e.g., M. Gell-Mann and M. Lévy, "The axial vector current in beta decay," *Nuovo Cimento 14* (1960), pp. 705–25.

15　P. W. Higgs, "Spontaneous Symmetry Breaking," in R. L. Crawford and R. Jennings, eds., *Phenomenology of Particles at High Energies*, 14th Scottish Universities' Summer School in Physics, 1973 (New York: Academic Press, 1974), pp. 529–52.

16　Y. Nambu, "Gauge invariance, vector-meson dominance, and spontaneous symmetry breaking," in L. Brown, M. Dresden, and L. Hoddeson, eds., *Pions to Quarks* (Cambridge University Press, 1989), pp. 639–42.

17　Nambu and Jona-Lasinio, "Dynamical model."

18　J. Goldstone, "Field theories with 'superconductor' solutions," *Nuovo Cimento 19* (1961), pp. 154–64.

19　J. Goldstone, A. Salam, and S. Weinberg, "Broken symmetries," *Phys. Rev. 127* (1962), pp. 965–70.

20　E.g., M. Baker and S. L. Glashow, "Spontaneous breakdown of elementary particle symmetries," *Phys. Rev. 128* (1962), pp. 2462–71.

21　P. W. Anderson, "Plasmons, gauge invariance, and mass," *Phys. Rev. 130* (1963), pp. 439–42; J. Schwinger, "Gauge invariance and mass," *Phys. Rev. 125* (1962), pp. 397–8.

22 P. W. Higgs, "Broken symmetries, massless particles and gauge fields,"
   Phys. Lett. 12 (1964), pp. 132–3; P. W. Higgs, "Broken symmetries and
   the mass of gauge vector bosons," *Phys. Rev. Lett. 13* (1964), pp. 508–9;
   F. Englert and R. Brout, "Broken symmetry and the mass of gauge
   vector bosons," *Phys. Rev. Lett. 13* (1964), pp. 321–3; G. S. Guralnik,
   C. R. Hagen, and T. W. B. Kibble, "Global conservation laws and
   massless particles," *Phys. Rev. Lett. 13* (1964), pp. 585–7;
   T. W. B. Kibble, "Symmetry breaking in Abelian gauge theories,"
   Phys. Rev. 155 (1967), pp. 1554–61.

23 S. Weinberg, "A model of leptons," *Phys. Rev. Lett. 19* (1967), pp.
   1264–6; A. Salam, "Weak and electromagnetic interactions," in N.
   Svartholm, ed., *Elementary Particle Theory* (New York: Wiley, 1968),
   pp. 367–77.

24 A general reference and bibliography to early development is R. Brout,
   Phase Transitions (New York: W. A. Benjamin, 1963).

25 A nice review is K. Wilson, "The renormalization group: critical
   phenomena and the Kondo problem," *Rev. Mod. Phys. 47* (1975), pp.
   773–840.

26 There is confusion in the literature concerning the word *classical.*
   Sometimes classical means molecular field and Landau theory, known to
   be valid for $d > 4$. Sometimes it means classical, as opposed to quantum
   field theory at finite temperature, as in the present context.

27 Goldstone, "Field theories."

28 Ibid.; Goldstone, Salam, and Weinberg, "Broken Symmetries."

29 See Brout, *Phase Transitions.*

30 Ibid.

31 The standard reference is V. de Alfaro, S. Fubini, G. Furlan, and
   C. Rossetti, *Currents in Hadron Physics* (Amsterdam: North-Holland,
   1973).

32 O. Penrose and L. Onsager, "Bose–Einstein condensation and liquid
   helium," *Phys. Rev. 104* (1956), pp. 576–84.

33 N. Bogoliubov, "On the theory of superfluidity," *Journal of Physics
   (USSR) 11* (1947), pp. 23–32.

34 For the many-body problem it is possible to work in the canonical
   ensemble, thus conserving N. SBS is then replaced by elevating certain
   correlations to a special role. This approach does not change local
   properties.

35 S. W. Hawking, "Black hole explosions?" *Nature 248* (1974), pp. 30–1.

36 W. Heisenberg and H. Euler, "Folgerungen aus der Diracschen Theorie
   des Positrons," *Z. Phys. 98* (1936), pp. 714–32; J. Schwinger, "On gauge
   invariance and vacuum polarization," *Phys. Rev. 82* (1951), pp. 664–79.

37 P. W. Anderson, "Random-phase approximation in the theory of
   superconductivity," *Phys. Rev. 112* (1958), pp. 1900–16; Y. Nambu,
   "Quasi-particles and gauge invariance in the theory of
   superconductivity," *Phys. Rev. 117* (1960), pp. 648–63.

38 Englert and Brout, "Broken symmetry."

39 Higgs, "Broken symmetries."

40 F. Englert, R. Brout, and M. F. Thiry, "Vector mesons in presence of
   broken symmetry," *Nuovo Cimento 43* (1966), pp. 244–57.

41 S. Weinberg, *Proceedings of the 14th Solvay Conference* (Brussels, 1966).

42 Ibid.

43 S. Glashow, J. Iliopoulos, and L. Maiani, "Weak interactions with lepton-hadron symmetry," *Phys. Rev. D2* (1970), pp. 1285–92.

44 S. Glashow, "Partial-symmetries of the weak interactions," *Nucl. Phys. 22* (1961), pp. 579–88.

45 Weinberg, "A model of leptons"; Salam, "Weak and electromagnetic interactions."

46 K. Wilson, "Confinement of quarks," *Phys. Rev. D10* (1974), pp. 2445–59.

47 For a review see D. Gross, Chapter 11; A. M. Polyakov, "Interaction of Goldstone particles in two dimensions. Applications to ferromagnets and massive Yang–Mills fields," *Phys. Lett. 59B* (1975), pp. 79–81.

48 J. M. Kosterlitz and D. J. Thouless, "Ordering, metastability and phase transitions in two-dimensional systems," *J. Phys. C6* (1973), pp. 1181–1203.

49 T. H. Berlin and M. Kac, "The spherical model of a ferromagnet," *Phys. Rev. 86* (1952), pp. 821–35; H. E. Stanley, "Spherical model as the limit of infinite spin dimensionality," *Phys. Rev. 176* (1968), pp. 718–22.

50 L. Onsager, "Electric moments of molecules in liquids," *J. Am. Chem. Soc. 58* (1936), pp. 1486–93.

51 R. Brout and H. Thomas, "Molecular field theory, the Onsager reaction field and the spherical model," *Physics 3* (1967), pp. 317–29.

52 The following version is due to R. Brout, "From asymptotic freedom to quark confinement," *Nucl. Phys. B310* (1988), pp. 127–40.

53 Wilson, "Confinement of quarks."

54 S. Coleman and E. Weinberg, "Radiative corrections as the origin of spontaneous symmetry breaking," *Phys. Rev. D7* (1973), pp. 1888–1910.

55 A. A. Belavin, A. M. Polyakov, A. S. Schwartz, and Y. S. Tyupkin, "Pseudoparticle solutions of the Yang–Mills equations," *Phys. Lett. 59B* (1975), pp. 85–7; G. t'Hooft, "Computation of the quantum effects due to a four-dimensional pseudoparticle," *Phys. Rev. D14* (1976), pp. 3432–50.

56 M. Gell-Mann, "Progress in Elementary Particle Theory, 1950–1964," in L. M. Brown et al., eds., *Pions to Quarks* (New York: Cambridge University Press, 1989).

57 Y. Nambu and G. Jona-Lasinio, "Dynamical model of elementary particles based on an analogy with superconductivity, I," *Phys. Rev. 122* (1961), pp. 345–58 and "Dynamical model of elementary particles based on an analogy with superconductivity, II," *Phys. Rev. 124* (1961), pp. 246–54.

58 Goldstone, "Field theories"; Goldstone, Salam, and Weinberg, "Broken symmetries."

59 Anderson, "Plasmons."

60 A. Klein and B. W. Lee, "Does spontaneous breakdown of symmetry imply zero-mass particles?" *Phys. Rev. Lett. 12* (1964), pp. 266–8.

61 W. Gilbert, "Broken symmetries and massless particles," *Phys. Rev. Lett. 12* (1964), pp. 713–14.

62 P. W. Higgs, "Broken symmetries, massless particles."

63 J. Schwinger, "Gauge invariance and mass," *Phys. Rev. 125* (1962), pp. 397–8 and "Gauge invariance and mass, II," *Phys. Rev. 128* (1962), pp. 2425–9.

64 Ibid.

65 Higgs, "Broken symmetries and the masses"; Englert and Brout, "Broken Symmetry."

66 P. B. Littlewood and C. M. Varma, "Gauge-invariant theory of the dynamical interaction of charge-density waves and superconductivity," *Phys. Rev. Lett. 47* (1981), p. 811. R. Sooryakumar and M. V. Klein, "Raman scattering by superconducting-gas excitations and their coupling to charge-density waves," *Phys. Rev. Lett. 45* (1980), p. 660.

67 Guralnik, Hagen, and Kibble, "Global conservation laws."

68 P. W. Higgs, "Spontaneous symmetry breakdown without massless bosons," *Phys. Rev. 145* (1966), pp. 1156–63.

69 S. L. Glashow, "Towards a unified theory: Threads in a tapestry," *Rev. Mod. Phys. 52* (1980), pp. 539–43.

70 Y. Nambu, "Symmetry breakdown and small-mass bosons," in E. C. G. Sudarshan, et al., eds., *The Past Decade in Particle Theory* (New York: Gordon and Breach, 1973), pp. 33–54; "Superconductivity and particle physics," *Physica 126B & C* (1987), pp. 328–34; "Gauge invariance, vector-meson dominance and spontaneous symmetry breaking," in L. M. Brown, M. Dresden, and L. Hoddeson, eds., *Pions to Quarks* (Cambridge University Press, 1989), pp. 639–42; "Concluding Remarks," *Phys. Rep. 104* (1984), pp. 237–58; "Mass formulas and dynamical symmetry breaking," in A. Das, ed., *From Symmetries to Strings: Forty Years of Rochester Conferences: A Symposium to Honor Susumu Okubo in His 60th Year* (Singapore: World Scientific, 1991), pp. 1–12; "Dynamical symmetry breaking," in M. Suzuki amd R. Kubo, eds., *Evolutionary Trends in Physical Sciences* Vol. 57, Springer Proceedings in Physics, (Berlin: Springer Verlag, 1991), pp. 51–66; Brown and Cao, "Spontaneous breakdown of symmetry"; L. Radicati, "Remarks on the early development of the notion of symmetry breaking," in M. G. Doncel, et al., eds., *Symmetries in Physics (1600–1990)*, Proc. 1st International Meeting on the History of Scientific Ideas, Catalonia (Univ. Autonoma de Barcelona, 1987), pp. 197–207; E. Farhi and R. Jackiw, eds., *Dynamical Gauge Symmetry Breaking* (Singapore: World Scientific, 1982).

71 P. Curie, "Sur la symétrie dans les phénomènes physique, symétrie d'un champ électrique et d'un champ magnétique," *J. Phys. 3* (1894), pp. 393–415.

72 P. Weiss, "L'hypothèse du champ meléculaire et la propriété ferromagnétique," *J. Phys. 6* (1907), pp. 661–90.

73 Dürr, Heisenberg, et al., "Zur Theorie der Elementarteilchen."

74 W. Heisenberg, "Zur Theorie des Ferromagnetismus," *Z. Phys. 49* (1928), pp. 619–36.

75 H. A. Jahn and E. Teller, "Stability of polyatomic molecules in degenerate electronic states. I-orbital degeneracy," *Proc. Roy. Soc. A161* (1937), pp. 220–35.

76 J. B. Bardeen, L. N. Cooper, and R. Schrieffer, "Microscopic theory of superconductivity," *Phys. Rev. 106* (1957), pp. 162–4.

77 Baker and Glashow, "Spontaneous breakdown."

78 J. Bardeen, "Gauge invariance and the energy gap model of superconductivity," *Nuovo Cimento 5* (1957), pp. 1766–8; P. W. Anderson, "Coherent excited states in the theory of superconductivity: Gauge invariance and the Meissner effect," *Phys. Rev. 110* (1958), pp. 827–35; "Random-Phase Approximation in the Theory of

Superconductivity," *Phys. Rev. 112* (1959), pp. 1900–16; G. Rickayzen, "Meissner effect and gauge invariance," *Phys. Rev. 111* (1958), pp. 817–21.

79 Bogoliubov, "A new method in the theory of superconductivity"; J. G. Valatin, "Comments on the theory of superconductivity," *Nuovo Cimento 7* (1957), pp. 843–57.

80 L. Tonks and I. Langmuir, "Oscillations in ionized gases," *Phys. Rev. 33* (1929), pp. 195–210.; D. Bohm and E. P. Gross, "Theory of plasma oscillations. A. Origin of medium-like behavior," *Phys. Rev. 75* (1949), pp. 1851–64, and "Theory of plasma oscillations. B. Excitations and damping of oscillations," *Phys. Rev. 75* (1949), pp. 1864–76; D. Bohm and D. Pines, "A collective description of electron interactions. I. Magnetic interactions," *Phys. Rev. 82* (1951), pp. 625–34, and "II. Collective vs. individual particle aspects of the interactions," *Phys. Rev. 85* (1952), pp. 338–53.

81 M. L. Goldberger and S. B. Treiman, "Decay of the pi meson," *Phys. Rev. 110* (1958), pp. 1178–84.

82 Y. Nambu, *Ninth International Conference on High Energy Physics*, Kiev, USSR (Moscow: Academy of Science USSR, 1962) Vol. 2, pp. 121–2; "Axial vector current conservation in weak interactions," *Phys. Rev. Lett. 4* (1960), pp. 380–2; "A 'superconductor' model"; "Dynamical theory of elementary particles suggested by superconductivity," *Proc. 1960 International Conference on High Energy Physics in Rochester* (1960), pp. 858–66; Y. Nambu and G. Jona-Lasinio, "Dynamical model of elementary particles."

83 V. G. Vaks and A. I. Larkin, "On the application of the methods of the superconductivity theory to the problem of the masses of elementary particles," in *Proc. of the 1960 Annual International Conference on High Energy Physics in Rochester*, p. 873.

84 Goldstone, "Field theories."

85 P. A. M. Dirac, "Is there an aether?" *Nature 168* (1951), pp. 906–7.

86 F. Bloch, "Zur Theorie des Ferromagnetismus," *Z. Phys. 61* (1930), pp. 206–19; "Zur Theorie des Austauschproblems und der Remenanzerscheinung der Ferromagnetika," *Z. Phys. 74* (1932), pp. 295–35.; T. Holstein and H. Primakoff, "Field dependence of the intrinsic domain magnetization of a ferromagnet," *Phys. Rev. 58* (1940), pp. 1098–1113.

87 Ginzburg and Landau, "On the theory of superconductivity."

88 Schwinger, "Gauge invariance"; Anderson, "Plasmons"; Higgs, "Broken symmetries, massless particles," and "Broken symmetries and the mass"; Englert and Brout, "Broken symmetry."

89 M. Gell-Mann and M. Levy, "The axial vector current in beta decay," *Nuovo Cimento 16* (1960), pp. 705–25; K. Nishijima, "Introduction of a neutral pseudoscalar field and a possible connection between strangeness and parity," *Nuovo Cimento 11* (1959), pp. 698–710; F. Gürsey, "On the symmetries of strong and weak interactions," *Nuovo Cimento 16* (1960), pp. 230–40; F. Gürsey, "On the structure and parity of weak interaction currents," *Ann. Phys. 12* (1960), pp. 91–117.

90 Y. Nambu and D. Lurié, "Chirality conservation and soft pion production," *Phys. Rev. 125* (1962), pp. 1429–36. Y. Nambu and E.

522   L. M. Brown, R. Brout, T. Y. Cao, P. Higgs, Y. Nambu

Shrauner, "Soft pion emission induced by electromagnetic and weak interactions," *Phys. Rev. 128* (1962), pp. 862–8.

91 R. Sooryakumar and V. Klein, "Raman scattering"; C. A. Balseiro and L. M. Falicov, "Phonon Raman scattering in superconductors," *Phys. Rev. Lett. 45* (1980), pp. 662–5; Littlewood and Varma, "Gauge-invariant theory"; M. A. Littlewood and C. M. Varma, "Amplitude collective modes in superconductors and their coupling to charge-density waves," *Phys. Rev. B26* (1982), pp. 4883–93.

92 S. Raby, S. Dimopolous and L. Susskind, "Tumbling gauge theories," *Nuc. Phys. B169* (1980), pp. 373–83.

93 Y. Nambu, "BCS mechanism, quasi-supersymmetry and fermion masses," in Z. A. Ajduk, et al., *New Theories in Physics, Proc. XI Warsaw Symposium on Elementary Particle Physics* (Singapore: World Scientific, 1989), pp. 1–10; "Quasi-supersymmetry bootstrap symmetry breaking, and fermion masses," in M. Bando, et al., eds., *New Trends in Strong Coupling Gauge Theories* (Singapore: World Scientific, 1989), pp. 3–11; V. Miransky, M. Tanabashi, and K. Yamawaki, "Is the t quark responsible for the mass of W and Z bosons?" *Mod. Phys. Lett. A4* (1989), pp. 1043–53; V. Miransky, M. Tanabashi, and K. Yamawaki, "Dynamical electroweak symmetry breaking with large anomalous dimension and t quark condensate," *Phys. Lett. B221* (1989), pp. 177–83; W. A. Bardeen, C. T. Hill, and M. Lindner, "Minimal dynamical symmetry breaking of the standard model," *Phys. Rev. D41* (1990), pp. 1647–60.

94 Y. Nambu, "Fermion-boson relations in BCS-type theories," *Physica 15D* (1985), pp. 147–51; "Mass formulas"; "Dynamical Symmetry Breaking."

95 Y. Nambu, "Dynamical symmetry breaking."

Part six

The Discovery of Quarks and Gluons

29

Early Baryon and Meson Spectroscopy Culminating in the Discovery of the Omega-Minus and Charmed Baryons

NICHOLAS SAMIOS

Born New York City, 1932; Ph.D., 1957 (physics), Columbia University; Director, Brookhaven National Laboratory; high-energy particle physics (experimental).

The era of studying particle resonance production in the mesonic and baryonic domain was truly exciting and productive. As one looks back, the most important findings occurred in a relatively short time period – roughly 1958–1964, with the preliminaries in the 1950s and lots of details in the 1970s and 1980s. This period of intense activity had many characteristics among which are the following:

1. Accelerators came into their own. Previous productive work occurred in cosmic rays, but now came the Cosmotron, Bevatron, AGS, and PS machines, all contributing important physics results.
2. There was strong interplay between experiment and theory. Global symmetry, the Sakata model, Pais–Piccioni conjecture, Treiman–Yang angle, Jackson angle, Lee–Yang inequalities (and of course, the Gell-Mann–Nishijima, Gell-Mann–Okubo mass formulas), all attest to this close relationship.
3. The early experimental results – even with low statistics – were usually correct. As you will see, the discovery of the ρ, K^*, φ, and η just popped out. On the other hand, one had to use some caution, for some of the early indications could be misleading, a case in point being the τ spin-parity, where Robert Oppenheimer cautioned Jay Orear not to bet on horses.
4. As data accumulated, a few incorrect results emerged – some of a major nature, which required large efforts in time and money to correct.

I begin by discussing the Barkas Table, the November 1957 version. It is worth noting that this earliest of compilations is very short – 16 entries. This was the time of emerging resonances, with Fermi, who in some respect started it all with the Δ, commenting, "Young man, if I could remember the names of these particles, I would have been a botanist." And that was 1954! Indeed, this period was characterized by a plethora of particles with a wide range of masses and properties.

The table contains three leptons (ν, e, μ), two mesons (π, K), and several baryons (p, n, Λ, Σ, Ξ). The spins of most of the particles were known; however, no parities are noted. Most of these particles were found in cosmic rays, in contrast to the later discoveries, which essentially all occurred at accelerator facilities. Even in those early days the neutral counterparts of some of the charged particles (Σ^0, Ξ^0, π^0) were found at accelerators.[1]

The modern era of resonances can be attributed to the work of Fermi and collaborators with their analysis of pion–nucleon phase shifts.[2] The onset of this whole new area of investigation, namely resonance production, is illustrated by the proton–proton total cross section, which changes abruptly from a rapidly decreasing behavior to flat and then to increasing in the region of a few GeV. The two major methods of investigating these structures were formation and production experiments. These are diagramatically illustrated in Fig. 29.1, where a variety of projectiles and exchanges are utilized to produce an assortment of resonances, both mesonic and baryonic.

The relevant accelerators involved in these investigations were the Cosmotron, Bevatron, AGS, and PS. The formation experiments were mainly executed by counter techniques, and some of the names associated with this activity are Cool–Piccioni, Kerth, Lindenbaum–Yuan, and Wenzel. The production experiments were mainly the province of bubble chambers. Their sizes were first measured in inches – 12 in., 15 in., 20 in., 30 in., 40 in., 72 in., 80 in. – as well as in metric units – 80 cm, 1 m, 2 m – and they used a variety of liquids: hydrogen, deuterium, propane, and freon. Their sizes increased dramatically with the advance of wide-angle optics and the 7 ft, 12 ft, 15 ft Gargamelle and BEBC came into being. The pioneers in the development of these chambers were Luis Alvarez, Ralph Shutt, Robert Palmer, Jack Steinberger, Bernard Gregory, Charles Peyrou, William Fowler, Joe Ballam, and Andre Lagarrigue. The masses, widths, and spin-parities of a variety of resonances were found via this powerful technique. The productivity of the formation technique is exemplified by the large number of N, Δ, Λ, and Σ resonances that are evident as bumps in total cross-section measurements. In particular the $\Delta(1238)$, $N(1510)$, and $N(1680)$ are

Formation Experiments

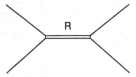

Production Experiments

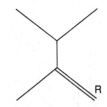

Fig. 29.1. Generalized Feynman diagrams for formation and production experiments.

unambiguous and prominent as well as the $\Delta(1900)$, $Y_0^*(1520)$ [later called the $\Lambda(1520)$], and $Y_0^*(1815)$.

The general technique of production experiments was equally fruitful in uncovering mesonic as well as baryonic resonances. In this manner the $\rho(780)$, $Y_1^*(1385)$ [i.e., $\Sigma(1385)$], and $K^*(885)$ were found in a short period in 1961.[3] It is astonishing to realize how nearly all the early evidence for resonances turned out to be correct. The discoveries of the $\omega(785)$ and $\eta(550)$, shown in Fig. 29.2, are interesting case studies.[4] In the production experiment of Pevsner et al., not only is the η clearly evident, but the ω is even clearer. Contrast this with the ω discovery via $\bar{p}p$ annihilation experiments: most mesons are produced in this reaction, but combinatorial problems reduce the signal.

Table 29.1 gives a snapshot of the resonant states as of August 1961. The number of states has increased, and their isospins are well established; however, their spin, parity, and charge-conjugation quantum numbers are poorly known. In particular, I vividly recall the heated debates (between Adair and Tripp) on the $Y^*(1385)$ spin, whether it was $\frac{1}{2}$ or $\frac{3}{2}$.

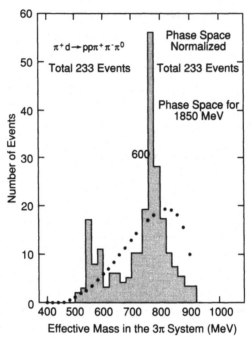

Fig. 29.2. Discovery of the $\eta(550)$ and $\omega(785)$ in the 1961 production experiment of Pevsner, et al.[5]

It is astonishing to realize how much debate and how many heated discussions ensued on issues that are not even appreciated today. A case in point is the $\Lambda\Sigma$ relative parity. There is a famous *Physical Review* letter by Nambu and Sakurai where they claim:[6]

> We wish to point out that, although even $\Lambda\Sigma$ parity has been tacitly assumed by many theoreticians, the available experimental data are suggestive of odd, rather than even, $\Lambda\Sigma$ parity. We can see this in the following eightfold way: ...

They then proceed to present eight reasons – all of which turned out to be wrong!

The experimental activities to determine the parities of particles were extensive and varied. In the particular case of the Σ^0, this involved possible cusp effects in the cross section for associated strange-particle production, and details in $\Sigma(1520)$ production and decay, which was elegantly determined via the decay $\Sigma^0 \to \Lambda^0 e^+ e^-$. Note that the Ξ parity had yet to be measured!

Table 29.1. *Possible resonances of strongly interacting particles (as of August 1961)*

	Mass (MeV)	Half-width $\Gamma/2$ (MeV)	Spin 1	Spin and parity j	Orbital wave	Products	Decay properties Branching fraction	α^j (MeV)	k (Mev/c)
ρ	750	$\cong 50$	1	1-	P	$\pi\pi$	100%	480	350
N	790	$\cong < 15$	0	1-		3π	100%	510	—
K^*	885	$\cong 8$	1/2 ?	?	?	$K+\pi$	100%	292	282
	1238	$\cong 45$	3/2	3/2+	P	$N+\pi$	100%	163	234
N^*	1510	$\cong 30$	1/2	3/2-	d	$N+\pi$	?	435	449
	1680	$\cong 50$	1/2	5/2+	$f+?$	$N+\pi$	?	605	567
	1900	$\cong 100$	3/2	?	?	?	?	—	
	1380	$\cong 25$	1	?	?	$\left\{\begin{array}{l}\Lambda^0+\pi\\\Sigma^0+\pi\end{array}\right.$	96% / 4%	130 / 54	205 / 122
Y^*	1405	$\cong 10$	0	?	?	$\left\{\begin{array}{l}\Sigma^0+\pi^0\\\Lambda^0+2\pi\end{array}\right\}$	100%	79 / 20	133
	1525	$\cong 20$	0	$\geq 3/2$	?	$\left\{\begin{array}{l}\Sigma+\pi\\\Lambda^0+2\pi\\K+p\end{array}\right.$	4 only 1 this ? ratio known	199 / 130 / 89	271 / — / 246
	1815	$\cong 60$	0	$\geq 3/2$	?	many	—	—	

This was the existing situation when the Gell-Mann–Okubo (GMO) mass formula was introduced in 1962.[7] Among the mesons, the members of the pseudoscalars (0^-) that were known were the pion and the kaon, with the GMO formula predicting a singlet mass of 600 MeV (the η was found at 550 MeV and the η' at 960 MeV). With respect to the vector family, utilizing the ω and ρ, the GMO formula placed the K^* at a mass of 780 MeV, to be compared with the experimental value of 890 MeV – not too good. The ϕ had yet to be found. Among the baryons the $\frac{1}{2}^+$ members were well known, the N, Λ, Σ, and Ξ (with the previously mentioned parity controversy), and the GMO formula was reasonably satisfied. Other possible multiplets involving the Δ, and other Σs, were too fragmentary to be useful at this time. This information is summarized in Table 29.2.

Table 29.2. *Performance of the Gell-Mann–Okubo mass formula in 1962*

$$M = a + bY + c\left(\frac{Y^2}{4} - I(I+1)\right) \quad Y = S + B$$

Bosons:	$\left(m(K)\right)^2 = \frac{3}{4}\left(m(\eta)\right)^2 + \frac{1}{4}\left(m(\pi)\right)^2$		
Baryons:	$\frac{1}{2}\left(m(N) + m(\Xi)\right) = \frac{3}{4}m(\Lambda) + \frac{1}{4}m(\Sigma)$		
0^{-+}	π, K	$m(\eta)_{th} = 600$ MeV	$m(\eta)_{ex} = 550$ MeV
1^{--}	ρ, ω	$m(K^*)_{th} = 780$ MeV	$m(K^*)_{ex} = 890$ MeV
$\frac{1}{2}^+$	N, Λ, Σ, Ξ	$1130 \approx 1135$	
$\frac{3}{2}^+$	$\Delta(1238), Y^*(1385)$		

By the summer of 1962 the ϕ and the $\Xi^*(1530)$ were found. As before, the early indications of these two resonances held up.[8] Not only were the masses and widths reliably determined, but the anomalously low rate for $\phi \to \rho\pi$ was noted and reasonably measured. This observation had a major influence on the systematics of strong decay and, in particular, on suppression mechanisms such as the Zweig rule. The discovery of the ϕ (see Fig. 29.3) as an additional vector meson transformed this 1^- octet to a nonet with perfect mixing. In the pseudoscalar family, it required several years before the η' was found, giving this 0^- nonet a rather complex mixing pattern.[9]

The finding of the $\Xi(1530)$ and its announcement at the 1962 CERN conference had an enormous consequence. At this meeting Gell-Mann made the dramatic pronouncement of the Eightfold Way and the expectation of a $J^P = \frac{3}{2}^+$ decimet comprised of the $\Delta(1238)$, $\Sigma(1385)$, $\Xi(1530)$, and a predicted singlet, the Ω^- with a mass pattern of equal spacing, so that the mass of the missing partner was predicted to be approximately 1675 MeV![10] This was the famous SU(3) symmetry, which I first learned of as a remark, after a theoretical session at this Rochester conference.[11] In fact, I remember that Gell-Mann and I had a very illuminating discussion immediately following his famous remark, where he ended by writing on a paper napkin the preferred production reaction,

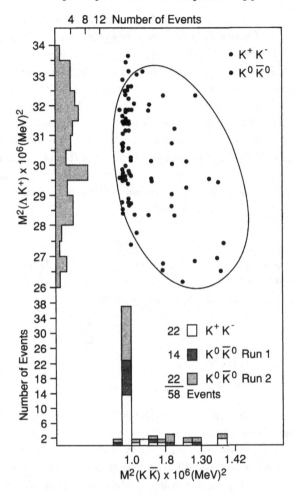

Fig. 29.3. Discovery of the $\phi(1020)$ in the 1962 Brookhaven experiment of Connolly, et al.[12]

namely $K^-p \to \Omega^- K^+ K^0$, the way the Ω^- was eventually produced and found.

The CERN conference provided an occasion for the presentation and discussion of a large volume and variety of experimental data as well as theoretical models and conjectures. The CERN PS and the Brookhaven AGS had come on in 1960, and their impact was clearly visible. With these larger accelerator facilities there came newer and more sophisticated detectors. Spark chambers, Cerenkov counters, and larger bubble chambers were being built and utilized. At Brookhaven Shutt and his group (of which I was a member) were busy designing and constructing the 80 in. hydrogen bubble chamber, with a useful volume of 1000 liters. Lively discussions were being conducted on the first physics to be attacked with such a device. It had not escaped our notice that the known catalogue of strange baryons was asymmetrical – more negative than positive strange particles. In our pursuit of finding and measuring the properties of such particles, it seemed evident that the K^- mesons were the beams of choice. As such we had proposed and were in the process of designing a 5 GeV/c separated K^- beam for the 80 in. chamber. That this single conjecture would be true was reinforced by an examination of the Gell-Mann–Nishijima formula

$$Q = I_3 + (B + S)/2,$$

where Q is the charge, I_3 third component of isospin, B baryon number, and S strangeness.

There was, therefore, room for particles with

$$S = -2 \qquad Q = I_3 - \frac{1}{2},$$

and more importantly, for

$$S = -3 \quad \text{with} \quad Q = I_3 - 1$$

in this latter case allowing for a triplet with $I = 1$ and $Q = -2, -1$, and 0 and a singlet $I = 0$ with $Q = -1$.

Murray's scheme focused our efforts and made running time more accessible. However, it should be noted that there was not theoretical unanimity – there were many other models being proposed – all having some validity. To illustrate this uncertainty, I quote from a 1963 paper by Oakes and Yang.[13]

> We have emphasized above some problems encountered in assigning the meson-baryon resonances to a pure multiplet in the octet symmetry scheme. In particular, we pointed out that the application of the mass formula to $N^*_{3/2}$, Y^*_1, $\Xi^*_{1/2}$ and omega minus, regarded as forming a pure tenfold multiplet, is without theoretical justification. However, equally spaced energy levels are always empirically worthy of attention, and the search for the omega minus should certainly be continued. We only emphasize that if the omega minus is found and if it does satisfy the equal-spacing rule, it can hardly be interpreted as giving support to the octet symmetry model, at least not without the introduction of drastically new physical principles.

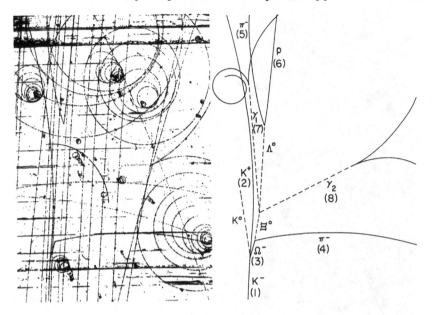

Fig. 29.4. Bubble-chamber photograph showing production and decay of the first Ω^- event observed.

After much hard work and difficulties, the first Ω^- event was found in 1964 on frame number 97025.[14] The observed reaction (Fig. 29.4) was:

$$K^- p \to \Omega^- K^+ (K^0)$$
$$\quad \hookrightarrow \Xi^0 \pi^-$$
$$\qquad \hookrightarrow \Lambda^0 \pi^0$$
$$\qquad\qquad \hookrightarrow \gamma \, \gamma$$
$$\qquad\qquad\quad \hookrightarrow e^+ \, e^-$$
$$\qquad\qquad\quad \hookrightarrow e^+ \, e^-$$
$$\qquad\qquad \hookrightarrow p \, \pi^-$$

with all particles observed except the K^0. It was a very unusual event, as the probability of both gamma rays from the π^0 materializing in the liquid hydrogen was less than 10^{-3}. Another striking feature of this event, not obvious to the nonexpert, is that the line of flight of the Λ^0 as determined by a straight line drawn between its vertex and crossover point of the Λ^0 misses the vertex of the negatively decaying particle. This immediately flagged this event as an Ω^- candidate, since a

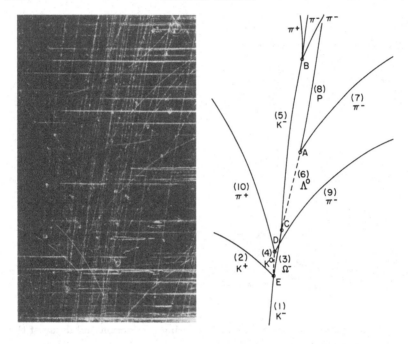

Fig. 29.5. The second Ω^- event observed at Brookhaven.

normal Ξ^- decay would not have this property. It is also true that the γ conversions were not observed during the original discovery of this event; the scanner, who happened to be me, missed them. However, they were found the next day while the event was being reexamined. The analysis was subsequently done in a few hours with the use of templates and rulers and resulted in the reconstruction of the correct π^0, Λ, Ξ^0 masses and the first value of the Ω^- mass, 1682 ± 12 MeV. The Ω^- mass as tabulated in 1992 is 1672 ± 0.3 MeV, a difference of 14 ± 12 MeV – not bad for the first experiment on a new detector in a new beam on such an important issue. With the precise mass value it is now evident that the 1954 Eisenberg event is not an Ω^-.[15] This is because either interpretation of the event $X^- \to \Lambda K^- + 4$ MeV or $X^- \to K^- \Sigma^0 + 4$ MeV yields mass values of 1613 MeV and 1690 MeV, respectively, inconsistent with the present mass value.

A second Ω^- event, shown in Fig. 29.5, was found soon thereafter. It was also somewhat unusual in that the K^- from the Ω^- decay also decayed in the visible volume of the bubble chamber, which should occur approximately 5% of the time. This reaction was

$$K^- \, p \to \Omega^- \; K^+ \; K^0$$
$$\quad\;\; \llcorner \;\; \llcorner \!\!\to \pi^+ \; \pi^-$$
$$\quad\; \llcorner\!\!\to \Lambda^0 \; K^-$$
$$\qquad\quad\; \llcorner \!\!\to \pi^- \; \pi^+ \; \pi^-$$
$$\qquad\; \llcorner \!\!\to p \; \pi^-$$

By late 1964 there were four Ω^- events that had been uncovered. By 1988 a Fermilab experiment accumulated 143,000 events, a formidable feat, but the spin and parity of this particle has yet to be measured.

In the next few years many of the low-lying multiplets among the mesons and baryons were uncovered.

$$
\begin{array}{lllcl}
\text{Mesons:} & \text{nonets} & J^{PC} & = & 0^{-+}, 1^{--}, 2^{++} \\[2ex]
\text{Baryons:} & \text{octets} & J^{P} & = & \frac{1}{2}^{+}, \frac{3}{2}^{-}, \frac{5}{2}^{+} \\[2ex]
& \text{decimets} & J^{P} & = & \frac{3}{2}^{+}, \frac{7}{2}^{+}
\end{array}
$$

Furthermore there was no multiplet higher than **1, 8, 10**; that is, no members of a $\overline{10}$ or **27** representation were identified. In particular, searches for a K^+K^+ and $\Xi^-\pi^+\pi^+$ resonance were negative. Such a spectroscopy led Gell-Mann and George Zweig to the concept of quarks, fractionally charged constituents.[16] This in turn led to the nonrelativistic quark model, in which all mesons were composed of a quark–antiquark pair $M = (q\bar{q})$ and baryons of three quarks $B = (qqq)$. All observed states could be described in this manner. The π^0 lifetime was also calculated utilizing this constituent quark model; agreement with experiment was reached only after including the factor of 3 due to color.

But spectroscopy has continued to this day, filling a book with more than 200 pages. The process has not been smooth, success alternating with failure. Among the minor mistakes we note the $K(725)$ and the zeta. Among the major faux pas one must list the split A_2 and the narrow $R, S, T,$ and U states.

The revolution of 1974 altered the simplicity of only three quarks – u, d, s – by the advent of a fourth quark, charm. The Glashow–Iliopoulos–Maiani mechanism required the existence of such a fourth quark to account for the low rate for certain K decays by introducing a cancellation.[17] The consequences for neutrino interactions of charmed particle production are illustrated in Fig. 29.6. Since the charm quark is coupled to the strange quark as $\cos^2 \theta_c$ and to the down quark as $\sin^2 \theta_c$

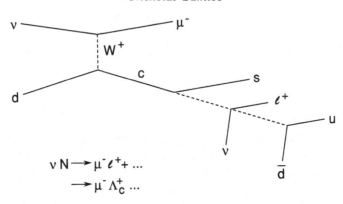

Fig. 29.6. Production of charmed baryons and opposite-sign dileptons in neutrino–nucleon scattering.

(Cabbibo angle), one would expect a reduced rate for the production of charmed baryons as well as a signature of opposite-sign dileptons. For the nonleptonic mode the signature would be the production of single strange particles, and that is precisely what was observed.

The first example of charmed baryons as well as bare charm production is shown in Fig. 29.7.[18] All tracks were identified by a unique occurrence or by kinematics. Among the positive tracks, all three are π^+ mesons: one decays, one interacts, and one has a δ ray. There are no missing neutrals, and the event was initiated by a 13 GeV neutrino. The event occurs in hydrogen; therefore this one event produces a doubly charged charm particle, the Σ_c^{++}, which in turn decays strongly into the lighter Λ_c^+. In effect, in this one event we see evidence for two charmed baryons. The reaction is

$$\nu\, p \rightarrow \mu^- \, \Sigma_c^{++}$$
$$\hookrightarrow \Lambda_c^+ \, \pi^+$$
$$\hookrightarrow \Lambda^0 \, \pi^+ \, \pi^+ \, \pi^-$$
$$\hookrightarrow p \, \pi^-$$

The mass of the Λ_c^+ was measured to be 2260 ± 20 MeV and the mass difference between the Σ_c^{++} and Λ_c^+ was 166 ± 15 MeV. This was the published value in April 1975, from the first experiment with the new 7 ft. chamber. Today's accepted value for the Λ_c^+ mass is 2285 MeV, or a difference from this first measurement of 25 ± 20 MeV. A reexamination of this first event when the systematics of the 7 ft. chamber were

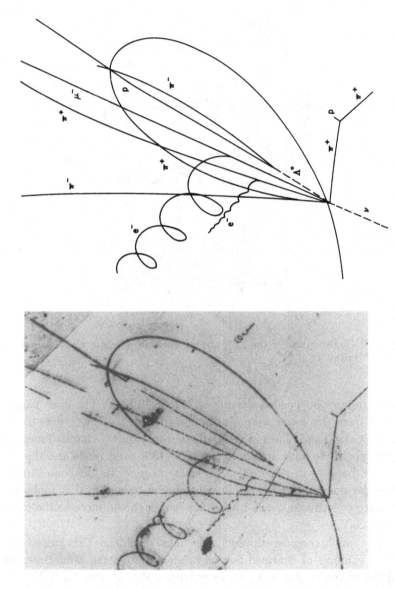

Fig. 29.7. Production of the Σ_c^{++} and Λ_c^+ by a neutrino–proton interaction in the 7 ft. Brookhaven bubble chamber.

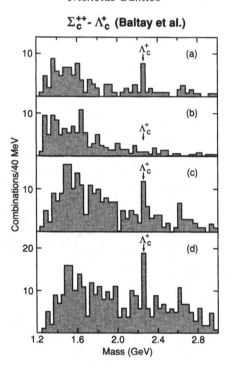

Σ_c^{++}- Λ_c^+ (Baltay et al.)

Fig. 29.8. Confirmation of the Σ_c^{++} and Λ_c^+ by a neon bubble chamber experiment of Baltay, et al.[19]

better understood yielded a value of 2275 ± 10 MeV. The mass values derived from this event were also in accord with the conjectures of De Rujula, Georgi and Glashow, and others where the 0^- charmed mesons were expected to have masses in the 1800–1860 mass range and the $\frac{1}{2}^+$ charmed baryons in the 2200–2300 mass range.[20] In fact, one can do first-order numerology for the heavier quarks by noting that replacing a strange constituent quark by a charm quark should increase the mass by ~ 1 GeV.

These simple considerations gave us added confidence that indeed this was an example of charmed-baryon production. At the same time, we had the pleasure of naming these particles, the Λ_c^+ and Σ_c^{++}.[21] This is the first instance of the use of this notation; we later presented the rationale for this particular naming of the charmed baryons, which is now the accepted nomenclature. In this publication of the first charmed baryon we concluded by noting[22] "that the signature $(\Delta S = -Q)$, rate, and decay-mass pattern are consistent with charmed-baryon production. All dynamical variables are normal under this hypothesis. In contrast other explanations involve extreme fluctuations and thus represent small probabilities, 3×10^{-5} or less. With the obvious caveat associated with one event, we find this observation to be strongly indicative of charmed-baryon production."

One can ask, in retrospect, why this discovery of charmed baryons was not immediately embraced by the high-energy physics community. An obvious answer is that it was only one event. But the Ω^- discovery consisted of one event, so why the difference? Unlike SU(3) the theoretical underpinning for charm was less firm. There was no spectroscopy and the mass predictions for charmed particles were not precise. Probably more important was the negative experimental findings by the SLAC–LBL detector at SPEAR. At first there seemed to be no excess production of strange particles at high energies, as would be expected if charmed particles were being produced. This was subsequently rectified, as pointed out by Gerson Goldhaber in Chapter 4.

This finding of charmed baryons was followed by a detailed investigation of neutrino interactions in neon, in particular, both leptonic and nonleptonic decays of charmed hadrons.[23] In this experiment it was verified that opposite-sign leptons were indeed examples of charmed-hadron production and decay via the GIM mechanism, by noting their association with strange particles and their detailed dynamics. A second charm event was also found, produced in deuterium, this time the Λ_c^+ decaying into $pK^-\pi^+$.[24] A verification of the existence of the Σ_c^{++} was supplied by the neon experiment (Fig. 29.8), where six events were found with a Σ_c^{++}–Λ_c^+ mass difference of 166 ± 3 MeV.[25]

Those years were a lot of fun – exciting, productive, and intellectually stimulating. How can one deny the pleasure of participating in the unraveling of a layer of Nature's secrets? How will it end? If the future imitates the past, then something new will emerge, not presently foreseen, to make our world exciting and stimulating. I close by acknowledging that whatever contributions I have made were due to the very pleasant, fruitful collaboration with outstanding and brilliant physicists; namely, J. Steinberger, M. Schwartz, J. Leitner, R. B. Palmer, and C. Baltay.

Notes

1 R. Plano, et al., "Systematics of Λ^0 and θ^0 decay," *Nuovo Cimento X, 5* (1957), pp.1700–15; L. W. Alvarez, et al., "Neutral cascade hyperon event," *Phys. Rev. Lett. 2* (1959), pp. 215–8 ; J. Steinberger, W. K. H. Panofsky, and J. Steller, "Evidence for the production of neutral mesons by photons," *Phys. Rev. 78* (1950), pp. 802–5.

2 H. L. Anderson, E. Fermi, et al., "Angular distribution of pions scattered by hydrogen," et al., *Phys. Rev. 91* (1953), pp. 155–68.

3 A. R. Erwin, et al., "Evidence for a π–π resonance in the $I = 1$, $J = 1$ state," *Phys. Rev. Lett. 6* (1961), pp. 628–30; M. Alston, et al., "Resonance in the $\Lambda\pi$ system," *Phys. Rev. Lett. 5* (1960), pp. 520–4; M. Alston, et al., "Resonance in the K–π system," *Phys. Rev. Lett. 6* (1961), pp. 300–2.

4 B. C. Maglić, et al., "Evidence for a $T = 0$ three-pion resonance," *Phys. Rev. Lett. 7* (1961), pp. 178–82; A. Pevsner, et al., "Evidence for a three-pion resonance near 550 MeV," *Phys. Rev. Lett. 7* (1961), pp. 421–23.

5 A. Pevsner, et al., "Evidence."

6 Y. Nambu and J. J. Sakurai, "Odd $\Lambda\Sigma$ parity and the nature of the $\pi\Lambda\Sigma$ coupling," *Phys. Rev. Lett. 6* (1961), pp. 377–80, erratum, p. 506.

7 M. Gell-Mann, "Symmetries of baryons and mesons," *Phys. Rev. 125* (1962), pp. 1067–84; "The Eightfold Way: A Theory of Strong Interaction Symmetry," Cal. Inst. Tech. Report No. CTSL-20 (1961), reprinted in M. Gell-Mann and Y. Neeman, eds., *The Eightfold Way* (New York: W. A. Benjamin, 1964), pp. 11–57; S. Okubo, "Note on unitary symmetry in strong interactions," *Prog. Theor. Phys. (Kyoto) 27* (1962), pp. 949–66.

8 L. Bertanza, et al., "Possible resonances in the $\Xi\pi$ and $K\bar{K}$ systems," *Phys. Rev. Lett. 9* (1962), pp. 180–3; G. M. Pjerrou, et al., "Resonance in the $(\Xi\pi)$ system at 1.53 GeV," *Phys. Rev. Lett. 9* (1962), pp. 114–17.

9 Goldberg, et al., "Existence of a new meson of mass 960 MeV," *Phys. Rev. Lett. 12* (1964), pp. 546–50; G. R. Kalbfleish, et al., "Observation of a nonstrange meson of mass 959 MeV," *Phys. Rev. Lett. 12* (1964), pp. 527–30.

10 M. Gell-Mann, "The Eightfold Way."

11 M. Gell-Mann, *Proc. of the International Conference on High Energy Physics* (Geneva: CERN, 1962), p. 805 (comment).

12 P. L. Connolly, et al., "Existence and properties of the ϕ meson," *Phys. Rev. Lett. 10* (1963), pp. 371–6.

13 R. J. Oakes and C. N. Yang, "Meson–baryon resonances and the mass formula," *Phys. Rev. Lett. 11* (1964), pp. 174–8.

14 V. E. Barnes, et al., "Observation of a hyperon with strangeness minus three," *Phys. Rev. Lett. 12* (1964), pp. 204–6.

15 Y. Eisenberg, "Possible existence of a new hyperon," *Phys. Rev. 96* (1954), pp. 541–43.

16 M. Gell-Mann, "A schematic model of baryons and mesons," *Phys. Lett. 8* (1964), pp. 214–15; G. Zweig, "An SU(3) model for strong interaction symmetry and its breaking," CERN Report No. 8181/Th 401 (January 1964); "An SU(3) model for strong interaction symmetry and its breaking: II," No. 8419/Th 412 (February 1964).

17 S. L. Glashow, J. Iliopoulos, and L. Maiani, "Weak interactions with lepton–hadron symmetry," *Phys. Rev. D 2* (1970), pp. 1285–92.
18 E. G. Cazzoli, et al., "Evidence for $\Delta S = -\Delta Q$ currents or charmed-baryon production by neutrinos," *Phys. Rev. Lett. 34* (1975), pp. 1125–28.
19 C. Baltay, et al., "Dilepton production"; "Confirmation."
20 M. K. Gaillard, B. W. Lee, and J. L. Rosner, "Search for charm," *Rev. Mod. Phys. 47* (1975), pp. 277–310.
21 N. P. Samios, "Charm baryons – observed in bubble chambers," in *Baryon 1980 Proceedings of the IVth International Conference on Baryon Resonances* (University of Toronto, 1980), pp. 309–16.
22 E. G. Cazzoli, et al, "Evidence," p. 1127.
23 C. Baltay, et al., "Dilepton production by neutrinos in neon," *Phys. Rev. Lett. 39* (1977), pp. 62–5; C. Baltay, et al., "Confirmation of the existence of the Σ_c^{++} and Λ_c^+ charmed baryons and observation of the decay $\Lambda_c^+ \to \Lambda\pi^+$ and $\bar{K}^0 p$," *Phys. Rev. Lett. 42* (1979), pp. 1721–4.
24 A. M. Cnops, et al., "Observation of the kaonic decay of the Λ_c^+ charmed baryon," *Phys. Rev. Lett. 42* (1979), pp. 197–200.
25 C. Baltay, et al., "Dilepton production"; "Confirmation."

30

Quark Models and Quark Phenomenology

HARRY LIPKIN

Born New York City, 1921; Ph.D., 1950 (physics), Princeton University; Professor of Physics at the Weizmann Institute of Science, Israel; School of Physics and Astronomy, Raymond and Beverly Sackler Faculty of Exact Sciences, Tel Aviv University, Tel Aviv, Israel; high-energy physics (theory).

I begin with a tribute to a great physicist who taught me how to think about quarks and physics in general, John Bardeen. A few sentences from John could often teach you more and give more deep insight than ten hours of lectures from almost anyone else. In 1966 when I began to take quarks seriously, I was unknowingly thinking about them in the language I had learned from John during two years at the University of Illinois, as quasi-particle degrees of freedom describing the low-lying elementary excitations of hadronic matter. Unfortunately I did not realize how much my own thinking had been influenced by John Bardeen until he was gone. I dedicate this paper to his memory.

Were quarks real? Quarks as real as Cooper pairs would have been enough. Quarks leading to anything remotely approaching the exciting physics of the BCS theory would have been more than enough. John always emphasized that Cooper pairs were not bosons, and that superconductivity was not Bose condensation. The physics was all in the difference between Cooper pairs and bosons. I was not disturbed when quarks did not behave according to the establishment criteria for particles. The physics might all be in the difference between quarks and normal particles. One had to explore the physics and see where the quark model led.

The arguments of the BCS critics that the theory was not gauge invariant did not disturb John; he knew where the right physics was. Similarly, the arguments criticizing quarks as nonrelativistic did not disturb me. The model had the right physics. It already in 1966 described so much experimental data not understood by any other model that it had to have the right physics. The formalism would come later and the basis of QCD was already published in 1966.[1] A model of colored quarks interacting with colored gauge bosons in the manner described by a non-Abelian gauge theory had so much of the right physics that it had to lead somewhere.[2] But there are none so blind as those who don't want to see.

A historical perspective

The history of this period can be characterized by repetition at successive levels of the conflict between "grand unification" and "compositeness" approaches to the structure of matter. Each stage began with the belief that the fundamental constituents of matter or "elements" were known. The experimental discoveries of too many elements led to attempts to unify the elements while still considering them as elementary, and to build them out of a smaller number of fundamental building blocks. In 1950 the nucleon and pion were considered the fundamental constituents of hadronic matter. Evidence for composite structure was resisted by the establishment, who sought to unify the large number of new "elementary" particles with concepts like nuclear democracy or higher symmetry, in which all particles were equally elementary. Today we have come full circle back to square one at a deeper level. All matter is constructed from quarks and leptons. The explanations of the large number of elementary objects using grand unification or compositeness have moved from the nucleon–pion level to the quark–lepton level.

The quark model developed very differently in the Eastern and Western Hemispheres. In the East the model was taken seriously from the beginning and supported by top establishment figures like N. N. Bogoliubov, Andrei Sakharov, Yakov Zeldovich, V. N. Gribov, W. Thirring, Giacomo Morpurgo, and Richard H. Dalitz. The Western approach was stated explicitly by Marvin L. Goldberger in introducing a colloquium speaker at Princeton in 1967. "A boy was standing on a street corner snapping his fingers and claiming that it kept the elephants away. When told that there had been no elephants around for many years, his response was 'You see! It works!' And now our speaker will talk about the quark model."

The approach of Galileo of studying Nature by experiments led Eastern physics to the conclusion: "The quark model works, and we do not understand it. Therefore it is interesting." Western theorists who seemed to have forgotten Galileo concluded: "The quark model works, but it contradicts the established dogma. Therefore it is heresy and witchcraft."

A true perspective requires distinguishing between dogma, phenomenology that contradicts established dogma but works, and phenomenology that contradicts established dogma but does not really work and is nonsense. The quark model really worked; it pointed the way toward future new ideas and a new and better understanding of the structure of matter. Two interesting examples in today's physics are high T_c superconductivity and cold fusion. Both surprised everybody when they were first announced. But high T_c really works and demands further investigation for a better understanding. Cold fusion is nonsense and does not work.

Israeli particle physics was at the crossroads between East and West, with roots in Moscow, Leningrad, and London. In 1967–68 when others referred to the quark model as witchcraft, a group of young junior faculty and postdocs named H. R. Rubinstein, Gabriele Veneziano, M. A. Virasoro, David Horn, Haim Harari, and Jonathan L. Rosner, who had come to Israel after spending time in the West, were putting the new quark model ideas together with accepted S-matrix Reggeism.[3] Thus began a new era in particle physics, then called duality, which laid the foundations for what is now called string theory.[4]

Weak and strong SU(3) – constituent and current quarks

Murray Gell-Mann pinpointed an important ingredient in understanding quarks: the difference between "weak" and "strong" SU(3) flavor algebras, which led to constituent and current quarks. Two independent breakthroughs were based on quarklike degrees of freedom. That color SU(3) had the right physics to describe strong interaction dynamics was already clear in 1966, with constituent quarks interpreted as quasiparticle degrees of freedom describing elementary excitations. But current quarks then only provided a mathematical basis for current algebra and were not seen as real physical pointlike objects until the quarkparton description of SLAC experiments. The relation between constituent and current quarks is expected to come somehow out of QCD, but it may well be as difficult as getting BCS theory out of the Lagrangian of QED.

Flavor symmetry and composite models

An early composite model of hadrons was the Fermi–Yang model of a pion as a bound nucleon–antinucleon pair. Its generalization by Sakata to include strange particles and a flavor symmetry generalized from isospin SU(2) to SU(3) was soon seen to be in conflict with experiment.[5]

The Eightfold Way of Gell-Mann and Ne'eman introduced an SU(3) flavor symmetry and a hadron classification from two different points of view. Gell-Mann's "weak SU(3)" began with the properties of the electroweak currents; Ne'eman's "strong SU(3)" with a non-Abelian gauge theory of strong interactions. Both used octet classifications for baryons and mesons with no theoretical explanation for the octet baryon classification nor any physical interpretation for the fundamental triplet. Goldberg and Ne'eman extended SU(3) to U(3) and included baryon number in a formulation constructing the baryon octet from three fundamental triplets carrying baryon number $\frac{1}{3}$.[6] Ne'eman also suggested that SU(3) was an exact symmetry of strong interactions broken by an additional "fifth interaction."[7] But the fundamental triplets of U(3) were presented only as an algebraic device and not as physical particles.

Today's accepted QCD is indeed a non-Abelian gauge theory with exact flavor symmetry for strong interactions and flavor symmetry broken by a completely different interaction outside of QCD (Higgs). This is just what Ne'eman proposed, but the basic degrees of freedom are completely different. The fundamental fermions and gauge bosons are not Ne'eman's baryon and vector meson octets but colored quarks and gluons, with more than three flavors and an additional color degree of freedom.

The "weak" and "strong" approaches to flavor symmetry go back to two very different lines of development of electroweak and strong interaction physics over the past 40 years. Electroweak physics is characterized by the "standard model syndrome," with most experiments either testing a standard model or looking for new physics beyond it. In 1945 the "standard model" was the quantum electrodynamics in Heitler's book and the Fermi theory of beta decay. Crises when the standard model appeared to be wrong were resolved by either revealing that crucial experiments were wrong or finding new concepts like parity nonconservation easily fit into the existing framework. The first indications of "physics beyond this standard model" arose from infinities in QED calculations and the Lamb-shift experiment, and in disagreements between measured beta-ray spectra and Fermi theory. The QED difficulties were solved by the new formulation of Feynman, Schwinger, and Tomonaga. The difficulties with beta-ray spectra went away after better experiments confirmed the Fermi theory. The development through various similar crises to modern electroweak theory was straightforward. It was always based on a field theory that explained low-energy phenomenology while revealing difficulties at higher energies. These were gradually solved without changing the low-energy phenomenology. Today's picture of the electron as a point particle with a Dirac magnetic moment surrounded by an electromagnetic field that gives QED corrections to $g - 2$ is essentially the same as the accepted model of the 1940s.

Hadron physics developed very differently with no sensible "standard model" until QCD. The original picture of elementary nucleons and pions with a Yukawa interaction failed to explain low-energy phenomenology and was soon discarded. Field theory was abandoned as useless for strong interactions and replaced by the analytic S-matrix, Reggeism, and the bootstrap. The nucleon and pion were first kept elementary, then made equally elementary with all other particles (nuclear democracy). Nothing worked, but the particle theory establishment clung to old dogma and refused to accept compositeness or revive field theory until forced by experimental data. Concepts now generally accepted – such as spontaneously broken symmetries, chiral symmetry, the unitary symmetry now called flavor-SU(3), quarks, and the color degree of freedom – were ridiculed by the reactionary establishment as they were dragged kicking and screaming along the path that eventually led to QCD. Today's picture of QCD proton structure and its magnetic moment bears no resemblance to accepted models of the 1940s, 1950s, and 1960s.

At the 1960 Rochester Conference I mentioned to Nambu that I had heard from Bardeen about his very interesting application of ideas from superconductivity to particle physics. Nambu said I was the only person at the conference who had expressed any interest in this work. At the 1962 Rochester conference in Geneva, the prediction that a particle later called the Ω^- should exist, already proposed in a paper by Glashow and Sakurai, was not considered important enough to be mentioned in any invited or contributed talk.[8] It was mentioned in a comment from the floor by Gell-Mann. The paper proposing the existence of quarks was accepted by *Physics Letters* only because it had Gell-Mann's name on it.[9] The editor said, "The paper looks crazy, but if I accept it and it is nonsense, everyone will blame Gell-Mann and not *Physics Letters*. If I reject it and it turns out to be right, I will be ridiculed."

$\bar{p}p$ *annihilation – first evidence for quarks*

Annihilation experiments performed shortly after the antiproton discovery gave results disagreeing with conventional model predictions.[10] A pion multiplicity of 5.3 ± 0.4 was found, much greater than the 2 or 3 predicted by statistical models, while e^+e^- pairs were not seen at the level predicted by QED from one-photon annihilation of a pointlike $\bar{p}p$ pair. Pions as quanta of a boson field could be created only after the annihilation of the positive and negative baryon number present in the initial state. No one considered the simple but unacceptably heretical explanation that both mesons and baryons were composite objects made of the same constituents carrying baryon number, rather than being elementary and completely different like photons and electrons, that no annihilation of baryon number was needed, and that constituents with opposite baryon number simply rearranged to form "positroniumlike" states with a multiplicity related to the number of constituents originally present. Shortly after the quark proposal, such a model showed that a rearrangement of the three quarks and three antiquarks in the proton and antiproton into three mesons gave the observed pion multiplicity.[11] A simple "back-of-the-envelope" calculation for pions produced from three s-wave $q\bar{q}$ pairs with the standard 3 : 1 statistical factor favoring the spin-triplet ρ that decays into two pions gives $3 \cdot \frac{3}{4} \cdot 2 + 3 \cdot \frac{1}{4} = \frac{21}{4} = 5.25$.

This quark-rearrangement model was ridiculed as nonsense when proposed in 1966.[12] The establishment prejudice against quarks even created serious difficulties for obtaining appointments and promotions for young people in our group. Deans and committees were influenced by pejorative comments in letters from well-known physicists about people who rush into print with such garbage.

Group theory

Until the discovery of the Ω^-, most particle physicists believed that group theory was useless for high-energy physics, thought of isospin as rotations in some three-dimensional space, and knew nothing about unitary groups. They therefore tried rotations in 4, 5, 6, 7, and 8 dimensions with fancy names like global symmetry, cosmic symmetry, and the like, before finding that the natural symmetry group to include the SU(2) $\times$ U(1) of isospin and strangeness was SU(3). Perhaps they called it the Eightfold Way because it took them eight years (1953–61) to find it.

Soon afterwards the pendulum swung and a flood of papers tried to include flavor SU(3) and space–time in a larger group and produced a number of fancy no-go theorems. I noted immediately that the physics underlying these fancy groups was completely crazy.[13] No sensible interaction could be invariant under transformations generated by operators acting nontrivially both in space–time and in an internal symmetry space. Translation invariance implies that a pion–nucleon scattering experiment at SLAC gives the same results when moved to Fermilab. Isospin invariance implies $\sigma(\pi^- p) = \sigma(\pi^+ n)$. But invariance under transformations acting in space–time like a translation and also transforming nontrivially under isospin can move a pion beam from a SLAC experiment to Fermilab, while leaving the nucleon target at SLAC. Any dynamics invariant under such transformations must obviously have no interactions, no bound states, and a continuous mass spectrum. However, no one paid attention to this kind of "low-brow phenomenology" and fancy theorems were published showing that nonsense is nonsense.

Static hadron properties in the quark model

The significance of quark-model predictions has been confused by model builders who produce an apparently large number of predictions from a specific model without noting that only two or three depend on the model and the rest all follow from model-independent symmetries like angular momentum, isospin, and SU(3). They get excellent but meaningless chi-squared fits to data. We avoid the pitfall by considering only those quark-model predictions not easily obtained in other ways, in particular, relations between mesons and baryons, and the determination of the values of parameters that are left free in SU(3).

The very early successes

The difference between the quark structures of the meson and baryon octets immediately explained striking regularities in the low-lying hadron spectrum not explained by SU(3) – for example, the baryon octets and decuplets, the meson nonets without the ninth baryon suggested by some SU(3) models, no meson decuplets, and the spin-parity quantum numbers $J^P = 0^-,\ 1^-, \frac{1}{2}^+,\ \frac{3}{2}^+$. Introducing U(3) rather than SU(3) and breaking SU(3) at the quark level by setting $m_s > m_u$ immediately gave the experimentally observed mass inequalities

$$M_\Xi > M_\Sigma \approx M_\Lambda > M_N; \quad M_\eta > M_{K^+} \approx M_{K^-} > M_\pi \quad (30.1)$$

instead of the bad baryon mass inequality that followed from using the same structure for baryon and meson octets,

$$M_\Lambda > M_N \approx M_\Xi > M_\Sigma. \quad (30.2)$$

These regularities still did not influence the establishment to take quarks seriously. Many open questions remained; for example, the reason for the decuplet classification for the spin-$\frac{3}{2}$ baryons, rather than octet or singlet, the reason for the Λ–Σ mass difference, and whether the next excited states were orbital excitations or states with additional $\bar{q}q$ pairs; that is, the so-called exotics.

The relevant degrees of freedom

Thirty years of experimental hadron spectroscopy have found an enormous number of hadronic states described as excitations of the spins and relative coordinates of the constituent quarks in the $\bar{q}q$ and qqq systems. Many additional degrees of freedom have been proposed for theoretical reasons – for instance, bags, strings, meson clouds, gluons, and a sea of $\bar{q}q$ pairs including strange quarks. But there has been so far no convincing evidence for hadronic states containing excitations of such degrees of freedom – for example, excitations describable as relative motion between the center-of-mass of the valence quarks and other constituents like a bag, cloud, or sea. Although the constituent quark is not believed to be an elementary pointlike object but rather a more complicated object with internal structure, there is so far no experimental evidence for low-lying excitations of this structure – that is, no evidence for "excited constituent quarks." Many model builders have attempted to introduce such additional degrees of freedom, either to satisfy theoretical prejudices or to obtain a "better fit" than the simple constituent quark model to certain experimental data. Any advantages claimed by these models must be scrutinized carefully before acceptance, and the absence of any observed low-lying excitations of such degrees of freedom must be explained.

SU(6) and the symmetric quark model

The great breakthrough in baryon spectroscopy was the application of $SU(6)$ symmetry with the unreasonable assumption that spin-$\frac{1}{2}$ quarks obeyed Bose statistics.[14] The contradiction was avoided by the introduction of parastatistics or an additional internal degree of freedom later called color.[15] Great progress was made in understanding the baryon spectrum without a fundamental understanding of statistics by the phenomenological "symmetric quark model," which classified the hadron spectrum according to the group $SU(6) \times O(3)$.[16] It described all baryons as three-quark states with wave functions satisfying Bose statistics and having orbital and radial excitations with quantum numbers qualitativelly described by a harmonic-oscillator shell model.[17] An enormous number of baryon resonances fit exactly into this simple potential model beginning with the $SU(6)$ 56 classification of the lowest baryons into a spin-$\frac{1}{2}$ flavor octet and a spin-$\frac{3}{2}$ decuplet, the first excited configuration being a 70 of $SU(6)$ with $L = 1$ and the second being a 56 with $L = 2$. But the overwhelming evidence repeatedly presented by Dalitz and others for this model was consistently dismissed by the establishment.[18]

The successful $SU(6)$ prediction of $-\frac{3}{2}$ for the ratio of the proton and neutron magnetic moments was again striking evidence for compositeness, since only a composite model gave a simple ratio for *total* moments. In other approaches adding Dirac and anomalous moments was like adding apples and oranges. The anomalous moment was a function of the strong interaction coupling constant; the Dirac moment was not. Meson magnetic moments were not measured directly, but the radiative magnetic dipole transition $\omega \to \pi\gamma$ is described by the same quark magnetic operators appearing in the proton moment. The successful prediction relating this transition to the proton magnetic moment again confirmed that mesons and baryons were made of the same quarks.[19]

The scale of the nucleon magnetic moments caused confusion since quark magnetic moments were expected to have the scale of the quark mass rather than the hadron mass, although detailed relativistic calculations of hadron properties by the Soviet group gave hadron moments at the right scale.[20] This confusion was resolved by noting that the effective mass appearing in the magnetic moment of a bound Dirac particle depends upon the Lorentz structure of the potential and its scale is set by the particle energy, not its mass, for a world scalar potential.[21] The relativistic calculations effectively assumed a world scalar potential.[22]

The prehistory of QCD

Andrei Sakharov was a pioneer in hadron physics who took quarks seriously already in 1966. He asked, "Why are the Λ and Σ masses different? They are made of the same quarks!"[23] His answer that the difference arose from a flavor-dependent hyperfine interaction led to relations between meson and baryon masses in surprising agreement with experiment.[24] He and his collaborators *anticipated* QCD by assuming a quark model for hadrons with a flavor-dependent linear mass term and a two-body interaction whose flavor dependence was all in a hyperfine interaction

$$v_{ij} = v_{ij}^0 + \vec{\sigma}_i \cdot \vec{\sigma}_j v_{ij}^{hyp}, \qquad (30.3)$$

where v_{ij}^0 is independent of spin and flavor, $\vec{\sigma}_i$ is a quark spin operator, and v_{ij}^{hyp} is a hyperfine interaction with different strengths but the same flavor dependence for qq and $\bar{q}q$ interactions. They obtained two relations between meson and baryon masses in surprising agreement with experiment.[25]

The mass difference between s- and u-quarks calculated in two ways from the linear term in meson and baryon masses showed that it costs exactly the same energy to replace a nonstrange quark by a strange quark in mesons and baryons, when the contribution from the hyperfine interaction is removed:

$$(m_s - m_u)_{bar} = \qquad M_\Lambda - M_N \qquad = 177\,\text{MeV}, \qquad (30.4)$$

$$(m_s - m_u)_{mes} = \frac{3(M_{K^*} - M_\rho) + M_K - M_\pi}{4} = 180\,\text{MeV}, \qquad (30.5)$$

where the subscripts u, d, and s refer to quark flavors. The flavor dependence of the hyperfine splittings calculated in two ways from meson and baryon masses gave the result

$$1.53 = \frac{M_\Lambda - M_N}{M_{\Sigma^*} - M_\Sigma} = \left(\frac{v_{ud}^{hyp}}{v_{us}^{hyp}}\right)_{bar} \qquad (30.6)$$

$$\left(\frac{v_{ud}^{hyp}}{v_{us}^{hyp}}\right)_{mes} = \frac{M_\rho - M_\pi}{M_{K^*} - M_K} = 1.61. \qquad (30.7)$$

This striking evidence that mesons and baryons are made of the same quarks and described by a universal linear mass formula with spin corrections in remarkable agreement with experiment was overlooked for amusing reasons and rediscovered only in 1978.[26] In that same year, 1966, Nambu derived just such a universal linear mass formula for mesons and baryons from a model in which colored quarks were bound into color singlet hadrons by an interaction generated by coupling the quarks to a non-Abelian SU(3) color gauge field, with spin effects neglected.[27]

The Nobel prize for QCD as a description of strong interactions might have been awarded to Sakharov, Zel'dovich, and Nambu. They had it all figured out in 1966: the Balmer formula, the Bohr atom, and the Schrödinger equation of strong interactions. All subsequent developments leading to QCD were just mathematics and public relations, with no new physics. But the particle physics establishment refused to recognize the beginnings of new physics and had to wait until new fancy names like color chromodynamics, confinement, and so on, were invented – together with a massive public-relations campaign. Then they claimed that they had discovered it all.

Color, confinement and large N_c

The color degree of freedom solved the quark-statistics problem for baryons and also provided answers to several puzzles previously unanswered. The observed hadron spectrum indicated that both qqq and $\bar{q}q$ interactions were attractive in all possible states of spin and parity. An antiquark should be attracted by the three quarks in a baryon to make a $qqq\bar{q}$ bound state. But there were no bound states with "exotic" quantum numbers that could not be made from the $q\bar{q}$ or qqq configurations. There was also the meson–baryon puzzle – why qqq and $\bar{q}q$ systems are bound but different. No simple meson-exchange model gave these properties.

In 1967 I noted that quarks would be confined in the limit where the number of colors was large, now called the large N_c limit.[28] The $\bar{q}q$ pairs were bound into mesons, the meson–meson interaction went to zero, the hadron spectrum was simply systems of noninteracting mesons, and free quarks would not be observed. At that time any heretical paper of this type would never have been accepted by a reputable refereed journal; I therefore put it into lecture notes. In 1972 I looked at saturation in toy models of nuclei and noted that a nucleon–nucleon isospin-exchange interaction produced by ρ exchange would bind only the deuteron and the isoscalar $N\bar{N}$ system, and that no higher-mass bound states would exist. This led naturally to replacing isospin SU(2) by color SU(3) and to a model with colored quarks interacting with a color-exchange potential to give the first explanation of the absence of exotics and the observed meson–baryon systematics as well as the relation between qq and $\bar{q}q$

potentials later used in all potential models treating both mesons and baryons.[29]

I was very excited to have found a simple explanation of so much hadron physics for which there was no other explanation, and wrote letters from Israel to several friends including Richard Feynman and Victor Weisskopf. Feynman never answered, and Weisskopf wrote that it was all very interesting but theorists would not like it because it was not renormalizable. This did not bother me; it rather reminded me of the criticisms of BCS as not being gauge invariant. Thinking along the lines of BCS, I was sure that I had found interesting physics and that the correct formalism would come later. In fact the discovery of asymptotic freedom came at the same time, and it is interesting to compare the situation in the summers of 1972 and 1973. In his summary talk at the 1972 Rochester Conference at Fermilab, Gell-Mann noted that the color degree of freedom was established from electroweak data, that strong interactions were still unsolved and would probably arise from exchanges of vector gluons. But there was no suggestion that color played any role in strong interactions. At the SLAC summer school in 1973, I was invited to talk about my work on "Quarks and colored glue," and Gross, Politzer, and Wilczek were talking about the great breakthrough of asymptotic freedom.

Someone called my attention to Nambu's old paper, the details of which I had forgotten, which had worked out the SU(3) algebra of this interaction but not investigated the spatial dependence or the implications for exotics.[30] In contrast with the behavior of some of my peers, I immediately rewrote my paper, giving Nambu full credit for the work that I had independently rediscovered, before submitting the paper for publication.

It is rather painful to note the disparaging and untrue criticism of my paper: "Recently this point has been given publicity by Lipkin, who treats, however, a Han–Nambu picture. ... We have rejected such a picture."[31] Gell-Mann is a great physicist whose work and ideas had a tremendous impact on the work and thinking of practically everyone attending this history conference, including myself. But the general consensus of those active in the field in 1973 is that there was nothing new nor original in this paper.[32] My paper treats only strong interactions, ignores electromagnetism and the possibility of integrally charged quarks and has nothing to do with Han-Nambu. This irrelevant red herring is discussed below. Their criticism is irrelevant nonsense.

Quark-model predictions for hadron reactions

Further evidence for a quark substructure of hadrons was found in the additive quark model for hadron reactions, the so-called ideal mixing pattern of vector and tensor mesons, a mysterious topological quark diagram selection rule (now called OZI), and peculiar systematics in the energy behavior of certain hadron total cross sections.[33]

The additive quark model, duality and dual resonance models

The simple additive quark model (AQM) of Levin and Frankfurt explained the ratio of $\frac{3}{2}$ between nucleon–nucleon and meson–nucleon scattering and again showed mesons and baryons to be made of the same quarks.[34] Further refinements included flavor dependence of the scattering amplitudes at the quark level.[35] That the total cross sections in channels now called exotic do not have the sharply decreasing behavior found in other channels was described in the AQM by attributing all the energy decrease to $\bar{q}q$ annihilation amplitudes.[36] The AQM was combined with a Regge picture attributing this energy behavior to exchange degeneracy of Regge trajectories by using the AQM to relate the couplings of hadrons to exchange-degenerate Regge trajectories.[37] The universality of additive quark couplings to mesons and baryons arose again and again in different contexts in these descriptions.

An S-matrix Regge approach beginning with finite-energy sum rules led to duality with the same states appearing both as s-channel resonances and t-channel exchanges and then to dual resonance models beginning with the Veneziano model.[38] Although this was not directly related to the quark model, it soon appeared that introducing the quark-model constraints on Reggeon couplings provided a powerful input with predictive power. Thus, for example, the absence of exotics as both resonances and t-channel exchanges led to the Okubo–Zweig–Iizuka (OZI) rule, while the exchange degeneracy and the dominance of the energy-dependent part of the cross section by $\bar{q}q$ annihilation led naturally to duality diagrams.[39] The energy-independent part of the cross section, later found to be slowly rising, was seen to be related to diffraction, described by Pomeron exchange, with a coupling given by the Levin–Frankfurt quark-counting recipe.

Neutral-meson mixing, OZI relations, and the November revolution

The first use of the additive quark model to obtain OZI relations for neutral mesons was the selection rule forbidding reactions such as[40]

$$\sigma(\pi^- p \to N\phi)$$

and its SU(3) rotation, and predicting the equality

$$\sigma(K^- p \to \Lambda\omega) = \sigma(K^- p \to \Lambda\rho) \tag{30.8}$$

The ρ^0 and ω mesons are produced in the reactions (30.8) only via their $u\bar{u}$ component and are thus produced equally.

An outstanding failure of a quantitative prediction of an OZI-forbidden process was the experimental discovery of the J/ψ by pure accident; no theorist had predicted the narrow width nor directed experimenters to look for these enormous signals. The big charm-search review paper by Gaillard, Lee, and Rosner predicted the vector charmonium state, overestimated its width by a factor of 30, and did not point out the striking signal of a very narrow resonance.[41] The very narrow width caused considerable confusion after the discovery of the J/ψ and was used as evidence against the charmonium interpretation. Feynman insisted that this "crazy Zweig rule" could not give such a large suppression, because it was violated by a two-step strong interaction process, where each step was allowed and perturbation theory was certainly not valid. There had to be some new symmetry principle with a new conserved quantum number.

This failure to understand the OZI rule led to overestimating the width by a factor of 30. The experimental $\phi \to \rho\pi$ width was used as input, and threshold effects were disregarded.[42] But the $\phi \to \rho\pi$ decay is dominated by the two-step transition $\phi \to K\bar{K} \to \rho\pi$ for which the OZI-allowed $K\bar{K}$ channel is open. The use of the experimental $\phi \to \rho\pi$ width as input can give only an upper bound for the width of the J/ψ decay where no OZI-allowed channel is open and the $D\bar{D}$ channel analogous to $K\bar{K}$ in $\phi \to \rho\pi$ is closed. The distinction between open on-shell and closed off-shell intermediate states is now known to be significant because the physically observable transitions to open on-shell channels are related by unitarity to the OZI-forbidden processes and because the amplitudes via on-shell intermediate states cannot be canceled by off-shell contributions.[43] But there still is no real answer to Feynman's argument against the narrowness of the J/ψ. Hand-waving arguments suggest that second-order processes are canceled by contributions from different intermediate states. But there is still no rigorous QCD argument supported by calculations.

The Gaillard, Lee, and Rosner paper contains a note attributed to me, suggesting e^+e^- collisions as the best place to look for charm, since the charge $+\frac{2}{3}$ gave a much larger relative cross section.[44] The most striking signal would be a large increase in the number of strange particles, since a charm quark would decay to a strange quark. Half of the hadronic events above charm threshold would contain strange particles. My argument was correct but the signal was not seen. At the 1975 Lepton–Photon Symposium, Haim Harari resolved the paradox by noting that the excess of strange particles was not observed because of the unexpected appearance near charm threshold of the tau lepton. The nonstrange hadrons from τ events compensated for the strangeness excess from charm decays. At that time the existence of the τ as well as the identification of the J/ψ as charmonium were still controversial.

Absence of free quarks and fractional charges

Much of the resistance of the particle physics establishment to the quark model was based upon their fractional charge and upon the failure of experimenters to find free quarks. Both arguments are red herrings.

Why there are no free quarks

Why should anyone expect to find free quarks? A so-called free electron is a very complicated object containing a cloud of virtual photons and e^+e^- pairs. The hydrogen atom is much more than a point electron and a point proton. The other constituents are observed in Lamb shift and other experiments. Theorists describe this complicated structure only by using infinite renormalizing constants. Pulling the hydrogen atom apart into an electron and a proton, each containing its own infinite cloud of junk, was possible because the vacuum polarization between the electron and proton was small when they were separated. The energy required to excite and ionize the hydrogen atom was less than the rest mass of an electron–positron pair by a factor of order 10^5.

But suppose the excitation energy of the first excited state of the hydrogen atom was more than double the mass of positronium. The excited states would decay almost immediately by emitting positronia and isolated electrons would not have been discovered. Hitting the electron with a photon having enough energy to move it far away from the proton would polarize the vacuum and create a string of electron–positron pairs, which would quickly recombine into neutral positronia. Atomic collisions could well produce "electron jets" of neutral atoms and positronia (and no free electrons). Free constituents would not be easily found for hadrons whose spectrum indicated a structure with the energy of the first excited state already greater than twice the pion mass. The energy required to move these constituents from their lowest orbit into the first excited orbit was already greater than double the rest mass of the lowest bound state. Thus pumping energy into the proton would simply create pions and other bound states. The forces and vacuum polarization created by trying to remove a quark from a proton were much too great to allow the quark to be removed like the electron from a hydrogen atom.

Already in the late 1960s the hadron spectrum suggested that hitting a quark produced a string of pairs and that the excitation spectrum looked like the spectrum of a string.[45] One does not have to invent fancy names like confinement and chromodynamics to understand this simple physics. But the establishment refused to budge from its reactionary position. The party line that nothing was more elementary than neutrons and protons was sacrosanct, and heretics were ridiculed.

Who needs integrally charged quarks?

The prejudice against fractional charge led to a number of proposals of models with integrally charged quarks and to a series of useless proposals for experiments to measure the quark charge. The basic fallacy in the arguments for and against integral charge is seen by noting that the electromagnetic current must have a color-octet component in all models with integrally charged quarks, and that all matrix elements of color octet operators vanish between color-singlet states. Thus all experiments involving only color-singlet hadrons can measure only the color-singlet component of the quark charge and will give the fractional charge.[46]

If quarks really have integral charge but color-octet hadrons exist only at the Planck mass, there is no way to observe the integral charge at reasonable energies and therefore no way to kill the integrally charged models. Looking for evidence for integrally charged quarks is useless far below the threshold for producing color-octet states. The only sensible answer to the proposal that quarks might have integral charge is "Who needs them?" Why bother shooting down such models? One can rephrase Pauli's remark about hidden variables: "Integrally charged quark models are like mosquitoes – the more you kill, the more there are."

Notes

1 Y. Nambu, "A systematics of hadrons in subnuclear physics," in A. de Shalit, H. Feshbach and L. Van Hove, eds., *Preludes in Theoretical Physics* (Amsterdam: North-Holland Publishing Company, 1966), pp. 133–42.

2 Ya. B. Zel'dovich and A. D. Sakharov, "The quark structure and masses of strongly interacting particles," *Yad. Fiz 4* (1966), pp. 395–406 (in Russian); *Sov. J. Nucl. Phys. 4* (1967), pp. 283–90 (English translation).

3 H. R. Rubinstein, A. Schwimmer, G. Veneziano, and M. A. Virasoro, "Generation of parallel daughters from superconvergence," *Phys. Rev. Lett. 21* (1968), pp. 491–5.

4 H. Harari, "Duality diagrams," *Phys. Rev. Lett. 22* (1969), pp. 562–5; J. L. Rosner, "Graphical form of duality," *Phys. Rev. Lett. 22* (1969), pp. 689–92; G. Veneziano, "An introduction to dual models of strong interactions and their physical motivations," *Phys. Rep. 9C* (1974), pp. 199–242.

5 C. A. Levinson, H. J. Lipkin, S. Meshkov, A. Salam, and R. Munir, "A reaction forbidden by the Sakata model of unitary symmetry," *Phys. Lett. 1* (1962), pp. 44–9.

6 H. Goldberg and Y. Ne'eman, "Baryon charge and R-inversion in the octet model," *Nuovo Cimento 27* (1963), pp. 1–5.

7 Yuval Ne'eman, "The fifth interaction: origins of the mass breaking asymmetry," *Phys. Rev. 134* (1964), pp. B1355–7.

8 S. L. Glashow and J. J. Sakurai, "The 27-fold way and other ways: symmetries of meson–baryon resonances," *Nuovo Cimento 25* (1962), pp. 337–54.

9 M. Gell-Mann, "A schematic model of mesons and baryons," *Phys. Lett. 8* (1964), pp. 214–15.

10 W. H. Barkas, et al., "Antiproton–nucleon annihilation process (Antiproton Collaboration Experiment)," *Phys. Rev. 105* (1957), pp. 1037–58.

11 H. R. Rubinstein and H. Stern, "Nucleon–antinucleon annihilation in the quark model," *Phys. Lett. 21* (1966), pp. 447–9.

12 Ibid.

13 Harry J. Lipkin, "Difficulties arising in the combination of internal symmetries and space–time," *Phys. Lett. 14* (1965), pp. 336–8 and "Group theory and theoretical physics," in J. E. Bowcock, ed., *Methods and Problems of Theoretical Physics* (Amsterdam: North-Holland Publishing Company, 1970), pp. 381–99.

14 F. Gürsey and L. A. Radicati, "Spin and unitary spin dependence of strong interactions," *Phys. Rev. Lett. 13* (1964), pp. 173–5.

15 O. W. Greenberg, "Spin and unitary spin dependence in a paraquark model of baryons and mesons," *Phys. Rev. Lett. 13* (1964), pp. 598–602; Y. Nambu, "A systematics of hadrons"; A. Tavkhelidze, in "Higher symmetries and composite models of elementary particles," and "Electromagnetic form factors in composite models of elementary

particles (relativistic models)," *High-Energy Physics and Elementary Particles* (Vienna: IAEA, 1965), pp. 753–62.

16 O. W. Greenberg, "Spin and unitary spin dependence"; P. Federman, H. R. Rubinstein, and I. Talmi, "Dynamical derivation of baryon masses in the quark model," *Phys. Lett. 22* (1966), pp. 208–9.

17 O. W. Greenberg, "Spin and unitary spin dependence"; G. Karl and E. Obryk, "On wave functions for three-body systems," *Nucl. Phys. B8* (1968), pp. 609–21.

18 R. H. Dalitz, "Excited nucleons and the baryonic supermultiplets," in Gordon L. Shaw and David Y. Wong, eds., *Proceedings of the 1967 Irvine Conference on Pion-Nucleon Scattering* (New York: John Wiley & Sons, 1969), pp. 187–207; "Symmetries and the strong interactions," in *Proceedings of the XIIIth International Conference on High Energy Physics* (Berkeley: University of California Press, 1967), pp. 215–32.

19 C. Becchi and G. Morpurgo, "Test of the nonrelativistic quark model for elementary particles: radiative decays of vector mesons," *Phys. Rev. 140B* (1965), pp. 687–90.

20 A. Tavkhelidze, "Higher symmetries," and "Electromagnetic form factors."

21 H. J. Lipkin and A. Tavkhelidze, "Magnetic moments of relativistic quark model of elementary particles," *Phys. Lett. 17* (1965), pp. 331–4.

22 A. Tavkhelidze, "Higher symmetries" and "Electromagnetic form factors."

23 Andrei D. Sakharov, *Memoirs* (New York: Alfred A. Knopf, 1990), p. 261.

24 Ya. B. Zeldovich and A. D. Sakharov, "The quark structure."

25 Ya. B. Zeldovich and A. D. Sakharov, "The quark structure"; A. D. Sakharov, "Mass formula for mesons and baryons with allowance for charm," *Piz. Zh. Eksp. Teor. Fiz. 21* (1975), pp. 554–7 (in Russian); *JETP Lett. 21* (1975), pp. 258–9 (English translation); "Mass formula for mesons and baryons," *Zh. Eksp. Teor. Fiz. 78* (1980), pp. 2112–15 (in Russian); *Sov. Phys. JETP 51* (1980), pp. 1059–60 (English translation); "An estimate of the coupling constant between quarks and the gluon field," *Zh. Eksp. Teor. Fiz. 79* (1980), pp. 350–3 (in Russian); *Sov. Phys. JETP 52* (1980), pp. 175–6 (English translation).

26 Harry J. Lipkin, "Magnetic moments of composite fermions," in Behram Kursunoglu and Arnold Perlmutter, eds., *Gauge Theories, Massive Neutrinos and Proton Decay* (New York: Plenum, 1981), pp. 359–76; A. D. Sakharov, private communication; Harry J. Lipkin, "A unified description of mesons, baryons and baryonium," *Phys. Lett. B74* (1978), pp. 399–403; "Collective phenomena and particle physics," *Annals of the New York Academy of Sciences 452* (1985), pp. 79–95, and *London Times Higher Education Supplement* (January 20, 1984), p. 17.

27 Y. Nambu, "A systematics of hadrons."

28 Harry J. Lipkin, "Particle physics for nuclear physicists," in C. de Witt and V. Gillet, eds., *Physique Nucléaire, Les Houches 1968* (New York: Gordon and Breach, 1969), pp. 585–67.

29 H. J. Lipkin, "Triality, exotics and the dynamical basis of the quark model," *Phys. Lett. 45B* (1973), pp. 267–71.

30 Y. Nambu, "A systematics of hadrons."

31 H. Fritzsch, M. Gell-Mann, and H. Leutwyler, "Advantages of the color

octet gluon picture," *Phys. Lett. 47B* (1973), pp. 365–8; H. J. Lipkin, "Triality, exotics."

32 H. Fritzsch, M. Gell-Mann, and H. Leutwyler, "Advantages."

33 S. Okubo, "φ-meson and unitary symmetry model," *Phys. Lett. 5* (1963), pp. 165–8; "Consequences of quark-line (Okubo–Zweig–Iizuka) rule," *Phys. Rev. D16* (1977), pp. 2336–52; G. Zweig, "An SU(3) model for strong interaction symmetry and its breaking," CERN Report No. 8419/TH412 (unpublished) (February 1964); "Fractionally charged particles and SU(6)," in A. Zichichi, ed., *Symmetries in Elementary Particle Physics* (New York: Academic Press, 1965), pp. 192–234; J. Iizuka, "A systematics and phenomenology of meson family," *Prog. Theor. Phys. Suppl. 37–38* (1966), pp. 21–34.

34 E. M. Levin and L. L. Frankfurt, "The quark hypothesis and relations between cross sections at high energies," *Pis. Zh. Eksp. Theor. Fiz. 2* (1965), pp. 105–7 (in Russian); *JETP Lett. 2* (1965), pp. 65–7 (English translation).

35 H. J. Lipkin and F. Scheck, "Quark model for forward scattering amplitudes," *Phys. Rev. Lett. 16* (1966), pp. 71–5.

36 H. J. Lipkin, " Quark models and high-energy scattering," *Phys. Rev. Lett. 16* (1966), pp. 1015–19.

37 H. J. Lipkin, "Some implications of the quark model for high energy scattering," *Z. Phys. 202* (1967), p. 414–24; "Exchange degeneracy, flat total cross-sections, and the absence of exotic resonances," *Nucl. Phys. B9* (1969), pp. 349–63.

38 G. Veneziano, "An introduction."

39 H. Harari, "Duality diagrams"; J. L. Rosner, "Graphical form of duality"; H. J. Lipkin, "Exchange degeneracy."

40 G. Alexander, H. J. Lipkin, and F. Scheck, "Neutral-meson production cross-sections and mixing angles in a quark model," *Phys. Rev. Lett. 17* (1966), pp. 412–16.

41 M. K. Gaillard, B. W. Lee, and J. L. Rosner, "Search for charm," *Rev. Mod. Phys. 47* (1975), pp. 277–310.

42 Ibid.

43 H. J. Lipkin, "Who understands the Zweig–Iizuka rule?" in J. Tran Thanh Van, ed., *New Fields in Hadronic Physics* (Orsay, France: Laboratoire de Physique Théorique et Particules Elémentaires, Université de Paris-Sud, 1976), Vol. I, pp. 327–60; "The OZI rule in charmonium decays above DD̄ threshold," *Phys. Lett. B179* (1986), pp. 278–80.

44 M. K. Gaillard, B. W. Lee, and J. L. Rosner, "Search for charm."

45 G. Veneziano, "An introduction."

46 Harry J. Lipkin, "What is the charge of a parton," *Phys. Rev. Lett. 28* (1972), pp. 63–6; "Rigorous results on measuring the quark charge below color threshold," *Phys. Lett. B85* (1979), pp. 236–40; "Color oscillations and measuring the quark charge," *Nuc. Phys. B155* (1979), pp. 104–14.

31

From the Nonrelativistic Quark Model to QCD and Back

GIACOMO MORPURGO

Born Florence, Italy, 1927; Laurea, 1948 (physics), University of Rome; Professor of Physics, University of Genoa; elementary particle physics (theory and experiment) and nuclear physics (theory).

Speaking of the birth, in 1969, of the parton model, David Gross wrote:[1] "From then on I was always convinced of the reality of the quarks, not *just as mnemonic devices for summarizing hadron symmetries, that they were then universally regarded to be*, but as physical pointlike constituents of the nucleon" (italics mine). In a letter of reply (note 1) I noted that while it is hard to predict how the notion of quarks will evolve, it is sure that – already since 1965 – their most productive description was a realistic one.

In a review article about the discovery of quarks,[2] Michael Riordan stated: "After several years of fruitless searches most particle physicists agreed that although quarks might be useful mathematical constructs, they had no innate physical reality as objects of experience." Again I disagree. For many people[3] trying to understand the remarkable developments of hadron spectroscopy, the quarks of the nonrelativistic quark model (NRQM)[4] were, already five years before partons, not a mathematical construct or a mnemonic device but something very realistic. I started a long experiment (from 1965 to 1982) to search for real free quarks[5] because of the *quantitative* results (well beyond group theory) that I had obtained with the NRQM.[6,7]

Of course in that period many theorists did not like the NRQM. As one example, at Vienna in 1968 my rapporteur talk (note 6) on the NRQM was inserted in the session on "Current Algebra." Many anecdotes might illustrate the split between the current algebraists, who mostly regarded quarks as mathematical objects, and the many people (note 3) working with the realistic quark model. But I omit them, noting only that in his reply (note 1) Gross recognized he had been too sweeping in his characterization of the universality of the opinion that quarks were just mnemonic devices. But, he added, that opinion was widely held.

Why did many theorists dislike the NRQM in the 1960s? The reasons were many; for example, some considered it as simply equivalent to abstract SU_6. But as I see it, the main reason was a misinterpretation of the NRQM due to a confusion between bare and dressed quarks. Of course partons are the simple, pointlike, almost massless bricks of hadrons. They can be identified with the current (or bare) quark fields. This is not so for the three constituent quarks in terms of which the structure of a proton is described in the NRQM; each of them is dressed (note 4) with its cloud of gluons and quark–antiquark pairs. This complexity is typical of any strong interaction, as stressed in note 4. If one interprets the NRQM as saying that in a proton there are just three bare quarks, as many did, then clearly the model is inconsistent. An example is g_A/g_V in the quark weak current (note 7), often indicated as a problem for the model. Clearly the current, in terms of bare quark fields, is $(1 + \gamma_5)$; but just because of this, we expect g_A/g_V to be different from unity ($\cong 0.74$) in the NRQM transition between dressed quarks (as in $N \to P$ beta–decay), because the vector current is not renormalized, while the axial vector is.

So in the NRQM the three quarks in a baryon must be seen as constituent quarks, each dressed and exchanging gluons with the others. The baryon is a superposition of infinite Fock states with infinitely many point quark–antiquark pairs and gluons. At the start, the quark mass I took was huge, their binding was due to Majorana forces, and "confinement" was due to the huge mass; but the Fock description was there from the start. Looking at the NRQM in this way (with three dressed quarks, thus with infinitely many bare quarks), the parton model does not conflict with a NRQM description, although it is obviously something entirely new. Unlike Riordan, I see the NRQM and the parton model as two different paths to QCD,[8] both implying realistic "quarks."

So much for the evolution from the NRQM to QCD. Now I move backwards from QCD to NRQM. Then one can discover something new, namely why the NRQM works fairly well *quantitatively*, a point that always struck me. I started this study of the relationship between field theory and NRQM based on the Fock description in 1968[9] and recently developed it further.[10] Here I must omit all the details. The method relates exactly a full-field calculation to a few-body one, using a unitary transformation between model and exact states, and integrating over the variables of the virtual quarks, antiquarks, and gluons. I will only illustrate a typical result, that of baryon magnetic moments (note 10a). As is well known, the simplest NRQM calculation of baryon magnetic moments is a two parameter calculation: One assigns to the $\mathcal{P}$ and $\mathcal{N}$ constituent quarks magnetic moments proportional to their charges and to the λ quark a smaller magnetic moment, due to its higher mass. One then calculates the expectation of the sum of the above magnetic moments on the spin-flavor factor of the NRQM wave function [the SU$_6$ factor]. Fixing the two above parameters from, say, P and Λ, one can predict the remaining five moments to better than 15 percent (Fig. 31.1). Now one may ask: If we could do a full exact QCD calculation, what would be the most general spin-flavor parametrization of the magnetic moments of the octet baryons? The answer is much simpler than one might guess. To first-order flavor breaking and omitting a term $O(10^{-2})$, the most general parameterization resulting from QCD is:

$$\vec{M} = (\mu + KS) \sum \left[(2/3)\vec{\sigma}^P - (1/3)\vec{\sigma}^N - (1/3)\vec{\sigma}^\lambda \right]$$
$$+ \tfrac{1}{3}(A + LQ) \sum \vec{\sigma}^\lambda + 2(FQ + HS + GQS)\vec{J} \tag{31.1}$$

Here μ, K, A, F, H, L, G are seven parameters, S is the strangeness, Q is the charge, and $\vec{J}$ the angular momentum ($J_z = \tfrac{1}{2}$) of the baryon being considered; the $\vec{\sigma}$'s are Pauli matrices. To obtain the magnetic moment M_B of baryon B, calculate $M_B = <W_B|M_z|W_B>$, where W_B is the spin–flavor factor of the NRQM wave function, the same as the SU(6) factor. I stress that although equation (31.1) looks like a typical NRQM expression, it is exact. A full QCD field calculation can give nothing else but this equation, which is exact (to first-order flavor breaking) and, being exact, is fully relativistic, although noncovariant. An explicit QCD calculation would only determine the coefficients in terms of α_s and the quark masses.

Because there are seven parameters in equation (31.1), the same as the number of measured moments, they are all determined. One can check that, indeed, the coefficients other than those (μ and A) that appear in the NRQM are at most 15 % of μ. The dominance of the terms of the NRQM is due to a decrease in the magnitude of terms of increasing complexity – a decrease related to the minimum number of gluons exchanged (note 10g).

Not only for the magnetic moments, but in all cases (compare note 10) the general exact parametrizations have a spin–flavor structure identical to those typical of the NRQM. To show this, no use is made of the magnitude of v/c. (v is the quark velocity in a hadron.) The gen-

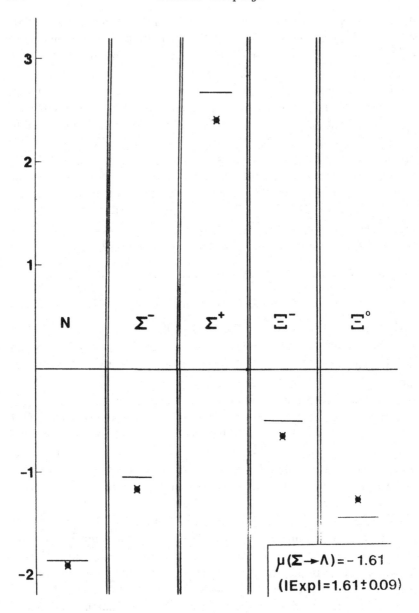

Fig. 31.1. The measured magnetic moments of the baryons compared with the values (solid lines) calculated with the NRQM. As stated in the text, the result of the NRQM calculation amounts to keep in equation (31.1) only the coefficients μ and A, and to fix them using as inputs the P and Λ moments.

eral parametrization is the result of a fully relativistic calculation (although expressed in non-covariant language) and, I repeat, the NRQM is fairly good quantitatively because it selects the dominant terms in this parametrization. It is amusing that the method solves some very old problems: it explains (note 10g and f) why certain "classical" formulas (like those of Gell-Mann–Okubo or of Coleman–Glashow) work so well. It also leads (note 10e) to a new octet-decuplet mass formula correct to one part per thousand and to several other predictions.[11]

Notes

1 D. J. Gross, "Asymptotic freedom," *Physics Today* (January 1987), pp. 39–44; G. Morpurgo, "Quarks realistic and naive," *Physics Today* (December 1987), p. 112 and D. Gross reply, pp. 112–13.

2 M. Riordan, "The discovery of quarks," *Science 256* (1992), pp. 1287–93.

3 R. H. Dalitz, "Symmetries and the Strong Interactions," in *Proc. 13th Int. Conf. High Energy Physics* (Berkeley: University of California Press, 1967), pp. 215–32; J. J. Kokkedee, *The Quark Model* (New York: W. A. Benjamin, 1969); A. Pickering, *Constructing Quarks* (Edinburgh: Edinburgh University Press, 1984), chapter 4 and refs. quoted therein.

4 G. Morpurgo, "Is a nonrelativistic approximation possible for the internal dynamics of 'elementary' particles?" *Physics 2* (1965), pp. 95–104.

5 For a complete account of the experiment see: M. Marinelli and G. Morpurgo, "Searches of fractionally charged particles in matter with the magnetic levitation technique," *Phys. Rep. 85* (1982), pp. 162–258; see also M. Marinelli and G. Morpurgo, "The electric neutrality of matter: a summary," *Phys. Lett. 137B* (1984), pp. 439–42.

6 G. Morpurgo, "The quark model," *Proc. 14th. Int. Conf. on High Energy Physics*, Vienna 1968 (Geneva: CERN, 1968), pp. 225–49.

7 G. Morpurgo, "Lectures on the quark model," in A. Zichichi, ed., *Theory and Phenomenology in Particle Physics*, Vol. A (New York: Academic Press, 1968), pp. 84–217, see especially p. 126.

8 For a graphical display of the two paths to QCD see, fig. 1.1, ref. 5, p. 166.

9 Ref. 7, Sections 8.2 and 8.3

10 G. Morpurgo, a) "Field theory and the nonrelativistic quark model: A parametrization of the baryon magnetic moments and masses," *Phys. Rev. D40* (1989), pp. 2997–3011; b) "Parametrization of the semileptonic baryon matrix elements," *Phys. Rev. D40* (1989), pp. 3111–24; c) "Field theory and the nonrelativistic quark model: A parametrization of the meson masses," *Phys. Rev. D41* (1990), pp. 2865–70; d) *Phys. Rev. D42* (1990), pp. 1497; e) *Phys. Rev. Lett. 68* (1992), pp. 139; f) *Phys. Rev. D45* (1992), pp. 1686; g) *Phys. Rev. D46* (1992), pp. 4068.

11 (Note added in proof.) See also: G. Dillon and G. Morpurgo, "The relation of constituent quark models to QCD: Why several simple models work 'so well'," *Phys. Rev. D53* (1996), p. 3754.

32

Deep-Inelastic Scattering
and the Discovery of Quarks

JEROME FRIEDMAN

Born Chicago, Illinois, 1930; Ph.D., 1956 (physics), University of
Chicago; Professor of Physics, Massachusetts Institute of Technology;
Nobel Prize in Physics, 1990; high-energy physics (experimental).

In 1961 Murray Gell-Mann and Yuval Ne'eman independently introduced a classification scheme, based on SU(3) symmetry, that placed
hadrons into families on the basis of spin and parity.[1] Like the periodic
table for the elements, this scheme had predictive as well as descriptive
powers. Hadrons that were predicted within this framework, such as the
Ω^-, were later discovered.

In 1964 Gell-Mann and George Zweig independently proposed quarks
as the building blocks of hadrons as a way of generating the SU(3) classification scheme.[2] When the quark model was first proposed, it postulated three types of quarks – up (u), down (d), and strange (s), having
charges $\frac{2}{3}$, $-\frac{1}{3}$, and $-\frac{1}{3}$, respectively; each of these was hypothesized to
be a spin-$\frac{1}{2}$ particle. In this model the nucleon (and all other baryons)
is made up of three quarks, and all mesons each consist of a quark and
an antiquark. For example, as the proton and neutron both have zero
strangeness, they are (u,u,d) and (d,d,u) systems, respectively. Though
the quark model provided the best available tool for understanding the
properties of the hadrons that had been discovered at the time, the
model was thought by many to be merely a mathematical representation of some deeper dynamics, but one of heuristic value. Among the
reasons for this assessment were the following: free quarks had not been
found, although they had been sought in numerous accelerator and cosmic ray investigations and in searches in the terrestrial environment;
there was a deep suspicion about the validity of their fractional charge
assignments; and the states in which the quarks were combined to form
baryons violated the Pauli exclusion principle. Despite these difficulties

there were a number of theorists who continued to apply the model to explain a wide range of hadronic phenomena.

The theory of hadron structure that was most widely accepted at the time was the bootstrap model, an approach based on S-matrix theory. This model, sometimes referred to as nuclear democracy, was based on the idea that there were no fundamental particles and that all hadrons are made up of one another. This picture was consistent with the low momentum transfer scattering seen in hadron–hadron interactions and with the observed "soft" electromagnetic form factors of the proton and neutron. Though the model was also used to derive an SU(3) hadronic symmetry, it could not provide the comprehensive description of multiplet structures that was given by the quark model.

The results from inelastic electron–nucleon scattering and later from neutrino–nucleon scattering played a pivotal role in resolving this dilemma by firmly establishing the quark model. These experiments demonstrated that proton and neutron are composite structures made up of spin-$\frac{1}{2}$ fractionally charged constituents.

More detailed descriptions of the deep-inelastic scattering program and its results are given in the written versions of the 1990 Nobel Lectures in Physics of Richard Taylor, Henry Kendall, and the present author.[3]

Early electron scattering experiments

In the latter half of 1967 a group of physicists from the Stanford Linear Accelerator Center and the Massachusetts Institute of Technology embarked on a program of inelastic electron–proton scattering after completing an initial study of elastic scattering with physicists from the California Institute of Technology.[4] This work was done on the newly completed 20 GeV Stanford Linear Accelerator. The initial purpose of the inelastic program was to study the electroproduction of resonances as a function of momentum transfer. It was thought that higher mass resonances might become more prominent when excited by virtual photons, and it was our intent to search for these at the very highest masses that could be reached. For completeness we also wanted to look at the inelastic continuum, since this was a new energy region that had not been previously explored. The proton resonances that we were able to measure showed no unexpected kinematic behavior.[5] Their transition form factors fell about as rapidly as the elastic proton form factor with increasing values of the invariant four-momentum transfer, q^2. However, we found some surprising features when we investigated the continuum region (now commonly called the deep-inelastic region).

The experiments consisted of measurements of spectra of inelastically scattered electrons, with only the scattered electrons detected, over a range of incident energies and scattering angles. A monochromatic beam from the linear accelerator was passed through a liquid hydrogen (and later deuterium) target and then through a series of monitors. The scattered electrons were momentum analyzed by one of three magnetic spectrometers installed in End Station A. In separate experiments the SLAC 20 GeV, 8 GeV, and 1.6 GeV spectrometers were used to cover different kinematic regions; however, most of the measurements were made with the two larger devices (see Fig. 32.1). Downstream of the magnetic elements of these spectrometers were placed scintillation counter hodoscopes that registered the momentum and scattering angle of each scattered electron. In conjunction with the hodoscopes there were particle identification counters that were employed to identify electrons amid a background of pions. These consisted of a gas Čerenkov counter, a total-absorption counter for electromagnetic cascades, and a few counters used to sample early shower development in the total-absorption counter.

The major experimental challenge in these measurements was to eliminate from the measured cross sections the effects of the radiation of photons that occur during the scattering process and during the electron's traversal of material before and after scattering. This required an intricate deconvolution procedure performed on the inelastic spectra measured at various incident energies for a given angle. As this procedure was based on an incomplete theoretical formulation and greatly taxed the computational facilities that were then available, this problem was of central concern to the group. Only after a number of comparisons between independent calculations, based on different approximations, were we convinced that we understood the systematic errors sufficiently well to have confidence in our results.

Early results

The first unexpected feature of these early results was that the deep-inelastic cross sections showed only a weak fall-off with increasing q^2.[6] When the experiment was planned, there was no clear theoretical picture of what to expect. The observations of Robert Hofstadter and his co-workers in their pioneering studies of elastic electron scattering from

Fig. 32.1. The three spectrometers used in the MIT–SLAC experiments. In the foreground is the 8 GeV spectrometer, while the 20 GeV spectrometer is behind it; at the extreme left, the 1.6 GeV spectrometer is just barely visible.

the proton showed that the proton had a size of about 10^{-13} cm and a smooth charge distribution.[7] This result, plus the theoretical framework most widely accepted at the time, suggested to our group when the experiment was planned that the deep-inelastic electron–proton scattering cross sections would fall rapidly with increasing q^2.

The yields we observed were a factor of 10 to 100 greater than we expected on the basis of a model of off-mass-shell photoproduction with the inclusion of the proton form factor. We were surprised and made extensive checks of our radiative corrections routine before we were convinced that our results were correct.

The weak dependence of the inelastic cross sections on momentum transfer for excitations well beyond the resonance region is illustrated in Fig. 32.2. The differential cross section divided by the Mott cross section, σ_{Mott}, is plotted as a function of the square of the four-momentum transfer, $q^2 = 2EE'(1 - \cos\theta)$, for constant values of the invariant mass of the recoiling target system, W, where $W^2 = 2M(E - E') + M^2 - q^2$.[8] The quantity E is the energy of the incident electron, E' is the energy of the final electron, and θ is the scattering angle, all defined in the laboratory system; M is the mass of the proton. The cross section is divided by the Mott cross section in order to remove the major part of the well-known q^2 dependence arising from the photon propagator. The q^2 dependence that remains is related primarily to the properties of the target system. Results from $10°$ are shown in the figure for each value of W. As W increases, the q^2 dependence appears to decrease. The striking difference between the behavior of the deep-inelastic and elastic cross sections is also illustrated in this figure, which shows the elastic cross section divided by the Mott cross section for $\theta = 10°$.

Scaling

The second surprising feature in the data, scaling, was found by following a suggestion of James Bjorken.[9] To describe the concept of scaling, one has to introduce the general expression for the differential cross section for unpolarized electrons scattering from unpolarized nucleons with only the scattered electrons detected.[10]

$$\frac{d^2\sigma}{d\Omega dE'} = \sigma_{Mott}\left[W_2 + 2W_1 \tan^2\frac{\theta}{2}\right].$$

The functions W_1 and W_2 are called structure functions and depend only on the properties of the target system. As there are two independent polarization states of the virtual photon, transverse and longitudinal, two such functions are required to describe this process. In general, W_1 and W_2 are expected to be functions of both q^2 and ν, where ν is the energy loss of the scattered electron. However, on the basis of models that satisfy current algebra, Bjorken conjectured that in the limit of q^2 and ν approaching infinity, the two quantities νW_2 and W_1 become functions only of the ratio $\omega = 2M\nu/q^2$; that is

$$2MW_1(\nu, q^2) \rightarrow F_1(\omega)$$
$$\nu W_2(\nu, q^2) \rightarrow F_2(\omega).$$

The scaling behavior of the structure functions is shown in Fig. 32.3, where experimental values of νW_2 and $2MW_1$ are plotted as a function of ω for values of q^2 ranging from 2 to 20 GeV^2. The data demonstrated scaling within experimental errors for $q^2 > 2$ GeV^2 and $W > 2.6$ GeV.

The dynamical origin of scaling was not clear at that time, and a number of models were proposed to account for this behavior and for the weak q^2 dependence of the inelastic cross section. Although most

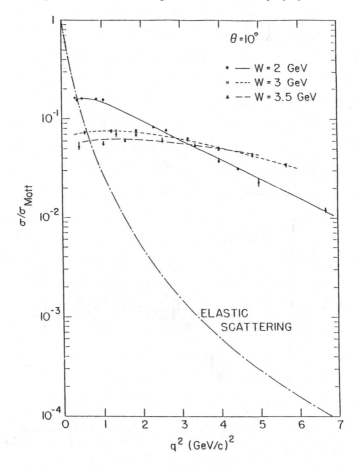

Fig. 32.2. $(d^2\sigma/dE'd\Omega)/\sigma_{\text{Mott}}$, plotted vs. q^2 for $W = 2, 3$ and 3.5 GeV at $\theta = 10°$. The lines drawn through the data are meant to guide the eye. Also shown is the cross section for elastic e–p scattering divided by σ_{Mott}, calculated for $\theta = 10°$ using the dipole form factor.

of these models were firmly embedded in S-matrix and Regge-pole formalisms, the experimental results caused some speculation regarding the existence of a possible pointlike structure in the proton. In his plenary talk at the Fourteenth International Conference on High Energy Physics, held in Vienna in 1968, where preliminary results on the weak q^2 dependence and scaling were presented, Wolfgang Panofsky reported that "theoretical speculations are focused on the possibility that these data might give evidence on the behavior of pointlike charged structures

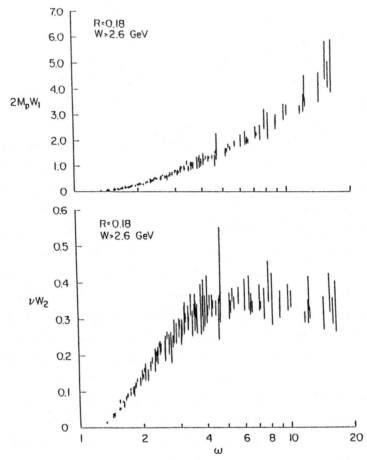

Fig. 32.3. The measured data for $2MW_1$ and νW_2 for the proton as functions of ω for $W > 2.6$ GeV, $q^2 > 1$ GeV/c^2, assuming $R = 0.18$.[11]

in the nucleon."[12] However, this was not the prevailing point of view at the time. This picture challenged the beliefs of most of the physics community, and only a small number of theorists took such a possibility seriously. One of these was Bjorken, who had proposed in his 1967 Varenna lectures that deep-inelastic electron scattering might provide evidence of elementary constituents.[13] Studying the sum-rule predictions derived from current algebra,[14] he stated, "We find these relations so perspicuous that, by an appeal to history, an interpretation in terms of elementary constituents is suggested."[15] In essence, Bjorken observed that a sum rule for neutrino scattering derived by Stephen Adler from the commutator of two time components of the weak currents led to an inequality for inelastic electron scattering,

$$\int_{q^2/2M}^{\infty} d\nu \left[W_2^p(\nu, q^2) + W_2^n(\nu, q^2) \right] \geq \frac{1}{2},$$

where W_2^p and W_2^n are structure functions for the proton and neutron respectively.[16]

This is equivalent to:

$$\lim_{E \to \infty} \left[\frac{d\sigma_{ep}}{dq^2} + \frac{d\sigma_{en}}{dq^2} \right] \geq \frac{2\pi\alpha^2}{q^4}.$$

The above inequality states that, as the electron energy goes to infinity, the sum of the electron–proton plus electron–neutron total cross sections (elastic plus inelastic) at fixed large q^2 is predicted to be greater than one-half the cross section for electrons scattering from a pointlike particle. Bjorken also derived a similar result for backward electron scattering.[17] These results were derived well before our inelastic measurements appeared. In hindsight, it is clear that they implied a pointlike structure of the proton and neutron and large cross sections at high q^2, but Bjorken's results made little impression on us at the time. Perhaps it was because his results were based on current algebra, which we found highly esoteric. Or perhaps we were very much steeped in the physics of the time, which suggested that hadrons were extended objects with diffuse substructures.

Nonconstituent models

The initial deep-inelastic measurements stimulated a flurry of theoretical work, and a number of nonconstituent models based on a variety of theoretical approaches were put forward to explain the surprising features of the data. One approach related the inelastic scattering to forward virtual Compton scattering, which was described in terms of Regge exchange using the Pomeranchuk trajectory, or a combination of it and nondiffractive trajectories.[18] Such models do not require a weak q^2 dependence, and scaling had to be explicitly inserted. Resonance models were also proposed to explain the data. Among these was a Veneziano-type model in which the density of resonances increases at a sufficiently rapid rate to compensate for the decrease of the contribution of each resonance with increasing q^2.[19] Another type of resonance model built up the structure functions from an infinite series of N and Δ resonances.[20]

None of these models was totally consistent with the full range of data accumulated in the deep-inelastic scattering program.

One of the first attempts to explain the deep inelastic scattering results employed the vector dominance model, which had been used to describe photon–hadron interactions over a wide range of energies.[21] This model, in which the photon is assumed to couple to a vector meson, which then interacts with a hadron, was extended using ρ-meson dominance to deep-inelastic electron scattering. It reproduced the gross features of the data in that νW_2 approached a single function of ω for ν much greater than M_ρ, the mass of the ρ-meson. The model also predicted that

$$R = \frac{\sigma_L}{\sigma_T} = \left(\frac{\varepsilon q^2}{M_\rho^2}\right)\left(1 - \frac{q^2}{2M\nu}\right),$$

where R is the ratio of σ_L and σ_T the photoabsorption cross sections of longitudinal and transverse virtual photons, respectively; and ε is the ratio of the vector meson–nucleon total cross sections for vector mesons with polarization vectors parallel and perpendicular to their direction of motion. Since the parameter ε is expected to have a value of about 1 at high energies, this theory predicted very large values of R for values of $q^2 \gg M_\rho^2$. The ratio R can be related to the structure functions in the following way:

$$R = \frac{W_2}{W_1}\left(1 + \frac{\nu^2}{q^2}\right) - 1.$$

The measurements of deep-inelastic scattering over a range of angles and energies allowed W_1 and W_2 to be separated and R to be determined experimentally. The measurements showed that R is small and does not increase with q^2. This eliminated the vector dominance model as a possible description of deep-inelastic scattering.

Various attempts to save the vector-meson dominance point of view were made with the extension of the vector-meson spectral function to higher masses, including approaches that included a structureless continuum of higher mass states.[22] These calculations of the generalized vector dominance model failed in general to describe the data over the full kinematic range.

The parton model

The constituent model that opened the way for a simple dynamical interpretation of the deep-inelastic results was the parton model of Richard

Feynman.[23] He developed this model to describe hadron–hadron interactions, in which the constituents of one hadron interact with those of another. These constituents, called partons, were identified with the fundamental bare particles of an unspecified underlying field theory of the strong interactions. He applied this model to deep-inelastic electron scattering after he had seen the early scaling results that were about to be presented at the Vienna Conference of 1968. Deep-inelastic electron scattering was an ideal process for the application of his model. In electron–hadron scattering the electron's interaction and structure were both known, whereas in hadron–hadron scattering neither the structures nor the interactions were understood at the time.

In this application of the model, the proton is conjectured to consist of pointlike partons from which the electron scatters. The model is implemented in a frame approaching the infinite momentum frame, in which the relativistic time dilation slows down the constituents to a nearly motionless state. The incoming electron thus "sees" and incoherently scatters from partons that do not interact with each other during the time in which the virtual photon is exchanged. In this frame the impulse approximation is assumed to hold, so that the scattering process is sensitive only to the properties and momenta of the partons. The recoil parton has some kind of final-state interaction in the nucleon, producing the secondaries emitted in inelastic scattering.

The parton model, with the assumption of pointlike constituents, automatically gave scaling behavior. The Bjorken scaling variable ω was seen to be the inverse of the fractional momentum of the struck parton, x, and νW_2 was shown to be the fractional momentum distribution of the partons, weighted by the squares of their charges.

In proposing the parton model, Feynman was not specific as to what the partons were. There were two competing proposals for their identity. Applications of the parton model identified partons with bare nucleons and pions, and also with quarks.[24] But parton models incorporating quarks had a glaring inconsistency. Quarks required strong final-state interactions to account for the fact that free quarks had not been observed in the laboratory. Before the theory of quantum chromodynamics (QCD) was developed, there was a serious problem in making the "free" behavior of the constituents during photon absorption compatible with this required strong final-state interaction. One of the ways to get out of this difficulty was to assign quarks very large masses, but this caused theoretical problems in constructing hadron structure from quarks. This problem was avoided in parton models employing bare nucleons and pions because the recoil constituents are allowed to decay into real particles when they are emitted from the nucleon.

Sidney Drell, Donald Levy and Tung-Mow Yan derived a parton model (in which the partons are bare nucleons and pions) from a canonical field theory of pions and nucleons with the insertion of a cutoff in transverse momentum.[25] A further development of this approach was a calculation by Tsung Dao Lee and Drell that provided a fully relativistic generalization of the parton model that was no longer restricted to an infinite momentum frame.[26] This theory obtained bound state solutions of the Bethe–Salpeter equation for a bare nucleon and bare mesons, and connected the observed scale invariance with the rapid decrease of the elastic electromagnetic form factors.

Bjorken and Emmanuel Paschos studied the parton model for a system of three quarks, commonly called valence quarks, embedded in a background of quark–antiquark pairs, often called the sea.[27] A more detailed description of a quark–parton model was later given by Julius Kuti and Victor Weisskopf.[28] Their model of the nucleon contained, in addition to the three valence quarks and a sea of quark–antiquark pairs, neutral gluons, which are quanta of the field responsible for the binding of the quarks. The momentum distribution of the quarks corresponding to large ω was given in terms of the requirements of Regge behavior.

Decisive tests of these models were provided by extensive measurements with hydrogen and deuterium targets that followed the early results, which were obtained using only hydrogen targets.

Measurements of proton and neutron structure functions

The first deep-inelastic electron scattering results were obtained in the period 1967–1968 from a hydrogen target with the 20 GeV spectrometer set at scattering angles of 6° and 10°.[29] By 1970 the proton data had been extended to scattering angles of 18°, 26°, and 34° with the use of the 8 GeV spectrometer.[30] The measurements covered a range of q^2 from 1 GeV^2 to 20 GeV^2, and a range of W^2 up to 25 GeV^2. By 1970 data had also been obtained at scattering angles of 6° and 10° with a deuterium target.[31] Subsequently, a series of matched measurements with better statistics and covering an extended range of q^2 and W^2 were done with hydrogen and deuterium targets, utilizing the 20 GeV, 8 GeV, and 1.6 GeV spectrometers.[32] These data sets provided, in addition to more detailed information about the proton structure functions, a test of scaling for the neutron. In addition, the measured ratio of the neutron

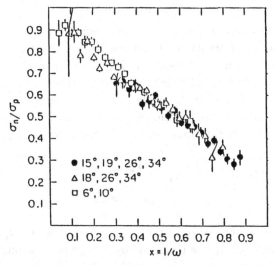

Fig. 32.4. Measured values of σ_n/σ_p, the ratio of the inelastic cross sections of the neutron and proton, as a function of x.[33]

and proton structure functions provided a decisive tool in discriminating among the various models proposed to explain the early proton results.

Neutron cross sections were extracted from measured deuteron cross sections using the impulse approximation along with a procedure to remove the effects of Fermi motion. The method used was that of William Atwood and Geoffrey West, with small modifications representing off-mass-shell corrections.[34] With the use of this method, the neutron cross section σ_n, free of the effects of Fermi motion, was obtained; from the measured proton cross section σ_p, the ratio σ_n/σ_p was determined.

The conclusions that were derived from the analysis of these extensive data sets were the following:

1. The deuteron and neutron structure functions showed the same approximate scaling behavior as the proton.

2. The values of R_p, R_n and R_d were equal within experimental errors.

3. The ratio of the neutron and proton inelastic cross sections falls continuously as the scaling variable x approaches unity ($x = 1/\omega$). From a value of about 1 near $x = 0$, the experimental ratio falls to about 0.3 in the neighborhood of $x = 0.85$, as shown in Fig. 32.4. These results put strong constraints on various models of nucleon structure.

Sum-rule results

A sum rule generally relates a weighted integral of a cross section (or of a quantity derived from it) and the properties of the interaction hypothesized to produce that cross section. Experimental evaluations of such relations thus provide a valuable tool in testing theoretical models. Sum-rule evaluations within the framework of the parton model provided an important element in identifying the constituents of the nucleon. Two important sum rules that were evaluated for neutrons and protons were:

$$I_1 = \int_1^\infty \nu W_2(\omega) \frac{d\omega}{\omega^2}$$

$$I_2 = \int_1^\infty \nu W_2(\omega) \frac{d\omega}{\omega}$$

where I_2 can be shown to be the weighted sum of the squares of the parton charges and I_1 the mean square charge per parton.[35] The sum I_2 is equivalent to a sum rule derived by Kurt Gottfried, who showed that for a proton that consists of three nonrelativistic pointlike quarks I_2^p equals 1 at a high q^2.[36] The experimental value of this integral when integrated over the range of the MIT–SLAC data gave:

$$I_2^p = \int_1^{20} \frac{d\omega}{\omega} \nu W_2^p = 0.78 \pm 0.04$$

where the integral was cut off for $\omega > 20$ because of insufficient information about R_p. Since the experimental values of νW_2 at large ω did not exclude a constant value (see Fig. 32.3), there was some suspicion that this integral might diverge. This would imply that in the quark model scattering occurs from an infinite sea of quark–antiquark pairs as ω approaches infinity.

The application of the quark–parton model allowed the weighted sum I_1 to be evaluated theoretically. If $u_p(x)$ and $d_p(x)$ are defined as the fractional momentum distributions of up and down quarks in the proton, then $F_2^p(x)$ is given by

$$F_2^p(x) = \nu W_2^p(x) = x \left[Q_u^2(u_p(x) + \bar{u}_p(x)) + Q_d^2(d_p(x) + \bar{d}_p(x)) \right]$$

where $\bar{u}_p(x)$ and $\bar{d}_p(x)$ are the distributions for anti-up and anti-down quarks, and Q_u^2 and Q_d^2 are the squares of the charges of the up and down quarks. The strange quark sea has been neglected.

Using charge symmetry it can be shown that

$$\frac{1}{2} \int_0^1 \left[F_2^p(x) + F_2^n(x) \right] dx$$

$$= \left[\frac{Q_u^2 + Q_d^2}{2} \right] \int_0^1 x \left[u_p(x) + \bar{u}_p(x) + d_p(x) + \bar{d}_p(x) \right] dx.$$

The integral on the right-hand side of the equation is the total fractional momentum carried by the quarks and antiquarks, which would equal 1 if they carried all the nucleon's total momentum. On this assumption the expected sum should equal

$$\frac{Q_u^2 + Q_d^2}{2} = \frac{1}{2} \left[\frac{4}{9} + \frac{1}{9} \right] = \frac{5}{18} = 0.28$$

The evaluations of the experimental sum from proton and neutron results over the entire kinematic range studied were

$$\frac{1}{2} \int \left[F_2^p(x) + F_2^n(x) \right] dx = 0.14 \pm 0.005.$$

Unlike I_2, the experimental value of I_1 was not very sensitive to the behavior of νW_2 for $\omega > 20$. The experimental value was about one-half the value predicted on the basis of a model of the proton having three valence quarks in a sea of quark–antiquark pairs. The Kuti–Weisskopf model, which included neutral gluons in addition to the valence quarks and the sea of quark–antiquark pairs, predicted a value of I_1 that was compatible with this experimental result.[37]

The difference $I_2^p - I_2^n$ was of great interest because it was presumed to be sensitive only to the valence quarks in the proton and the neutron. On the assumption that the quark–antiquark sea is an isotopic scalar, the effects of the sea cancel out in the above difference, giving $I_2^p - I_2^n = \frac{1}{3}$. Unfortunately, it was difficult to extract a meaningful value from the data because of the importance of the behavior of νW_2 at large ω. Extrapolating $\nu W_2^p - \nu W_2^n$ toward $\omega \to \infty$ for $\omega > 12$, with the asymptotic dependence $(1/\omega)^{\frac{1}{2}}$ expected on the basis of Regge theory, we obtained a rough estimate of $I_2^p - I_2^n = 0.22 \pm 0.07$. This was compatible with the expected value, given the error and the uncertainties in extrapolation.

The Bjorken inequality previously discussed, namely,

$$\int_{q^2/2M}^{\infty} d\nu \left[W_2^p(\nu, q^2) + W_2^n(\nu, q^2) \right] \geq \frac{1}{2}$$

was also evaluated. This inequality was found to be satisfied at $\omega \simeq 5$, demonstrating consistency with "pointlike" structure in the nucleon.

Identification of the nucleon constituents as quarks

The confirmation of a constituent model of the nucleon and the identification of the constituents as quarks took a number of years and was the result of continuing interplay between experiment and theory. By the time of the Fifteenth International Conference on High Energy Physics held in Kiev in 1970, there was an acceptance in some parts of the high-energy physics community of the view that the proton is composed of pointlike constituents. At that time we were reasonably convinced that we were seeing constituent structure in our experimental results, and afterwards our group directed its efforts to trying to identify these constituents and making comparisons with the last remaining competing models.

The electron-scattering results that played a crucial role in identifying the constituents of protons and neutrons or that ruled out competing models were the following:

Measurement of R

At the Fourth International Symposium on Electron and Photon Interactions at High Energies held in Liverpool in 1969, MIT–SLAC results were presented that showed that R was small and was consistent with being independent of q^2. The subsequent measurements, which decreased the errors, were consistent with this behavior.[38]

The experimental result that R was small for the proton and neutron at large values of q^2 and ν required that the constituents responsible for the scattering have spin-$\frac{1}{2}$, as was pointed out by Curtis Callan and David Gross.[39] These results ruled out pions as constituents but were consistent with the constituents being quarks or bare protons.

The ratio σ_n/σ_p

As discussed in a previous section, σ_n/σ_p decreased from 1 at about $x = 0$ to 0.3 in the neighborhood of $x = 0.85$. The ratio σ_n/σ_p is equivalent to W_2^n/W_2^p for $R_p = R_n$. In the quark model a lower bound of 0.25 is imposed on W_2^n/W_2^p. While the experimental values approached and were consistent with this lower bound, Regge and resonance models had difficulty at large x, as they predicted values for the ratio of about 0.6 and 0.7, respectively, near $x = 1$, while pure diffractive models predicted 1.0. The relativistic parton model in which the partons were associated with bare nucleons and mesons predicted a result for W_2^n/W_2^p that fell to zero at $x = 1$ and was about 0.1 at $x = 0.85$, in disagreement with our results.

A quark model in which up and down quarks have identical momentum distributions would give a value of $W_2^n/W_2^p = 0.67$. Thus the small value observed experimentally requires a difference in these distributions and quark–quark correlations at low x. To get a ratio of 0.25, the lower limit of the quark model, only a down quark from the neutron and an up quark from the proton can contribute to the scattering at the value of x at which the limit occurs.

Sum rules

As previously discussed, several sum-rule predictions suggested pointlike structure in the nucleon. The experimental evaluations of the sum rule related to the average squared charge of the constituents were consistent with the fractional charge assignments of the quark model provided that half the nucleon's momentum is carried by gluons.

Early neutrino results

Deep-inelastic neutrino–nucleon scattering produced complementary information that provided stringent tests of the above interpretation. Since charged-current neutrino interactions with quarks were expected to be independent of quark charges but were hypothesized to depend on the quark momentum distributions in a manner similar to electrons, the ratio of electron and neutrino deep inelastic scattering was expected to depend on the quark charges, with the momentum distributions canceling out.

That is,

$$\frac{\frac{1}{2}\int [F_2^{\nu p}(x) + F_2^{\nu n}(x)]\, dx}{\frac{1}{2}\int [F_2^{e p}(x) + F_2^{e n}(x)]\, dx} = \frac{2}{Q_u^2 + Q_d^2} = \frac{18}{5},$$

where $\frac{1}{2}(F_2^{\nu p}(x) + F_2^{\nu n}(x))$ is the F_2 structure function obtained from neutrino–nucleon scattering from a target having an equal number of neutrons and protons. The integral of this neutrino structure function over x is equal to the total fraction of the nucleon's momentum carried by the constituents of the nucleon that interact with the neutrino. This

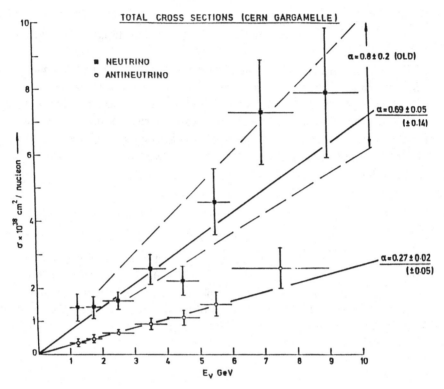

Fig. 32.5. The early Gargamelle measurements of neutrino–nucleon and antineutrino–nucleon cross sections as a function of neutrino energy.[40]

directly measures the fractional momentum carried by the quarks and antiquarks, because gluons are not expected to interact with neutrinos.

The first neutrino and antineutrino total cross sections were presented in 1972 at the Sixteenth International Conference on High Energy Physics held at Fermilab and the University of Chicago. The measurements were made at CERN using the large heavy-liquid bubble chamber Gargamelle. At this meeting Donald Perkins, who reported these results, stated that "... the preliminary data on the cross-sections provide an astonishing verification for the Gell-Mann/Zweig quark model of hadrons."[41]

These total cross section results, presented in Fig. 32.5, demonstrate a linear dependence on neutrino energy for both neutrinos and antineutrinos that is a consequence of Bjorken scaling of the structure functions in the deep-inelastic region. By combining the neutrino and antineutrino cross sections, the Gargamelle group was able to show that

$$\frac{1}{2} \int \left[F_2^{\nu p}(x) + F_2^{\nu n}(x) \right] dx =$$

$$\int x \left[u_p(x) + \bar{u}_p(x) + d_p(x) + \bar{d}_p(x) \right] dx = 0.49 \pm 0.07,$$

which confirmed the interpretation of the electron scattering results that suggested that the quarks and antiquarks carry only about half of the nucleon's momentum. When they compared this result with

$$\frac{1}{2} \int \left[F_2^{ep}(x) + F_2^{en}(x) \right] dx,$$

they found the ratio of neutrino and electron integrals was 3.4 ± 0.7 compared to the value predicted for the quark model, $\frac{18}{5} = 3.6$. This was a striking success for the quark model.

Within the next few years additional neutrino results solidified these conclusions. The results presented at the London Conference in 1974 demonstrated that the ratio $\frac{18}{5}$ was valid.[42] Figure 32.6, taken from Gargamelle data, shows a comparison of $F^{\nu N}(x)$ and $\frac{18}{5} F_2^{eN}(x)$, where $F_2^{\nu N}(x)$ and $F_2^{eN}(x)$ each represents an average of proton and neutron structure functions. In addition, the Gargamelle group evaluated the Gross–Llewellyn Smith sum rule for the F_3 structure function, which uniquely occurs in the general expressions for the inelastic neutrino and antineutrino nucleon cross sections as a consequence of parity violation in the weak interaction.[43] This sum rule states that

$$\int F_3^{\nu N}(x) dx = (\text{number of quarks}) - (\text{number of antiquarks}),$$

which equals 3 for a nucleon in the quark model. Obtaining values of $F_3^{\nu N}(x)$ from the differences of the neutrino and antineutrino cross sections, the Gargamelle group found the sum to be 3.2 ± 0.64, another significant success for the quark model.

General acceptance of quarks as nucleon constituents

After the 1974 London Conference, with its strong confirmation of the quark model, a general change of view developed with regard to the structure of hadrons. The bootstrap approach and the concept of nuclear democracy were in decline, and by the end of the 1970s the quark structure of hadrons became the dominant view for developing theory and planning experiments. A crucial element in this change was the general

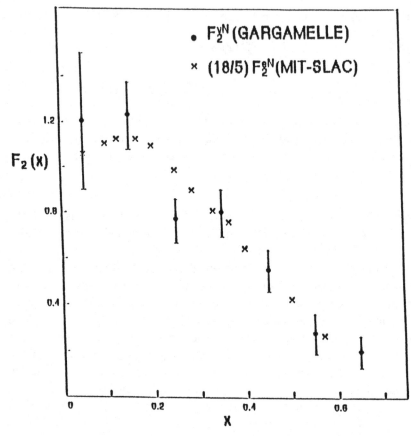

Fig. 32.6. Gargamelle measurements of $F_2^{\nu N}$ compared with $\left(\frac{18}{5}\right) F_2^{eN}$ calculated from the MIT–SLAC results.[44]

acceptance of QCD, which eliminated the last paradox, namely, why are there no free quarks?[45] The conjectured infrared slavery mechanism of QCD provided a reason to accept quarks as physical constituents without demanding the existence of free quarks. The asymptotic freedom property of QCD also provided a ready explanation of scaling, but logarithmic deviations from scaling were inescapable in this theory. While our later MIT–SLAC results had sufficient precision to establish small deviations from scaling, we did not have a wide enough energy range to verify their logarithmic behavior. This was later confirmed in higher energy muon and neutrino scattering experiments at FNAL and CERN. There were a number of other important experimental results reported in 1974 and in the latter half of the decade that provided further strong confirmations of the quark model. Among these were the discovery of charmonium, and its excited states, investigations of the total cross section for $e^+e^- \rightarrow hadrons$, and the discoveries of quark jets and gluon jets.[46] The quark model, with quark interactions described by QCD, became the accepted basis for the structure of hadrons. This picture, which is one of the foundations of the Standard Model, has not been contradicted by experimental evidence in the intervening years.

Notes

1 M. Gell-Mann, "The eightfold way: a theory of strong interaction symmetry," Caltech Synchrotron Laboratory Report No. CTSL-20 (1961); Y. Ne'eman, "Derivation of strong interactions from a gauge invariance," *Nucl. Phys. 26* (1961), pp. 222–9.

2 M. Gell-Mann, "A schematic model of baryons and mesons," *Phys. Lett. 8* (1964), pp. 214–15; G. Zweig, "An SU(3) model for strong interaction symmetry and its breaking," CERN preprint 8182/TH 401 (1964); "An SU(3) model for strong interaction symmetry and its breaking: II," CERN preprint 8419/TH 412 (1964).

3 R. E. Taylor, "Deep inelastic scattering: the early years," in *Les Prix Nobel 1990* (Stockholm: Almqvist & Wiksell, 1991), also in *Rev. Mod. Phys. 63* (1991), pp. 573–95; H. W. Kendall, "Deep inelastic scattering: experiments on the proton and the observation of scaling," in *Les Prix Nobel 1990*, also in *Rev. Mod. Phys. 63* (1991), pp. 597–614; J. I. Friedman, "Deep inelastic scattering: comparison with the quark model," in *Les Prix Nobel 1990*, also in *Rev. Mod. Phys. 63* (1991), pp. 615–27. Excerpts from the last publication are used in the present article.

4 The following physicists participated in the inelastic electron scattering experiments described in this paper: W. B. Atwood, E. Bloom, A. Bodek, M. Breidenbach, G. Buschhorn, R. Cottrell, D. Coward, H. DeStaebler, R. Ditzler, J. Drees, J. Elias, J. Friedman, G. Hartmann, C. Jordan, H. Kendall, M. Mestayer, G. Miller, L. Mo, H. Piel, J. Poucher, C. Prescott, E. M. Riordan, L. Rochester, D. Sherden, M. Sogard, S. Stein, R. Taylor, D. Trines, and R. Verdier. For additional acknowledgments, see J. I. Friedman, H. W. Kendall, and R. E. Taylor, "Deep inelastic scattering: acknowledgments," *Les Prix Nobel 1990*, also in *Rev. Mod. Phys. 63* (1991), p. 629. For the elastic scattering experiment, see D. H. Coward, et al., "Electron–proton elastic scattering at high momentum transfer," *Phys. Rev. Lett. 20* (1968), pp. 292–5.

5 W. K. H. Panofsky, "Low q^2 electrodynamics, elastic and inelastic electron (and muon) scattering," in J. Prentki and J. Steinberger, eds., *Proceedings of 14th International Conference on High Energy Physics* (Geneva: CERN, 1968), pp. 23–39. The experimental report, which I presented, is not published in the conference proceedings. It was, however, produced as a SLAC preprint and submitted as paper No. 563.

6 E. D. Bloom, et al., "High-energy inelastic e–p scattering at 6° and 10°," *Phys. Rev. Lett. 23* (1969), pp. 930–4; M. Breidenbach, et al., "Observed

behavior of highly inelastic electron–proton scattering," *Phys. Rev. Lett.* *23* (1969), pp. 935–41.

7 R. W. McAllister and R. Hofstadter, "Elastic scattering of 188-MeV electrons from the proton and the alpha particle," *Phys. Rev. 102* (1956), pp. 851–6.

8 The Mott scattering cross section $\sigma_{Mott} = (e^4/4E^2)(\cos^2 \frac{\theta}{2} / \sin^4 \frac{\theta}{2})$.

9 J. D. Bjorken, "Asymptotic sum rules at infinite momentum," *Phys. Rev. 179* (1969), pp. 1547–53; in a private communication, Bjorken told the MIT–SLAC group about scaling in 1968.

10 S. D. Drell and J. D. Walecka, "Electrodynamic processes with nuclear targets," *Ann. Phys. (NY) 28* (1964), pp. 18–33.

11 G. Miller, et al., "Inelastic electron–proton scattering at large momentum transfers and the inelastic structure functions of the nucleon," *Phys. Rev. D5* (1972), pp. 528–44.

12 W. K. H. Panofsky, "Low q^2 electrodynamics," p. 37.

13 J. D. Bjorken, "Current algebra at small distances," in J. Steinberger, ed., *Proceedings of the International School of Physics "Enrico Fermi," Course XLI* (New York: Academic Press, 1968), pp. 55–81.

14 M. Gell-Mann, "Symmetries of baryons and mesons," *Phys. Rev. 125* (1962), pp. 1067–84. For a review of current algebra, see: J. D. Bjorken and M. Nauenberg, "Current algebra," *Ann. Rev. Nucl. Sci. 18* (1968), pp. 229–64.

15 J. D. Bjorken, "Current algebra at small distances," p. 56.

16 S. L. Adler, "Sum rules giving tests of local current commutation relations in high-energy neutrino reactions," *Phys. Rev. 143* (1966), pp. 1144–55; J. D. Bjorken, "Inequality for electron and muon scattering from nucleons," *Phys. Rev. Lett. 16* (1966), pp. 408–10.

17 J. D. Bjorken, "Inequality for backward electron and muon nucleon scattering at high momentum transfer," *Phys. Rev. 163* (1967), pp. 1767–9.

18 H. D. Abarbanel, M. L. Goldberger, and S. B. Treiman, "Asymptotic properties of electroproduction structure functions," *Phys. Rev. Lett. 22* (1969), pp. 500–2; H. Harari, "Inelastic electron–nucleon scattering and the Pomeranchuk singularity," *Phys. Rev. Lett. 22* (1969), pp. 1078–81; "Duality, quarks, and inelastic electron–hadron scattering," *Phys. Rev. Lett. 24* (1970), pp. 286–90; T. Akiba, M. Sakuraoka, and T. Ebata, "Deep inelastic e–p scattering and a Reggeization scheme for photon processes," *Lett. Nuovo Cimento 4* (1970), pp. 1281–4; H. Pagels, "Regge model for inelastic lepton–nucleon scattering," *Phys. Rev. D3* (1971), pp. 1217–21; J. W. Moffat and V. G. Snell, "Regge model with scale-invariance for nucleon Compton scattering, electroproduction, and electron–positron annihilation," *Phys. Rev. D3* (1971), pp. 2848–61.

19 P. V. Landshoff and J. C. Polkinghorne, "The scaling law for deep inelastic scattering in a new Veneziano-like amplitude," *Nucl. Phys. B19* (1970), pp. 432–44.

20 G. Domokos, S. Kovesi-Domokos, and E. Shonberg, "Direct-channel resonance model of deep-inelastic electron scattering, I. Scattering on unpolarized targets," *Phys. Rev. D3* (1971), pp. 1184–90; "II. Spin dependence of the cross section," *Phys. Rev. D3* (1971), pp. 1191–5.

21 J. J. Sakurai, "Vector-meson dominance and high-energy electron–proton inelastic scattering," *Phys. Rev. Lett. 22* (1969), pp. 981–4.

22 For a review of the vector dominance and generalized vector dominance models, see T. H. Bauer, R. E. Spital, D. R. Yennie, and F. M. Pipkin, "The hadronic properties of the photon in high-energy interactions," *Rev. Mod. Phys. 50* (1978), pp. 261–436.

23 R. P. Feynman, "Very high-energy collisions of hadrons," *Phys. Rev. Lett. 23* (1969), pp. 1415–17; "The behavior of hadron collisions at extreme energies," in C. N. Yang, et al., eds., *Proceedings of the Third International Conference on High Energy Collisions* (New York: Gordon and Breach, 1969), pp. 237–58.

24 S. Drell, D. J. Levy, and T. M. Yan, "Theory of deep-inelastic lepton–nucleon scattering and lepton-pair annihilation processes. I," *Phys. Rev. 187* (1969), pp. 2159–71; and "II. Deep-inelastic electron scattering," *Phys. Rev. D1* (1970), pp. 1035–38; N. Cabbibo, G. Parisi, M. Testa, and A. Verganelakis, "Deep inelastic scattering and the nature of partons," *Lett. Nuovo Cimento 4* (1970), pp. 569–74; J. D. Bjorken and E. A. Paschos, "Inelastic electron–proton and γ–proton scattering and the structure of the nucleon," *Phys. Rev. 185* (1969), pp. 1975–82; J. Kuti and V. F. Weisskopf, "Inelastic lepton–nucleon scattering and lepton pair production in the relativistic quark–parton model," *Phys. Rev. D4* (1971), pp. 3418–39; P. V. Landschoff and J. C. Polkinghorne, "Partons and duality in deep inelastic lepton scattering," *Nucl. Phys. B28* (1971), pp. 240–52.

25 S. Drell, D. J. Levy, and T. M. Yan, "Theory of deep-inelastic," and "II. Deep-inelastic electron."

26 T. D. Lee and S. D. Drell, "Scaling properties and the bound-state nature of the physical nucleon," *Phys. Rev. D5* (1972), pp. 1738–63.

27 J. D. Bjorken and E. A. Paschos, "Inelastic electron–proton."

28 J. Kuti and V. F. Weisskopf, "Inelastic lepton–nucleon scattering."

29 E. D. Bloom, et al., "High-energy inelastic"; M. Breidenbach, et al., "Observed behavior."

30 G. Miller, et al., "Inelastic electron–proton scattering."

31 J. S. Poucher, et al., "High-energy single-arm inelastic e–p and e–d scattering at 6° and 10°," *Phys. Rev. Lett. 32* (1974), pp. 118–21.

32 A. Bodek, et al., "Comparisons of deep-inelastic e–p and e–n cross sections," *Phys. Rev. Lett. 30* (1973), pp. 1087–9; "The ratio of deep-inelastic e–n and e–p cross sections in the threshold region," *Phys. Lett. 51B* (1974), pp. 417–20; "Experimental studies of the neutron and proton electromagnetic structure functions," *Phys. Rev. D20* (1979), pp. 1471–1552; E. M. Riordan, et al., "Extraction of $R = \sigma_L/\sigma_T$ from deep inelastic e–p and e–d cross sections," *Phys. Rev. Lett. 33* (1974), pp. 561–4; "Tests of scaling of the proton electromagnetic structure functions," *Phys. Lett. 52B* (1974), pp. 249–52; W. B. Atwood, et al., "Inelastic electron scattering from hydrogen 50° and 60°," *Phys. Lett. 64B* (1976), pp. 479–82.

33 A. Bodek, "Comparisons of deep inelastic"; "The ratio of deep inelastic"; "Experimental studies."

34 W. B. Atwood and G. B. West, "Extraction of asymptotic nucleon cross sections from deuterium data," *Phys. Rev. D7* (1973), pp. 773–83; A. Bodek, "Comment on the extraction of nucleon cross sections from deuterium data," *Phys. Rev. D8* (1973), pp. 2331–4.

35 C. G. Callan and D. J. Gross, "Crucial test of a theory of currents," *Phys. Rev. Lett. 21* (1968), pp. 311–13.
36 K. Gottfried, "Sum rule for high-energy electron-proton scattering," *Phys. Rev. Lett. 18* (1967), pp. 1174–77.
37 J. Kuti and V. F. Weisskopf, "Inelastic lepton–nucleon scattering."
38 A. Bodek, et al., "Comparisons of deep-inelastic"; "The ratio of deep-inelastic"; "Experimental studies"; E. M. Riordan, et al., "Extraction of $R = \sigma_L/\sigma_T$"; "Tests of scaling."
39 C. G. Callan and D. J. Gross, "High-energy electroproduction and the constitution of the electric current," *Phys. Rev. Lett. 22* (1969), p. 156.
40 Ibid.
41 D. H. Perkins, "Neutrino interactions," in J. D. Jackson, A. Roberts, and R. Donaldson, eds., *Proceedings of the XVI International Conference on High Energy Physics* (Batavia, IL: National Accelerator Laboratory, 1972), Vol. 4, pp. 189–247, on p. 189.
42 M. Haguenauer, "'Gargamelle' experiment," pp. IV-95–100; F. Sciulli, "Caltech–Fermilab experiment," pp. IV-105–9; D. C. Cundy, "Neutrino physics," pp. IV-131–48; all in J. R. Smith, ed., *Proceedings of the XVII International Conference on High Energy Physics* (Chilton, UK: Science Research Council, 1974).
43 D. J. Gross and C. H. Llewellyn Smith, "High-energy neutrino-nucleon scattering, current algebra and partons," *Nucl. Phys. B14* (1969), pp. 337–47.
44 M. Haguenauer, "'Gargamelle' experiment," p. IV-96.
45 D. J. Gross and F. Wilczek, "Ultraviolet behavior of non-Abelian gauge theories," *Phys. Rev. Lett. 30* (1973), pp. 1343–5; H. D. Politzer, "Reliable perturbative results for strong interactions," *Phys. Rev. Lett. 30* (1973), pp. 1346–9.
46 J. J. Aubert, et al., "Experimental observation of a heavy particle J," *Phys. Rev. Lett. 33* (1974), pp. 1404–6; J. E. Augustin, et al., "Discovery of a narrow resonance in e^+e^- annihilation," *Phys. Rev. Lett. 33* (1974), pp. 1406–8. For a compendium of reprints and references covering the discoveries of the J/Ψ and its excited states, see R. N. Cahn and G. Goldhaber, *The Experimental Foundations of Particle Physics* (Cambridge: Cambridge University Press, 1989), pp. 257–78. For a review of these early results, see R. F. Schwitters and K. Strauch, "The physics of e^+e^- collisions," *Ann. Rev. Nucl. Sci. 26* (1976), pp. 89–149. G. Hanson, et al., "Evidence for jet structure in hadron production by e^+e^- annihilation," *Phys. Rev. Lett. 35* (1975), pp. 1609–12. For referenced reviews of the early gluon jet data, see P. Duinker and D. Luckey, "In search of gluons," *Comm. Nucl. Part. Phys. 9* (1980), pp. 123–39; P. Söding and G. Wolf, "Experimental evidence on QCD," *Ann. Rev. Nucl. Part. Sci. 31* (1981), pp. 231–93.

33

Deep-Inelastic Scattering: From Current Algebra to Partons

JAMES BJORKEN

Born Chicago, Illinois, 1934; Ph.D., 1959 (physics), Stanford University; Staff Physicist at the Stanford Linear Accelerator Center; high-energy physics (theory and experiment).

I begin with a disclaimer: what follows is subjective recollection, with no serious attempt of setting down an objective history. I also limit the scope of my remarks to the period roughly from 1966 to 1971. This period can in turn be divided in two parts – BF (Before Feynman) and AF (After Feynman).

Before Feynman

The climate in the beginning of this period was very different from now. David Gross has quite accurately and eloquently described it in Chapter 11, and I need not elaborate it very much here again. Field theory for the strong and weak interactions was not trusted. The emphasis was on observables, in close analogy to the Heisenberg matrix mechanics that heralded the golden age of quantum mechanics in the late 1920s. Local fields for strongly interacting particles were simply too far away from observations to be regarded as reliable descriptive elements. It was Murray Gell-Mann's great contribution to identify the totality of the matrix elements of electroweak currents between hadron states as operationally defined descriptive elements, upon which one could base a phenomenology with a lot of predictive power.

As did matrix mechanics, Gell-Mann's current algebra allowed the construction of sum rules based upon equal-time commutation relations of the electroweak currents with each other.[1] The idea was picked up by Sergio Fubini and his collaborators, who greatly extended what Gell-Mann had started, and then by Stephen Adler and William Weisberger, who produced one of the most important and celebrated results of the period.[2]

Adler went on to explore with great thoroughness the consequences of the current-algebra/sum-rule approach for neutrino reactions. His work provided a most important basis for what was to follow when the ideas were applied to electroproduction. However, most of the extant neutrino data and ancillary electroproduction phenomenology was at relatively low energy. While there were sum rules, derived by Adler, for what is now known as the deep-inelastic region of high energies and momentum transfer, most of his – and others' – attention was consumed by the region for which data existed.[3]

At this time I was at SLAC, in the midst of the construction of the first truly geographical accelerator, and it was natural to concentrate my efforts on the electron-scattering opportunities the new machine presented, especially since many of my close friends and colleagues were preparing to do those experiments. At the outset it was clear to the wise heads at Stanford that the SLAC linac should be an ideal tool for observing the instantaneous charge distribution inside the proton via inelastic scattering. There was a long tradition at Stanford of using inelastic electron–nucleus scattering and sum-rule techniques to study the constituent nucleons.[4] The extension of such ideas to search for constituents of the nucleon itself was natural to consider. I recall Leonard Schiff, in a colloquium devoted to the first announcement of Project M to the physics department as a whole, describing this opportunity – in particular showing that the energy- and momentum-transfer scales were more than adequate for seeing the insides of the proton. The problem of applying these ideas to the electron-scattering program was not conceptual but technical: for the first time control of highly relativistic kinematics was essential, and this was not easy to do. It would take some time to realize that one had to fully commit to the opposite extreme of highly relativistic motion in order to retrieve the non-relativistic intuition.

Local current algebra techniques were relativistic and provided quite solid ground for getting going. Adler's sum rules admitted corollaries for the electroproduction channel, and new techniques for developing asymptotic sum rules (now superseded by the operator-product expansion methodology) appeared that also allowed new insights into electroweak radiative corrections (they could be shown to diverge).[5] All of this machinery had a pointlike flavor to it. The most trustworthy results were those that were valid for free pointlike particles; indeed, the acid test for a good sum rule was that it worked in the case of free particles. And the locality assumption of the charge and current densities, and of their commutation relations, was literally axiomatic, doubted by few theorists for good reason. It was this locality assumption that led to these "pointlike" consequences. Finally, in this period Richard Feynman, Gell-Mann, and George Zweig boldly conjectured free-field behavior for the commutators of the space components of the electroweak currents.[6] This was venturing out onto thin ice, but their hypothesis led to strong predictions, for better or worse. Remarkably, their guess has essentially survived to the present day.

Nevertheless the sum-rule/current-algebra machinery was not sufficient for the local SLAC problem. The most important new consideration was to think about the actual size and shape of the well-known structure functions as a function of their by now well-known arguments ν and q^2. (The kinematic formalism had been around for quite a while, although those who had digested it and gotten it in their bones were a very small subculture.)[7]

Why was this next step necessary? It was important because one needed to know the ranges of ν needed for convergence of the sums and which values of q^2 were necessary for the asymptotic sum rules to be relevant experimentally. It would be of little use to have all these formal results were there not the expectation that they would be applicable at the SLAC energy scale.

For me there was another motivation, which came from Utah. There I learned, thanks to a series of most pleasant visits, of the underground neutrino detector being built by Jack Keuffel and his group. It was an impressively big enterprise. After seeing it installed under the Park City ski area in all its glory, the Fermilab neutrino detectors built later looked disappointingly small. The Utah detector included magnetic analysis of the upward-going muons emerging from the rock into the device. The energy scale went up to a TeV, so it was appropriate to try to apply the consequences of the Adler sum rules and all that to their upcoming experimental program. However, to do that again required detailed assumptions of the behavior of the structure functions with respect to ν and q^2. Jumping ahead a little in the story, it turned out that the scaling assumption led to a strong prediction of roughly equal numbers of muons per factor 10 in momentum, something that could be checked with quite small statistics and that was sensitive to the pointlike assumption (scaling behavior), which was the input. As it turned out, their results appeared only shortly before the Fermilab neutrino program got under way, and the number of observed muons was only five (although one was above 100 GeV momentum).[8]

And before going on, I here want to acknowledge the great debt I owe to Jack. He was a most talented physicist, from whom I learned an enormous amount of physics. And not only physics. It was physics on the way up the lifts at Alta, and even more important things in deep powder on the way down. It is regrettable that he was taken from us so early on.

In thinking about the size and shape of the structure functions, the only real technique available was common sense – by which I mean basically guesswork: what option looked most reasonable? It was a method that I greatly distrusted at the time, although by now I'm willing to use it, for better or worse, more freely.[9] But formal training does not encourage such thinking. Theoretical physics is traditionally an if–then linear logical exercise, at least when it gets to the publication level, not to mention the pedagogical level. In this case the "if" was often not too credible, and there was very little way to get to the "then."

My common-sense guess of the structure functions was not so bad. Threshold behavior at large x I guessed to be $(1 - x)^3$, using Bloom–Gilman duality ideas, developed because of the insistent queries of Jerry Friedman, who wanted to understand the interplay of what I was talking about in the continuum with what was going on in the resonance region.[10] The small-x region was controlled by Regge asymptotics, a subject for which much expertise was available locally, thanks to the presence of Fred Gilman and Haim Harari.[11] The value of ν at a given q^2 for attaining this Regge region could be guessed using the kinematics of coherent production, something that was a specialty of Sam Berman and Sid Drell.[12] The answer was $x \ll \frac{1}{3}$. The region of moderate x was therefore the region in which to expect saturation of the sum rules – the valence region. This in turn implied (approximate) scaling. The area under the structure function was determined by the sum rule. Finally spin-$\frac{1}{2}$ fields building the electroweak currents seemed most reasonable, and so I guessed the longitudinal structure function would be unimportant. This led to the correct expectation for the y-distribution as well as a cartoon of the structure function (Fig. 33.1) which is fine – except for the normalization, a factor 2 to 3 too large.

In addition, all this was consistent with a simple constituent picture. I had some of the infinite-momentum ideology, but lacked a lot also.[13] And in any case, I felt the constituent picture to be a very unreliable and naive viewpoint compared to the purportedly more solidly based

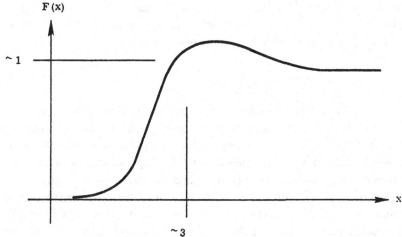

Fig. 33.1. A sketch of the conjectured structure function I made in 1967 prior to the SLAC experiments.

picture using sum rules, Regge asymptotics, and the like – despite the fact that pointlike constituents could be more easily explained to the experimentalists. I also looked at the problem using formal methods, one of which was published, others of which were discarded.[14] But though they had the patina of respectability, I think it fair to say that they played a lesser role than what I have described.

While I now look back with considerable satisfaction at all this, I must most emphatically add that at the time I didn't have much confidence in what was basically a lot of guesswork. Was this in fact legitimate theoretical physics? It was not clear at all. And to compound the situation, there was no solid basis even for the sum rules – there were loopholes of a technical nature in all those relevant to the deep inelastic region. If there was a fixed pole at $J = 1$ in a certain unobservable sub-amplitude for forward virtual Compton scattering, Adler's sum rule died. And the asymptotic sum rules rested on the fragile assumptions of Feynman, Gell-Mann, and Zweig. I not only worked on the scaling ideas but also published a paper with Richard Brandt detailing what might be done were all deep-inelastic structure functions to vanish in the scaling limit.[15]

Indeed the existing theoretical climate was strongly conditioned by what I called VD – vector dominance. It basically anticipated a short-distance behavior that was softer by a power of q^2. It was most elegantly canonized by the field-algebra formalism of T. D. Lee, Steven Weinberg, and Bruno Zumino.[16] Later on it was carried forward vigorously by Jun Sakurai.[17] So there was no imperative, for me or any other theorist at the time (BF), that the scaling limit need be nontrivial.

After Feynman

The story of the deep-inelastic experiments has been told by now in many places. As the raw data from the experiment started rolling in, there were leaks into the theory group. So those of us who were interested in this topic had some idea which way the wind was blowing. However, there was uncertainty regarding the effect electromagnetic radiative corrections would have on the final numbers; these were a major concern to us all.[18] Nevertheless, I kept my own tally of what I thought F_2 was doing, on a map of the ν, q^2 plane (assuming the longitudinal scattering to be small).

Finally Henry Kendall came by with a batch of more official data, and I knew what the next step should be. Henry went away, had his Transcendental Revelation, and returned with considerable excitement.[19] I must say I never had such a moment, and felt at the time Henry overreacted a little. For me everything moved slowly and steadily, one small advance after another, over a time span of years. Anyway, it was not long before Pief Panofsky presented this analysis in Vienna.[20]

Up to this point the business of guessing the size and shape of the structure function was a quite lonely enterprise. To this day I do not know if anyone else was even working on this problem. Interest in it grew after Pief's talk, but to my recollection remained at best moderate until Feynman entered the picture. Thereafter it became more widespread than the current algebra enterprise that preceded these developments.

The story of Feynman's visit to SLAC has also been recounted in many places.[21] I can add here very little, since I was away except for the last few hours he was around. Our communication was confused; he didn't have the perspective on the problem I described above. From my point of view, the way he described the infinite-momentum constituent picture so familiar now was somewhat foreign, and seemingly naive. Retrospectively, there was nothing naive about it. I was hampered by my own flawed version of the constituent viewpoint, where for half of the argument I would use infinite-momentum thinking, and for the other half retreat to the proton rest frame.[22]

Feynman energized the theorists as well as the experimental community. A variety of hypotheses were put forward to interpret the data.

Main lines of development, other than the parton model (which had more than one incarnation), included light-cone current algebra, vector dominance, and diffraction dominance.[23]

Almost immediately after the scaling results, it was recognized by Boris Ioffe and independently by others – including Feynman – that at small x the important regions of space–time contributing to the deep-inelastic phenomenon were near the light cone, but involved large longitudinal distances in the laboratory frame.[24] This evolved into an elaborate extension of the original Gell-Mann current algebra, one that served as a rather model-independent, albeit cumbersome, descriptive tool.[25] It is impressive how this very respectable approach has largely disappeared from sight in favor of the simple, yet less rigorous, parton approach. Nevertheless, in the AF period it enjoyed great popularity.

Also important to mention here are the beautiful, deep contributions of Ken Wilson and others on the operator-product expansion, which together with the light-cone formalism provided a solid theoretical basis for dealing with structure functions and their moments.[26] It provided many of the tools appropriate for the subsequent developments of asymptotic freedom and scaling violations, as described in Chapter 11.

As mentioned already, the main feature of the vector dominance approach was a predicted nonscaling of the structure functions. The apparent scaling in the data was blamed on a large longitudinal contribution at small x.

Diffraction dominance exploited the fact that the shape of the measured structure function looked rather unimpressive.[27] No quasi-elastic peak, which might have been anticipated from historical examples of electron-nucleus scattering, appeared in the data, and it was a reasonable hypothesis that only "Pomeron exchange" contributed in the scaling limit. The test of this idea was electron–neutron scattering, which should have been identical to electron–proton scattering.

With time the data decided everything. Nevertheless, the parton picture, both for me and for Feynman, remained a precarious matter. It took a long time before confidence in it became strong.[28] For me, the biggest problem was the size of the structure function. I suppose that I had a hard time shrugging off my early guess.[29] The small value of F_2 nowadays is attributed to the fact that half the nucleon momentum is carried by neutral gluons. I was very slow on picking this up; an early, clear statement was provided by Kuti and Weisskopf.[30]

Putting gluons in the parton distribution was suggestive of a field theoretic origin of the strong force. On this I – and I believe Feynman as well – remained agnostic for a long time.[31]   There were strong warnings that renormalizable field theories and deep-inelastic scaling did not peacefully coexist.[32] This did not bother me because I had no commitment to a field theory of the strong force – that was my upbringing. And during this period string theories came into vogue: maybe the strong interactions were mediated by strings. The parton content of a string was a very negotiable concept.

The question of quark quantum numbers for the partons was one that was settled once and for all by neutrino data. From my point of view, quarks were natural from the start, because they helped make F_2 small; even with the fractional charge it wasn't small enough for me.

Some final remarks

This has been a very personal view of a subject in which the real contributions do not belong to me. I am grateful for having had the opportunity to contribute, however small the contribution has been. It is my credo that technological advances drive the progress in experimental physics and that experiments in turn drive the theory. Without those ingredients, the most brilliant theoretical constructs languish worthlessly. There is in my opinion no greater calling for a theorist than to help advance the experiments. It is not an easy thing to do. In the case of the deep-inelastic experiments, I feel my real contribution was to do this. Even with the advantage of close personal friendships, it was not easy to generate commitment to an enterprise that led into such unfamiliar territory. Since then I have found it no easier to do, and lately considerably harder. Nevertheless I am not quite ready to give up.

Notes

1 M. Gell-Mann, "Symmetries of baryons and mesons," *Phys. Rev. 125* (1962), pp. 1067–84; "The symmetry group of vector and axial vector currents," *Physics 1* (1964), pp. 63–75.

2 S. Fubini, G. Furlan, and A. Rossetti, "A dispersion theory of symmetry breaking," *Nuovo Cimento 40* (1965), pp. 1171–93; "Nucleon magnetic moments and photoproduction sum rules," *Nuovo Cimento 43A* (1966), pp. 161–70; S. Adler, "Calculation of the axial-vector coupling constant renormalization in β-decay," *Phys. Rev. Letters 14* (1965), pp. 1051–5; W. Weisberger, "Renormalization of the weak axial-vector coupling constant," *Phys. Rev. Lett. 14* (1965), pp. 1047–51.

3 S. Adler, "Sum rules giving tests of local current commutation relations in high-energy neutrino reactions," *Phys. Rev. 143* (1966), pp. 1144–55. For reviews of current algebra, see S. Adler and R. Dashen, *Current*

Algebras and Applications to Particle Physics (New York: W. A. Benjamin, 1968); also J. Bjorken and M. Nauenberg, "Current algebra," *Ann. Rev. Nucl. Sci. 18* (1968), pp. 229–64.

4 S. Drell and C. Schwartz, "Sum rules for inelastic electron scattering," *Phys. Rev. 112* (1958), pp. 568–79.

5 J. D. Bjorken, "Inequality for electron and muon scattering from nucleons," *Phys. Rev. Lett. 16* (1966), p. 408; J. D. Bjorken, "Applications of the chiral $U(6) \otimes U(6)$ algebra of current densities," *Phys. Rev. 148* (1966), pp. 1467–78; K. Johnson and F. Low, "Current algebras in a simple model," *Prog. Theor. Phys. Suppl. 37–38* (1966), pp. 74–93.

6 R. P. Feynman, M. Gell-Mann, and G. Zweig, "Group $U(6) \otimes U(6)$ generated by current components," *Phys. Rev. Lett. 13* (1964), pp. 678–80.

7 L. Hand, "Experimental investigation of pion electroproduction," *Phys. Rev. 129* (1963), pp. 1834–46; S. Drell and J. D. Walecka, "Electrodynamic processes with nuclear targets," *Ann. Phys. (N.Y.) 28* (1964), pp. 18–33.

8 H. Bergeson, G. Cassiday, and M. Hendricks, "Neutrino-induced muons deep underground," *Phys. Rev. Lett. 31* (1973), pp. 66–70.

9 While this is somewhat off the subject, let me offer an example of common-sense thinking – it is an argument that the e^+e^- ratio R should neither vanish nor become infinite at high energies: Using CVC we relate R to W decay. If R is very small then the leptonic branching ratio of W is nearly 100%, too good to be true from an experimental standpoint. If R is very large, hadroproduction in e^+e^- collisions is enormous, again too good to be true.

10 E. Bloom and F. Gilman, "Scaling, duality, and the behavior of resonances in inelastic electron–proton scattering," *Phys. Rev. Lett. 25* (1970), pp. 1140–3.

11 F. Gilman and H. Harari, "Strong-interaction sum rules for pion-hadron scattering," *Phys. Rev. 165* (1968), pp. 1803–29.

12 J. Bjorken, "Theoretical ideas on inelastic electron and muon scattering," *Proceedings of the 1967 International Symposium on Electron and Photon Interactions at High Energies* (Stanford, California, September 5–9, 1967), pp. 109–27; S. Berman and S. D. Drell, "Speculations on the production of vector mesons," *Phys. Rev. 133B* (1964), pp. 791–801.

13 For some reason I thought in terms of an effective mass for the constituent as seen in the rest frame, not the momentum fraction at infinite momentum. See J. D. Bjorken, "Current algebra at small distances," in J. Steinberger, ed., *Proceedings of the International School of Physics "Enrico Fermi," Course 41, 1967, "Selected Topics in Particle Physics"* (New York: Academic Press, 1968), pp. 55–81.

14 J. D. Bjorken, "Asymptotic sum rules at infinite momentum," *Phys. Rev. 179* (1969), pp. 1547–53. This work built on the contributions of M. Cornwall and R. Norton, "Current commutators and electron scattering at high momentum transfer," *Phys. Rev. 177* (1969), pp. 2584–6. The only complaint I have had regarding Feynman is the following: In his book *Photon–Hadron Interactions* (Reading, Mass.: W. A. Benjamin, 1972) he discusses these formal methods, including one involving the DGS (Deser–Gilbert–Sudarshan) and Dyson representations for current

commutators. He comments (p. 186), "I believe it was this argument with the Dyson representation which either led Bjorken to his scaling hypothesis, or helped confirm his suspicions of its truth." Not so, and quite the opposite. I did know that material, but found that no matter what the data turned out to be, a weight function could be found that was consistent with it. In other words, the method led to no experimental implications for deep-inelastic processes. I was unhappy that Feynman would think I would rely on such a formal methodology.

15 J. D. Bjorken and R. Brandt, "Minimal current algebra," *Phys. Rev. 177* (1969), pp. 2331–6.

16 T. D. Lee, S. Weinberg, and B. Zumino, "Algebra of fields," *Phys. Rev. Lett. 18* (1967), pp. 1029–32.

17 J. J. Sakurai, "Vector-meson dominance and high energy electron-proton inelastic scattering," *Phys. Rev. Lett. 22* (1969), pp. 981–4.

18 See for example Fig. 4 in note 15.

19 "This author ... recalls wondering how Balmer may have felt when he saw, for the first time, the striking agreement of the formula that bears his name with the measured wavelengths of the atomic spectra of hydrogen," H. W. Kendall, "Deep inelastic scattering: experiments on the proton and the observation of scaling," *Rev. Mod. Phys. 63* (1991), pp. 597–614.

20 W. K. H. Panofsky, "Low q^2 electrodynamics, elastic and inelastic electron (and muon) scattering," in J. Prentki and J. Steinberger, eds., *Proceedings of the 14th International Conference on High Energy Physics* (Geneva: CERN, 1968), pp. 23–39.

21 M. Riordan, *The Hunting of the Quark* (New York: Simon & Schuster, 1987), pp. 148–54. See also J. I. Friedman, "Deep inelastic scattering: Comparison with the quark model," *Rev. Mod. Phys. 63* (1991), pp. 615–27; H. W. Kendall, "Deep inelastic scattering"; R. E. Taylor, "Deep inelastic scattering: The early years," *Rev. Mod. Phys. 63* (1991), pp. 573–95, and references therein.

22 J. D. Bjorken and M. Nauenberg, "Current algebra."

23 J. Bjorken and E. Paschos, "Inelastic electron–proton and γ–proton scattering and the structure of the nucleon," *Phys. Rev. 185* (1969), pp. 1975–82. S. Drell, D. Levy and T. M. Yan, "Theory of deep-inelastic lepton-nucleon scattering and lepton-pair annihilation processes. I.," *Phys. Rev. 187* (1969), pp. 2159–71; P. Landshoff and J. Polkinghorne, "Partons and duality in deep inelastic lepton scattering," *Nucl. Phys. B28* (1971), pp. 240–52.

24 B. L. Ioffe, "Space-time picture of photon and neutrino scattering and electroproduction cross section asymptotics," *Phys. Lett. 30B* (1969), pp. 123–25.

25 H. Fritzsch and M. Gell-Mann, "Light cone current algebra," in E. Gotsman, ed., *Proceedings of the International Conference on Duality and Symmetry in Hadron Physics* (Tel Aviv: Weizmann Science Press, 1971).

26 K. Wilson, "Non-Lagrangian models of current algebra," *Phys. Rev. 179* (1969), pp. 1499–1512; S. Ciccariello et al., "Broken scale invariance and symptotic sum rules," *Phys. Lett. 30B* (1969), pp. 546–8; G. Mack, "Universality of deep inelastic lepton–hadron scattering," *Phys. Rev. Lett. 25* (1970), pp. 400–1.

27 H. Abarbanel, M. Goldberger, and S. Treiman, "Asymptotic properties of

electroproduction structure functions," *Phys. Rev. Lett.* *22* (1969), pp. 500–2; H. Harari, "Inelastic electron–nucleon scattering and the Pomeranchuk singularity," *Phys. Rev. Lett.* *22* (1969), pp. 1078–81.

28 In 1972 Feynman wrote, "We have built a very tall house of cards making so many weakly based conjectures one upon the other and a great deal may be wrong." R. P. Feynman, *Photon–Hadron Interactions*, p. 268.

29 As late as 1972 I was still questioning the Adler sum rule: J. D. Bjorken and S. F. Tuan, "Is the Adler sum rule for inelastic lepton–hadron processes correct?" *Comm. of Nucl. Part. Phys.* *5* (1972), pp. 71–8.

30 J. Kuti and V. Weisskopf, "Inelastic lepton-nucleon scattering and lepton pair production in the relativistic quark–parton model," *Phys. Rev. D4* (1971), pp. 3418–39.

31 See for example the discussion remarks of Feynman at Cornell in 1971: N. B. Mistry, ed., *Proceedings of the 1971 International Conference on Electron and Photon Interactions at High Energies* (Ithaca, NY: Cornell Laboratory for Nuclear Studies, 1972), pp. 296–7.

32 S. Adler and W. K. Tung, "Breakdown of asymptotic sum rules in perturbation theory," *Phys. Rev. Lett.* *22* (1969), pp. 978–81; R. Jackiw and G. Preparata, "T products at high energy and commutators," *Phys. Rev.* *185* (1969), pp. 1929–40.

34

Hadron Jets and the Discovery of the Gluon*

SAU LAN WU

Born Hong Kong; Ph.D., 1970 (physics), Harvard University; Enrico Fermi Professor of Physics at the University of Wisconsin, Madison; high-energy physics (experimental).

Quarks were proposed in 1964 by Murray Gell-Mann and George Zweig to explain the multiplet structure of the observed hadrons.[1] Their experimental observation occurred a few years later, through inelastic electron–nucleon scattering at SLAC by Jerome Friedman, Henry Kendall, Richard Taylor, and their collaborators.[2]

Quarks have had many successes. For example, since the discovery of the upsilon in 1977 by Leon Lederman and his collaborators, there are five known quarks: u, d, c, s, and b.[3] Thus, for e^+e^- energies above the $b\bar{b}$ threshold but much below the rest mass of the Z boson, the production of all quark pairs is expected to be given by the Feynman diagram of Fig. 34.1(a), leading to the result

$$\frac{\sigma(e^+e^- \to u\bar{u}, d\bar{d}, c\bar{c}, s\bar{s}, b\bar{b})}{\sigma(e^+e^- \to \mu^+\mu^-)}$$

$$= 3\left[\left(\frac{2}{3}\right)^2 + \left(\frac{1}{3}\right)^2 + \left(\frac{2}{3}\right)^2 + \left(\frac{1}{3}\right)^2 + \left(\frac{1}{3}\right)^2\right]$$

$$= \frac{11}{3}, \tag{34.1}$$

where the overall common factor 3 is due to the number of colors for each of the five quarks. This is in good agreement with experimental results on total hadronic production cross sections.

In spite of these successes, the produced quark pairs of fractional charge have never been observed directly. This dilemma leads to the

* Work supported in part by Department of Energy contract DE-AC02-76ER00881.

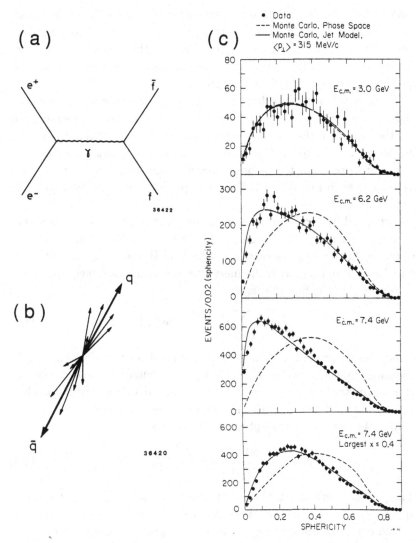

Fig. 34.1. (a) Feynman diagram for $e^+e^- \to f\bar{f}$, where the fermion f may be a quark or a lepton. (b) Schematic drawing for e^+e^- annihilation in the quark model – production of a $q\bar{q}$ pair followed by hadronization. (c) Sphericity distributions at various center-of-mass energies for hadronic events in which three or more particles are detected. These experimental data agree with the jet model. [Source, Hanson, et al., note 4.]

working hypothesis, due mainly to James Bjorken and Richard Feynman, that somehow the quarks turn into a group of hadrons through strong interactions. Another way of stating this hypothesis is that quarks are "confined," so that in the final state the quarks must combine with other quarks or antiquarks to form hadrons.

Independent of the mechanism for turning quarks into hadrons, the hadrons are expected to retain some memory of the quark momentum. In other words, if the quarks are produced in the x-direction, the resulting hadrons are expected to have, on the average, larger momenta in the x-direction than the y- or z-directions, especially at high energies, as illustrated in Fig. 34.1(b). From this point of view, the occurrence of jets is natural.

Motivated by such considerations, the jet structure in e^+e^- annihilation was first sought and found at SPEAR by the SLAC–LBL collaboration, with the analysis carried out by Gail Hanson.[4] Their method is based on an analogy with the inertia tensor in classical mechanics. For each event they define the tensor

$$T_{\alpha\beta} = \sum_i (\delta_{\alpha\beta} p_i^2 - p_{i\alpha} p_{i\beta}), \qquad (34.2)$$

where the summation is over all detected particles and α and β refer to the three spatial components of each particle momentum $\vec{p}_i$. Since $T_{\alpha\beta}$ is a symmetrical tensor, it can be diagonalized to give the eigenvalues $\lambda_1, \lambda_2,$ and λ_3, together with the normalized eigenvectors $\hat{n}_1, \hat{n}_2,$ and $\hat{n}_3$. If $\lambda_1 \geq \lambda_2 \geq \lambda_3$, then the sphericity S is defined as

$$S = \frac{3\lambda_3}{\lambda_1 + \lambda_2 + \lambda_3} = \frac{3(\sum_i p_{\perp i}^2)_{min}}{2\sum_i p_i^2}. \qquad (34.3)$$

The jet structure is established by studying the energy dependence of sphericity. The experimental results are given in Fig. 34.1(c), showing, as the center-of-mass energy increases, progressive deviation from the Monte Carlo simulation based on phase space. Jet structure is clearly evident at the center-of-mass energy of 7.4 GeV. The spin-$\frac{1}{2}$ behavior of quarks was verified by the observation of an azimuthal asymmetry in inclusive hadron production by Roy Schwitters and his collaborators, and also by the jet-axis angular distribution integrated over the azimuthal angle by Hanson and her collaborators.[5]

At the higher energies of PETRA and PEP, the jet structure can be seen much more directly; indeed, no analysis using $T_{\alpha\beta}$ is necessary, and jets are clearly identified by the naked eye on an event-by-event basis.

PETRA (Positron–Electron Tandem Ring Accelerator) is located in the German national laboratory Deutsches Elektronen-Synchrotron (DESY), in a suburb west of Hamburg, Germany. Established in 1959 under the direction of Willibald Jentschke, this laboratory has played a crucial role in the reemergence of Germany as one of the leading countries in physics.

The proposal for the project to construct PETRA was submitted to the government of the German Federal Republic in November 1974.[6] Due to the dedicated efforts of Herwig Schopper, the director of DESY at that time, approval was granted one year later. Shortly thereafter, on January 27, 1976, the foundation "stone," actually an aluminum vacuum chamber, was laid. Under the direction of Gustav Voss, the construction of PETRA proceeded very rapidly. The electron beam was first stored on July 15, 1978, more than nine months earlier than originally scheduled. In September 1978, collisions were first observed; a month later, three detectors, PLUTO, MARK J, and TASSO, were moved into place. On November 18, 1978, the first hadronic event was observed by PLUTO, at a center-of-mass energy of 13 GeV. JADE was moved into the beam in February 1979, and CELLO was moved in to replace PLUTO in March 1980.

At the time PETRA began operation, there were several theoretically proposed particles, including the gluon and the weak vector bosons W and Z.[7] The gluon of quantum chromodynamics is its Yang–Mills non-Abelian gauge particle.[8] Thus the strong interactions between quarks are mediated by the gluon, in much the same way as the electromagnetic interactions between electrons are mediated by the photon.

Shortly after the discovery of quarks, the inelastic electron scattering experiment at SLAC gave first evidence that there is something inside nucleons besides the three quarks. As explained in detail in this volume by Friedman, the inelastic structure functions W_1, W_2, and so on, are functions of two variables, ν and q^2, where ν is the energy lost by the electron and q^2 is the square of the invariant mass of the virtual photon emitted by the electron and absorbed by the target particle (the proton, for example). Bjorken scaling means that, instead of these two variables, for deep-inelastic processes there is actually only one variable $\omega = 2M\nu/q^2$, where M is the proton mass. If Bjorken scaling were exact, then there would be a sum rule relating an integral of W_2 to the sum of the squares of the quark charges in the proton, which equals $\frac{1}{3}$. However, experimental measurement of this structure function W_2 gave a value of only 0.16 to the integral in the sum rule.[9] Within the quark–

parton model, this discrepancy could be explained by postulating that the three quarks in the proton carry only about half of its total momentum, with the other half carried by something else, perhaps gluons.

In the years following the pioneering MIT–SLAC experiment, efforts were made to determine not only the quark distribution function, but also the gluon distribution function in nucleons. The very extensive neutrino scattering data from the BEBC and CDHS collaborations at CERN made it feasible to determine these distribution functions by comparison with what was expected from QCD, and it was found that the gluon distribution function was not small.[10] This information about the gluon is interesting but indirect, similar to the evidence for the Z provided by observation of the $\mu^+\mu^-$ asymmetry in electron–positron annihilation.[11] This summarizes what was known about the gluon at the time of PETRA's turn-on.

In 1977 I moved from MIT to the University of Wisconsin to become a faculty member, and at the same time joined the TASSO (Two-Arm Spectrometer SOlenoid) collaboration at PETRA. Besides taking an active part in the construction of the Čerenkov counters for the TASSO detector, I thought a lot about what physics to do at PETRA. Among the particles that were theoretically expected but not yet observed experimentally, the most interesting ones included the gluon, the W and the Z. They were especially interesting because they were new gauge particles. Up to this point, the only known gauge particle was the photon, predicted by Einstein in 1905 and observed by Compton in 1923.[12] Thus no new gauge particle had been discovered for more than fifty years. Furthermore, theoreticians predicted that these new gauge particles are fundamentally different from the photon in that they have self-interactions. Whichever way one looked at them, they were exciting objects.

Since PETRA did not have enough energy to produce the W or the Z, the only realistic possibility was the gluon.[13] One possible effect of the gluon would be the broadening of a quark jet in two-jet events as the center-of-mass energy increases. While this jet broadening is indeed an interesting effect, I felt that the discovery of the gluon required direct observation.

Since the gluon is the gauge particle for strong interactions, the simplest way to produce it is by the gluon *bremsstrahlung* process

$$e^+e^- \to q\bar{q}g \tag{34.4}$$

analogous to the photon *bremsstrahlung* process $e^+e^- \to \mu^+\mu^-\gamma$. Since the gluon, similar to the quark, was expected to hadronize into a jet, this process (Eq. 34.4) should lead to a three-jet event. How could I find such three-jet events?

One of the first worries was whether the e^+e^- center-of-mass energy at PETRA was sufficiently high to produce events where the three jets are clearly separated. No convincing arguments could be given one way or the other, but I was encouraged by my rough estimate that three times the SPEAR energy might be sufficient.[14] Since PETRA was expected to exceed this $3 \times 7.4 \sim 22$ GeV very soon, I decided to proceed on the assumption that the energy was high enough.

At that time several of my collaborators in TASSO were interested in the broadening of the jet as mentioned above. Although I was not aware of it then, being a newcomer to e^+e^- colliding-beam physics, both the jet broadening and the three-jet events had been discussed theoretically by John Ellis, Mary K. Gaillard, and Graham Ross.[15] The general feeling at that time, shared by both the theoreticians and many experimentalists, was that "the first observable effect should be a tendency for the two-jet cigars to be unexpectedly oblate.... Eventually, events with large p_T would have a three-jet structure."[16] I was fortunate in that my rough estimate steered me to the search for gluons through three-jet events as soon as PETRA became operational.

Even with the assumption that the PETRA energy was high enough to produce three-jet events with clearly separated jets, I made a number of false starts until I realized the power of the following simple observation. By energy-momentum conservation, the two jets in $e^+e^- \to q\bar{q}$ must be back-to-back. Similarly, the three jets in $e^+e^- \to q\bar{q}g$ must be coplanar. Therefore, the search for the three jets can be carried out in the *two-dimensional* event plane, the plane formed by the momenta of $q, \bar{q}$, and g. Figure 34.2 shows a page of my notes written in June 1978.

The importance of this observation is not due simply to the reduction of dimensionality, but to the qualitative difference between vectors in three-dimensional spaces and those in two-dimensional spaces. If you have a number of vectors in a three-dimensional space, there is no natural way to order them. By contrast, vectors in a plane can be naturally ordered cyclically. Thus, if the polar coordinates of N vectors $\vec{q}_j$ are (q_j, θ_j), then these $\vec{q}_j$ can be relabeled such that

$$0 \leq \theta_1 \leq \theta_2 \leq \theta_3 \leq \ldots \leq \theta_N < 2\pi. \tag{34.5}$$

With this cyclic ordering, the $\vec{q}_j$'s can be split up into three sets of *contiguous* vectors, and these three sets are to be identified as the three jets.

S.L. Wu
a June 1978.

Analysis of Jets

1. Two opposite jet
2. Two non-opposite jets
3. Planar case
4. Three jets

P9
4-9

The analysis of 3 jet events

A. With suitable χ^2 cuts, pick out the events from the region marked "Planar Distribution". For these events, T_3 is relatively small. Let $\hat{n}_3$ be ~~eigenvalue~~ eigenvector corresponding to T_3, and we project out this $\hat{n}_3$ component.

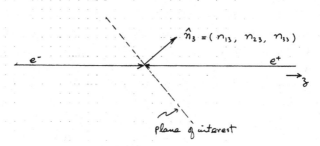

$$\hat{n}_3 = (n_{13}, n_{23}, n_{33})$$

e^- ——————— e^+

$\rightarrow z$

plane of interest

We choose n_{33} to be positive

Let the X-axis be determined by projecting the x-axis to the $\hat{n}_1 - \hat{n}_2$ plane.

$$\hat{X} = \frac{\hat{z} - \hat{n}_3 (\hat{x} \cdot \hat{n}_3)}{|\hat{x} - \hat{n}_3 (\hat{x} \cdot \hat{n}_3)|}$$

~~and~~ and the Y-axis by orthogonality

$$\hat{Y} = \hat{n}_3 \times \hat{X} = \frac{\hat{n}_3 \times \hat{z}}{|\hat{x} - \hat{n}_3 (\hat{x} \cdot \hat{n}_3)|}$$

Fig. 34.2. A page from my logbook notes of June 1978 on three-jet analysis.

There are of course a number of ways of carrying out this splitting, and, with suitable restrictions, the one with the smallest average transverse momentum is chosen as the best approximation to the correct way of identifying the three jets.

Once this basic point is realized, the search for three-jet events proceeds as follows. First, using either the tensor (Eq. 34.2) or equivalently the momentum tensor $M_{\alpha\beta} = \sum_j p_{j\alpha} p_{j\beta}$, the event plane is determined as the plane with the smallest transverse momentum. Then all the measured momenta of the produced particles are projected into this event plane. Using Eq. 34.5 above, rearrange these projected momenta $\vec{q}_j$ into a cyclic order. Split them into three contiguous sets and for each of these sets define a two-dimensional analogy of the momentum tensor above; let $\Lambda^{(\tau)}$ be the larger eigenvalue and $\hat{m}^{(\tau)}$ the corresponding normalized eigenvector. Since each jet can contain particles in one direction only, the signs of $\hat{m}^{(\tau)}$ can be chosen so that $\vec{q}_j \cdot \hat{m}^{(\tau)} > 0$ for each j in the corresponding set. For each way of splitting into three contiguous sets, calculate the sum of these three largest eigenvalues:

$$\Lambda(N_1, N_2, N_3) = \Lambda^{(1)} + \Lambda^{(2)} + \Lambda^{(3)}. \tag{34.6}$$

This $\Lambda(N_1, N_2, N_3)$ is maximized over all allowed ways of splitting into three contiguous sets. This maximizing partition gives the three jets, and the corresponding $\hat{m}^{(1)}, \hat{m}^{(2)}$, and $\hat{m}^{(3)}$ yield the directions of the jet axes.

In short, the event plane is used to put the projections of the measured momenta of the produced particles into cyclic order. For each way of splitting into three contiguous sets, the sum of the larger eigenvalues corresponding to the two-dimensional momentum tensors for the three sets is evaluated. The particular splitting with the largest value of this sum corresponds to the smallest average momentum transverse to the three axes and is therefore chosen as the way to identify the three jets. It is then straightforward to study the various properties of the jets, such as the average transverse momentum of each jet in three-jet events, and compare them with the corresponding properties in two-jet events.

This procedure has a number of desirable features. First, all three jet axes are determined, and they are in the same plane. This feature makes it easy to display any three-jet event, simply by projecting the observed momenta into this plane. Second, particle identification is not needed, since there is no Lorentz transformation. Third, the computer time is moderate even when all the measured momenta are used. Finally, it is not necessary to have the momenta of all the produced particles; it is only necessary to have at least one momentum from each of the three jets. Thus, for example, the procedure works well even when no neutral particles are included.

This last advantage is important, and it is the reason why this procedure is a good match to the TASSO detector at the time of PETRA turn-on. For the purpose of using this procedure, the most important part of the detector is the large drift chamber, which had a sensitive length of 3.23 m with inner and outer diameters of 0.73 and 2.44 m. There were 15 layers, nine with the sense wires parallel to the axis of the chamber and six with the sense wires oriented at an angle of approximately ±4°. These six layers made it possible to measure not only the transverse momenta of the produced charged particles but also their longitudinal momenta. This drift chamber was designed and constructed under the direction of Björn Wiik.

My procedure of identifying the three jets, programmed with the help of my postdoc Georg Zobernig, was ready before the turn-on of PETRA in September 1978. Shortly thereafter, I showed the procedure and the program to Wiik, and he was very excited about them. I presented my analysis method in a TASSO collaboration meeting and later had it published.[17]

In April 1979 there was a rumor that the PLUTO collaboration had discovered gluon jets from their DORIS data on upsilon decays. This group employed the method of energy flow, using the information from all the observed particles, including not only charged particles but also neutral particles. In this way, the thrust axis was determined. A suitable line-up according to the thrust axes allowed a summation over the events and led to a polar plot of energy flow. Their polar plot, Fig. 34.3, showed "a clear three-prong pattern," which was the basis of a rumor that PLUTO had observed the decay of the upsilon into three gluons.

However, shortly thereafter the PLUTO collaboration carried out for comparison a Monte Carlo calculation of this energy-flow plot assuming a phase-space distribution. It was found that the Monte Carlo energy-flow plot also showed a comparable three-prong pattern. Thus this three-prong pattern could not be taken as evidence for gluon jets.[18]

When we had obtained data for center-of-mass energies of 13 GeV and 17 GeV, Zobernig and I looked for three-jet events. It was not until just before the Bergen Conference in June 1979 that we started to obtain data at the center-of-mass energy of 27.4 GeV. Zobernig and I found one clear three-jet event from a total of 40 hadronic events at this center-of-mass energy. This first three-jet event of PETRA, as seen

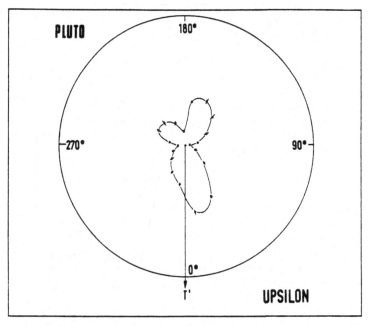

Fig. 34.3. Energy distribution from upsilon decay observed by the PLUTO Collaboration at the DORIS e^+e^- storage ring. Although the distribution shows a clear three-pronged pattern, it cannot be considered as evidence for the decay of the upsilon into three gluons. (See text and note 18.)

in the event plane, is shown in Fig. 34.4. When this event was found, Wiik had already left Hamburg to go to Bergen, so I took my display of this event to show him at his house near the city. It turned out that Ellis was there too; after seeing my event, he described this event as "gold-plated." During this weekend, I also telephoned Günter Wolf, the TASSO spokesman, at his home in Hamburg and told him of the finding. Wiik showed the event in his plenary talk and referred to me for questions. Donald Perkins took this offer and challenged me to show him all forty TASSO events. I did so, and, after we had spent some time together studying the events, he was convinced.

The following is quoted from Wiik's talk:[19]

If hard gluon bremsstrahlung is causing the large $p_\perp$ values in the plane then a small fraction of the events should display a three jet structure. The events were analyzed for a three jet structure using a method proposed by Wu and Zobernig.... A candidate for a three jet event, observed by the TASSO group at 27.4 GeV, is shown in Fig. 21 viewed

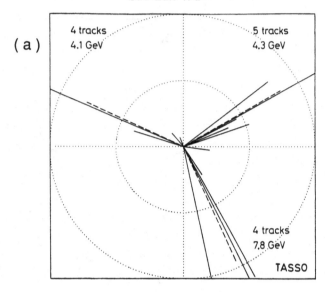

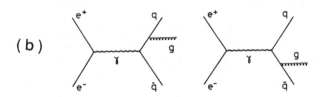

Fig. 34.4. (a) The first three-jet event from electron-positron annihilation, as viewed in the event plane. It has three well-separated jets. [Source, Wiik, note 19, and Wu and Zobernig, note 20.] (b) Feynman diagrams for the gluon *bremsstrahlung* process $e^+e^- \rightarrow q\bar{q}g$.

along the $\hat{n}_3$ direction. Note that the event has [three clear well separated jets] and is just not a widening of a jet.

As soon as I returned from Bergen, I wrote a TASSO note (Fig. 34.5) with Zobernig on the observation of this three-jet event.[20] Both in Wiik's talk and in this TASSO note, this three-jet event was already considered to be due to the hard gluon *bremsstrahlung* process (Eq. 34.4). As seen from Fig. 34.4, this first three-jet event had three clear, well-separated jets, and was considered to be more convincing than a good deal of statistical analysis. Indeed, before the question of statistical fluctuations

TASSO Note No. 84
26.6.1979

From Sau Lan Wu and Haimo Zobernig

On: A three-jet candidate (run 447 event 13177)

We have made a three jet analysis to all the hadronic candidates (43 events for $\Sigma|P_i| \geq 9$ GeV) of the May 1979 data at $E_{cm} = 27.4$ GeV using our method described in DESY 79/23 (A method of three jet analysis in e^+e^- annihilation).

Fig. 1 gives the triangular plot of the normalized eigenvalues Q_1, Q_2 and Q_3 ($Q_1 \leq Q_2 \leq Q_3$) of the momentum matrix

$$M_{\alpha\beta} = \sum_j (P_{j\alpha} \; P_{j\beta})$$

(See equation (1) and Fig. 1 of DESY 79/23). We find two three jet candidates

        run 447   event 13177

        run 439   event 12845

We then display each event on the 3 planes
        plane 1: normal to $\hat{n}_1$, the normalized eigenvector
        corresponding to Q_1. $\sum_i |P_{i\perp}|^2$

with respect to this plane is minimized.

        plane 2: normal to $\hat{n}_2$, the normalized eigenvector
        ~~to $\hat{n}_2$~~ corresponding to Q_2

        plane 3: normal to $\hat{n}_3$, the normalized eigenvector
        corresponding to Q_3.

Fig. 2 displays the three jet candidate (run 447 event 13177) on planes 1, 2, 3.

Fig. 3 displays plane 1 of this event in a blow up scale.

The axis for each of the three jets are found. Given the axes and $\Sigma|P_i|$ of each jet, the total energy of each jet is determined assuming the mass of each quark (or gluon) is zero.

Fig. 4 displays the event run 439 event 12845. This event looks like two charged jets and one neutral jet.

Fig. 34.5. The first page of TASSO Note No. 84, by Wu and Zobernig [note 20].

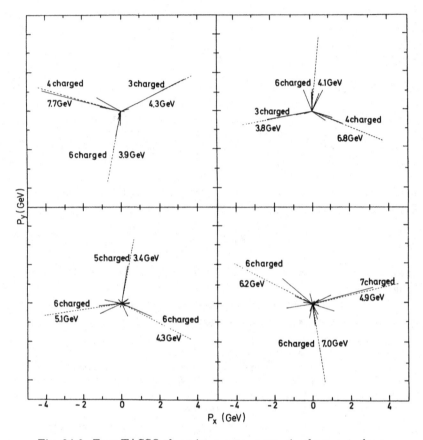

Fig. 34.6. Four TASSO three-jet events as seen in the event plane.

could be seriously raised, events from $E_{cm} = 27.4$ GeV rolled in and we found a number of other three-jet events.

Less than two weeks after the Bergen Conference, four of the TASSO three-jet events, as shown in Fig. 34.6, were shown by Paul Söding of DESY and TASSO at the European Physical Society Conference in Geneva.[21] Comparisons of event shape distributions with the QCD predictions were also included. Fig. 34.7 gives several plots of the various transverse momentum distributions. The first one is the distribution of

$$< p_\perp^2 >_{out} = \frac{1}{N} \sum_j (\vec{p}_j \cdot \hat{n}_1)^2 \qquad (34.7)$$

(the square of the momentum component normal to the event plane averaged over the charged particles in one event), while the second one

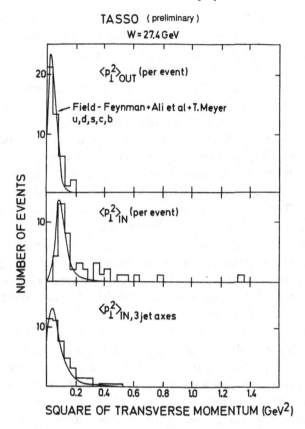

Fig. 34.7. Distribution of the average squared transverse momentum component out of the event plane (top), and in the event plane (middle), for the early TASSO events at the center-of-mass energy of 27.4 GeV (averaging over charged hadrons only). The curves represent $q\bar{q}$ jets without gluon *bremsstrahlung*. Comparison of these distributions gives evidence that broadening (compared to $q\bar{q}$ jets) occurs in one plane. The bottom figure shows $< p_\perp^2 >$ per jet when three jet axes are fitted, again compared with the two-jet model. [Source, Söding, note 21.]

is that of

$$< p_\perp^2 >_{in} = \frac{1}{N} \sum_j (\vec{p}_j \cdot \hat{n}_2)^2 \qquad (34.8)$$

(the square of the momentum component in the event plane and perpendicular to the sphericity axis averaged the same way). A comparison of these two plots shows that the major difference is the absence of a tail for $< p_\perp^2 >_{out}$ and the presence of one for $< p_\perp^2 >_{in}$. Since three-jet events tend to have a small $< p_\perp^2 >_{out}$ but a much larger $< p_\perp^2 >_{in}$, this distribution of $< p_\perp^2 >_{in}$ shows a continuous transition from two-jet events to three-jet events. Also shown in Fig. 34.7 is $< p_\perp^2 >_{in, \ 3 \ jet \ axes}$, which is defined the same way as Eq. 34.8 but, for each jet, the jet axis found by my method is used. The absence of a tail and the similarity to the first distribution means that the jets in three-jet events are similar to those in two-jet events, justifying the use of the same word "jet" in both cases. At the time of the EPS Conference in Geneva, no other experiment at PETRA was mentioning anything about three-jet events.

On July 31, 1979, there were presentations by each of the PETRA experiments at the open session of the DESY Physics Research Committee. Again only TASSO (represented by Peter Schmüser of the University of Hamburg) gave evidence for three-jet events.

With these three-jet events, the question was: what are the three jets? Since quarks are fermions, and two fermions (electron and positron) cannot become three fermions, it immediately followed that these three jets could not all be quarks and antiquarks. In other words, a new particle had been discovered.

Second, since this new particle, similar to the quarks, also hadronizes into a jet, it cannot be a color singlet. Color singlets, such as the pion, the kaon, and the proton, either leave a track (if charged), give an energy deposition in a calorimeter, or decay into well-defined final states, but do not metamorphose into jets. Therefore, the appearance of three-jet events meant that the carrier of strong forces, unlike the photon, was not an Abelian gauge particle (which must be colorless).

For these reasons, it was readily accepted by most of the high-energy physicists that the three-jet events were due to

$$e^+ e^- \to q \bar{q} g. \tag{34.9}$$

Motivated by the result of TASSO, by the end of August the other three collaborations at PETRA began to have their own three-jet analyses ready. At the Lepton–Photon Conference held at FNAL in late August 1979, all four PETRA experiments gave more extensive data,[22] thereby providing important confirmation of the discovery of the gluon by TASSO. In a period of three months, between August 29 and December 7, these more extensive data were submitted for publication by TASSO,[23] MARK J,[24] PLUTO,[25] and JADE,[26] in that order. The gluon remains one of the most interesting discoveries from PETRA.[27]

TASSO

The presentation at the FNAL Conference at the end of August 1979 was TASSO's fourth public announcement of three-jet events (after the Bergen Conference, the EPS–Geneva Conference, both in June 1979, and the DESY Physics Research Committee in July 1979). The published paper (note 23), which was received for publication by *Physics Letters* on August 29, 1979, gives more extensive TASSO data, and the results are, of course, more accurate. They confirm, but are in no way different from, what was presented above. The data were not only from center-of-mass energies of 27.4 GeV, but also from 27.7 GeV, 30 GeV, and 31.6 GeV. This paper concluded that:

> The planar events exhibit three axes, the average transverse momentum of the hadrons with respect to these axes being 0.30 GeV/c.... The data are most naturally explained by hard noncollinear bremsstrahlung $e^+e^- \to q\bar{q}g$.

MARK J

The presentation at the FNAL Conference at the end of August 1979 was MARK J's first public announcement of three-jet events, two months after that of TASSO at Bergen. Its paper (note 24) was received for publication by *Physical Review Letters* on August 31, 1979. Its analysis was based on the method of energy flow, similar to that used by the PLUTO collaboration at DORIS for the upsilon decay.[28] This paper stated that:

> In conclusion, we have shown that the energy flow of hadronic events from e^+e^- interactions can be described in terms of QCD.... The energy distribution of the events with thrust < 0.8 and oblateness > 0.1 shows three distinct jet structures.

Their energy-flow plot is shown in Fig. 34.9(a).

After the presentation by the MARK J collaboration at the FNAL Conference, Heinrich Meyer of the PLUTO collaboration made the following important point. Drawing on his personal experience of the PLUTO/DORIS three-jet analysis (Fig. 34.3), he pointed out the difficulty of using the energy-flow method to distinguish a phase-space distribution from a three-jet distribution.[29] In order to compare with the MARK J result, Meyer showed (for PETRA energies), a Monte Carlo simulation of the energy-flow plot [Fig. 34.9(b)] assuming a phase-space

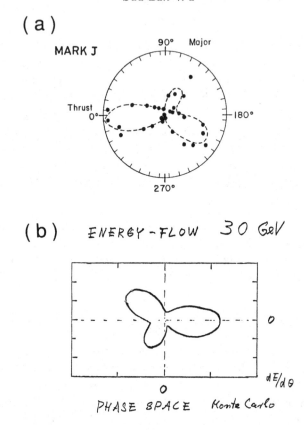

Fig. 34.8. (a) Energy distribution in the event plane as defined by the thrust and the major axes, shown by the MARK J Collaboration at the FNAL Conference in 1979. The energy value is proportional to the radial distance. The superimposed dashed line represents the distribution calculated with use of the $q\bar{q}g$ model [Source: D. P. Barber, et al., note 22, p. 18.] (b) Energy distribution shown by H. Meyer after the MARK J presentation. It was obtained by a Monte Carlo simulation using a phase-space model. [Source: Note 29]

distribution. It was evident that there was essentially no difference between the Mark J data and the Monte Carlo phase-space distribution. In connection with this comparison, the MARK J paper states, that "Phase-space distribution will show three nearly identical lobes due to the method of selection used."[30]

While at the time of the FNAL Conference in 1979 there was not enough data to give an unambiguous interpretation of the MARK J energy-flow result (phase space vs. three jets), this vital point was resolved by the MARK J collaboration two years later.[31] In this study, a much larger number of events, higher energies, and an improved energy-flow method with very different event-selection criteria were used. The result was that both phase space and $q\bar{q}g$ showed three-lobe structures but $q\bar{q}g$ gave a better agreement with the data.

PLUTO

The presentation at the FNAL Conference at the end of August 1979 was also PLUTO's first public announcement of three-jet events. Its paper (note 25) was received for publication by *Physics Letters* on September 13, 1979. The method of analysis used by the PLUTO Collaboration has features of both that of TASSO and that of MARK J. It began with the thrust T, but also used the generalization of thrust to the triplicity T_3 of Siegmund Brandt and Hans-Dieter Dahmen.[32] The value of this triplicity T_3 ranges between $T_3 = 1$ for a perfect three-jet event and $T_3 = 0.65$ for a perfectly spherical event. Since three-jet events are concentrated in a band of large triplicity, while two-jet events have a large thrust, PLUTO chose three-jet events by $T < 0.8$ and $T_3 > 0.9$. With this selection, PLUTO observed 48 three-jet events, in good agreement with the $q\bar{q}g$ model but not with the $q\bar{q}$ model without a gluon. Their data at center-of-mass energies between 27.4 and 31.6 GeV led to the conclusion:

> Again at high energies the data are consistent with $q\bar{q}g$ only. We conclude from this observation that indeed a three-jet event structure develops with increasing energy as predicted by the $q\bar{q}g$ model.

JADE

The presentation at the FNAL Conference was also JADE's first public announcement of three-jet events. Their method of analysis presented at FNAL is similar to the one used by TASSO. JADES's paper (note 26) was received for publication by *Physics Letters* on December 7, 1979. They concluded that:

> Both the qualitative nature of these planar, three-jet events and the rate at which they occur are in good agreement with what is expected from the process $e^+e^- \rightarrow q\bar{q}g$. This strongly suggests that gluon *bremsstrahlung* is the origin of the planar, three-jet events.

Conclusions

In summary, since two fermions cannot turn into three fermions, the experimental observation of three-jet events in e^+e^- annihilation, first accomplished by the TASSO collaboration in June 1979 and confirmed by the other collaborations at PETRA two months later, implies the discovery of a new particle. Similar to the quarks, this new particle hadronizes into a jet, and therefore cannot be a color singlet. These three-jet events are most naturally explained by hard noncollinear *bremsstrahlung* $e^+e^- \rightarrow q\bar{q}g$.

It is nevertheless nice to have a measurement of the spin of the gluon, which, being a gauge particle, must necessarily have spin 1. The amount of data collected at PETRA up to the time of the FNAL Conference was not quite enough for this goal, but the TASSO collaboration carried out this spin determination as soon as there were enough data.[33] Not surprisingly, the spin of the gluon was found to indeed be 1. Shortly thereafter, this result for the spin of the gluon was first verified by the PLUTO collaboration and then by the other collaborations.[34]

Thus the 1979 discovery of the second gauge particle, the gluon, occurred more than fifty years after that of the photon. This particle is also the first Yang–Mills non-Abelian gauge particle, that is, a gauge particle with self-interactions. Four years later, in 1983, the second and the third non-Abelian gauge particles, the W and the Z, were discovered at the CERN proton–antiproton collider by the UA1 and UA2 collaborations.[35]

Notes

1 M. Gell-Mann, "A Schematic Model of Baryons and Mesons," *Phys. Lett.* *8* (1964), pp. 214–215; G. Zweig, "An SU₃ Model for Strong Interaction Symmetry and its Breaking," CERN Reports No. 81821 TH401 (January 17, 1964) and 1891 TH412 (February 21, 1964).

2 E. D. Bloom, et al., "High-Energy Inelastic e–p Scattering at 6° and 10°," *Phys. Rev. Lett.* *23* (1969), pp. 930–934; M. Breidenbach, et al., "Observed Behavior of Highly Inelastic Electron–Proton Scattering," *Phys. Rev. Lett.* *23* (1969), pp. 935–939.

3 S. W. Herb, et al., "Observation of a Dimuon Resonance at 9.5 GeV in 400-GeV Proton–Nucleus Collisions," *Phys. Rev. Lett.* *39* (1977), pp. 252–255; W. R. Innes, et al., "Observation of Structure in the Υ Region," *Phys. Rev. Lett.* *39* (1977), pp. 1240–1242 (erratum p. 1640). Recent Fermilab experiments (1995) show there is a sixth quark, too.

4 G. Hanson, et al., "Evidence for Jet Structure in Hadron Production by e^+e^- Annihilation," *Phys. Rev. Lett.* *35* (1975), pp. 1609–1612.

5 R. F. Schwitters, et al., "Azimuthal Asymmetry in Inclusive Hadron Production by e^+e^- Annihilation," *Phys. Rev. Lett.* *35* (1975), pp. 1320–1322. Hanson, et al., "Evidence for Jet Structure."

6 Deutsches Elektronen-Synchrotron, "PETRA – a proposal for extending the storage-ring facilities at DESY to higher energies" (Hamburg: DESY, November 1974); Deutsches Elektronen-Synchrotron, "PETRA – updated version of the PETRA proposal" (Hamburg: DESY, February 1976).

7 S. L. Glashow, "Partial-Symmetries of Weak Interactions," *Nucl. Phys. 22* (1961), pp. 579–588; S. Weinberg, "A Model of Leptons," *Phys. Rev. Lett. 19* (1967), pp. 1264–1266; A. Salam, "Weak and Electromagnetic Interactions," in M. Svartholm, ed., *Proceedings of the Eighth Nobel Symposium* (New York: Wiley, 1968), pp. 367–377; S. L. Glashow, J. Iliopoulos, and L. Maiani, "Weak Interactions with Lepton–Hadron Symmetry," *Phys. Rev. D2* (1970), pp. 1285–1292.

8 C. N. Yang and R. L. Mills, "Conservation of Isotopic Spin and Isotopic Gauge Invariance," *Phys. Rev. 96* (1954), pp. 191–195.

9 J. Friedman, "Deep-Inelastic Experiments and the Discovery of Quarks," Chapter 32, this volume.

10 P. C. Bosetti, et al., "Analysis of Nuclear Structure Functions in CERN Bubble Chamber Neutrino Experiments," *Nucl. Phys. B142* (1978), pp. 1–28; J. G. H. de Groot, et al., "QCD Analysis of Charged-Current Structure Functions," *Phys. Lett. 82B* (1979), pp. 456–460; J. G. H. de Groot, et al., "Inclusive Interactions of High-Energy Neutrinos and Antineutrinos in Iron," *Z. Phys. C: Part. Fields 1* (1979), pp. 143–162.

11 CELLO Collaboration, H. J. Behrend, et al., "Measurement of the Reaction $e^+e^- \to \mu^+\mu^-$ for $14 \leq \sqrt{s} \leq 36.4$ GeV," *Z. Phys. C: Part. Fields 14* (1982), pp. 283–288; JADE Collaboration, W. Bartel, et al., "Observation of a Charge Asymmetry in $e^+e^- \to \mu^+\mu^-$," *Phys. Lett. 108B* (1982), pp. 140–144; MAC Collaboration, E. Fernandez, et al., "Electroweak Effects in $e^+e^- \to \mu^+\mu^-$ at 29 GeV," *Phys. Rev. Lett. 50* (1983), pp. 1238–1241; MARK J Collaboration, B. Adeva, et al., "Measurement of Charge Asymmetry in $e^+e^- \to \mu^+\mu^-$," *Phys. Rev. Lett. 48* (1982), pp. 1701–1704; MARK II Collaboration, M. E. Levi, et al., "Weak Neutral Currents in e^+e^- Collisions at $\sqrt{s} = 29$ GeV," *Phys. Rev. Lett. 51* (1983), pp. 1941–1944; PLUTO Collaboration, Ch. Berger, et al., "Measurement of the Muon Pair Asymmetry in e^+e^- Annihilation at $\sqrt{s} = 34.7$ GeV," *Z. Phys. C: Part. Fields 21* (1983), pp. 53–57; TASSO Collaboration, R. Brandelik, et al., "Charge Asymmetry and Weak Interaction Effects in $e^+e^- \to \mu^+\mu^-$ and $e^+e^- \to \tau^+\tau^-$," *Phys. Lett. 110B* (1982), pp. 173–180.

12 A. Einstein, "Über einen die Erzeugung und Verwandlung des Lichtes betreffenden heuristischen Gesichtspunkt" (On a Heuristic Point of View Concerning the Production and Transformation of Light), *Annalen der Physik 17* (1905), pp. 132–148; A. H. Compton, "The Total Reflexion of X-Rays," *Philosophical Magazine 45* (1923), pp. 1121–1131.

13 The mass limits $m_w > 40$ GeV and $m_z > 80$ GeV were given in Weinberg, "A Model of Leptons."

14 This very rough estimate, carried out in 1978, went as follows. Consider the most favorable kinematic situation where the three jets make angles of 120° with each other. Beginning with the assumption that the invariant mass of each pair of jets is given by the SPEAR energy, i.e., 7.4 GeV, then the energy of each jet is about 4.3 GeV. The total energy of the three jets is thus 13 GeV, which must be further increased because each jet has to be narrower than the SPEAR jets. This additional factor

is estimated to be $180°/120° = 1.5$, leading to about 20 GeV. But the phase space for this most favorable kinematic region is very small, and hence a further increase of at least 10% or 20% is called for, resulting in a factor of 3 over the SPEAR energy of 7.4 GeV.

15 J. Ellis, M. K. Gaillard, and G. G. Ross, "Search for Gluons in e^+e^- Annihilation," *Nucl. Phys. B111* (1976), pp. 253–271; erratum, *B130* (1977), p. 516.

16 Ibid. p. 270.

17 Sau Lan Wu and Georg Zobernig, "A Method of Three-Jet Analysis in e^+e^- Annihilation," *Z. Phys. C: Part. Fields 2* (1979), pp. 107–110.

18 This PLUTO result was reported in the *CERN Courier* as a news item under Around the Laboratories: "DESY – Three jets in upsilon decay," *CERN Courier* (English edn., May 1979), pp. 108–109. A month later, this news item was withdrawn. *CERN Courier* (English edn., June 1979), p. 148, stated that "Later it became clear that the evidence is inconclusive, as other mechanisms can reproduce the three-jet pattern seen in the upsilon decays." One of the "other mechanisms" is a phase-space distribution.

19 B. H. Wiik, "First Results from PETRA," in A. Haatuft and C. Jarlskog, eds., *Proceedings of Neutrino 79* (Bergen, June 18–22, 1979), Vol. 1, pp. 113–154, on pp. 127–128.

20 Sau Lan Wu and Georg Zobernig, "A Three-jet Candidate (run 447, event 13177)," TASSO Note No. 84 (June 26, 1979).

21 P. Söding, "Jet Analysis," in *Proceedings of the European Physical Society International Conference on High Energy Physics* (Geneva, 27 June–4 July 1979), Vol. 1, pp. 271–281.

22 D. P. Barber, et al., "The First Year of MARK-J at PETRA," in T. B. W. Kirk and H. D. I. Abarbanel, eds., *Proceedings of the 1979 International Symposium on Lepton and Photon Interactions at High Energies* (Batavia, Illinois: Fermi National Accelerator Laboratory, August 23–29, 1979), pp. 3–18; Ch. Berger, "Results from the PLUTO Experiment on e^+e^- Reactions at High Energies," in ibid., pp. 19–33; TASSO Collaboration, "TASSO Results on e^+e^- Annihilation Between 13 and 31.6 GeV and Evidence for Three Jet Events," in ibid., pp. 34–51; JADE Collaboration, "First Results from JADE," in ibid., pp. 52–69. Since these experiments were run simultaneously, the amounts of data were similar and there is not much difference in their statistical significance.

23 TASSO Collaboration, R. Brandelik, et al., "Evidence for Planar Events in e^+e^- Annihilation at High Energies," *Phys. Lett. 86B* (1979), pp. 243–249.

24 MARK J Collaboration, D. P. Barber, et al., "Discovery of Three-Jet Events and a Test of Quantum Chromodynamics at PETRA," *Phys. Rev. Lett. 43* (1979), pp. 830–833.

25 PLUTO Collaboration, Ch. Berger, et al., "Evidence for Gluon Bremsstrahlung in e^+e^- Annihilation at High Energies," *Phys. Lett. 86B* (1979), pp. 418–425.

26 JADE Collaboration, W. Bartel, et al., "Observation of Planar Three-Jet Events in e^+e^- Annihilation and Evidence for Gluon Bremsstrahlung," *Phys. Lett. 91B* (1980), pp. 142–147.

27 See, for example, Sau Lan Wu, "e^+e^- Physics at PETRA – the First Five Years," *Phys. Rep. 107* (1984), pp. 59–324.

28 *CERN Courier,* "DESY – Three jets." See note 18.

29 H. Meyer, "Measurements of the Properties of the Υ Family from e^+e^- Annihilation," in T. B. W. Kirk and H. D. I. Abarbinel, eds., *Proceedings of the 1979 International Symposium on Lepton and Photon Interactions at High Energies,* pp. 214–27, on p. 225.

30 D. P. Barber, et al., "Discovery," p. 833, footnote 5.

31 MARK J Collaboration, D. P. Barber, et al., "Measurement of Hadron Production and Three-Jet Event Properties at PETRA," *Phys. Lett. 108B* (1982), pp. 63–66.

32 S. Brandt and H.-D. Dahmen, "Axes and Scalar Measures of Two-Jet and Three-Jet Events," *Z. Phys. C: Part. Fields 1* (1979), pp. 61–70.

33 TASSO Collaboration, R. Brandelik, et al., "Evidence for Spin-1 Gluon in Three-Jet Events," *Phys. Lett. 97B* (1980), pp. 453–458.

34 PLUTO Collaboration, Ch. Berger, et al., "A Study of Multi-Jet Events in e^+e^- Annihilation," *Phys. Lett. 97B* (1980), pp. 459–464.

35 UA1 Collaboration, G. Arnison, et al., "Experimental Observation of Isolated Large Transverse Energy Electrons with Associated Missing Energy at $\sqrt{s} = 540$ GeV," *Phys. Lett. 122B* (1983), pp. 103–116; "Experimental Observation of Lepton Pairs of Invariant Mass around 95 GeV/c^2 at the CERN SPS Collider," *Phys. Lett. 126B* (1983), pp. 398–410; UA2 Collaboration, M. Banner, et al., "Observation of Single Isolated Electrons of High Transverse Momentum in Events with Missing Transverse Energy at the CERN $\bar{p}p$ Collider," *Phys. Lett. 122B* (1983), pp. 476–485; UA2 Collaboration, P. Bagnaia, et al., "Evidence for $Z^0 \rightarrow e^+e^-$ at the CERN $\bar{p}p$ Collider," *Phys. Lett. 129B* (1983), pp. 130–140.

Part seven

Personal Overviews

35

Quarks, Color, and QCD

MURRAY GELL-MANN

Born New York City, 1929; Ph.D., 1951 (physics), Massachusetts Institute of Technology; Robert A. Millikan Professor of Physics Emeritus at the California Institute of Technology; Professor, Santa Fe Institute; Nobel Prize in Physics, 1969; elementary particle physics (theory).

The story of my early speculations about quarks begins in March 1963, when I was on leave from Caltech at MIT and playing with various schemes for elementary objects that could underlie the hadrons. On a visit to Columbia, I was asked by Bob Serber why I didn't postulate a triplet of what we would now call SU(3) of flavor, making use of my relation $\underline{3} \times \underline{3} \times \underline{3} = \underline{1} + \underline{8} + \underline{8} + \underline{10}$ to explain baryon octets, decimets, and singlets. I explained to him that I had tried it. I showed him on a napkin (at the Columbia Faculty Club, I believe) that the electric charges would come out $+\frac{2}{3}, -\frac{1}{3}, -\frac{1}{3}$ for the fundamental objects. During my colloquium that afternoon, I mentioned the notion briefly, but meanwhile I was reflecting that if those objects could not emerge to be seen individually, then all observable hadrons could still have integral charge, and also the principle of "nuclear democracy" (better called "hadronic egalitarianism") could still be preserved unchanged *for observable hadrons*. With that proviso, the scheme appealed to me.

I didn't get a great deal of time to work on these "kworks," as I thought of them (not yet having spotted the word "quarks" in *Finnegans Wake*) until September, 1963, when I returned to Caltech.

While I was away, my student George Zweig had completed his final oral, with Richard Feynman replacing me, and had left during the summer for CERN, where he conceived his ideas about "aces." We did not overlap, so we had no discussions about quarks and were in ignorance of each other's work. I did talk with Viki Weisskopf, then Director General of CERN, in the early fall by telephone between Pasadena and Geneva, but when I started to tell him about quarks he said, "this is a transatlantic phone call and we shouldn't waste time on things like that." I presume, therefore, that there was no further discussion of my ideas at CERN that year.

Naturally, my first concern about the scheme was the matter of statistics. The quarks, viewed as *constituents* of the low-lying baryon states, were most simply interpreted as being in a symmetrical state of space, spin, and isotopic spin or SU(3) coordinates. But Fermi–Dirac statistics would predict the opposite. Thus, from the beginning I wanted some weird statistics for the quarks, and I associated that requirement ("quark statistics") with their not emerging to be viewed individually. But what kind of weird statistics was involved?

Back at Caltech, I tried various speculations about the statistics. I had heard, at MIT, about the idea of parafermions and wondered, naturally, whether quarks might be parafermions of rank 3. Yuval Ne'eman, who was visiting Caltech that academic year, recalls that he accompanied me to the library to look up formulae relating to such objects, to see whether certain combinations of three of them would behave like fermions in a symmetrical state of space, spin, and isotopic spin or SU(3) coordinates. He says that we consulted an article that contained a mistake or misprint and therefore missed the construction, although, in fact, three such parafermions would yield, among other things, a state behaving like a fermion, symmetrical in the other variables.

Thus I left the matter of quark statistics for a later time and wrote up the triplet idea during the fall of 1963 (using the spelling "quark" and a reference to Joyce) in a brief submission to *Physics Letters*, emphasizing the "current quark" aspects more than the "constituent quark" aspects. I employed the term "mathematical" for quarks that would not emerge singly and "real" for quarks that would. In the letter, to illustrate what I meant by "mathematical," I gave the example of the limit of infinite mass and infinite binding energy.

Later, in my introductory lecture at the 1966 International Conference on High Energy Physics in Berkeley, I improved the characterization of mathematical quarks by describing them in terms of the limit of an infinite potential, essentially the way confinement is regarded today. Thus what I meant by "mathematical" for quarks is what is now generally thought to be both true and predicted by QCD. Yet, up to the present, numerous authors keep stating or implying that when I wrote that quarks were likely to be "mathematical" and unlikely to be "real," I meant that they somehow weren't there. Of course I meant nothing of the kind.

During the 1960s and 1970s, numerous experiments were undertaken to search for real quarks. Peter Franken had the idea that quarks in sea water would be concentrated by oysters, and he would phone me at midnight to tell me of the progress of his experiments on those molluscs, which he ground up and checked for spectral lines that might be emitted by atoms containing real quarks. William Fairbank at Stanford thought at times that he had candidate events for real quarks. Others were looking for real quarks in the cosmic radiation. If they existed, then some fractionally charged particle would be stable, and that stable object could have practical applications, not only possible catalysis of nuclear fusion, but presumably others as well. I used to say that real quarks would lead to a quarkonics industry. But I have always believed instead in mathematical quarks, ones that do not emerge singly to be observed or utilized.

I did not want to call such quarks "real" because I wanted to avoid painful arguments with philosophers about the reality of permanently confined objects. In view of the widespread misunderstanding of my carefully explained notation, I should probably have ignored the philosopher problem and used different words.

The prescription that I frequently recommended for studying quarks was current algebra, for example as abstracted from a useful but obviously wrong field theory of quarks with a single neutral gluon. It was in that connection that I mentioned the recipe that Valentine Telegdi attributed to Escoffier, in which pheasant meat is cooked between two slices of veal and the veal is then thrown away. The veal was the incorrect single gluon theory, not field theory in general.

Later on, in the late 1960s, Bjorken gave some further, approximate generalizations of the quark current algebra, arriving at what was a kind of impulse approximation for suitable high energy, high momentum-transfer inclusive processes such as deeply inelastic collisions of electrons and nucleons. This was what was renamed by Feynman the "parton model." But acknowledgment that the "partons" were just quarks (and antiquarks and sometimes gluons) came slowly, and even when some authors began to refer to "quark partons," it was still often implied that they were somehow different from current quarks treated in a particular approximation. The approximation in question is, of course, one that is justified when the strength of the interaction between colored particles, such as quarks, becomes weaker and weaker at shorter and shorter distances.

Let us now return to the matter of quark statistics. After the appearance of my letter, Oscar Greenberg suggested that the quarks might be parafermions of rank 3. He did the mathematics correctly and showed that fermionic baryons would result. However, he did not, so far as I know, go on to require that the baryons with bizarre statistics be suppressed, leaving only fermions.

Likewise, when Moo-Young Han and Yōichirō Nambu suggested the existence of what we called "color" at Caltech, they did not consider the suppression of color nonsinglets. In fact, in a special twist, they assigned different electric charges to different colors, in such a way that all the quarks and antiquarks would have integral charges $+1, 0$, or -1. That way color nonsinglets would not only exist but would be excited by the electromagnetic interaction.

In both the Greenberg and Han–Nambu schemes, three quarks in a *symmetrical* configuration with respect to space, spin, and isospin or SU(3) coordinates would yield a *fermion* composite, but other kinds of baryons would also exist, in one case with outlandish statistics, and in the other case with non–singlet color.

Since I was always convinced that quarks would not emerge to be observed as single particles ("real quarks"), I never paid much attention to the Han–Nambu model, in which their emergence was supposed to be made plausible by giving them integral charges. However, it is a pity that I missed a follow-up article by Nambu. It appeared in the 1966 Festschrift volume devoted to my old teacher, Weisskopf, in which friends and admirers celebrated his sixtieth birthday. As one of his students, I should, of course, have contributed, but my habit of procrastination proved – as it had many times before – to be an obstacle, and I didn't produce an article for the volume in time for publication. I was so ashamed of that situation that I never opened the book.

In his 1966 paper Nambu pointed out how a color octet vector interaction between quarks would serve to lower the energy of the color singlet configuration of three quarks relative to octet and decimet representations, thus explaining the observation at modest energies of baryon configurations symmetrical in the other variables. However, sticking to the Han–Nambu point of view, he did not try to abolish the color non–singlet representations, and in his scheme they would appear at higher energies.

If I had seen Nambu's article, I might have concluded that his color octet vector interaction should be utilized without the Han–Nambu idea of integral charges and with confined ("mathematical") quarks. I might then have made progress in the direction of QCD.

Instead, that sort of insight was delayed until 1971. It was in that year, on the day of the earthquake that shook the Los Angeles area, that Harald Fritzsch arrived at Caltech. (In memory of that occasion, I left the pictures on the wall askew, until they were further disturbed by the 1987 earthquake.)

In the fall of 1971, both Fritzsch and I moved to the vicinity of Geneva to spend the year working at CERN. William Bardeen was there with us. We took up again the questions involved in putting the quarks in the known baryon states into symmetrical configurations of the usual variables. First, we reexamined parastatistics, getting the arithmetic right this time, and saw at once that there would occur, for three para-quarks of rank 3, one regular fermion pattern, in which the quarks would have a symmetrical wave function in the usual variables, as well as three patterns with unconventional statistics. If we somehow abolished those three extra ones, then baryons would all be fermions, and the configurations assigned to the known baryon states would be all right. Similarly, quark–antiquark states could be restricted to bosons.

Looking at color, we knew that the requirement of color singlets only would accomplish the same thing. Moreover, we discovered that each of the four patterns formed by three paraquarks of rank 3 could be identified with a single component of one of the representations $\underline{1}$, $\underline{8}$, $\underline{8}$, and $\underline{10}$ of color SU(3) arising from the product $\underline{3} \times \underline{3} \times \underline{3}$. The fermion composed of three paraquarks of rank 3 is the color singlet, and the states of unconventional statistics are just single components of the color octet and decimet representations.

Thus it was borne in on us that the situation was very simple. Para-quarks with prohibition of unconventional statistics would yield baryons and mesons in agreement with observation and nothing crazy. Colored quarks with prohibition of nonsinglet color would yield a mathematically equivalent situation. We concluded that this must, in fact, be the right way to describe Nature.

That fall, everyone was talking about the demonstration, by means of anomalies, that the elementary perturbation theory formula for $\pi^\circ \to \gamma\gamma$ decay, more or less as given by Steinberger in 1950, when he was briefly a theorist at the Institute for Advanced Study in Princeton, must be correct to a good approximation (small pion–mass squared) provided the right elementary fermions are put into the diagram. With quarks, the decay rate comes out right if the factor 3 for color is inserted. In fact, one then gets the same result that Steinberger got by putting in the neutron and proton, because $1^2 - 0^2 = 3\left[(\frac{2}{3})^2 - (-\frac{1}{3})^2\right]$. We concluded that colored quarks (presumably with suppression of color nonsinglets) were correct.

Likewise the ratio R measured by the SLAC–LBL experiment on the cross section for e^+e^- annihilation is, to a good approximation, just the sum of the squares of the charges of the fundamental fermions, other than the electron, that are relevant up to the given energy. One gets a contribution $1^2 = 1$ from the muon, and, in the case of color, a contribution $3[(\frac{2}{3})^2 + (-\frac{1}{3})^2 + (-\frac{1}{3})^2] = 2$ from u, d, and s quarks. We noted that above the charm threshold one should add to the sum 3 an additional term $3[\frac{2}{3}]^2 = \frac{4}{3}$, giving $\frac{13}{3}$. We did not, of course, know about the contributions from the third family, starting with a term $1^2 = 1$ for the tau lepton.

Burton Richter of SLAC tried, in his Stockholm lecture, to bury this essentially correct prediction in a mess of irrelevant numbers, so as to make fun of theorists, but in fact *he* had been claiming that the ratio didn't level off at all, while we theorists subscribing to hidden color had the contribution per family correct.

In fact, color used in this way was not very well received in all circles. Some theorists at SLAC laughed at it, and, adding insult to injury, contrasted colored quarks with what they called "Gell-Mann and Zweig quarks"!

Fritzsch and I went on to construct, during the winter of 1971–72, a field theory of colored quarks and gluons. Although we were still ignorant of Nambu's 1966 work, we knew that it would be a good idea to have a Yang–Mills theory based on gauged SU(3) of color, with perfect conservation of color and perfect gauge invariance. Such a theory would be renormalizable and also compatible with the electroweak theory, with quark masses supplied in both cases by the same "soft-mass" mechanism. (I had advertised for such a mechanism, by the way, in my Caltech report of January 1961 on the Eightfold Way.) One of the most important virtues of describing the strong interaction by means of a Yang–Mills theory based on color, with electromagnetic and weak charges not color-dependent, is that the strong and electroweak Yang–Mills theories utilize separate internal variables and so do not clash with each other. Back in 1961, I had had to abandon my speculations about a flavor SU(3) Yang–Mills theory for the strong interaction, mainly because of such a clash.

Although enthusiastic about the beauty of this theory, we hesitated a bit in endorsing it in print, for three reasons:

1. We were worried about how to generate a nonzero trace of the stress–energy–momentum tensor in the limit of zero quark masses. We knew that such a nonzero trace was needed: the mass of the nucleon, unlike the masses of the lowest pseudoscalar mesons, had to be non-vanishing in that limit and scale invariance had to be broken. Somewhere there was a source of mass that would hold up as quark masses vanished.

Even without the explicit dimensional transmutation later demonstrated by Sidney Coleman and Erick Weinberg, it was easy to show that in such a theory the trace could be nonvanishing in the limit. John Ellis had been a visitor at Caltech during 1969–70, and he had lectured there on the possible generation of an anomalous trace, yielding what he called – appropriately for that era – POT, for partially zero trace. If I had remembered his work, I would not have troubled myself about the generation of mass from no mass.

2. We understood that some form of string theory (in terms of which the Veneziano model had just been reformulated) was the embodiment of the bootstrap, and in those days, of course, the bootstrap idea was thought to apply to hadrons alone rather than to all the elementary particles. Thus we thought at times that perhaps the Yang–Mills field theory of colored quarks and gluons ought to be replaced by some kind of related string theory. (Of course, it turns out that QCD structures behave like bags and, when they are elongated, somewhat like strings, but those are approximate and derived features of the theory, not fundamental ones.)

3. We didn't understand what was causing the suppression of color nonsinglets (i.e., the confinement of color or the "mathematical" character of quarks and gluons). We didn't know that it would follow from the color SU(3) Yang–Mills theory itself.

In the summer of 1972, I presented a paper on behalf of Fritzsch and myself at the International Conference on High Energy Physics, at a session chaired by David Gross. I discussed these ideas and, in the spoken version, emphasized the possibility of a Yang–Mills theory of SU(3) of color coupled to colored quarks, as well as the alternative possibility that fundamental colored strings might somehow be involved. In preparing the written version, unfortunately, we were troubled by the doubts just mentioned, and we retreated into technical matters.

Fritzsch and I continued to work on the Yang–Mills theory and its implications, and the next summer, together with H. Leutwyler, we composed in Aspen a letter on "The Advantages of the Color Octet Gluon Model."

We compared the Yang–Mills color octet gluons with the "throwaway" model theory with a single neutral gluon. We pointed out that the octet theory overcame, in the case of color, a fundamental difficulty noticed by Lev Okun in the singlet model, namely invariance of the interaction under $SU(3n)$, where n is the number of flavors.

We went on to discuss the asymptotic freedom that had recently been pointed out in the Yang–Mills theory by David Politzer and by Gross and Frank Wilczek (and also by Gerard 't Hooft, who may not have fully appreciated its significance). They used the "renormalization group" method that Francis Low and I had developed for QED. In connection with that method, he and I had shown that for a charged particle (the electron) in QED, the weight function of the propagator did not have to be positive, whereas for the neutral photon it did have to be, with the result that the strength of the force carried by the photon had to increase from the infrared (where it is the renormalized charge squared) toward the ultraviolet. In QCD, the gluon is itself charged, and the positivity of the weight function can therefore be violated, so that the strength of the coupling need not vary in the same way as for electromagnetism. In fact, the relevant parameter in perturbative Yang–Mills theory has the opposite sign to that of the corresponding parameter in QED, and so there is asymptotic freedom in the ultraviolet and the possibility of a confining potential in the infrared.

If such a confining potential does result, then that explains, as Gross and Wilczek remarked, why color nonsinglets are eliminated from the spectrum of particles that emerge singly; and that in turn explains why quarks are "mathematical."

Finally, we discussed the question of whether there is, in the limit of vanishing quark masses, conservation of the flavor-singlet axial vector current, which threatens to yield four light pseudoscalar mesons instead of three for $SU(2)$ of flavor, and nine instead of eight for $SU(3)$ of flavor, contrary to fact. That was an old preoccupation of mine:

$$\partial_\mu \mathcal{F}_{0\mu}^5 \overset{?}{\to} 0 \text{ as masses} \to 0.$$

The theory seemed at first to have that difficulty;

BUT there is an anomaly proportional to $g^2 \epsilon_{\mu\nu\kappa\lambda} G_{\mu\nu}^A G_{\kappa\lambda}^A$, where G is the Yang–Mills field strength;

BUT the anomaly term itself is the divergence of a current $\mathcal{J}_\mu^5$, so

$$\partial_\mu(\mathcal{F}_{0\mu}^5 - \mathcal{J}_\mu^5) \to 0 \text{ as masses} \to 0,$$

which seems to revive the difficulty in another form;

BUT $\mathcal{J}_\mu^5$ is not gauge-invariant, so the difficulty does not seem serious.

Then there were two more BUTs (fortunately an even number!) that we did not cover in the letter:

BUT the *charge* $\int \mathcal{J}_0^5 d^3x$ does appear to be gauge-invariant, with

$$\frac{d}{dt}(\int \mathcal{F}_{00}^5 d^3x - \int \mathcal{J}_0^5 d^3x) \to 0 \text{ as masses} \to 0,$$

apparently gauge-invariant;

BUT, as was soon shown by Alexander Polyakov et al. and 't Hooft, in connection with instantons, this charge is locally but not globally gauge-invariant, so in fact there is no problem of a fourth or ninth light pseudoscalar boson.

Meanwhile, the asymptotic freedom of QCD not only suggested that there could be a corresponding "infrared slavery" that would yield confinement of colors, but at the same time it gave directly an explanation of the so-called parton model, which amounted to assuming that quarks (and antiquarks and gluons) had a weakened interaction at short distances or large momentum transfers. Again, Gross and Wilczek discussed this important point in their paper.

The theory had many virtues and no known vices. It was during a subsequent summer at Aspen that I invented the name quantum chromodynamics, or QCD, for the theory and urged it upon Heinz Pagels and others. Feynman continued to believe, for a while, that the "parton" picture was something other than an approximation to QCD, but finally, at the Irvine meeting in December of 1975, he admitted that it was nothing else but that.

The mathematical consequences of QCD have still not been properly extracted, and so, although most of us are persuaded that it is the correct theory of hadronic phenomena, a really convincing proof still requires more work. It may be that it would be helpful to have some more satisfactory method of truncating the theory, say by means of collective coordinates, than is provided by the brute-force lattice gauge theory approximation!

36

The Philosopher Problem

PAUL TELLER

Born Chicago, Illinois, 1943; Ph.D., 1969 (philosophy) MIT; Professor, Department of Philosophy, University of California at Davis, philosophy of science.

Professor Murray Gell-Mann told us how, in 1963, in a submission to *Physics Letters*, he "employed the term 'mathematical' for quarks that would not emerge singly and 'real' for quarks that would." Three years later he offered an improved "characterization of mathematical quarks by describing them in terms of the limit of an infinite potential, essentially the way confinement is regarded today. Thus what I meant by 'mathematical' for quarks is what is now generally thought to be both true and predicted by QCD." But in using the term "mathematical" Professor Gell-Mann got himself into some hot water, for "up to the present, numerous authors keep stating or implying that when I wrote that quarks were likely to be 'mathematical' and unlikely to be 'real,' I meant that they somehow weren't there. Of course, I meant nothing of the kind."

How did Gell-Mann get himself into this little predicament? "I did not want to call [confined] quarks 'real' because I wanted to avoid painful arguments with philosophers about the reality of permanently confined objects. In view of the widespread misunderstanding of my carefully explained notation, I should probably have ignored the philosopher problem and used different words."

At the conference Gell-Mann told us about the doctor's prescription he kept posted in his office admonishing him not to debate philosophers, suggesting that his choice of the word "mathematical" was his effort to follow the prescription. In this case the medicine may have turned out to be worse than the ill it was meant to cure.

I want to touch on the "philosopher problem," what it was and where we are with it today. I take it that the "philosopher problem" refers to the attitude crudely summarized by saying that "if we can't see it, can't see it under any circumstances, then it isn't real." We might quibble about whether this attitude is most appropriately called the "philosopher problem," since in the nineteenth century it was largely advocated by physicists skeptical, for this reason, about the reality of atoms. But with the early twentieth century work of Einstein and Perrin the physics community largely accepted atoms. Possibly philosophers have been slower, and if so we might fairly use the term "philosopher problem."

But the important point is that for most philosophers as well as physicists, the whole matter has become a dead issue. In recent decades philosophers have very much taken to heart the point that when we see anything, we do so indirectly. (What goes here for perception goes in the same way for anything called "observation," "detection," and the like.) When we see chairs, tables, and any sort of middle-sized macroscopic objects, we do so only via the good offices of a flood of photons, massaged by a lot of optics, interaction with the neurons in the retina, data manipulation in the optic nerve and further cortical processing, until whatever neural processes that ensue count as perception.

Now, given that our perception of ordinary chairs and tables is this indirect, what grounds could we have for denying reality (in whatever sense chairs and tables are real) to something we see only slightly more indirectly? Most philosophers are then quite delighted to slide the slippery slope all the way down to quarks. From this point of view the issue of confinement is perfectly irrelevant.

Here is another way to make the point. The "If we can't see it, it isn't real" principle is based on a false dichotomy. It is tacitly assumed that sometimes we see things directly, as when I plainly see the table now before me, and sometimes we detect the presence of an object only through indirect evidence. People who hold the suspect principle acknowledge that we often detect the presence of real objects only indirectly. But, they insist, indirect detection is always susceptible to alternative interpretations. If we can never check up on indirect detection with direct perception, our claims to reality of the detected objects can never be more than conjectural. And conjecture, it is finally suggested, isn't a solid enough basis for claims that the thing or things *really* exist.

There is a lot in this line of argument to which one ought to object. But we need not fuss with the details, since the whole argument collapses with the collapse of the false dichotomy. I am not concerned with philosophers (not so many of us these days) who worry whether chairs and tables are real. In whatever sense you think chairs and tables are real, and once you appreciate the indirectness of our perception of these things, the greater indirectness of seeing smaller things is not going to be an in-principle reason for thinking the smaller things are not real in the same sense. Of course, as the chain involved in indirect perception gets longer, the chance of error may increase; the chances that we have been fooled about quarks may well be larger than the (vanishingly small, if you like merely "philosophers' ") chance we are systematically fooled about chairs and tables. But the *kind* of reason we have for thinking that quarks are real differs in degree, not in kind, from the *kind* of reason we have for thinking that chairs and tables are real, always with the same sense for the word "real." The fact that we can "see" quarks only with the aid of the "microscope" of deep-inelastic scattering experiments in no way shows that quarks are not real in whatever sense applies to chairs and tables seen with the aid of the instruments that Nature gave us at birth.

I don't know about 1963. But today there is no need to hide behind the word "mathematical."

37

Should We Believe in Quarks and QCD?*

MICHAEL REDHEAD

Born London, England, 1929; Ph.D. (mathematical physics), University College, London; Professor of History and Philosophy of Science at the University of Cambridge, England; philosophy of science.

There are two questions I want to address in this chapter. First, what is the evidential status of entities such as quarks and theories such as quantum chromodynamics, or QCD? In particular, is there a special problematic associated with just *these* entities and *this* theory?

But that leads to the second question of a more general nature: What is the evidential status of any theoretical entities and their properties and relations as encoded in some area of theoretical discourse? The second question touches on a central concern of general philosophy of science. But let me start with the first question.

Quarks first came into the physics vocabulary via the fundamental representation of the SU(3) symmetry introduced into hadronic physics in 1964 by Murray Gell-Mann and George Zweig. The actual known particles were associated with higher-dimensional representations of the symmetry, such as the octet, the original Eightfold Way. The quarks were at first a somewhat shadowy substratum for building up the particles actually observed in Nature (in particular the famously predicted Ω^-). I say shadowy because one could, for example, abstract from the quarks an algebra of currents, take this algebra seriously and discard the quarks – throwing away the ladder after making the ascent, so to speak. But then, in the late 1960s, came the deep-inelastic electron scattering experiments at SLAC, the verification of Bjorken scaling, and its immediate interpretation in terms of pointlike constituents, the parton model of the nucleons. It was then a small step to identify the partons, which in a sense one could directly "see," with the highly conjectural quarks.

---

* This paper is based on material prepared for the Tarner Lectures delivered in Cambridge, England, in February 1993 under the title "From Physics to Metaphysics."

But with the quarks came the theory of quark interactions, the color degrees of freedom, the gluon fields, and the whole apparatus of non-Abelian gauge theory in the now familiar Standard Model, augmenting the electroweak theory of Steven Weinberg and Abdus Salam with the quantum chromodynamics of strong interactions.

And there were immediate successes in terms of empirical predictions, quantitatively verified departures from crude Bjorken scaling, the production of jets, and so on. So did physicists believe in the theory? (I will come to philosophers later.) Well, not exactly. It was not that the theory was *empirically refuted,* far from it, but there were *theoretical* puzzles that made it unattractive as the demonstrable "last word" in the theory of strong interactions.

First there are all the general puzzles about understanding and interpreting quantum mechanics. These problems, however, are generally brushed under the carpet, so to speak, by most physicists, but dirt under the carpet is still dirt, I would submit. Then there was the generally unsatisfactory business of infinite renormalizations. Most physicists regarded a renormalized theory as some sort of "effective theory," hiding the detail of the "true" theory behind renormalized parameters, whose values were to be taken from experiment.

Next there was a sense of ad hocery in the number of adjustable parameters in the Standard Model, and the curious role of the Higgs particle in the electroweak sector. Then physicists were tempted by the Holy Grail of grand unification, combining the leptons and quarks in a single scheme. Grand unified theories generally predict the instability of the proton, via the interconvertability of quarks into leptons. This has not so far been observed, but many physicists still expect that it is an allowed process, although on a very long time scale. To that extent they do not believe crude QCD as the final theory.

Finally, there is of course the whole question of incorporating gravitation in a Theory of Everything, and the recent surge of enthusiasm for superstring theories.

So the question "Do physicists believe in QCD?" is rather like asking "Do physicists believe in classical mechanics?" The answer is yes for certain limited purposes of theoretical modeling of phenomena, but not in the sense that it is a serious candidate for being dead right – the final answer in strong interaction physics.

But what about the quarks themselves? There is often thought to be a special problem here associated with the phenomenon of quark confinement. In the past the "real" has been probed by the "manifest."

Electrons, atoms, nucleons, and so on could be dealt with singly in their free state, and then everything explained by an elaborate *Aufbau* principle, putting the single elements together. This is the classical method of analysis, of understanding complex wholes in terms of their simple constituents. But in a sense the quarks are a sort of counterexample, since they cannot be separated from their partners.

But this stress on making real entities manifest is a somewhat crude rendering of what we mean by manifest. The deep-inelastic scattering experiments manifest the quarks just as surely as holding them, one at a time, in the hollow of one's hand, so to speak. "Direct" observation is actually pretty "indirect," as far as particle physics is concerned. We "see" particles by actually seeing what they can do, producing tracks in bubble chambers, firing off spark chambers, and so on.

Let us now turn to what a philosopher might say on reading the preceding paragraphs, which are supposed to represent the views of physicists. I will therefore turn to my second question, which has a much broader focus than just quarks and QCD.

In what sense should one believe in science at all? There's a broad spectrum of what I may call "isms and schisms" in answering such a question that fill the pages of philosophical monographs and journals. At one end of the spectrum there are the relativists, the antirealists, the social constructivists, the irrationalists. At the other end are the realists, the objectivists, the champions of scientific rationality – and there's pretty well every shade in between. For example, there are the positivists, who are generally happy to be realists at the level of direct macroscopic observation, but – drawing a sharp observational–theoretical distinction – treat the theoretical machinery just as a calculational device or instrument, if you will, for connecting observational input in what amounts to a black-box approach to theoretical physics. To be inquisitive, seeking to get out the metaphysical screwdriver, lever off the lid, and see what is going on inside, is rigorously prohibited.

But much of recent philosophy of science has served to throw doubt on any sharp distinction between observation and theory (the slogan here is the "theory-ladenness" of observation). If one is moved by these arguments, one will either reject metaphysical realism even at the level of observation and move to a position such as pragmatism, where the slogan becomes: "It is true for you if it works for you." But given certain aims we may still try to retain a sense of rationality in how best to achieve those aims, *or* one may be driven in the opposite direction, interpreting the theoretical machinery as well as the observations realistically.

Let me now sketch the extremes. But I shall stress that the compromise positions generally tend to be unstable, and it is very easy to be driven to one of the extremes, if one looks at the matter dispassionately. So what are the arguments of the out-and-out relativist, put in the proverbial nutshell? First the relativist denies that there is an objective fact of the matter about any area of discourse, whether it be natural science, ethical questions, or even logic and mathematics. There is no Archimedean point, no God's-eye perspective, from which truth in the sense of correspondence with what is actually the case makes any sense. And any claim to grasp "reality" as it is in itself, the Kantian *ding an sich,* is just a metaphysical conceit.

How, ask the relativists, could we achieve knowledge of this sort – either by reason, which they dismiss as ridiculous, or on the basis of empirical observation, which they say is conditioned by theoretical presuppositions? So conditioned, in fact, that it can provide no sure foundations for knowledge in the old-fashioned sense of knowing the objective truth about things, what they are, how they behave, and so on.

Everything is relativized to purely subjective opinions, or at best intersubjective agreement conditioned not by the world and its scientific investigation, for example, but by socioeconomic factors and ideologies. Truth comes out as coherence, or perhaps whatever makes us feel good, never correspondence with what is in fact the case. For example, relativists can well say religion is true in their sense, but not because there actually is a God, or a moral law, and it is the same with science. "Quarks exist" is true just because someone believes that quarks exist, but *never* because (surprise, surprise) quarks do in fact exist. There are two conclusions that one can reach with this line of thought. Either one knows nothing – one is just a skeptic in the famous Pyrrhonist tradition of antiquity – or, surprisingly enough, one knows anything and everything that one has an opinion or a belief about, because that is what a relativist means by "knowing" something.

Of course, relativism has some apparently curious features. I know that fairies live at the bottom of my garden, if that is what I, or perhaps my local community, believe – even if nobody else does. For the relativist, since there is no robust notion of truth, there is also no notion of error, of being wrong. To be wrong in their Pickwickian sense is just to disagree with someone else. But even that is a bad way of putting it. You are both right in your own terms and from your own point of view if Jack says there are fairies at the bottom of his garden, and Jill says there are not. Now why should we think that relativism is true?

Well, in what sense of "true"? Relativists often speak as though the truth of relativism is the one thing they can know in the old-fashioned sense, but since they don't admit the old-fashioned sense, does the truth of relativism just dissolve into a matter of mere opinion? But then why do they bother with arguments? Of course, the relativists regard their position as the height of postmodernist sophistication, but in fact the whole position teeters toward incoherent absurdity.

Now let us look at the other extreme, realism, which roughly denies everything that the relativist asserts. There *is* an objective world, quite distinct from us and our musings and imaginings, where quarks either do exist or do not exist. We may never come to know decisively which is the case, but experimental evidence can be adduced to bear on the question – to provide degrees of support or confirmation, for the claim that quarks exist. Do we know indubitably the evidential basis itself? No, not for sure, but again we can make reasonable estimates about the reliability of the experimental reports, based on the usual procedures of testing and calibration.

This all sounds much closer to what the physicists said, but have we really grasped the nettle of the *ding an sich,* of knowing how things really are? I believe that my old mentor Karl Popper had the right approach to this problem. We conjecture how things are, we are never in a position to know for sure whether we are right, but the conjectures are not made in a purely fanciful or speculative way, for they are subject to evidential control by the techniques of experimental science, by observation, by inspecting the world. Popper of course emphasized the negative control of refutation, and one might want to allow some more positive sense of support or confirmation. But there is *some* control, the world kicks back, we cannot just make it up any way that pleases us. We don't construct quarks; we actually assess evidence for the conjecture that they do actually exist.

Now back to the compromises? Well, once you go soft on the notion of truth, you have started on a slippery slope. You may try to hang on to some middle ground such as pragmatism, with its talk of truth as the eventual consensus of some group of ideal enquirers, but how can we tell what makes an ideal enquirer? Could it be, cynically, just someone who ultimately comes to agree with you yourself? But the pull to extreme relativism is really impossible to resist. So the trick of not turning into a relativist is not to allow the first, subtly alluring move, of going soft on truth. For quarks it may not matter so much, but in everyday life I believe it really does make a difference whether we believe in medical science as against witchcraft and spells, and I know for sure which jetliner I want to travel in, the realist's not the relativist's!

So let us start again at the other end, and ask whether we should be realists about matters relating to the macroscopic world of everyday life, as the positivists would allow. And if the answer seems to be yes, let us make the reverse slide, if I can put it like that, from a robust realism about tables and chairs, to a definitely more conjectural realism, but realism all the same, about quarks and QCD.

I believe the physicist's gut feeling for these things is probably right (that is another conjecture for you) and I totally reject the apparently liberal, open-minded, and egalitarian but ultimately destructive doctrines of the relativists and social constructivists. And if that opinion is imputed to my socioeconomic environment, or my prenatal experiences in my mother's womb, I will simply respond: "poppycock."

So I have nailed my colors to the mast as a realist. But there are a number of issues that still need attending to:

1. Underdetermination: there may be two or more quite distinct metaphysical accounts of the nature of reality, which have exactly the same empirical consequences. So how could experiments ever be brought to bear in selecting one account of reality rather than another?

   An example that arises in classical physics is field versus particle accounts of the ultimate nature of matter. Consider the simplest question. How does a lump of matter get from A to B? The particle account says that a substantial lump of matter made of particles just moves across from A to B. But the field theorist would say that a force field of impenetrability has changed its configuration from one concentrated around A to one concentrated around B.

   Although two theories may be impossible to distinguish empirically, they may function quite differently from the point of view of heuristics – that is, one metaphysical approach may lead in a very natural manner to a succession of empirically testable theories, while the other may be relatively barren from the heuristic point of view. This is well illustrated by the great fertility of quantum field theory developments in the 1960s as compared with the rival S-matrix program.

2. Realism has often been attacked on the grounds that there is a significant lack of convergence in the history of theoretical physics, which (so the argument runs) is characterized by discontinuity rather than any continuously cumulative progression. But I believe that detailed his-

torical analysis often reveals much more continuity than one suspects, at any rate at the level of structure rather than ontology.

3. Realism seems to require some adequate notion of truth-likeliness or verisimilitude. Given that our theories are often discarded, can one nevertheless make sense of an approach to the truth? This is a very thorny technical problem in philosophy of science that hinges on the question: What is a theory really about?

   For example, consider an astronomical theory that predicts the number of planets P and the number of days in the week D. Suppose it gets both these numbers wrong, but gets $P + D$ right? Should this count in assessing whether the theory has gotten closer to the truth? This is a question to which no totally satisfactory answer has been given. Intuitively $P + D$ is not an interesting or significant quantity to get right, but how can we rule it out on purely logical grounds?

4. This problem points to the question: Are we supposed to be realists about every aspect of a theory? In the case of many theories in mathematical physics, the answer seems to be clearly no. The physical content is often embedded in a wider mathematical structure that itself has no physical referent. An example of such a situation might be the analytic S-matrix, where axioms are introduced controlling the behavior of the physical quantities continued analytically to the complex plane. More relevant to our purpose is the role of gauge transformations in a theory where the physically significant quantities are gauge invariant.

   Such developments certainly encourage the view that mathematical physics is just pieces of mathematics, a black box that is only connected to the world via its empirical predictions. But because some parts of a mathematical theory do not have physical correlates, it does not follow that we should be instrumentalists about the whole structure.

5. I have already referred to the problems about the interpretation of quantum mechanics. Recent developments particularly associated with the work of the late John Bell stress the problems of reconciling quantum mechanics with local realism. But these issues are a good deal more subtle than many physicists will allow. Realism certainly seems to require some form of nonlocality, but this may not be of a form that makes the whole program of realistically interpreting quantum mechanics necessarily inconsistent with relativity theory.

Thus there are certainly problems associated with the realist position. But these problems merely demonstrate that philosophy of physics is no more a finished enterprise than physics itself. They provide us with a challenge for further endeavor.

38

A Historical Perspective on the Rise of the Standard Model

SILVAN SCHWEBER

Born Strasbourg, France, 1928; Ph.D., 1952 (physics), Princeton University; Professor of Physics at Brandeis University; high-energy physics (theory) and history of science.

The establishment of the Standard Model marked the attainment of another stage in the attempt to give a unified description of the forces of nature. The program was initiated at the beginning of the nineteenth century by Oersted and Faraday, the "natural philosophers," who, influenced by *Naturphilosophie*, gave credibility to the quest and provided the first experimental indication that the program had validity. Thereafter, Maxwell constructed a model for a unified theory of electricity and magnetism, providing a mathematical formulation that was able to explain much of the observed phenomena and to make predictions of new ones. With Einstein the vision became all-encompassing. In addition, Einstein advocated a radical form of theory reductionism. For him the supreme test of the physicist was "to arrive at those universal elementary laws from which the cosmos can be built up by pure deduction." A commitment to reductionism and a desire for unification animated the quest for the understanding of the subnuclear domain – and success in obtaining an effective representation was achieved by those committed to that vision.

The formulation of the Standard Model is one of the great achievements of the human intellect – one that rivals the genesis of quantum mechanics. It will be remembered – together with general relativity, quantum mechanics, and the unravelling of the genetic code – as one of the outstanding intellectual advances of the twentieth century. But much more so than general relativity and quantum mechanics, it is the product of a communal effort.[1] That it could not have been accomplished without the genius of the experimenters and engineers who invented the technologies and built the accelerators, detectors, and computers, designed the experiments, and analyzed the data goes without saying. That it required technical competence, imagination, powers of concentration, and perseverance on the part of the theorists that matched those of Wolgang Pauli, Erwin Schrödinger, and Werner Heisenberg is also true. But merely analyzing the technical dimensions of the community only highlights the necessary components that made success possible. They are not sufficient.

My chapter focuses on some of the other dimensions that made the high-energy physics enterprise possible. Its success owes much to the effectiveness of its proponents in the councils of state. For the most part these were men who had been transformed by World War II. Thus my chapter is concerned with three generations: the generation that came of age with quantum mechanics; the generation trained by these men; and the generation that joined the ranks of the community at the end of the 1960s and the early 1970s. In the first are to be found, among others: Eduardo Amaldi, Robert Bacher, Hans Bethe, Enrico Fermi, Ernest Lawrence, J. Robert Oppenheimer, Isadore I. Rabi, Victor Weisskopf – the outstanding scientists and visionaries who after World War II made high energy the dominant field in physics.[2] They were also statesmen who could interface between the scientific and the political realms, and their political and diplomatic efforts within the councils of state made possible the construction of the laboratories and the requisite subsequent funding. They were committed to a tradition that had been molded by Niels Bohr, and high-energy physics for them was also a vehicle for international scientific cooperation.

Most of the physicists who contributed to the creation and establishment of the Standard Model are members of the generation that was trained after World War II. Among those trained by members of the first generation after the war are Murray Gell-Mann, Steven Weinberg, Sheldon Glashow, Abdus Salam, and many others. The third consists of the physicists coming of age in the late 1960s, and early 1970s: Gerhard 't Hooft, David Gross, Howard Georgi, Roman Jackiw, David Politzer, Thomas Applequist, Helen Quinn, and Roy Schwitters, to name a few. The mythic figures for all three generations were Bohr and Einstein. Einstein gave the community an intellectual vision, that of unification. Bohr gave it a vision of an intellectual community.

My chapter is organized as follows: Part I deals with the wider context. It looks at World War II and the post-World War II period and examines some of the factors that made possible the ascendancy and commanding status of high-energy physics in the United States and Western

Europe. Part II looks at the President's Science Advisory Committee (PSAC) and the General Advisory Committee (GAC) of the Atomic Energy Commission, their activities in shaping the high-energy program, and the justification that were given for obtaining support for high energy. It also considers the values that shaped "science policy" in the period under consideration. Part III deals with the relation between high-energy and condensed-matter physics and with some of the intellectual transformations that were brought about by the Standard Model. The epilogue briefly addresses issues of community.

I. The wider context

World War II was responsible for a large-scale scientific revolution that was brought about by the plethora of novel devices and instruments developed principally by physicists. Many of these devices had been introduced before the war but in a relatively primitive state and on an individual basis. It is the scale on which these devices and instruments become available that transforms the stage. The new institutions that were created by the war – for instance, OSRD, NDRC – introduced a novel contractual system between the government and universities and private industries that channeled most of the research and development carried out during the war. This contract and granting system became the mechanism through which research at universities would be funded after the war.

World War II also initiated a revolution in management science and military and industrial planning. Cybernetics, information theory, game theory, and operations research – disciplines born during the war – were particularly influential in these developments.[3] The consolidation during World War II of an engineering approach that became known as systems engineering owes much to the activities of physicists at the Radiation Laboratory at MIT, the Applied Physics Laboratory at Johns Hopkins, and the Jet Propulsion Lab at Caltech. They there molded a new role for themselves in which they did not merely function as experts to be tapped to solve problems defined by administrators and military personnel. They made themselves part of the procedure that assessed, defined, and reassessed the problems themselves. And in the process they were not content merely to build better radar sets or better proximity fuses in Cambridge and Baltimore, but they went out to the battlefields and determined and designed weapons systems that would be most effective in given situations. They were so successful in all that they undertook that they came to be asked to help devise strategy, and to map out bombing raids and plan submarine campaigns for maximum effectiveness.[4] Systems analysis – the "whole problem" approach that is presumably sensitive to the interdependence of the component parts of the system – became the appellation for the conceptual framework that had been wrought by the physicists working on the problems of research, development, and manufacture of radar at wartime laboratories, particularly the Rad Lab at MIT. It became emblematic of the rationalized, quantitative way to plan and implement large-scale projects with well-defined ends such as the design of the Nike system or of Project Apollo, whose mission was to land a man on the moon.[5]

The patterns of interaction between civilian scientists and the military that were developed by OSRD, NCRD, NACA, the OR groups, at the Rad Lab, Los Alamos, and elsewhere became the model for the postwar committees that advised the armed forces on weapons development and on tactical and strategic matters. This new framework for expert advice to the armed services by civilian scientists – principally physicists or former physicists – made scientists a new elite, with physicists at the top of the pecking order. The proliferation of advisory committees was indicative of the new dependence of the state on its scientific community. A new relationship between science and the state emerged. The basic assumptions of the alliance had been spelled out by Vannevar Bush in his influential *Science: The Endless Frontier*, in which he also outlined how the compact was to be implemented with the creation of a National Science Foundation.

The United States had won the war because of its industrial productive capacity and its technological innovations. The stimulus for technological innovation is scientific knowledge. Furthermore, "A nation which depends upon others for its new basic scientific knowledge will be slow in its industrial progress and weak in its competitive position in world trade, regardless of its mechanical skill."[6] Scientific knowledge would automatically generate progress provided the government would not interfere, except for dispensing the necessary subventions. To guarantee the vitality of its industrial and technological enterprises the United States must invest in research and development. And "[t]he simplest and most effective way in which Government can strengthen industrial research is to support basic research and to develop scientific talent."[7]

Basic research was indeed lavishly supported – at first, by ONR and the AEC – and scientific talent was generated. The single most important criterion that determined research support was excellence, which implied that the peaks among the institutions of higher learning grew yet higher. Similarly reflecting the commitments of Vannevar Bush, James B. Conant, Karl T. Compton, and Richard Tolman – the wartime leadership – to a meritocratic and elitist view of education, excellence, rather than equality, determined the support of graduate studies.

The Cold War cemented the wartime relationship. National security and national prestige became the major determinants for both the size and the pace of growth of the governmental budgets supporting research and development in general, and the physical sciences in particular. The United States saw itself in an international technological competition with the Soviet Union: national security and national prestige demanded that it remain the frontrunner in that race and that it preserve its supremacy in all the major fields of science. It could maintain its leadership position by supporting the best scientists and by providing maximum educational opportunities to everyone capable of benefiting from it. The physical sciences, which were seen as a national resource and as the catalysts for technological innovation, were mobilized in the interest of national security, and whatever support could not be justified in terms of national security was justified in terms of national prestige. The maintenance of the preeminent standing of the United States in the international rivalry became the justification for the support of those basic sciences whose relevance to technology was not immediately apparent. High-energy physics was one of the primary beneficiaries of that policy.

CERN

The funding required to support a program of experimental high energy physics using accelerators implied that support had to be obtained at the national or supranational level. Success in establishing a large-scale program therefore required a convergence of interests.

In the United States, the confluence of the intellectual interests of the nuclear physicists and those of the government in maintaining the preeminent position in the sciences made it possible after World War II for the high-energy program to take off on its exponential growth during the next two decades. International rivalry was the driving force. In Western Europe the confluence of a different set of interests made a large high-energy physics laboratory feasible after World War II. CERN became one of the means to implement the vision of an economically and politically integrated Western Europe that had been advanced by a number of politicians. CERN could be established because conditions were favorable for regional unification, and in turn CERN helped make the political environment more conducive to further regional integration. It was clearly important for Western Europe to rebuild its scientific infrastructure and to create a base of scientific expertise at the level attained by the United States during the war if it was to compete successfully in world markets and regain its political standing in the new world order. A cooperative scientific enterprise requiring large resources – both financial and technical – that were beyond the economic means and technical capabilities of any single Western European nation would be the ideal vehicle to implement this policy. Political reality made it obvious that cooperation would be impossible in atomic energy matters, given their important security and economic components. Indeed, the program of research that was chosen – high-energy physics – was one that did not interfere with the industrial and military interests of the individual nations, and thus obvious sources of conflicts could be avoided.

The alliance between scientific and diplomatic interests was consolidated in the administrative structure of CERN. Its commitment to excellence was made possible by the manner in which member states accepted the administration of the research center under the guidance of the scientific community. The success of the enterprise owes much to the fact that decisions regarding scientific research and industrial procurement have been relatively free from pressures from member states. The models adopted originated in the lessons learned from the experiences at the wartime U.S. laboratories – in particular, the Rad Lab at MIT and Los Alamos.[8]

Priorities in foreign policy, not purely scientific considerations, have been most important motivations for member states to support the organization since the 1950s. For the member nations the laboratory has a meaning beyond the strictly scientific activities it performs. CERN is an emblem of European unification and helped Western Europe achieve a scientific eminence that would not have been possible otherwise.

Thus in Europe high-energy physics escaped becoming a showcase of national capabilities, the "mascot of national greatness" that it was sometimes represented to be in the United States.[9]

Periodization

The history of high-energy physics in the United States from the end of World War II till the late 1970s can readily be fitted into a periodization that is useful when trying to understand the growth of science and technology after World War II.[10]

The first period lasted from 1945 to the mid-1950s. An unquestioned faith in science is perhaps one of its most striking features. It was Science per se, as a valid human activity, rather than the institutions of science that was supported and the public's trust in Science was such that its management was left to the new elite. During this period the scientific community obtained an unusual degree of autonomy in the procedures by which research grants were allocated.[11] Physics rode the crest of the wave, and physicists reaped the benefits of their wartime contributions. Physicists, and in particular the nuclear physicists who had worked at Los Alamos, reached the apex of their influence during this period.

During the second period, which ranged roughly from the mid-1950s to the mid-1960s, the implementation of the program that Bush had outlined became rationalized, in Max Weber's sense. The Berlin blockade and the Korean War had convinced the American public of the reality of a technological and scientific race with the USSR, and of the importance of winning it. Sputnik jolted the nation into implementing the rationalization of planning at a much quicker pace. The Cold War and the Korean conflict had created a national security state committed to "making the world safe for democracy." This national dedication required large investments in the development and manufacture of new weaponry and in the military establishment. Only a constantly expanding economy could sustain the effort. How to keep the economy growing so as to guarantee full employment and an ever-increasing GNP became the cardinal question for economists in the beginning of the 1960s.[12] In 1947 the Steelman Report had given the following answer:[13]

> Only through research and more research [in the basic sciences] can we provide the basis for an expanding economy, and continued high levels of employment.

In the late 1950s the panacea was to be continual technological innovation since the latter would secure constant economic growth. Moreover, it was argued that the urban problems that were becoming apparent and the social and economic obstacles encountered by the poorer classes, particularly blacks and other minorities, would be resolved and overcome more or less automatically by rapid economic growth and full employ-

ment. Economic growth would create higher-paying and more satisfying jobs, and the process would secure political stability. The support
of science now became partially justified in terms of its essential contribution to technological innovation.[14] The spectacular success of the
science-based industries near Stanford and MIT – Hewlett-Packard, Var-
ian, DEC, Motorola – were proof of the correctness of the assumptions.

The creation of the post of Science Adviser to the President and of
PSAC immediately after the launching of Sputnik in the fall of 1957
should be seen as trying to make more effective and expeditious the
mechanism for "nonpolitical" science and technology advice at the highest level of government. Sputnik accelerated and intensified a planning
procedure that had been in place but did not alter the assumptions under
which the planning was being effected. PSAC's role was to help implement the national strategy, especially in matters of "science in policy."[15]
Thus in 1962 a PSAC-appointed panel issued a report, *Technology and
Economic Prosperity*, that addressed the issue of economic growth in
which it recommended new steps to strengthen the U.S. technological
leadership and to increase productivity.

When PSAC was established in 1957 the pressing issues before it were
primarily related to national security, what Rabi had called "science
in policy": the missile gap, the technical issues involved in a nuclear-
weapons test ban, surprise attack, evaluations of choices of weapon systems for development.[16] In the early 1960s, policy questions regarding
defense issues were given over to the National Security Council. As its
role in security matters decreased PSAC acquired greater influence in
nonmilitary areas such as the space program and education. The problems addressed by PSAC became "policy for science" questions such as:
setting priorities for high-energy physics, astronomy, computers, geology, chemistry; establishing a balance among these various disciplines;
and the balance between pure and applied research in these areas.

One of the first items on PSAC's agenda was a review of the state of
scientific education in the United States and the status of high-energy
physics. PSAC's position was essentially the same as that outlined in
the Bush report. It too believed that United States leadership would be
protected as long as the economy kept expanding, basic research throve,
and universities maintained strong graduate programs in the sciences.
The PSSC high school physics course and the Berkeley physics courses
for undergraduate science majors are examples of some of the impressive
coordinated efforts that were stimulated by PSAC to strengthen educa-

tion in the sciences. I will turn to the reports of the PSAC panels on high-energy physics in Part II.

During this second period, systems analysis – undoubtedly, one of the most influential contributions made by the physicists during World War II – reached the peak of its influence. The stimulus for "planning-programming-budgeting" and "cost-effectiveness," the procedures that Robert McNamara instituted at the Pentagon, came from systems analysis; it was part of the same rationality. The effectiveness of that approach was demonstrated in the management of NASA. Systems analysis was *de rigueur* at RAND, and the same was true at the Institute of Defense Analysis (IDA), which was paradigmatic of the involvement of scientists in this rationalization of military planning. Later it became a focal point in the attack by students on scientists' participation in the Vietnam War.[17]

Nor was the systems approach limited to situations where the applied sciences and engineering factors were the most important components. In an influential book published in 1971, Alice Rivlin, who as assistant secretary for planning and evaluation in the Department of Health, Education and Welfare, oversaw the applications of planning-programming-budgeting methods there during the 1960s, urged that the systems approach be applied more generally to the solutions of social ills.[18] Similarly, Murray Gell-Mann, in an address at the dedication of the new physics building at the University of Santa Barbara in 1970, indicated that "we need something like what is called systems analysis, to take into account all the factors" when addressing social and environmental problems. But recognizing that systems analysis had in the past reduced people to personnel and wild creatures to resources, had set to zero anything that was hard to quantify, and often had been used "to justify unwise decisions in the field of national security," Gell-Man called for a "a systems analysis with heart ... to recommend to society a set of incentives for the humanely rational use of technology."[19]

The third period stretches from the mid-1960s till the mid-1970s. Already at the beginning of the 1960s, budgetary strains were beginning to appear. The increased numbers of scientists that the implementation of the Bush agenda had produced, the increased cost of laboratory science, and the effects of the funding for NASA, NIH, and other big projects made it apparent that some criteria had to be imposed on the allocation of the finite resources. What these criteria ought to be constituted the debates over "science policy."[20] It was obvious that a rational and comprehensive science policy would deeply influence the course of scientific development through governmental decisions in funding. At one extreme was Michael Polanyi's conception of the republic of science that had complete autonomy over its activities.[21] The somewhat more "realistic" notion that resources should be allocated to diverse uses in accordance with the importance of those uses gained support within the scientific community.

During the early 1960s, Congress began to scrutinize science and technology appropriations and to inject itself into the funding procedures. Thus the 1964 Daddario report recommended greater uniformity of geographical distribution in the allocation of grants; its justification, "[the] need to build geographic base of capabilities." During this same period, pressure was also mounting – from Congress, the President, DoD – to shift the emphasis of the funding from basic research to more applied and more immediately useful research.[22] The Daddario–Kennedy amendment did precisely this for the NSF in 1968. It "changed the National Science Foundation in both form and substance" by authorizing it to fund applied research as well as basic research. The amendment also designated the social sciences as eligible for support, and required the NSF to submit to Congress an annual report on the state of U.S. science.[23] In 1970 William Koch, the director of the AIP, noted that physics "from being regarded as a science desperately needed for national survival and prestige" has become placed in "a more conventional social context, with new priorities."[24]

This changing context made it clear that the justification for the support of high-energy physics needed to be reanalyzed. Because basic science, and high-energy physics in particular, then and now, have only a relatively small political constituency, rational arguments rather than political pressure have to be invoked to support their claim for high ranking in the budget. The rational arguments usually consist in claims about the benefits that flow from the research activities. Depending on the period and the wider context, what are deemed the benefits has varied; however, economic benefits are always part of the reasons. I shall look at that story in Part III.

During this period, which includes the beginning of United States involvement in the Vietnam War, that disenchantment with science set in among the public at large, and particularly so among young people. The Vietnam War focused the discontent and exacerbated the deep uneasiness. Daily, home television screens made explicit the destructive potential of science and technology, and the scale on which they could contribute to human suffering and cause the depredation of the environment. The public had already been alerted to the deleterious effects ascribed to technology – environmental pollution, automobile death tolls – by the publication of Rachel Carson's *Silent Spring* and of Ralph Nader's *Unsafe at Any Speed*. Carson in 1962 had indicated how pesticides, the product of chemistry for better living, had led to the deterioration of the environment. Nader in 1965 had shown how a technology can be misused by large corporations in their quest for immediate profits. Similarly the sluggish growth of the economy undermined the belief that all problems could be solved by rational planning or by technological fixes. The war crystallized these feelings and led to a massive antiscience movement. Its roots sprang from seeing the physical sciences as the main source of technology, identifying science with technology, and blaming the bad effects of technology on science. Science became identified with war and nationalist competition.[25]

Herbert Marcuse, who became the philosophical guru of the young rebels, gave a trenchant criticism of modern technological society with "its progressing transfer of power from the human individual to the technical or bureaucratic apparatus, from living to dead labor, from personal to remote control, from a machine (or group of machines) to a whole mechanized system." He found it regressive and dehumanizing and deeply deplored the fact that this transfer of power had allowed human beings to abdicate their moral responsibilities; for modern technological society had "released the individual from being an autonomous person: in work and in leisure, in his needs and satisfactions, in his thoughts and emotions."[26] He ascribed its inhumanity to the fact that:[27]

> the mathematical character of modern science determines the range and size of its creativity, and leaves the nonquantifiable qualities of *humanitas* outside the domain of exact science. The mathematical propositions about nature are held to be *the* truth about nature, and the mathematical conception and project of science are held to be the only "scientific" one. This notion amounts to claiming universal validity for a specific historical theory and practice of science and other modes of knowledge appear as less scientific and therefore as less exactly true. Or to put it more bluntly: after having removed the non-quantifiable qualities of man and nature from scientific method, science feels the need for redemption by coming to terms with the "humanities."

Marcuse clearly found a resonance in the younger generation. The young elite in the upper middle class suburban high schools and at the elite colleges and research universities rebelled the most. They saw the system of education at their schools as a mechanism to exclude the poor and the blacks from full participation in society; science as the source of industrial innovations that led to further deterioration of the environment and more lethal weapons; and rationalized planning as one further step toward Orwell's world of 1984.

The upheaval was momentous. One of the most perceptive contemporary analyses of this tumultuous period was given by Don K. Price in his presidential address before the AAAS in December 1968. Price noted that the student rebellion was "the first radical international movement for two or three centuries ... that does not have material progress as its aim." Far from proposing to enlist science and technology to improve the material welfare of the poor, the student rebellion rejected technological progress as a goal. It accepted Marcuse's premise that science, by its intrinsic nature, had reduced itself to an inhumane mode of thought and that technology had become an engine of oppression. Since the industrial system and the polity were based upon science and technology, they had to be overthrown.[28]

By the end of the decade it was apparent that the United States, for all of its advanced technology, sophisticated weaponry, and overstocked nuclear arsenal, could not subdue a technologically backward, but resourceful and determined North Vietnam. Andrew Hacker, a political scientist at the University of Rochester, wrote of "the end of the American era."[29]

From 1967 to 1975 federal R&D funding became dramatically reduced.[30] At the end of the period available funds were down by some 20% (in real terms) from their peak. The cutbacks resulted in substantial unemployment among scientists and engineers. Support for research in the physical sciences was similarly affected; federal funding dropped by some 15% from its peak. Far fewer young Americans enrolled for graduate education in the physical sciences, and the Ph.D. production of Americans in these areas plummeted and has not recovered since.

The end of this third period could be taken to be January 1973. Shortly after being inaugurated president of the United States for a second term, Richard Nixon disbanded PSAC by accepting the pro forma resignation of its members. He also discontinued the post of Science Advisor to the President and transferred the civilian functions of OST to the Director of the NSF and the military functions to the National Security Council.[31]

II. The justifications for high-energy physics

In a letter written in May 1943 near the end of his stay in London, to where he had gone to learn about British wartime scientific activities, John von Neumann feelingly described his visit to Oswald Veblen, his friend and colleague at the Institute for Advanced Study:[32]

> I think that I see clearly that the best course for me at present is to concentrate on Ordnance work, and the Gas Dynamical matters connected therewith. I think that I have learned here a good deal of experimental physics, particularly of the Gas Dynamical variety, and that I shall return a better and impurer man. I have also developed an obscene interest in computational techniques. I am looking forward to discussing these matters with you. I really feel like proselytizing – even if I am going to tell you only things which you have known much longer than I did.

Von Neumann's description of his encounter with the intellectual challenges of the war applies equally well to the physicists who worked in the wartime laboratories: They had indeed become better scientists if impurer men.[33] When the hostilities ceased, many of them, particularly the nuclear physicists who had worked at the Metallurgical Laboratory and at Los Alamos, sought ways to become once again purer men and purer scientists. Guaranteeing and demonstrating the peaceful uses of atomic energy was one avenue for redemption. The efforts to secure civilian control over the nuclear technology entailed intense political lobbying and was an affirmation of the atomic scientists' commitment to the international character of scientific knowledge, for they knew that military control would make international control impossible. A second avenue to purity was unraveling the secrets of nature at the subnuclear level. For many physicists the wartime experience had reinforced the notion that only pure physics – physics for physics' sake – was "basic" or "fundamental" physics and "good" physics.[34] High-energy physics offered fertile ground for both purification and "good" physics.

Most of the members of the GAC and of PSAC, and many of those serving on the advisory committee to ONR – the bodies that decided on the support of high-energy physics after World War II – were nuclear physicists who had been associated with Los Alamos. Their support of high-energy activities was important for the growth of high-energy physics. In fact, these men – Bacher, Rabi, Oppenheimer, Lawrence, Fermi and so on – were some of the most convincing advocates of high-energy physics, and the spectacular flowering of the field owes much to their effectiveness as proponents.

High-energy physics prospered during the 1950s and 1960s, and many of the best and brightest young physicists went into the field. The various reports on high-energy physics issued by the panels convened by either PSAC or the AEC map the growth of the field in the United States. In each the justification given for support reflects the changing context.

In the fall of 1958, a year after its establishment, and again in December 1960, PSAC and the General Advisory Committee of the AEC convened a panel to assess the state of high-energy physics. As would be expected in the aftermath of Sputnik these panels recommended the expansion of activities and greater support for the field.[35] In 1963, a third panel was convened under the joint sponsorship of PSAC and the GAC "to assess future needs in the field of high-energy accelerator physics."[36] It issued its report on May 10, 1963, and recommended:

1. the construction of a 200 BeV machine at Berkeley;
2. the construction of storage rings at Brookhaven;
3. a feasibility study of a 600–1000 BeV machine;.
4. the construction of a high-intensity alternating gradient 12.5 BeV machine in Madison, Wisconsin.

The panel stated that recent progress in the study of elementary particle physics had clearly exhibited the direction to be followed and that major advances could be achieved by continued intensive efforts in the field. It emphasized that it believed that its recommendations were implementable and "likely to yield far reaching results." It also indicated that its recommendations were based on "internal needs of the discipline" and that the needs for accelerators in other fields of science and technology were not considered. The geopolitical assumptions that underlay its recommendations were the same as before: "The scientists of the US, native and foreign born, have led the world in high-energy physics." The United States has maintained leadership in high-energy research as a result of the government's willingness to subsidize a broad range of activities, including the building and support of accelerators. "[I]t is essential that the United States maintain its leading position in this area of research which ranks among our most prominent scientific undertakings."

Let me quote some of the panel's statements for the technical justification for its recommendations. They are of interest because they clearly indicate that belief in unification was widespread; that a commitment to reductionism was prevalent; and that in the eyes of the panel high-energy physics was the most fundamental field contributing to a deeper understanding of the nature of matter:

> The laws of the behavior of the elementary particles (together with the laws of the universe itself) underlie and determine the principles of physics and provide the ultimate basis for all natural science.
>
> The central theme of science is the reduction of many different phenomena to a simple set of principles through which the known facts are understood and new results predicted. This process culminates in elementary particle physics and cosmology, where one seeks unifying concepts that embrace all phenomena.
>
> The Challenge [sic] of the future will be in the unification of all the [four] kinds of forces in one coherent set of simple laws. Such an objective is necessarily a difficult one, but the way has been prepared by the entire history of physics.

Although the report mentions possible technological by-products (such as the design and construction of higher-powered klystrons), the panel stressed that the importance of the high-energy accelerator program lay in pushing technology to its limits. Its challenging technical problems engaged some of the most "inventive and resourceful scientists" who form "a reservoir of inventive energy and broadly based scientific and engineering skill from which leadership can be drawn for other scientific enterprises." High-energy physics offers "a unique training ground for some of our most creative people."[37]

The panel did note some clouds on the horizon. Although the basic aim of high-energy physics was accepted by the scientific community, the activities of the field had not caught the imagination of the public. It encouraged a more organized effort by the AEC and by the high-energy scientists to explain the meaning and the extent of "this highly successful US activity both at home and abroad."

The Weinberg–Weisskopf–Anderson debate

Although many of the recommendations of PSAC's 1963 High Energy panel were implemented – though not necessarily at the sites that the panel suggested – the report did not go unchallenged. In the face of the growing demands on an R&D budget for science that was beginning to level off, how to make the choices became the focus of an intense debate during the 1960s. Science administrators, practicing scientists, economists and philosophers all entered the fray, and as a result the literature on science policy grew enormously. I want to focus on three papers that shaped the thinking of the physics community. These papers by Alvin Weinberg, Victor Weisskopf, and Philip Anderson staked out interesting positions, and their content is still of relevance today.

In an influential article that appeared in *Minerva* in the winter of 1963, Alvin Weinberg, the director of Oak Ridge National Laboratory, posed the following questions: When research funds are allocated to science, how shall the choices be made among different, often incommensurable, fields of science that compete with one another for the allocations; for example between high-energy physics and molecular biology, or between oceanography and metallurgy? How shall the choices among the different institutions that make claims on these governmental allocations – universities, government laboratories, and industry – be made? Weinberg was particularly interested in seeing whether he could formulate a scale of values that might help establish priorities among the very large fields of science, particularly between different branches of basic science such as molecular biology and high-energy physics. He proposed that expenditures for scientific activities be ranked according to the intrinsic and instrumental value of the activities to be supported by the expenditures, and recommended that usefulness and relevance to neighboring fields of science be given top priority. For Weinberg the word "fundamental" in basic science was equivalent to "relevance to neighboring areas of science." His proposal for the criterion of scientific merit was that, other things being equal, "that field has the most scientific merit which contributes most heavily to and illuminates most brightly its neighboring scientific disciplines."[38]

To implement these views, Weinberg suggested, scientific panels that judged how much money should be allocated to a given branch of science rather than to another should include representatives of neighboring branches of science. Furthermore, when deciding whether to fund that field, one would also have to answer the questions: Does it have technological merit? Does it have social merit, that is, does it have relevance to human welfare and the values of man? Weinberg was of course aware that the latter are intrinsically more difficult questions to answer.

Weinberg then went on to analyze whether, on the basis of these criteria, molecular biology and high-energy physics ought to be funded. Molecular biology passed with flying colors. High-energy physics, on the other hand, was found wanting. Its original major task had been to understand the nuclear forces. "In this it has only been modestly successful; instead, it has opened an undreamed-of world of strange particles.... The field has no end of interesting things to do, it knows how to do them, and its people are the best." "Yet," Weinberg went on, "I would be bold enough to argue that, at least by the criteria that I have set forth – relevance to the sciences in which it is embedded, relevance to human affairs, and relevance to technology – high-energy physics rates poorly." More specifically, Weinberg noted that "the world of subnuclear particles seems remote from the rest of the physical sciences.... I know of few discoveries in ultra-energy physics which bear strongly on the rest of science.... As for its bearing on human welfare and technology, I believe it is essentially nil. These two low grades would not bother me if high-energy physics were cheap. But it is terribly expensive.... especially [in] those brilliant talents who could contribute so ably to other fields which contribute much more to the rest of science and to humanity than does high-energy physics. On the other hand, if high-energy physics could be made a vehicle for international cooperation.... between East and West.... the expense of high-energy physics would become a virtue."[39] Weinberg would then be willing to give the enterprise a slightly higher grade in social merit.

The challenge Weinberg posed to the scientific community was to prove its worth to society. This was particularly difficult for high-energy physics, precisely because it is more reductionist, more esoteric, as well as more expensive than most of the other basic sciences.

In a second article published the following year, Weinberg turned to the broader question: "What criteria can society use in deciding how much it can allocate to science as a whole rather then to competing activities such as education, social security, foreign aid and the like?"[40] In particular, Weinberg was concerned with those fields of *basic* research that were very expensive, very remote from any applied scientific problems, and that were being "pursued primarily because the researchers find the science intensely interesting, often because the findings in these fields are likely to illuminate neighbouring branches of *basic* science" (emphasis added). His answer was that basic science in fields clearly relevant to applied science (such as biology vis-à-vis medicine) be viewed as an overhead charge on that particular applied science and that the purest basic science – such as high-energy physics – be viewed as an overhead charge on the society's entire technical enterprise, that is, that its budget be a fixed percentage of the overall budget devoted to these activities.

Weinberg disseminated his views to the physics community in an article in *Physics Today* in March 1964. His challenge to high-energy physics could not go unheeded, coming from a respected member of the physics community and the highly regarded administrator of one of the largest government laboratories, with ready access to the highest echelons of the government's decision-making elite. The confrontation came at a time when the justification for new and expensive large-scale science was being critically analyzed by those making contending claims and by members of Congress who felt accountable when voting large appropriations for projects that seemingly had no relevance to their constituencies. Expensive and apparently "useless" projects in high-energy physics and astronomy would clearly come under special scrutiny. Astronomy had the advantage of being connected to "space" and therefore benefited from the post-Sputnik emphasis on projects connected to space science. High-energy physics, "after having ridden on the coattails of nuclear energy for a number of years," no longer enjoyed such unquestioned support.[41]

There were many responses.[42] I shall limit myself to that coming from within the physics community.

The Yuan report

In January 1965 the Brookhaven National Laboratory issued a remarkable booklet entitled "Nature of Matter: Purposes of High Energy Physics." It was edited by Luke C. L. Yuan, the chairman of the Super High Energy Physics Committee at Brookhaven that for the previous three years had been exploring the design and experimental programs for a 600–1000 BeV accelerator.[43] The aim of the pamphlet was to dispel the misunderstandings of the objectives of high-energy physics, "not only among the general public, but also among the scientific community as a whole" and justify the building of a "super high energy" accelerator. Some 5000 copies were printed by the government and distributed to the members of the Congress and high-ranking members of the executive branch. It was part of an active campaign to obtain the endorsement of the scientific community for the building of such a machine.

The Research Division of the U.S. Atomic Energy Commission, then the principal source of governmental support of high-energy physics, had "encouraged and supported" the project. J. R. Oppenheimer, who ten years earlier had been dishonored by the AEC, was asked to write the foreword to the publication. This could be interpreted as an act of penitence by the AEC, but the Commission was also relying on one of the most charismatic personalities within the physics community to unify it and have it back the funding of a big new accelerator.

In the pamphlet, some of the most distinguished theoretical physicists working in the United States at the time expounded their views concerning the purposes of high-energy physics.[44] All the statements were fairly short, averaging two or three pages. The first one was by Bethe. Although he was no longer active in high-energy physics and had been working on low-energy nuclear physics for the previous ten years, Bethe had contributed importantly to all the fields of physics and was one of the most admired and widely respected physicists – in both the physics community and within government circles. He was, therefore, in an ideal position to answer Weinberg. His paper set the tone:[45]

> High energy physics is undoubtedly today the frontier of physics. The discoveries in this field of study contribute most to the advance of our fundamental understanding of nature.

Bethe went on to indicate that when he started his career in physics just after the formulation of quantum mechanics in 1925–7, "it was astonishing" how quickly every problem in atomic physics, then the frontier of physics, yielded to theoretical treatment. "Physicists were spoiled by this period of amazing success of a single theoretical approach." Quantum mechanics furthermore explained the chemical bond and gave an understanding of the solid state. "Solid state," he continued, "is still a very fruitful field, giving many important advances and new insights into the working of the nonrelativistic Schrödinger equation for complicated systems. However, one can hardly claim that it advances our *fundamental* understanding of nature" (emphasis in the original). The frontier during the 1930s and 1940s was in nuclear physics, and the task was to establish the forces between nucleons and to calculate the quantum states of nuclei under these forces. Neither of these tasks was completed as yet "and much interesting work remains to be done."

Addressing the issue of relevance that had been central to Weinberg, Bethe noted that, "Particle physics, or high-energy physics, is different from atomic and nuclear physics in being far removed from our daily experience." It is easy to justify work in atomic and solid-state physics as it deals with and explains so much of the world in which we live. These fields are also of great technical and practical importance, having yielded such devices as masers, lasers, and transistors. "In nuclear physics, the practical application of nuclear power and atomic weapons is too well known to need discussion." But "[n]o such practical application has appeared or is likely to appear, for particle physics. Indeed the processes observed in particle physics may not occur in nature outside the laboratory to any important degree." How can the "fascination" of high-energy physics then be justified? Bethe gave three reasons:

1. particle physics is the most basic field of knowledge in the physical world;
2. high-energy physics will give the basis for the theoretical treatment of nuclear physics – which *is* related to the world as we know it;
3. the very difficulty of the theory and the conceptual challenges the field poses make it one of the most demanding human endeavors.

Additionally, the difficulty of the theory requires it to be closely linked to experiments.

It was thus not surprising to Bethe that particle physics had attracted "the most ambitious and the best brains among the young physicists." He concluded with the statement: "I believe that particle physics deserves the greatest support among all the branches of our science because it gives the most fundamental insight."

Gary Feinberg, in his answer to why the government should support high-energy physics, made the following moving statement:[46]

> Each human society excels at a small number of the many activities that people carry out. Our own society is preeminent at large scale technological and scientific projects, such as the building of high energy accelerators. It is therefore an expression of the highest spirit of our culture to carry on with the task we have begun, the exploration of nature to all its limits. Indeed, it may well be judged that this spirit is our greatest contribution to the human outlook. High energy physics is clearly one of the subjects on the frontier of such exploration.

Schwinger stressed unification. The goal of the high-energy theorist is not merely to find an organizing principle for subnuclear particles, "a new periodic table of the elements, interesting and important as that may be. Rather we are groping toward a new concept of matter, one that will unify and transcend what are now only understood as separate and unrelated aspects of natural phenomena." And with the characteristic hubris of the physicist he noted that "The scientific level of any period is epitomized by the current attitude toward the fundamental properties of matter. The world view of the physicist sets the style of the technology and the culture of the society, and gives direction to future progress." Writing four years before the first lunar landing, Schwinger asked, "Would mankind now stand on the threshold of the pathway to the stars without the astronomical and mechanical insights that marked the beginning of the scientific age?" He concluded his brief statement with the following remarks:[47]

> [O]ne should not overlook how fateful a decision to curtail the continued development of an essential element of the society can be. By the Fifteenth century, the Chinese had a mastery of ocean voyaging far beyond anything existing in Europe. Then in an abrupt change of intellectual climate, the insular party at court took control. The great ships were burnt and the crews disbanded. It was in those years that small Portuguese ships rounded the Cape of Good Hope.

Steven Weinberg based his justification for "Why build accelerators?" on reductionism:

> I believe that such questions must be built on the assumption that nature has absolute laws of great simplicity, from which all the sciences flow in a hierarchy.

Phenomena at one level are to be explained on the basis of physical sciences up in the hierarchy – with elementary particle physics and cosmology being the sciences at the apex of the hierarchy.[48]

> The discovery that next moves us closer to the ultimate laws of nature will almost certainly be made in one (or hopefully both) of these fields. For this reason particle physics and cosmology have an intrinsic interest not shared with any other science... [W]e are interested in [them] because it brings us as close as now possible to the absolute logical structure of the universe. It is a pity that new accelerators and telescopes happen to be expensive, but not to build them would mean that science must renounce the highest of its objectives, the discovery of the laws of nature. Instead of feuding with one another for public favor, it would be fitting for scientists to think of themselves as members of an expedition sent to explore an unfamiliar but civilized commonwealth whose laws and customs are dimly understood. However exciting and profitable it may be to establish themselves in rich coastal cities of biochemistry and solid state physics, it would be tragic to cut off support to the parties already working their way up the river, past portages of particle physics and cosmology, toward the mysterious inland capital where the laws are made.

Victor Weisskopf's defense of high-energy physics opened a line of argumentation that gained wide currency for a while.[49] He discerned two kinds of researches in the development of twentieth-century science, which he called intensive and extensive research. "Intensive research goes for the fundamental laws, extensive research goes for the explanation of phenomena in terms of known fundamental laws.... Solid-state physics, plasma physics, and perhaps also biology are extensive. High-energy physics and a good part of nuclear physics are intensive." There is always much more activity in extensive fields than in intensive fields. There are thus two dimensions to basic research. "The frontier of science extends all along a long line from the newest and most modern intensive research, over the extensive research spawned by the intensive research of yesterday, to the broad and well developed web of extensive research activities based on intensive research of past decades....Given these definitions it is clear that high-energy physics is still mostly intensive in character." Furthermore, "each part of this scientific frontier is of importance, [and] it would be most dangerous to neglect some parts relative to others."[50]

It is the very fact that high-energy physics research is the frontier of intense research that implies that at present the field leads to very little extensive research; it is also responsible for the fact that it attracts so large a proportion of the brightest scientists and that its cost per scientist is so much higher than in many other parts of the scientific frontier.[51] It is precisely its intensiveness that compels and justifies its support.

Weisskopf was willing to concede that further progress in "biology or in solid-state physics" was possible without any further research in high energy. But the spirit and the character of the scientific enterprise would be changed if basic questions that could be answered would be left unanswered – and this would adversely affect all fields of science. Moreover, Weisskopf believed that if this were to occur it would greatly harm the education of young scientists. Furthermore, it was improbable that high-energy physics would in fact deprive other fields of science of skilled manpower. It is by its very nature a limited field. Competition is heavy; success is rare and depends more often than not on luck and opportunity. Many of the best scientific brains avoid this field because of the narrow choice of activities.

Two years later in an article in *Physics Today* Weisskopf softened his stand somewhat and pointed to nuclear physics as a field that successfully straddles the intensive–extensive demarcation.[52] Nuclear structure has "the enviable position to be in between" these two extreme positions, and should please both "extensivists" and "intensivists." But the thrust of the original position had not really changed. The opening statement of the 1967 report of the High Energy Physics Advisory Panel of the AEC that was chaired by Weisskopf reiterated that position:[53]

High energy physics is one of the main fronts of science and an essential part of our scientific effort. It tries to establish the fundamental laws of physics, which are at the base of all we know about matter. It searches for the laws governing the four fundamental interactions – nuclear, electromagnetic, weak and gravitational – with the final aim of unifying these interactions by finding some common origin.... Apart from seeking an understanding of "interactions" between particles, high energy physics seeks to find reasons for the existence of the particles themselves. Why is matter made of nucleons and electrons?

Drawing on the Weisskopf model, the report indicated that

Since the late 1940s high-energy physics has played the role played by atomic physics in the first quarter of the 20th century and by nuclear in the second quarter. It is the present frontier in the ongoing study of matter. As such, it is an essential part of physics education. Excluding it or relegating it to a minor role would deprive science education of a most essential feature: the quest for fundamental laws and the urge to know more about new and unknown phenomena. Like atomic and nuclear physics in earlier periods, high-energy physics has the characteristics of a frontier area. The excitement of penetrating into the unknown attracts a large number of bright students, so this field plays a relatively large part in the training of physics PhD's.

Sensitive to the criteria that Alvin Weinberg had proposed, the panel's report conceded that there have been relatively few applications of the discoveries of high-energy physics to other sciences and technologies. But it noted that it was typical for a field with completely new phenomena that its connections with other sciences develop at a later stage. It also pointed out that

It is no coincidence that the greatest advances in man's knowledge of the basic nature of matter have always been made in countries which are also the leaders economically and industrially, such as England in the nineteenth century, Germany in the early 20th century, and the United States in the last 40 years. There is a causal relation in either direction: advanced industry creates the means of research, and basic research creates the knowledge and atmosphere of daring inquiry which is necessary for advances in modern technology.

Anderson's challenge

Both Weisskopf's assumptions and Steven Weinberg's reductionism were questioned in 1972 by Philip Anderson. One of the foremost condensed-matter physicists, he challenged the radical theoretical reductionist view held by the majority of elementary-particle physicists and questioned the validity of the intensive–extensive model that Weisskopf had advanced. His was not only an attack on the philosophical position of the high-energy physicists; it also challenged their dominance within the physics community and within the councils of state. Anderson asserted that[54]

> [T]he reductionist hypothesis does not by any means imply a "constructionist" one: The ability to reduce everything to simple fundamental laws does not imply the ability to start from those laws and reconstruct the universe. In fact, the more the elementary particle physicists tell us about the nature of the fundamental laws, the less relevance they seem to have to the very real problems of the rest of science, much less to those of society. The constructionist hypothesis breaks down when confronted with the twin difficulties of scale and complexity.

Anderson believes in emergent laws. He holds the view that each level has its own "fundamental" laws and its own ontology.[55] But it is not enough to know the "fundamental" laws at a given level. It is the solutions to equations, not the equations themselves, that provide a mathematical description of the physical phenomena. Emergence refers to properties of the solutions. The properties of solutions are not readily apparent from the equations. Moreover, the behavior of a large and complex aggregate of "elementary" entities is not to be understood "in terms of a simple extrapolation of the properties of a few particles." Although there may be suggestive indications of how to relate one level to another, it is next to impossible to *deduce* the complexity and novelty that can emerge through composition.[56] The study of the new behavior at each level of complexity requires research that Anderson believes "to be as fundamental in its nature as any other." Although one may array the sciences in a roughly linear hierarchy, according to the notion: the elementary entities of the science X obey the laws of the science Y one step lower; it does not follow that science X is "just applied science Y." The elementary entities of condensed-matter physics obey the laws of elementary particle physics, but condensed-matter physics is not just applied "elementary-particle physics," nor is chemistry applied many-body physics, pace Dirac. In his article Anderson sketched how the theory of "broken symmetry" helped explain the shift from quantitative to qualitative differentiation in condensed-matter physics and why the constructionist converse of reductionism broke down.

In fact, developments in quantum field theory, in particular the use of renormalization group methods and the promulgation of the concept of "effective field theories," gave strong support to Anderson's views.[57]

III. The changed intellectual scene

Advances in particle physics, and, in particular, the establishment of the Standard Model and the subsequent attempts to unify the electroweak and strong interactions in grand unified theories, resulted in exciting interdisciplinary activities in cosmology and elementary-particle physics. Thus the cosmological abundance of helium-4 fixes an upper bound on the number of flavors (quark varieties) in models that have symmetry between quarks and leptons. The striking success of the inflationary models in accounting for the flatness and horizon problems gives some credence to the approach – but a word of caution is in order since the validity of the Higgs mechanism will remain open until a Higgs particle is discovered and its properties determined.

I want to focus on one aspect of these developments to suggest that these advances have been responsible for yet another important transformation of physics.

The introduction of the idea of broken symmetry into elementary-particle physics can be compared to the breaking of the spheres by Copernicus and Kepler. It totally transformed the way of describing and understanding elementary-particle interactions.[58] When the dynamics of symmetry breaking at zero temperature were understood, it was natural to ask whether the broken symmetries of elementary-particle physics would be restored by heating the system to a sufficiently high temperature, in the same way as the rotational invariance of a ferromagnet is restored by heating it above its Curie temperature. Kirzhnits and Linde indicated that this was the case in field theories with broken global symmetries: starting from a spontaneously broken ground state, heating produces a phase transition, and at high temperatures the effects of the spontaneously broken symmetry disappear.[59] Dolan and Jackiw, as well as Steven Weinberg, calculated the critical temperature in both gauge and nongauge field theories and showed how finite-temperature effects in renormalizable quantum field theories can restore a symmetry that is broken at zero temperature.[60] Furthermore, Weinberg noted that the existence of such phase transitions had important cosmological implications in "big bang" cosmologies. These were elaborated by Sidney Coleman, Alan Guth, Linde, and others.[61]

The qualitative features of even the simplest model are sufficient for the point I want to make. At the very high temperatures such as occur in the very early Universe in big bang cosmologies, for most gauge theories one is in the symmetric phase with the expectation value of the Higgs field equal to zero, corresponding to the only minimum of the effective potential for the Higgs field. As the Universe expands and the temperature decreases, the effective potential develops additional minima. At first these correspond to false vacua, but at still lower temperature they become the true vacua. The state with the expectation value of the Higgs field equal to zero is now the false vacuum. Coleman showed that quantum fluctuations could create bubbles with one of the true vacua on their inside. Such a bubble would grow, and upon thermalization it would result in a large region of heated true vacuum. That region will be described by the theory with spontaneously broken symmetry.[62]

If there is validity to this picture, then an interesting shift has taken place in accounting for the regularities of the physical world – much closer in spirit to the way lawfulness is interpreted in biology. It would imply that evolution, that is, history, would be a component in physical explanation. There are obvious caveats in that the scenario hinges on the Higgs mechanism and the inflationary cosmology is not without its problems, the cosmological constant being one of them.[63] It is, of course, the case that astrophysics brought a historical perspective to physics, starting with Immanuel Kant's, William Herschel's, and Simon Laplace's nebular hypotheses. Modern astrophysics deals with the life histories of stars and galaxies, nucleosynthesis, and other such evolutionary processes. Although we have come to appreciate that, at some stage in the development of the Universe, Mendeleev's table consisted of only two elements – rather than the present 100 or so – nucleosynthesis and stellar evolution were explained in terms of laws of Nature that were thought fixed and immutable. It is the structures that evolved, not the "fundamental" laws. Normally, we don't think of Maxwell's equations or Schrödinger's equation as having evolved or having a history.

To state that QED is the low-energy effective field theory approximation of the Weinberg–Glashow–Salam theory of electroweak interactions still casts the result within the traditional framework. To state that the Weinberg–Glashow–Salam theory of electroweak interaction is also the result of an evolutionary, historical process clearly adds a new dimension to the explanatory scheme.[64]

If there is legitimacy to this view, it would resonate with a position taken toward the end of the last century by Charles Saunders Peirce, the great American philosopher who was one of the founders of pragmatism. Peirce had as deep an insight into Darwin's theory of evolution as anyone in the nineteenth century and was profoundly affected by it. Like Darwin himself Peirce asked: How did natural selection evolve? The best Darwin could do was to suggest that perhaps maximalizing the amount of life on the surface of the earth is the global principle that drives the evolution of life on the planet, and gives rise to natural selection. In the *Descent of Man*, to give an example of how regularity "evolved," Darwin traced the "evolution" of sexual selection from the lowest sexually reproducing life forms to human beings. Had he read the reprint Mendel sent him he could have asked: "How did Mendel's laws evolve?" Clearly there was a time in the history of life on earth when there were no Mendelian laws.

Peirce took the position that to accept absolute, immutable, invariable laws makes the regularities ultimate and thus closes the door to the possibility of ever explaining them or explaining how there came to be as much regularity in the Universe as there is.[65]

> To suppose universal laws of nature capable of being apprehended by the mind and yet having no reason for their special forms, but standing inexplicable and irrational, is hardly a justifiable position. Uniformities are precisely the sort of facts that need to be accounted for. That a pitched coin should sometimes turn up heads and sometimes tails calls for no particular explanation; but if it shows up heads every time, we wish to know how this result has been brought about. Law is *par excellence* the thing that wants a reason.
>
> Now the only possible way of accounting for the laws of nature and for uniformity in general is to suppose them results of evolution. This supposes them not to be absolute, not to be obeyed precisely. It makes an element of indeterminacy, spontaneity, or absolute chance in nature.

For Peirce evolution and "absolute" chance were fundamental. I want to stress the historical – evolutionary – component.

Interestingly, in the late 1960s Giuseppe Cocconi had arrived at a somewhat similar position as a result of his confrontation with the diversity and complexity encountered in the world of elementary particles.[66] Cocconi had pondered whether insights obtained from the nonexact disciplines could be of help in developing further understanding of the exact and simple world of physics. He contrasted "complex" disciplines such as evolutionary biology and geology, with the "simple" and exact disciplines, such as physics and astronomy. As Wigner had done before, he pointed out that the exact sciences are characterized by the possibility of isolating the systems they deal with and the fact that these systems require only a small number of relevant variables for the description of phenomena. From these features ensue the possibility of mathematization and prediction. "Complex disciplines are those obeying laws that evolve historically," and they defy mathematization precisely because of the historical dimension. History is inseparable from the idea of the arrow of time, a concept absent in all fundamental laws of physics that can be expressed mathematically.

Cocconi then asked: "What is the consequence of evolution, of the arrow of time, in establishing the fundamental laws, the exact and simple laws of quantum physics?"[67] He observed that in the exact sciences complexity arose when the simplest reactions involved in the systems under consideration became endothermic, that is, when energy has to be supplied to the evolving medium – and that the same was true for the "so-called elementary particles." Drawing on the parallel between the molecular complexity and elementary particle complexity, Cocconi speculated that

> It would be appealing to think that in the realm of high energies, situations could develop similar to those for molecules, and that subtle and apparently insignificant details of some interactions could have unimaginable and radical consequences in the historical evolution of matter.... If such a possibility really exists, our conception of the physical world would be greatly affected. The "immutable" laws of physics could become as ephemeral as those of organic life, "immutable" only for observations limited in space, and even more exotic, the evolution of these laws would depend on history, a history that has followed a path that, to a great extent must have been determined by chance. Another kind of life, the life of the physical world, would then be developing around us, in parallel with what we are accustomed to call real life, that on earth, of the organic world.

Epilogue: Crisis and community

In my chapter I have tried to cull out some of the salient features of the postwar period and of the 1960s and 1970s, in order to indicate some of the roots of the current sense of crisis that permeates physics.[68] Some of the arguments to justify the activities of high-energy physicists that were advanced in these decades are the same ones as were being marshalled in support of the SSC. Then and now they range from national prestige and utility to pointing out that scientific advance is one measure of a civilization and to warning that a society entirely preoccupied with its own problems will lose its spirit and vitality. Similarly, the sense of crisis in the culture at large has many of the same components; it is what has been called the postmodern crisis, and Vietnam highlighted it. It is the crisis engendered by the critical confrontation with the legacy of the Enlightenment and its enshrinement of Reason, a confrontation that was initiated by Nietzsche at the turn of the century. For us in the United States the need for reassessment comes at a time when we have to face the hidden costs of the waging of the Cold War – a conflict that has left us almost bankrupt economically and in need of finding new bonds to hold the nation together since military expenditures and national security can no longer be relied upon to do so.

I believe that high-energy/elementary-particle physics (and I here include the activities of string theorists), astrophysics, and cosmology have a special role to play, precisely because of what the proponents of these fields after World War II called their purity: their remoteness from everyday phenomena and their seeming lack of relevance in utilitarian matters. There must be a part of the scientific enterprise that does not respond easily to the customary demand for relevance. It has become very clear that the demand for relevance for economic, technological, security and other instrumental ends can easily become a source for corruption of the scientific process. Moreover, relevance can also refer to other, more exalted ends. Elementary-particle physics, astrophysics, and cosmology are among the few remaining areas of science where the advancement of the field is determined *internally*, on the basis of its own intellectual agenda, its experimental findings, and its own intrinsic conceptual structure. Particle physics and cosmology have not been "stabilized" and may never be. These communities have thus a special role – and a special responsibility – as a *community* committed to the Bohrian and Peircian vision of seeking truth. They are communities committed to rationality – but not instrumental rationality – for whom communication inheres in their very being, but ones that also believe in a basic ontology of the world and affirm that it is possible to decipher the logical structure of the physical universe. Most importantly, they are the guarantors that one of the most exalted of human aspirations, which is "to be a member of a society which is free but not anarchical," can indeed be realized. In his closing remarks at the centennial celebration of the National Academy of Sciences in 1963, Rabi pointed to this aspect of science as one of its greatest attractions. Let me conclude by quoting Rabi's depiction of this "free but not anarchical" community:[69]

Members of this community possess an inner solidity which comes from a sense of achievement and an inner conviction that the advance of science is important and worthy of their greatest effort. This solidity comes in a context of fierce competition, strongly held conviction, and differing assessments as to the value of one achievement or another. Over and above all this too human confusion is the assurance that with further study will come order and beauty and a deeper understanding.

Acknowledgments

I am particularly indebted to Harvey Brooks for allowing me to study the notes he took on the meetings of PSAC and COSPUP while he was a member of these committees. They were particularly helpful in drafting part I of the paper. I also had the benefit of valuable discussions with Tian Yu Cao.

Notes

1 One only has to think of the frequency of both large and small conferences. Among the large ones: the Rochester conferences, the Moriond meetings, the NSF regional conferences; small conferences and workshops held in Marseilles, Amsterdam, CERN, etc. Most of these smaller conferences issued mimeographed notes that were circulated to the community. In addition, there were many summer schools : Brandeis, Boulder, Erice, Les Houches, Cargese, and numerous NATO-sponsored ones, at which the most recent advances were presented in a coherent and critical fashion. Furthermore, the preprint method of communication made available to most members of the community the results of research as it was completed, well in advance of official publication.

2 In a talk delivered at Harvard on May 15, 1992, on "What is fundamental Physics?" David Nelson characterized high energy physics as the "flagship" of physics during the 1950s and 1960s. It is an apt metaphor. The dominance of high energy resonated with a tradition in physics that was prominent since Rutherford's time, which maintained that the search for the fundamental constituents of matter is the fundamental enterprise of physics. Until the 1970s, the more or less accepted view – and the one usually disseminated to the public at large – was that the history of the advances of physics since the beginning of the century was the progressive understanding of the properties of matter: from atoms, molecules, and solids, to nuclei, to the subatomic entities that account for the fundamental processes in nature. This view also became part of the lore that the physics community impressed on its students, as attested by such books as Eywind Wichmann's *Quantum Physics* (New York: McGraw Hill, 1971) – the fourth volume of the Berkeley physics series for the training of undergraduate physics majors – which was published in the late 1960s. For many of the theorists, however, it is not the search for the "elementary constituents" of matter that is fundamental, but rather how they help us understand why the world is the way it is and how it came to be thus.

3 M. Fortun and S. S. Schweber, "Scientists and the State: The Legacy of World War II," *Social Studies of Science 23* (1993), pp. 595–642 and "The Legacy of World War II: Physicists and Operations Research," in

G. Gavroglu, ed., *Recent Developments in the Historiography of Science* (Dordrecht: Kluwer, 1993), pp. 151–83.

4 I. I. Rabi, *Science: The Center of Culture* (New York: The World Publishing Company, 1970).

5 The activities of the RAND Corp. (an acronym for Research and Development) and of the MITRE Corp. (MIT R&D) after the war became paradigmatic of the approach.

6 Vannevar Bush, *Science: The Endless Frontier. A Report to the President on a Program for Postwar Scientific Research* (Washington, D.C., July 1945), reprinted 1960 by the National Science Foundation (Washington, D.C., July 1960), p. 10. As Harvey Brooks has noted: "The implicit message of the Bush report seemed to be that technology was essentially the application of leading-edge science and that, if the country created and sustained a first-class science establishment based primarily in the universities, new technologies for national security, economic growth, job creation, and social welfare would be generated almost automatically without explicit policy attention to all the other complementary aspects of innovation." Harvey Brooks, "What is the National Agenda for Science, and How Did It Come About?," *American Scientist 75* (1987), pp. 511–17, esp. p. 512. For a somewhat different reading of this text, see David A. Hollinger, "Free Enterprise and Free Inquiry: The Emergence of Laissez-Faire Communitarianism in the Ideology of Science in the United States," *New Literary History 21* (1990), pp. 897–919. Hollinger puts great emphasis on the commitment to individualism on the part of some of the leading spokesmen for the scientific enterprise, in particular Millikan and Hale. From my reading of the leading practitioners of science during the thirties – K. T. Compton, J. C. Slater, J. R. Oppenheimer, etc. – I would temper somewhat this assessment since "cooperation" and "coordination" is certainly part of their rhetoric. See also Alan Needell, "From Military Research to Big Science: Lloyd Berkner and Science Statesmanship in the Postwar Era," in Bruce Hevly and Peter Galison, eds., *Big Science: The Growth of Large Scale Research* (Stanford: Stanford University Press, 1992), pp. 290–311.

7 Bush, *Science: The Endless Frontier*, p. 21.

8 In fact, the idea of a large regional collaborative laboratory was suggested by Rabi. See Armin Hermann, John Krige, and Dominique Pestre, *History of CERN*. Vol. 1, (Amsterdam: Elsevier Science Publishers, 1987), and E. Amaldi, "The History of CERN during the early 1950's" in L. M. Brown, M. Dresden, and L. Hoddeson, eds., *Pions to Quarks* (Cambridge University Press, 1989); Paulo Bilyk, "CERN," paper for a graduate seminar in the Department of History of Science at Harvard University, May 1991. For an account that focuses on CERN as the realization of a vision of an international science embodied in the internationalism of a laboratory, see Robert Jungk, *The Big Machine* (London: Andre Deutsch, 1969).

9 Lew Kowarski, "Reflexions sur la Science," in Gabriel Minder, ed., *New Forms of Organization in Physical Research after 1945* (Geneva: Institut des Hautes Etudes Internationales, 1978), pp. 230–7. Incidentally, Kowarski noted that international science tends to be big, and that big science tends to be international.

10 This periodization was made by Harvey Brooks in the lectures he gave

during the 1970s. Harvey Brooks, "National Science Policy and Technological Innovation," in Ralph Landau and Nathan Rosenberg, eds., *The Positive Sum Strategy: Harnessing Technology for Economic Growth* (Washington: National Academy Press, 1987); Harvey Brooks, "What Is the National Agenda for Science." See also Don K. Price, *The Scientific Estate* (Cambridge: Harvard University Press, 1964); Don K. Price, *America's Unwritten Constitution* (Baton Rouge: Louisiana State University Press, 1983); Bruce L.R. Smith, *American Science Policy since World War II* (Washington: The Brookings Institute, 1990) gives a valuable overview and insightful account of science and technology policy in the United States until the late 1980s.

11 This was certainly the case for basic research – research to advance knowledge and understanding – supported by the ONR, the AEC and the NSF.

12 One of John F. Kennedy's promises in his 1960 presidential campaign was that he would overcome the stagnancy of the economy and get the country moving again.

13 President's Scientific Research Board, *Science and Public Policy: Administration for Research* (Washington: Government Printing Office, 1947), Vol. 1, p. 4. This three-volume report became known as the Steelman Report.

14 From 1945 till the mid-1960s, there was a steady growth of funding for research and development and in the size of the technical community. Toward the end of the 1960s, the fraction of the GNP devoted to R&D was roughly twice what it had been at the peak of military and atomic weapon R&D toward the close of World War II.

15 See I. I. Rabi. *Science and Public Policy, The Joseph Wunsch Lecture 1963* (Haifa: The Technion, 1963), reprinted in I. I. Rabi, *Science: The Center of Culture* (New York: World Publishing Co., 1970), pp. 75ff. H. Brooks, "The Scientific Advisor," in Robert Gilpin and Christopher Wright, eds., *Scientists and National Policy Making* (New York: Columbia University Press, 1964), pp. 73–96.

16 PSAC acted a body of "wise men" that critically analyzed the problems before them and aimed at reaching an intellectual consensus about the issues involved. It appointed panels and commissioned reports to help them in their deliberations, but it did not endorse or reject the recommendations presented in the reports. Many of the original members of PSAC were personally known to the President, had extensive experience in rocketry, radar, and atomic weaponry, and had served on high-level advisory committees. While addressing these concerns, PSAC helped establish the notion that disarmament is an aspect of national security. In its reports it stressed that some measure of safety may be gained through arms reduction and by limits on the testing of nuclear weapons instead of the continual increase of offensive capabilities.

17 The non-profit IDA had been created in the early 1960s to assist the Secretary of Defense in synthesizing policy. It did so by drafting critical reports that presented all views, evaluated all possible choices, and were as "objective" as possible. Ironically, its creation had been firmly opposed by "officers of the most powerful and independent segment of American bureaucracy, the career military services, supported by [their] military

clients, who disapproved on principle of any not-for-profit corporation."
D. K. Price, "Purists and Politicians," *Science 163* (1969), pp. 25–31.

18 Alice Rivlin, *Systematic Thinking for Social Action* (Washington: Brookings Institute, 1971).

19 "Like beauty or diversity or the irreversibility of change in the case of the environment; [and] like privacy for the individual." M. Gell-Mann, "How Scientists Can Really Help," *Physics Today 28/4* (May 1971), pp. 23–5. It might be recalled that Gell-Mann's speech was given shortly after the first heart transplant was performed by Dr. Christian Barnard.

20 See Edward Shils, ed., *Criteria for Scientific Development: Public Policy and National Goals* (Cambridge: MIT Press, 1968).

21 For a summary, see Michael Polanyi, "The Republic of Science: Its Political and Economic Theory," *Minerva I:1* (Autumn 1962), 54–73.

22 Some influential members of Congress and many high government officials were of the opinion that basic research is an expensive luxury that should be supported only if likely to yield an immediate spinoff in terms of practical applications for industry, medicine, or national defense. Thus, in 1966 the DoD underwrote Project Hindsight, whose assignment was to analyze the assumptions that underlay the support of science since World War II. Its report challenged the simplistic notion that commercialization followed directly from research and questioned the relevance of basic research. It suggested that agencies place much greater emphasis on applied research. Chalmers W. Sherwin and Raymond Isenson, *First Interim Report on Project Hindsight* (Office of the Director, Defense Research and Engineering, 1966).

23 John T. Wilson, *Academic Science, Higher Education and the Federal Government: 1950–1983* (Chicago: University of Chicago Press, 1983), pp. 57–59.

24 H. W. Koch, "An Age of Change," *Physics Today 27:1* (January 1970), pp. 27–32.

25 For a succinct expression of the state of siege that scientists felt under see the convocation address of the president of the National Academy of Sciences which he delivered at Hebrew University on June 5, 1971. Philip Handler, "In Defence of Science," the Hebrew University of Jerusalem, 1971. For an analysis of the anti-science currents of the 1970s, see *Daedalus 103* (Summer 1974) and the Ciba Foundation Symposium, *Civilization & Science* (Amsterdam: Associated Scientific Publishers, 1972); also Harvey Brooks, "The Technology of Zero Growth," *Daedalus 102* (Fall 1973), pp. 129–52.

26 Herbert Marcuse, "The Individual in the Great Society," in Bertram M. Gross, ed., *A Great Society?* (New York: Basic Books, Inc., 1968), pp. 58–80, esp. p. 62.

27 Ibid., p. 74. For a similar viewpoint see the sharp criticism of science and technology by Lewis Mumford, in Lewis Mumford, *Pentagon of Power* (New York: Harcourt Brace, 1970). Note in particular his censure of the Megamachine – the name he gave to the technological–economic complex that he saw molding and ruling the lives of people. Mumford claimed that by focusing only on empirically observed phenomena, scientists had become "absolute lawgivers," quantifiability had become the test for truth, and importance attached only to those things that could be measured and quantified.

28 On March 3 and 4, 1969, workshops and panel discussions were held at
MIT to discuss the crisis. Jonathan Allen, ed., *March 4. Scientists,
Students and Society* (Cambridge: MIT Press, 1970). The situation was
such that V. F. Weisskopf, at the time chairman of the physics
department at MIT, felt the need to justify basic research. See his
editorial, "The Need for Basic Research," in *Science 167* (1970), p. 935.

29 Andrew Hacker, *The End of the American Era* (New York: Athenaeum,
1970). "Thus far corporate America has escaped open attack because the
victims of new technology do not yet outnumber its beneficiaries."

30 In January of 1966 the *New York Times* reported that "For the first time
in its history, NASA has been allocated a budget appropriation that
represents a decrease from that of the previous year.... Like NASA the
AEC received a smaller total appropriation than last year. At a briefing
... a space agency official who called the budget stringent said the cut
also reflected the war in Vietnam and the nation's commitment to health
and antipoverty programs at home." Harold Schmeck, "NASA budget
cut $163-million; First trim in its 8-year history," *New York Times*,
January 25, 1966. Although the estimates showed a decrease in money to
be spent for development – e.g., the production cost of large space
rockets – it indicated an increase for spending in a wide range of scientific
endeavors including manned spaced flight, water desalination, and studies
in high-energy physics.

31 Already in December 1972 Nixon had sent to Congress a Reorganization
Plan that abolished the Office of Science and Technology (OST) in the
Executive Office of the President. There is little doubt that the abolition
of PSAC and the post of Science Advisor was a vindictive political act.
Nixon had put the full weight of his office behind the deployment of an
antiballistic missile system and in support of the Supersonic Transport
plane (SST). However, the military effectiveness of the ABM system was
seriously questioned by PSAC. Moreover, former Science Advisors – not
members of PSAC at the time – made public their reservations in
testimony before Congress in 1969. In addition, several PSAC members
were on the board of the Federation of American Scientists, which
launched an active campaign against ABM. In the SST matter, Richard
Garwin, the chairman of the SST panel of PSAC, testified and
campaigned publicly against the program. He was careful to emphasize
that he was speaking in a personal capacity and had come to his position
on the basis of information fully available in the public domain.

A subsequent challenge under the Freedom of Information Act by an
environmental action organization forced the White House to release the
PSAC panel report that questioned both the economic feasibility and the
benignness of the environmental impact of the SST. Most of the content
of the panel report was known to Congress when it rejected the program.
The political activism of PSAC members and of the scientific community
in general clearly rankled Nixon. It should also be recalled that scientists
and engineers had organized the political organization "Scientists and
Engineers for Johnson–Humphrey" in 1964, and some of them became
strong supporters of "Scientists and Engineers for McCarthy" in 1968.
PSAC members faced a real dilemma when trying to balance the need for
the confidentiality of their advice – which is given in such a way that the
President can arrive at a decision in situations where other factors may

be of equal if not greater importance to him – and public pressure to make the advice available – as exemplified by the Freedom of Information Act and the Federal Advisory Committee Act (which requires that Committee meetings be open to the public except where matters of national security are involved). David Z. Beckler, "The Precarious Life in the White House," in G. Holton and William A. Blanpied, eds., *Science and Its Public: The Changing Relationship* (Dordrecht: D. Reidel Publishing Co., 1976).

32 J. von Neumann to Oswald Veblen, May 21, 1943. Oswald Veblen papers, Library of Congress, Box 15, von Neumann File. (It was during this visit that von Neumann learned about explosive lenses and the British activities in operations research.) The notion of purity was clearly of great significance to von Neumann after World War II. On the occasion of a dinner honoring Robert Kent in the early 1950s, he commented:

It was through him [Kent] that I was introduced to military science, and it was through military science that I was introduced to applied sciences. Before this I was, apart from some lesser infidelities, essentially a pure mathematician, or at least a very pure theoretician. Whatever else may have happened in the meantime, I have certainly succeeded in losing my purity.

From 1937 on von Neumann was a consultant for the Ballistic Research Lab (BRL) at Aberdeen Proving Grounds, Maryland. He was introduced to the activities of the BRL by Veblen, who had worked there during World War I. In 1937 the research program at the BNL was revitalized in view of the deteriorating situation in Europe. In 1940 the Army asked von Neumann to be on the board that was to review BRL, where Robert Kent was one of the senior administrators (von Neumann Papers, Library of Congress).

33 The emphasis on gender is to highlight the fact that very few women were involved. What this implies about the community is the subject of another investigation. The rhetoric of purity after World War II merits a thorough inquiry. Lewis Strauss once quipped that there were three kinds of scientists: pure, applied, and political. "Pure" science and "pure" mathematics were thought to be "fun." A different set of rules was thought to apply to their performance as compared to applied science.

34 The idea that work in the borderline between physics and chemistry, metallurgy and other such "applied" fields, was beneath the dignity of the true physicist had been given currency by Pauli in the 1930s. His correspondence with Heisenberg during that decade contains many remarks deprecating solid-state physics.

35 On May 17, 1959, the White House issued "An Explanatory Statement on Elementary Particle Physics" and "A Proposed Federal Program in Support of High Energy Accelerator Physics," the two documents that the panel appointed by PSAC and GAC had drafted. The panel consisted of Emanuel Piore, chairman; Jesse Beams, Hans A. Bethe, Leland Haworth, and Edwin McMillan. It is interesting to note that the introduction to the "Explanatory Statement" stated that

Elementary particle physics ... is concerned with phenomena remote from our immediate and familiar surroundings ... and proceeds, not with any view towards useful application, but by pursuing discovery for its own sake. It is the very heart of modern physics, and the product of many centuries of effort to understand the universe.

The panel recommended the construction of the Stanford linear electron accelerator, continued support of the activities of the Midwestern Universities Research Association to design an accelerator, and the exploration of the possibility of building an accelerator at Oak Ridge. More specifically, it recommended an increased level of funding by the government to reach approximately $135 million annually by the fiscal year 1963. In fact, by 1963 there were some 900 Ph.D.s active in the field, and 92% of the U.S. support came from the AEC.

36 Jerome Weisner was the chairman of PSAC at the time, and Manson Benedict was chair of the GAC. The membership of the panel was as follows: Norman Ramsey, Harvard, chair; Philip Abelson, Carnegie Institution of Washington; Owen Chamberlain, University of California, Berkeley; Murray Gell-Mann, Caltech; Edwin Goldwasser, University of Illinois; T. D. Lee, Columbia; W. K. H. Panofsky, Stanford; E. M. Purcell, Harvard; Frederick Seitz, NAS; J. H. Williams, University of Minnesota.

37 Most of these themes had first been advanced by John Archibald Wheeler in an influential paper read before the National Academy of Sciences on November 17, 1945, at a symposium on "Atomic Energy and its Implications." The intent of his presentation was "to survey the outstanding problems [of elementary particle physics] and to discuss possible lines of investigation." One of the specific questions Wheeler considered was the relationship between the elementary particles. It is in his lecture that one finds for the first time explicit reference to the four kinds of interaction between the "elementary particles" and to the differing strength of the coupling constants used in their field-theoretic description. Furthermore, he stressed that "the task of reducing our experience to order includes ... the creative function of assimilating the fruits of the collaboration [between theorists and experimenters] into a unified view of matter." He ended his talk with the following remarks:

On men like these will depend our future in war and peace. They will make for us new tools of defense in time of danger. They will leaven our applied science, our technology, our industry, and our intellectual life in the days of peace. Their qualities of mind and heart are the prize. We must seek out our able young men, outfit them, and send them forward to work their way through the unknown, not only because the land is rich, but most of all because participation in this great Odyssey will develop men of the kind on whom our future as a nation depends. There is no other way.

John Archibald Wheeler, "Problems and Prospects in Elementary Particle Research," *Proc. Am. Philos. Soc. 90* (1946), pp. 36–47.

38 Alvin M. Weinberg, "Criteria for Scientific Choice," *Minerva I:2* (1963), pp. 159–171. Reprinted in Edward Shils, ed., *Criteria for Scientific Development: Public Policy and National Goals* (Cambridge: MIT Press, 1968).

39 The possibility of international cooperation in high-energy physics, and in particular the possibility of a joint US–USSR cooperative effort in

building a 1000 BeV machine was discussed at the Atoms for Peace meeting in Geneva on April 2, 1963.

40 Alvin Weinberg, "Criteria for Scientific Choice II: The Two Cultures," *Minerva III:1* (1964), pp. 3–14. Reprinted in Shils, ed., *Criteria*. The two cultures Weinberg was referring to were basic and applied science.

41 V. F. Weisskopf, "In Defense of High Energy Physics," in Luke C. L. Yuan, ed., *Nature of Matter: Purposes of High Energy Physics* (Upton, NY: Brookhaven National Laboratory, 1965), p. 24.

42 See for example the articles written by philosopher Stephen Toulmin that are reprinted in Shils, ed., *Criteria*.

43 Luke C. L. Yuan, ed., *Nature of Matter: Purposes of High Energy Physics* (Upton, NY: Brookhaven National Laboratory, 1965). Similar feasibility studies were being carried out at Argonne, Berkeley Radiation Laboratory, SLAC, and at several other universities.

44 The most notable exceptions were Richard Feynman and Murray Gell-Mann.

45 H. A. Bethe, "High Energy Physics," in Yuan, *Nature of Matter*, pp. 9–11.

46 G. Feinberg, "The Future of High Energy Physics," in Yuan, *Nature of Matter*, pp. 12–14.

47 J. Schwinger, "The Future of Fundamental Physics," in Yuan, *Nature of Matter*, p. 23.

48 S. Weinberg, "Why Build Accelerators?," in Yuan, *Nature of Matter*, pp. 71–73. The metaphor of navigators and explorers had been introduced by Wheeler in his NAS address in 1945. His source was the coded telephone conversation that Arthur Compton had with James Conant on December 2, 1942, in which he informed him of Fermi's successful experiment to control a nuclear chain reaction. Compton reported that "The Italian navigator has discovered America.... and Columbus finds the natives are friendly." In his speech, Wheeler called for the exploration of the new continent of elementary particles that lay beyond the shores of nuclear physics.

49 See for example, H. Koch, "An Age of Change."

50 V. F. Weisskopf, "In Defense of High Energy Physics," in Yuan, *Nature of Matter*, pp. 24–27.

51 It was estimated that the average cost of a Ph.D. in high-energy physics was close to $350,000, much higher than the cost of a Ph.D. in other areas of physics or in chemistry.

52 Victor F. Weisskopf, "Nuclear Structure and Modern Research," *Physics Today 20:5* (May 1967), pp. 23–32. Weisskopf's article included a famous diagram in which he plotted "extensiveness vs. intensiveness." The three steps that physics had taken in its study of matter divided the vertical intensive axis: atomic physics, nuclear physics, high-energy physics – which Weisskopf labeled mesonic physics. A line drawn at 45° indicated the progress of science. In this diagram the lower the field is in the intensive development (atomic physics), the more it was extended in the extensive direction.

53 "The Status and Problems of High Energy Physics Today – A Report of the High Energy Physics Advisory Panel of the AEC," *Science 161* (July 1968), pp. 11–19. Established in January 1967, the Panel was composed of the following physical scientists: Rodney Cool (Brookhaven); Earle C.

Fowler (Duke); Leon Lederman (Columbia); Edward J. Lofgren (Lawrence Radiation Laboratory); George E. Pake (Washington University); W. K. H. Panofsky (Stanford); Robert Sachs (Argonne); Keith R. Symon (Wisconsin); Robert Walker (Caltech); Robert R. Wilson (Cornell); C. N. Yang (Stony Brook); V. F. Weisskopf (MIT), chairman. By 1967 the impact of the Vietnam War on R&D budgets was considerable. The Panel noted that the FY 1968 funding for high energy was 40% below the projections that the AEC had made in its policy statement of 1965. Thus the future growth of high energy was at stake. A subcommittee of the Panel consisting of Weisskopf, Panofsky, Wilson, and Yang discussed the report with PSAC on June 20, 1967. The Panel's recommendations were as follows:

1. that a substantial increase in the annual operating budget for high-energy physics be followed by moderate growth.

2. that authorization be given for the building of the electron–positron ring at SLAC and the 4.2 meter bubble chamber at Brookhaven.

3. that authorization for the construction of the 200 BeV accelerator be granted in FY 1969. In addition, the Panel encouraged and recommended support for US–USSR collaborative research at Serpukhov.

54 P. W. Anderson, "More Is Different," *Science 177* (1972), pp. 393–6.

55 Translated into the language of particle physicists, Anderson would say each level has its effective Lagrangian and its sets of (quasi-stable) particles. In each level the effective Lagrangians – the "fundamental" description at that level – is the best we can do. When particle theorists explore the consequences of the $SU(3) \times SU(2) \times U(1)$ Standard Model, their work is not more fundamental than what Anderson does exploring the consequences of the Schrödinger theory for condensed-matter physics. Note that in this case, positivism and pragmatism lead to intellectual democracy.

56 The formulation of the strong interactions as a non-Abelian gauge theory can be considered a "fundamental" description. But to go from that description to an effective chiral Lagrangian to describe low-energy pion–nucleon scattering, or to deduce from it the binding energy of the deuteron and explain why it is so small, presents enormous difficulties that have not as yet been overcome. This reflects the fact that ascertaining the properties of the solutions of the "fundamental" equations – to say nothing of obtaining actual solutions – is an extremely difficult mathematical task involving delicate limiting procedures (judging from some of the exact results obtained for phase transitions in various dimensions).

57 S. Weinberg, "Phenomenological Lagrangians," *Physica 96A* (1979), pp. 327–340; H. Georgi, "Effective Quantum Field Theories," in Paul Davies, ed., *The New Physics* (Cambridge: Cambridge University Press, 1989), pp. 4446–57. The history of these changes is analyzed in T. Y. Cao and S. S. Schweber, "The Conceptual Foundations and Philosophical Aspects of Renormalization Theory," *Synthese 97* (1993), pp. 33–108.

58 As is well known, dynamical symmetry breaking was brought into elementary particle physics from many-body physics. See L. M. Brown and Tian Yu Cao, "Spontaneous breakdown of Symmetry: Its Rediscovery and Integration into Quantum Field Theory," *Hist. Stud. Phys. Biol. Sci. 21* (1991), pp. 211–35.

59 D. A. Kirzhnits and A. D. Linde, "Macroscopic Consequences of the Weinberg Model," *Phys. Lett. 42B* (1972), pp. 471–4.

60 L. Dolan and R. Jackiw, "Symmetry Behavior at Finite Temperature," *Phys. Rev. D9* (1974), pp. 3320–41. S. Weinberg, "Gauge and global Symmetries at High Temperature," *Phys. Rev. D9* (1974), pp. 3357–77.

61 See for example L. F. Abbott and So-Young Pi, eds., *Inflationary Cosmology* (Singapore: World Scientific, 1986); Andrei Linde, *Particle Physics and Inflationary Cosmology* (New York: Harwood Academic Press, 1990), and A. Guth and P. Steinhardt, "The Inflationary Universe," in Paul Davies, ed., *The New Physics*, pp. 34–60.

62 S. Coleman, "Fate of the False Vacuum: Semiclassical Theory," *Phys. Rev. D15* (1977), pp. 2929–36.

63 See for example Larry Abbott, "The Mystery of the Cosmological Constant," *Scientific American 256:5* (May 1988), pp. 106–13.

64 Note that the viewpoint posited does not exclude the possibility that there exists a "fundamental" unified theory, which can be represented at various energy scales by "effective" field theories. The present situation with respect to symmetries may also be analogous to the relation between special and general relativity. In the special theory of relativity, space–time is a rigid stage unaffected by the matter and energy present – as was the case in Newtonian physics. General relativity dynamicized space–time, with the geometry of space–time becoming a response to the energy–momentum density present. The analogy is the following: before the Standard Model, symmetries were likewise thought of as "rigid." Gauge field theories have allowed the possibility to "dynamicize" the symmetries.

65 C. S. Peirce, "The Architecture of Theories," in *Chance, Love and Logic. Philosophical Essays* (New York: Harcourt Brace, 1923), p. 162.

66 G. Cocconi, "The Role of Complexity in Nature," in M. Conversi, ed., *Evolution of Particle Physics* (New York: Academic Press, 1970), pp. 81–7.

67 Cocconi observed that when history and the arrow of time were not recognized as operating in the biological sciences, the living world was considered static, ruled by fixed laws – except for acts of creation that defied delineation. He considered physics to be in the same state: "We have some immutable laws ... which all mechanical facts obey ... and far back in time, ten billion years ago, some big event ... put all this world into being."

68 Some of the causes of the sense of crisis are analyzed in S. S. Schweber, "Physics, Community and the Crisis in Physical Theory," *Physics Today* (1993). In high-energy physics the disparity in the time scales for constructing new accelerators and detectors and that for creating novel theories is partly responsible for the unease in that community. There have essentially been no new experimental results – the impetus for advances in the field – since the early 1980s, when the electroweak theory was confirmed with the discovery of the Ws and the Z^o, except for the corroboration of the number of families in the 1980s and the strong evidence for the existence of the top quark in 1994–5. Particle physics theorists are divided into various camps – with string theorists representing the tradition of trying to seek unifying (unitary) theories and with many of them resonating to Dirac's contention that theories

ought to be "beautiful." But since their efforts thus far are so removed from experimental relevance, they are branded mathematicians by the more phenomenologically inclined theorists committed to effective field theory approaches. And everyone seems to agree that some of the most exciting aspects of high-energy physics are to be found in astrophysics. Nor has the condensed matter physics community been immune to the sense of crisis. The excitement generated by the solution of the problem of phase transitions has abated. The problems of explaining high temperature superconductivity have proven more refractory than initially anticipated. The departure of a number of distinguished practitioners to such fields as biophysics and neural networks has not escaped notice. The splits between the various subfields of physics were sharpened during the 1980s partly because of shrinking budgets. The end of the Cold War in the late 1980s and early 1990s has exacerbated the sense of crisis. In the spring of 1992, the *New York Times* related that some 800 applicants had applied for a single tenure-track position at Amherst College. *Physics Today* in March 1992 reported on the sense of despair that characterizes the atmosphere of the meetings of the AIP. See John M. Rowell, "Condensed Matter Physics in a Market Economy," *Physics Today 45:5* (May 1992), pp. 40–7. These suggest that the discipline is facing a situation as difficult as that during the early 1930s at the height of the Depression. It is likely that the discipline will shrink sharply in size over the next decade or so.

69 I. I. Rabi. "Science in the Satisfaction of Human Aspiration," in *The Scientific Endeavor. Centennial Celebration of the National Academy of Sciences* (New York: The Rockefeller Institute Press, 1963). The views of the republic of science that Rabi was promulgating found their most forceful expression in Don K. Price, *The Scientific Estate* (Cambridge: Belknap Press of Harvard University Press, 1965). They also resonated with the views of science that had been advanced by Robert K. Merton, Rabi's colleague at Columbia. See Robert K. Merton, "Science and the Social Order," *Philosophy of Science 5* (1938), pp. 321–7; "A Note on Science and Democracy," *Journal of Legal and Political Sociology 1* (1942), pp. 115–26; *Social Theory and Social Structure*, 2d ed. (Glencoe, Ill.: The Free Press, 1957). See also Michael Polanyi, "The Republic of Science: Its Political and Social Significance," *Minerva 3* (1964), pp. 455–76. For a most thoughtful and insightful overview, see Yaron Ezrahi, *The Descent of Icarus* (Cambridge: Harvard University Press, 1990).

Index